Computational Materials Science: Theory and Applications

Computational Materials Science: Theory and Applications

Edited by
Lily Chen

WILLFORD PRESS
www.willfordpress.com

Published by Willford Press,
118-35 Queens Blvd., Suite 400,
Forest Hills, NY 11375, USA

ISBN: 978-1-68285-426-6

Cataloging-in-Publication Data

Computational materials science : theory and applications / edited by Lily Chen.
    p. cm.
Includes bibliographical references and index.
ISBN 978-1-68285-426-6
1. Materials science--Data processing. 2. Materials--Computer simulation. 3. Materials--Data processing.
I. Chen, Lily.
TA404.23 .C66 2018
620.11--dc23

For information on all Willford Press publications
visit our website at www.willfordpress.com

Contents

Preface

This book has been an outcome of determined endeavour from a group of educationists in the field. The primary objective was to involve a broad spectrum of professionals from diverse cultural background involved in the field for developing new researches. The book not only targets students but also scholars pursuing higher research for further enhancement of the theoretical and practical applications of the subject.

Computational materials science is a fast growing field. It involves computational tools for solving problems related to materials science. Different mathematical models are used for developing a better understanding of material structures and properties. Most research done in this field focuses on the behavior of materials at varied levels. This book on computational materials science is a collective contribution of a renowned group of international experts. It aims to serve as a resource guide for students and experts alike and contribute to the growth of the discipline.

It was an honour to edit such a profound book and also a challenging task to compile and examine all the relevant data for accuracy and originality. I wish to acknowledge the efforts of the contributors for submitting such brilliant and diverse chapters in the field and for endlessly working for the completion of the book. Last, but not the least; I thank my family for being a constant source of support in all my research endeavours.

Editor

The origin of electronic band structure anomaly in topological crystalline insulator group-IV tellurides

Zhen-Yu Ye[1,2], Hui-Xiong Deng[2], Hui-Zhen Wu[1], Shu-Shen Li[2], Su-Huai Wei[3] and Jun-Wei Luo[2]

Group-IV tellurides have exhibited exotic band structures. Specifically, despite the fact that Sn sits between Ge and Pb in the same column of the periodic table, cubic SnTe is a topological crystalline insulator with band inversion, but both isovalent GeTe and PbTe are trivial semiconductors with normal band order. By performing first-principles band structure calculations, we unravel the origin of this abnormal behaviour by using symmetry analysis and the atomic orbital energy levels and atomic sizes of these elements. In group-IV tellurides, the s lone pair band of the group-IV element is allowed by symmetry to couple with the anion valence p band at the L-point, and such s–p coupling leads to the occurrence of bandgap at the L-point. We find that such s–p coupling is so strong in SnTe that it inverts the band order near the bandgap; however, it is not strong enough in both GeTe and PbTe, so they remain normal semiconductors. The reason for this is the incomplete screening of the core of the relatively tight-binding Ge 4s orbital by its 3d orbitals and the large atomic size and strong relativistic effect in Pb, respectively. Interestingly, we also find that the rhombohedral distortion removes the inversion symmetry and the reduced s–p coupling transforms the α-SnTe back to a normal semiconductor. Our study demonstrates that, in addition to spin–orbit coupling, strain and interface dipole fields, inter-orbital coupling is another effective way to engineer the topological insulators.

INTRODUCTION

Topological insulators (TIs) are a group of materials with interesting electronic properties. They can exhibit insulating state in the bulk, but support spin-momentum locked gapless states at the boundaries. For three-dimensional TIs,[1,2] the metallic topological surface states induced by spin–orbit coupling (SOC) are protected by time-reversal symmetry, known as Kramer's pair. Time-reversal symmetry forbids the elastic backscattering of Kramer's pair surface states, and the dispassion-less transport of the topological surface states is robust against nonmagnetic weak disorder. Topological crystalline insulators[3] (TCIs) are new topological phases of matter in which the topological surface states are protected by crystal symmetries,[4] such as the mirror symmetry of the crystal with respect to {110} plane in the rocksalt (RS) structure of semiconductor SnTe, instead of time-reversal symmetry in TIs. The SnTe-like TCI state is characterised by a new topological invariant called the mirror Chern number, instead of the Z_2 invariant in general TIs.

The RS structure IV–VI narrow band gap semiconductors such as GeTe, SnTe and PbTe and their alloys have been extensively explored in developing infrared optoelectronic devices[5–8] and as high-performance thermoelectric materials.[9] Besides their wide range of applications, the IV–VI materials also exhibit exotic electronic structures. For example, SnTe was recently found to be the prototype of TCIs. The topological surface states of SnTe and related IV–VI group alloys $Pb_{1-x}Sn_xTe$ and $Pb_{1-x}Sn_xSe$ have been theoretically predicted by performing the first-principles calculations[4] and experimentally observed by carrying out angle-resolved photoemission spectroscopy,[10–13] scanning tunnelling microscope,[14,15] as well as transport measurements.[16,17] Surprisingly, although SnTe is a well-known TCI, both of its lighter isovalent GeTe and heavier isovalent PbTe counterparts are normal insulators. This non-monotonic behaviour of IV–Te compounds is quite unexpected and the origin has not been fully explained.

In this paper, performing first-principles calculations, we unravel the origin of the anomalies by studying the electronic structures of GeTe, SnTe and PbTe. We find that in the cubic (RS, β-phase) the (low lying) cation s lone-pair state, which are valence electrons that are not shared with another atom, couples strongly with (high lying) anion valence p state that leads to band inversion in SnTe, but the coupling is not large enough for GeTe due to its low 4s orbital energy and for PbTe due to its large Pb atom size and relativistic effects, so they remain to be normal semiconductors. Surprisingly, we find that the strong admixture of cation p valence state and anion p conduction state in rhombohedral phase, which does not have inversion symmetry, leads to loss of band inversion and makes the low-temperature phase α-SnTe a normal semiconductor. This raises an interesting question about the validity of previous theoretical[4] and experimental[10–17] studies about the topological behaviour of SnTe, suggesting that more studies are needed to clarify this issue.

RESULTS AND DISCUSSION

Figure 1 shows the calculated band structures of GeTe, SnTe and PbTe in ideal RS structure. They share common features over a large energy range of 20 eV, despite some systematic changes from compound to compound arising from the chemical shift of cations, changes in lattice constant, as well as spin–orbit splitting. Specifically, in the displayed energy window, all the conduction and valence band levels at high symmetry k-points occur in the

[1]Department of Physics, State Key Laboratory for Silicon Materials, Zhejiang University, Hangzhou, China; [2]State Key Laboratory of Superlattices and Microstructures, Institute of Semiconductors, Chinese Academy of Sciences, Beijing, China and [3]National Renewable Energy Laboratory, Golden, CO, USA.
Correspondence: J-W Luo (jwluo@semi.ac.cn)

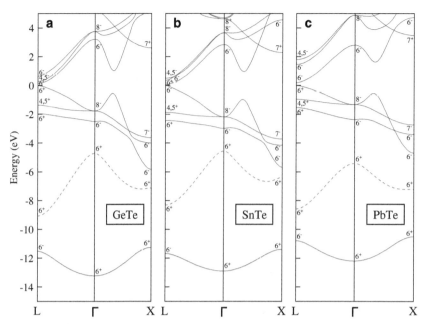

Figure 1. Calculated band structures of cubic (**a**) GeTe, (**b**) SnTe and (**c**) PbTe. The energy zero sets at the top of the valence band.

same order, except near the fundamental band gap, which is at the L point. II–VI semiconductors (e.g., CdTe) are usually more stable in the zinc-blende structure and IV–VI semiconductors (e.g., PbTe) are usually more stable in RS structure, although both have similar face-centred cubic BZ. Unlike conventional II–VI semiconductors where the cation s band becomes the lowest conduction band and is unoccupied, the large electronegativity of the group-IV elements results in their cation s band (dashed lines in Figure 1) being occupied and below the anion p band, forming the so-called lone-pair s band. Because of this, II–VI semiconductors usually have conduction band minimum (CBM) state occurring at Γ point with mostly cation s orbital character, whereas IV–VI semiconductors usually have CBM state at L point with mostly cation p orbital character. The s–p coupling between cation s band and anion valence p band is forbidden at the high symmetry Γ point but is allowed at the L point. In IV–VI compounds, the occupied lone-pair s band is low lying. The s–p level repulsion pushes the valence p band up remarkably at L point in energy, making it the valence band maximum (VBM) over the whole BZ. Thus, IV–VI semiconductors often have a direct band gap at the L point despite the fact that all II–VI compounds have direct band gap at the Γ point.[18] In the following, we will discuss in more detail the band structures and chemical trends of GeTe, SnTe and PbTe.

Figure 2 shows the band structures of RS SnTe calculated with and without SOC. We first examine the effects of the lone-pair s band on the band structure of IV–VI semiconductors by neglecting the SOC, as shown in Figure 2a. At Γ point, the anion p orbital valence band belongs to the triply degenerate Γ_{15} irreducible representation (single group notation[19] and chosen the cation sites as origin) but the low-lying cation lone-pair s band belongs to the Γ_1 irreducible representation, thus their coupling is strictly forbidden by symmetry since only states with the same symmetry can couple to each other.[20] But the valence band Γ_{15} state can couple with the high-lying cation p derived conduction band Γ_{15} state, pushing the valence band state down and the conduction band state up in energy. At X point, the lone-pair s band has X_1 symmetry, whereas the three-fold degenerate valence p orbital bands split into two bands with non-degenerate X_4' symmetry and doubly degenerate X_5' symmetry, respectively, and thus their coupling to the lone-pair band is also forbidden. Like at Γ point,

the p–p coupling pushes the valence band (X_5' state) down and the conduction band (X_5' state) up in energy. Therefore, at both Γ and X points the low-lying lone-pair s band could not directly influence the valence p band, but the p–p coupling leads to a level repulsion that lowers the valence p band and raises the conduction p band. However, at the L point, the valence anion p orbital bands split into a doubly degenerate L_3 state and a non-degenerate L_1 state, and the lone-pair s band also has the L_1 symmetry, leading to inter-band coupling between p orbital L_1 state and lone-pair s orbital L_1 state and resulting in a strong level repulsion. The emergent s–p coupling at L point significantly pushes the L_1 branch of the valence p states upward to a much higher energy than ones at Γ point, making it the VBM. On the other hand, the p–p coupling, which is allowed at Γ and X points, is forbidden by symmetry at L point as shown in Figure 2a, and thus the conduction band energy at the L point is low. The conduction bands at L point are further pushed down by high-lying conduction bands (with anion d character) of the same symmetry, making it the CBM. The SOC lowers the crystal symmetry and further lifts the degeneracy of p orbital bands, as shown in Figure 2b, and may introduce additional inter-band coupling. However, the SOC introduced inter-band coupling is usually weak compared with the crystal field induced coupling unless the coupled states are nearly degenerate. Therefore, the general band structure is unaffected by including SOC. This explains why the IV–VI RS compounds have a direct band gap occurs at the L point.

Note that for conventional II–VI semiconductors in zinc-blende structure, the cation s state is unoccupied and forms the conduction band. The lowest conduction s band has different symmetry at Γ point but same symmetry at X and L points as the valence p derived states. The allowed s–p coupling between conduction band and valence band at X and L points repels substantially the conduction band upwards and valence band downwards. This s–p coupling thus makes most of the zinc-blende II–VI semiconductors direct bandgap, having the CBM at the Γ point and the VBM at the Γ point.

We now turn to the abnormal trend of band ordering of the fundamental band gap among group-IV tellurides. Figure 1 shows that for GeTe and PbTe, the ordering of the bands at L point is normal in the sense that they have a cation p derived L_6^- (double group notation) CBM state and an anion p derived L_6^+ VBM state.

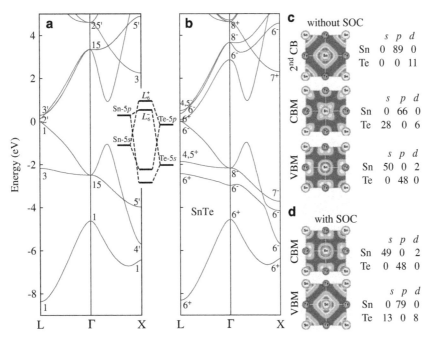

Figure 2. Calculated electronic band structures of SnTe (**a**) without SOC and (**b**) with SOC. The wave function squares as well as atomic orbital components (in percentage) of the CBM and VBM states are given in **c** and **d** for calculation without SOC and with SOC, respectively. Inset shows the schematic plot of s–p coupling in SnTe.

Table 1. Calculated band gaps at $L[E(L_6^-) - E(L_6^+)]$ and $\Gamma[E(\Gamma_6^-) - E(\Gamma_8^-)]$ points for GeTe, SnTe and PbTe at the experimental lattice constants a

	a (Å)	$E_g(L)$ (eV)		$E_g(\Gamma)$ (eV)	E_{cation} (eV)		r_{cation} (Å)
		Calc.	Expt.		E(s)	E(p)	
GeTe	5.996 (ref. 21)	0.17	0.2 (ref. 21)	4.99	−11.93	−4.05	1.52
SnTe	6.300 (ref. 4)	−0.11	−0.185 (ref. 4)	5.05	−10.78	−3.87	1.72
PbTe	6.428 (ref. 22)	0.20	0.19 (ref. 22)	4.13	−12.04	−3.50	1.81

The calculated band gaps are compared with available experimental data. Calculated s and p valence atomic orbital energies (in eV) of Ge, Sn and Pb are listed. The last column gives the atomic radii (in Å) for these atoms.

However, for SnTe this band order is reversed, known as the Dimmock reversal,[21] despite the fact that Sn element is in the middle of Ge and Pb in the group 14 column of the periodic table. Specifically, the CBM is a L_6^+ state and the VBM is a L_6^- state, as shown in Figures 1 and 2b. It is usually expected that physical properties will change monotonically as the atomic number of the isovalent elements changes. For example, the band gaps of ZnTe, CdTe and HgTe are reduced from 2.4 to 1.5 eV and to −0.3 eV.[18] Table 1 shows the calculated band gaps of GeTe, SnTe and PbTe, which are in good agreement with experimental results. Such band reversal can also be illustrated by examining the charge density distribution and the atomic orbital characters[22] of the VBM and CBM states. Figure 3 shows clearly a switch of wavefunctions between CBM and VBM states from GeTe to SnTe and a switch back from SnTe to PbTe. This switch back indicates that the band gap order of SnTe is opposite to both of its lighter (GeTe) and heavier (PbTe) isovalent neighbours. By projecting the CBM and VBM wavefunctions onto the atomic orbitals, we find in Figure 3 that the CBM of GeTe and PbTe and the VBM of SnTe are cation p predominated with minor contributions from anion s and anion d orbitals, whereas the VBM of GeTe and PbTe and the CBM of SnTe are mainly from anion p with remarkable contributions from cation s. This is consistent with the wave function distribution of CBM and VBM among GeTe, SnTe and PbTe. Therefore, the CBM symmetry of both GeTe and PbTe is L_6^- and their VBM symmetry

is L_6^+. On the other hand, the order is reversed for the SnTe, i.e., the CBM is L_6^+ and VBM is L_6^-. Because of the band inversion, SnTe was recently found to be a TCI, but GeTe and PbTe are normal semiconductors.

The anomalous band gap ordering among GeTe, SnTe and PbTe can be explained as follows. Although the Sn element sits between Ge and Pb elements in the group 14 column of the periodic table, the energy of Sn 5s orbital is 1.15 and 1.26 eV higher than the Ge 4s and Pb 6s orbitals, respectively, and thus more delocalised in real space and closer in energy to the high-lying Te 4p orbital (Table 1). The abnormally low Ge 4s orbital in energy is due to the incomplete screening of the nucleus by the 3d orbitals, whereas the low Pb 6s orbital energy is due to the large relativistic effect of Pb.[23] The high energy and delocalised character of Sn 5s orbital lead to stronger s–p coupling in SnTe than in other two compounds. In fact, the s–p coupling (Figure 2a inset) in SnTe is so strong that the L_6^+ (Figures 1b and 2b) state is pushed up even above the L_6^- state and leads to the band inversion. Such s–p coupling also appears in the other two compounds, but it is not strong enough to leading to band inversion in GeTe and PbTe.

Besides the cation s atomic energy levels, there are several other factors that also have consequences for the band ordering. First, the cation atomic radius increases in the order $r(Ge) < r(Sn) < r(Pb)$ (Table 1) so are the bond lengths in the IV–Te compounds.

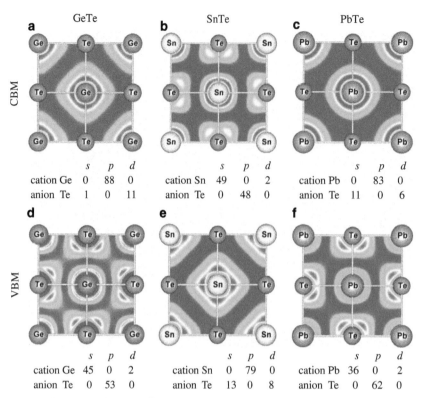

Figure 3. Wave function square (or charge distribution) of CBM and VBM states that occur at the L point for cubic (**a** and **d**) GeTe, (**b** and **e**) SnTe and (**c** and **f**) PbTe. The site and angular momentum projected wave function characters within the projector augmented wave spheres of the CBM and VBM states in percentage are listed for each state.

The shorter the bond length is, the larger the s–p repulsion could be. Among the three compounds, GeTe has the shortest bond length, so this factor will favour GeTe to have inverted band gap. Second, SOC is generally regarded as the main driving force to invert band order for three-dimensional TIs.[2] The SOC in Pb is the largest among the three cation elements as evidenced by the fact that the atomic p orbital spin–orbit splitting is 0.29, 0.71 and 2.4 eV for Ge, Sn and Pb,[24] respectively, and the spin–orbit splitting of the conduction band at the Γ point $\left[E(\Gamma_8^-) - E(\Gamma_6^-)\right]$ is 0.55, 0.81 and 2.1 eV in GeTe, SnTe and PbTe, respectively, which are in good agreement with the atomic trend. The spin–orbit splitting pushes the CBM down and VBM up and therefore will favour PbTe to having band inversion. Furthermore, the Ge 4p is 0.18 and 0.55 eV lower in energy than Sn 5p and Pb 6p orbitals, respectively. The lower energy of the cation p orbital also favors the band inversion. The net effect of the 'balancing acts' is that all the three IV–Te compounds have similar band structures with SnTe being band inverted.

It is interesting to see that the band gaps at Γ point $E_g(\Gamma_6^- - \Gamma_8^-)$ also show non-monotonic behaviour of the three compounds, but in the exactly opposite way, that is, $E_g(\Gamma_6^- - \Gamma_8^-)$ decreases from SnTe to GeTe to PbTe (Table 1). From Figure 1, it can be seen that $E_g(\Gamma_6^- - \Gamma_8^-)$ is between the bonding and anti-bonding states of anion p and cation p orbitals. SnTe has the largest $E_g(\Gamma_6^- - \Gamma_8^-)$ because Sn 5p orbital has higher energy than that of Ge 4p orbital energy. For PbTe, although Pb has even higher 6p orbital energy than Sn 5p, the larger atomic size and strong SOC of Pb reduces $E_g(\Gamma_6^- - \Gamma_8^-)$ of PbTe.

For GeTe and SnTe, the stable structure below the transition temperature T_c is actually the rhombohedral distorted a phase. The T_c is 670 K and 98 K, respectively, for GeTe and SnTe[25] in the low carrier concentration limit. For PbTe, the ferroelectric phase transition does not occur at low temperature.[26] The lattice parameters of the primitive rhombohedral cell[27–30] are $a_r = 4.293$

Å, $a_r = 58.032°$ for GeTe and $a_r = 4.475$ Å, $a_r = 59.879°$ for SnTe. The RS-to-rhombohedral ferroelectric phase transition is accompanied by a relative shift τ (0.028 for GeTe and 0.014 for SnTe) of the cation and anion sublattices along the cubic $<111>$ direction that breaks the space inversion symmetry. The four degenerate L points in RS BZ split into three L points with symmetry C_s and one Z point with symmetry C_{3v} in the rhombohedral BZ.

The calculated direct energy band gap at L point is 0.59 eV for a-GeTe (rhombohedral structure), much larger than 0.17 eV for β-GeTe (cubic RS structure) due to atom displacement in rhombohedral structure, which lowers the crystal symmetry by lifting the inversion symmetry and leads to reduced s–p coupling and admixture between the cation p valence state and anion p conduction state at L point, hence opens up a large gap between valence and conduction states. For a-SnTe, the reduced s–p coupling and strong admixture between cation p valence state and anion p conduction state leads to loss of band inversion and makes the material a normal semiconductor with a band gap of 0.20 eV (Figure 4). It is interesting to notice that if we assume there is no atom displacement in a-SnTe (i.e., $\tau = 0$), the character of VBM and CBM is same as in the β-SnTe and the band gap is still inverted.

The topological phase transition induced by the structure phase transition in SnTe is consistent with theoretical results in ref. 28. It is believed that the T_c depends strongly on the carrier concentration and the T_c can be much lower when there is high carrier concentration.[31] Although some of the reported experiments[10–17] were conducted at a temperature below the low carrier concentration T_c limit, topological phase transition induced by structure phase transition have not been observed to our knowledge. Further experimental confirmation is called for to test our prediction that there is a topological phase transition between RS β-SnTe and rhombohedral a-SnTe.

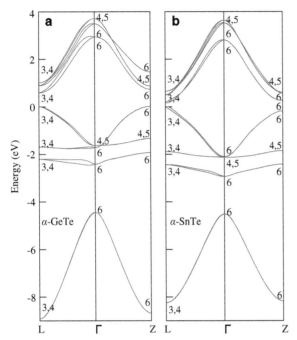

Figure 4. Calculated energy band structure of (**a**) GeTe and (**b**) SnTe in rhombohedral structure.

Conclusions

We have investigated the electronic structure of group-IV tellurides GeTe, SnTe and PbTe using first-principles calculation. The abnormal trend in their electronic band structures, i.e., a band inversion in SnTe but not in GeTe and PbTe, is explained by symmetry analysis and the atomic energy levels and sizes. We show that the low-lying cation s lone-pair band state couples strongly with the high-lying anion valence p band states at L point and this s–p coupling pushes the anion valence p band up to become the VBM. The Ge 4s orbital energy is lower than Sn 5s orbital energy due to the incomplete screening of the nucleus by the 3d orbitals. The Pb 6s orbital energy is lower than Sn 5s orbital energy due to the large relativistic effect of Pb atom. The highest cation s orbital energy of Sn among the three cation atoms, leading to the largest s–p coupling in SnTe with respect to GeTe and PbTe, makes SnTe band gap (at the L point) inverted, thus explains the peculiar chemical trends in the band structure among the three compounds. We also demonstrate that internal atom displacement in α-GeTe and α-SnTe induces strong admixture between CBM and VBM and reduces s–p coupling, thus increases the band gap. The α-SnTe is found to be a normal semiconductor because of reduced s–p coupling and admixture of p orbitals, which requires further experimental verification. Our study illustrates that besides reported SOC,[2] strain,[2] alloying[32] and interface dipole field,[33] inter-level coupling can also be an effective pathway to design TIs. Specifically, we could choose different atoms with different atomic orbital energies, apply pressure or strain to change the atom–atom bond length, regulate the charge state (Coulomb U) of an atom to control inter-level coupling.

COMPUTATIONAL METHODS

The electronic structure and total energy are calculated using first-principles density functional theory as implemented in the Vienna Ab initio simulation package.[34] The local density approximation[35,36] is used to deal the correlation effects whereas the modified Becke and Johnson exchange potential (termed TB-mBJ)[37] is adopted to improve the local density approximation description of band structures. The projector

augmented wave method[38,39] is used to treat the interaction between the valence electrons and the core electrons.[39] The valence configurations of Ge, Sn, Pb and Te atoms considered in the calculations are $3d^{10}4s^24p^2$, $4d^{10}5s^25p^2$, $5d^{10}6s^26p^2$ and $5s^25p^4$, respectively. A plane-wave expansion up to 400 eV is applied and a Γ-centred $8 \times 8 \times 8$ Monkhorst-Pack[40] k-mesh is used for the Brillouin zone (BZ) integration. Spin–orbit interaction, implemented in the Vienna Ab initio simulation package code following the approach of Kleinman and Bylander,[41] is included in the self-consistent calculations because it involves heavy elements of Ge, Sn, Pb and Te and has an important role in the topological behaviour of the compounds. The experimental lattice constants (Table 1) are used in TB-mBJ band structure calculation as suggested in ref. 42. to avoid uncertainty in the calculated band structure due to a small error in calculated lattice constants.[4] It yields band gaps of semiconductors with an accuracy comparable to GW methods but is computationally much less expensive. It is worth to note that the TB-mBJ approach remarkably improves over the original BJ approach for the description of band gaps[37] and the local density approximation error for p–p like band gap such as those in IV–Te compounds is expected to be small.[6]

ACKNOWLEDGEMENTS

This work was supported by Natural Science Foundation of China (No. 61290305, 11374259, 11374293 and 61474116). JWL was supported by the National Young 1000 Talents Plan. The work at NREL was funded by the US Department of Energy, under Contract No. DE-AC36-08GO28308.

COMPETING INTERESTS

The authors declare no conflict of interest.

REFERENCES

1 Hasan, M. Z. & Kane, C. L. Colloquium: topological insulators. *Rev. Mod. Phys.* **82**, 3045–3067 (2010).
2 Qi, X.-L. & Zhang, S.-C. Topological insulators and superconductors. *Rev. Mod. Phys.* **83**, 1057–1110 (2011).
3 Fu, L. Topological crystalline insulators. *Phys. Rev. Lett.* **106**, 106802 (2011).
4 Hsieh, T. H. *et al.* Topological crystalline insulators in the SnTe material class. *Nat. Commun.* **3**, 982 (2012).
5 Springholz, G. in *Molecular Beam Epitaxy: From Research to Mass Production* (ed. Henini M.) (Elsevier, 2013).
6 Wei, S.-H. & Zunger, A. Electronic and structural anomalies in lead chalcogenides. *Phys. Rev. B* **55**, 13605–13610 (1997).
7 Jin, S., Cai, C., Bi, G., Zhang, B., Wu, H. & Zhang, Y. Two-dimensional electron gas at the metastable twisted interfaces of CdTe/PbTe (111) single heterojunctions. *Phys. Rev. B* **87**, 235315 (2013).
8 Zhang, B. *et al.* Quantum oscillations in a two-dimensional electron gas at the rocksalt/zincblende interface of PbTe/CdTe (111) heterostructures. *Nano Lett.* **15**, 4381–4386 (2015).
9 Zhang, B. *et al.* Phonon blocking by two dimensional electron gas in polar CdTe/PbTe heterojunctions. *Appl. Phys. Lett.* **104**, 161601 (2014).
10 Tanaka, Y. *et al.* Experimental realization of a topological crystalline insulator in SnTe. *Nat. Phys.* **8**, 800–803 (2012).
11 Dziawa, P. *et al.* Topological crystalline insulator states in Pb$_{1-x}$Sn$_x$Se. *Nat. Mater.* **11**, 1023–1027 (2012).
12 Xu, S. Y. *et al.* Observation of a topological crystalline insulator phase and topological phase transition in Pb$_{1-x}$Sn$_x$Te. *Nat. Commun.* **3**, 1192 (2012).
13 Yan, C. *et al.* Experimental observation of dirac-like surface states and topological phase transition in Pb$_{1-x}$Sn$_x$Te (111) films. *Phys. Rev. Lett.* **112**, 186801 (2014).
14 Okada, Y. *et al.* Observation of dirac node formation and mass acquisition in a topological crystalline insulator. *Science* **341**, 1496–1499 (2013).
15 Zhang, D. *et al.* Quasiparticle scattering from topological crystalline insulator SnTe (001) surface states. *Phys. Rev. B* **89**, 245445 (2014).
16 Assaf, B. A. *et al.* Quantum coherent transport in SnTe topological crystalline insulator thin films. *Appl. Phys. Lett.* **105**, 102108 (2014).
17 Taskin, A. A., Yang, F., Sasaki, S., Segawa, K. & Ando, Y. Topological surface transport in epitaxial SnTe thin films grown on Bi$_2$Te$_3$. *Phys. Rev. B* **89**, 121302 (2014).
18 Madelung, O. *Semiconductors: Data Handbook* 3rd edn (Springer, 2004).
19 Koster, G. F., Dimmock, J. O., Wheeler, R. G. & Statz, H. *Properties of the Thirty-Two Point Groups* (MIT Press, 1963).

20 Harrison, W. A. *Electronic Structure and the Properties of Solids: The Physics of the Chemical Bond* (Dover, 1989).

21 Dimmock, J. O., Melngailis, I. & Strauss, A. J. Band structure and laser action in $Pb_xSn_{1-x}Te$. *Phys. Rev. Lett.* **16**, 1193–1196 (1966).

22 Dimmock, J. O. & Wright, G. B. Band edge structure of PbS, PbSe, and PbTe. *Phys. Rev.* **135**, A821–A830 (1964).

23 Phillips, J. C. *Bonds and Bands in Semiconductors* (Academic Press, 1973).

24 Herman, F., Kuglin, C. D., Cuff, K. F. & Kortum, R. L. Relativistic corrections to the band structure of tetrahedrally bonded semiconductors. *Phys. Rev. Lett.* **11**, 541–545 (1963).

25 Bussmann-Holder, A., Bilz, H. & Vogl, R. in *Dynamical Properties of IV-VI Compounds* (Springer, 1983).

26 Bate, R. T., Carter, D. L. & Wrobel, J. S. Paraelectric behavior of PbTe. *Phys. Rev. Lett.* **25**, 159–162 (1970).

27 Wdowik, U. D., Parlinski, K., Rols, S. & Chatterji, T. Soft-phonon mediated structural phase transition in GeTe. *Phys. Rev. B* **89**, 224306 (2014).

28 Plekhanov, E., Barone, P., Di Sante, D. & Picozzi, S. Engineering relativistic effects in ferroelectric SnTe. *Phys. Rev. B* **90**, 161108 (2014).

29 Goldak, J., Barrett, C. S., Innes, D. & Youdelis, W. Structure of alpha GeTe. *J. Chem. Phys.* **44**, 3323–3325 (1966).

30 Shaltaf, R., Durgun, E., Raty, J. Y., Ghosez, P. & Gonze, X. Dynamical, dielectric, and elastic properties of GeTe investigated with first-principles density functional theory. *Phys. Rev. B* **78**, 205203 (2008).

31 Iizumi, M., Hamaguchi, Y., Komatsubara, F. K. & Kato, Y. Phase transition in SnTe with low carrier concentration. *J. Phys. Soc. Jpn.* **38**, 443–449 (1975).

32 Yan, B. & Zhang, S.-C. Topological materials. *Rep. Prog. Phys.* **75**, 096501 (2012).

33 Zhang, D., Lou, W., Miao, M., Zhang, S.-C. & Chang, K. Interface-induced topological insulator transition in GaAs/Ge/GaAs quantum wells. *Phys. Rev. Lett.* **111**, 156402 (2013).

34 Kresse, G. & Furthmüller, J. Efficiency off ab-initio total energy calculations for metals and semiconductors using a plane-wave basis set. *Comput. Mat. Sci.* **6**, 15–50 (1996).

35 Ceperley, D. M. Ground state of the electron gas by a stochastic method. *Phys. Rev. Lett.* **45**, 566–569 (1980).

36 Perdew, J. P. & Zunger, A. Self-interaction correction to density-functional approximations for many-electron systems. *Phys. Rev. B* **23**, 5048–5079 (1981).

37 Tran, F. & Blaha, P. Accurate band gaps of semiconductors and insulators with a semilocal exchange-correlation potential. *Phys. Rev. Lett.* **102**, 226401 (2009).

38 Blöchl, P. E. Projector augmented-wave method. *Phys. Rev. B* **50**, 17953–17979 (1994).

39 Kresse, G. & Joubert, D. From ultrasoft pseudopotentials to the projector augmented-wave method. *Phys. Rev. B* **59**, 1758–1775 (1999).

40 Monkhorst, H. J. & Pack, J. D. Special points for Brillouin-zone integrations. *Phys. Rev. B* **13**, 5188–5192 (1976).

41 Kleinman, L. & Bylander, D. M. Efficacious form for model pseudopotentials. *Phys. Rev. Lett.* **48**, 1425–1428 (1982).

42 Kim, Y.-S., Marsman, M., Kresse, G., Tran, F. & Blaha, P. Towards efficient band structure and effective mass calculations for III-V direct band-gap semiconductors. *Phys. Rev. B* **82**, 205–212 (2010).

2

An efficient descriptor model for designing materials for solar cells

Fahhad H Alharbi[1,2], Sergey N Rashkeev[2], Fedwa El-Mellouhi[2], Hans P Lüthi[3], Nouar Tabet[1,2] and Sabre Kais[1,2]

An efficient descriptor model for fast screening of potential materials for solar cell applications is presented. It works for both excitonic and non-excitonic solar cells materials, and in addition to the energy gap it includes the absorption spectrum ($a(E)$) of the material. The charge transport properties of the explored materials are modelled using the characteristic diffusion length (L_d) determined for the respective family of compounds. The presented model surpasses the widely used Scharber model developed for bulk heterojunction solar cells. Using published experimental data, we show that the presented model is more accurate in predicting the achievable efficiencies. To model both excitonic and non-excitonic systems, two different sets of parameters are used to account for the different modes of operation. The analysis of the presented descriptor model clearly shows the benefit of including $a(E)$ and L_d in view of improved screening results.

INTRODUCTION

There has been a remarkable thrust towards developing cost-effective photovoltaics in the past two decades.[1–5] Different materials and device concepts have been deployed and the highest achieved conversion efficiency so far is 44.7% by quadruple junction using III–V materials.[6] As for the market, it is dominated by the conventional Si solar cells. Nonetheless, dye-sensitised solar cells, organic photovoltaics (OPV) and the recently emerged hybrid perovskite solar cells could become more cost effective and competitive if produced at large scale.[7]

In principle, an efficient single-junction solar cell can be made of any semiconducting material with an energy gap (E_g) ranging between 1.0 and 1.7 eV and with reasonable transport to allow the photogenerated carrier to be collected.[4,8–11] Hence, many organic and inorganic semiconductors have been used to make solar cells.[4,11] The selection was mostly based on known materials as, till recently, experimental data were the main source for screening materials for solar cells. Despite the rich data, this certainly limits the screening space. However, the sophisticated computational capabilities have provided an alternative route to explore new materials for solar cells much beyond the rich experimental data. There are many initiatives in this regard. Among the most noticeable ones is the Clean Energy Project at Harvard University.[12–14] It is a high-throughput discovery and design program for the next generation of OPV materials. By 2013, 2.3 million of organic molecules and polymers were analysed using > 150 million density functional theory calculations to assess their applicability for solar cells.[14]

The Clean Energy Project, like other initiatives,[15–17] is based on atomistic calculations, which are then fed into empirical descriptor models to assess the potential of the studied material for photovoltaics. The commonly used descriptor model, at least within the OPV community, is the one proposed by Scharber,[18] a one-parameter model based on the computed E_g, in which the open-circuit voltage (V_{oc}) is assumed to be a fixed reduction of E_g defined (by the Scharber model) as the difference between the highest occupied molecular orbital of the donor and the lowest unoccupied molecular orbital of the acceptor. The short-circuit current (J_{sc}) is estimated as a fraction of the current resulting from absorbing all incident photons above E_g, whereas the fill factor (FF) is set to a fixed value. Usually, the FF and the scaling parameter for J_{sc} are both set to a value of 0.65. Although these approximations for V_{oc} and FF appear reasonable, the assumption that all the photons above E_g are absorbed and a fraction of them is extracted as current is an extreme oversimplification. These assumptions ignore the inhomogeneity of the absorption spectrum. Furthermore, it assumes that the transport is highly efficient and that the diffusion length is much larger than the absorption length so that the detailed balance fraction of the photogenerated carriers can be collected. Practically, it is important to consider in more detail the absorption spectrum and the transport limitations.

In this paper, we propose a descriptor model where the absorption spectrum ($a(E)$) is obtained from the same electronic structure calculations used to determine E_g. E is the photon energy in eV, which is used as the unit for energy throughout in this paper. These are the only atomistic calculations needed here. In addition, the transport is characterised by the diffusion length (L_d), which is a measure for the mean distance that an excited carrier can cross through random diffusion before recombining. Calculating L_d needs lengthy calculations, which would make combinatorial screening prohibitively expensive. To avoid this, each material is given a value for L_d, which is characteristic for the family of compounds it belongs to.

The focus in this paper is on OPV as in the Scharber model. Nonetheless, the same model is applicable to other photovoltaics technologies. Yet, due to the slightly different modes of operation between excitonic (such as OPV) and non-excitonic (e.g., inorganic

[1]College of Science and Engineering, Hamad Bin Khalifa University, Doha, Qatar; [2]Qatar Environment and Energy research Institute (QEERI), Hamad Bin Khalifa University, Doha, Qatar and [3]Department of Chemistry and Applied Bioscience, ETH Zurich, Zurich, Switzerland.
Correspondence: FH Alharbi (falharbi@qf.org.qa)

semiconductor cells) solar cells, two distinct sets of parameters should be used. For non-excitonic solar cells, the binding energy of excitons is small and hence the exciton can be dissociated thermally or by potential gradient. On the other hand, in excitonic cells, where the binding energy is large, the heterojunction band offset is needed to dissociate excitons. Thus, a considerable additional loss in the voltage is unavoidable. Therefore, it is essential to make distinction between these two classes of cells.

We intend to apply our model for large-scale virtual screening of organic compounds, where the absorption spectra fluctuate considerably. Hence, we expect that by taking into account the details of their absorption spectrum, we will be able to better discriminate between candidate compounds. The initial validation analysis clearly shows the merit of including $a(E)$ and L_d in the descriptor model. Just as an example, the Scharber model suggests that copper phthalocyanine (CuPc) is better than the parent (2,4-bis[4-(N,N-diisobutylamino)-2,6-dihydroxyphenyl] squaraine) (SQ) donor. However, by including $a(E)$, the improved model shows that SQ should be more efficient than CuPc if the film thickness is < 100 nm, which is within the normal range of OPV donor thickness.[19,20]

THE PROPOSED DESCRIPTOR MODEL

As known, photovoltaics efficiency (η) is commonly expressed as

$$\eta = \frac{V_{oc} J_{sc} FF}{P_{in}}, \tag{1}$$

where P_{in} is the input power density. Certainly, many factors contribute to V_{oc}, J_{sc} and FF. The main factors are materials related. Yet, the device design, fabrication quality and operational conditions have major roles as well. For materials screening, it is reasonable to assume that the device design and quality are optimised. So, the merit of the material's potential for photovoltaic depends mainly on its optoelectronic properties. In this section, relations are proposed to link the materials properties to practical estimations for the maximum obtainable values for V_{oc}, J_{sc} and FF and hence the efficiency.

For OPV technology, the most efficient cells are bulk heterojunction (BHJ) devices. Conceptually, for materials screening, BHJ device requires multi-purpose multi-dimensional screening; i.e., a matrix of possible devices need to be screened based on a set of acceptors and a set of donors that fulfil the requirements to make a working solar cell. If the sets are small, then, the two-dimensional screening is possible. Otherwise, it can become intractable. So, most of the related large-scale screening is performed for single-layer OPV.[12,13,18] However, as BHJ devices allow for greater thickness than single-layer ones, the assumed thickness should be larger than the actual 'exciton' diffusion length (L_{Xd}), which is usually $\ll 100$ nm.[19–21] In this work, we follow the same track of materials screening for single-layer OPV, but assume that the thickness is equal to the nominal average of BHJ devices, i.e., around 100 nm.[19,20] For practical reasons, the focus is on finding a small set of promising donor materials, for which it will be later possible to find matching acceptors.

The proposed descriptor model parameters will be based on the best experimentally reported efficiencies for different organic, inorganic and organometallic materials. These data are tabulated in Tables 1–3. There are some better reported efficiencies; unfortunately, there are no details about the performance parameters of these cells. So, we limit the analysis to the best reported cells with full details. The used reference materials are:

Excitonic:SQ, DTS (5,5-bis(4-(7-hexylthiophen-2-yl)thiophen-2-yl)-[1,2,5]thiadiazolo[3,4-c]pyridine-3,3-di-2-ethylhexylsilylene-2,2'-bithiophene), CuPc, ZnPc (zinc phthalocyanine), DBP ((dibenzo([f,f]-4,4',7,7'-tetraphenyl)diindeno[1,2,3-cd:1',2',3'-lm] perylene), P3HT (poly(3-hexylthiophene-2,5-diyl), PTB7 (poly[[4,8-

Table 1. The efficiencies and cell performances for the most-efficient reported inorganic homojunction solar cells with full data

	E_g (eV)	L (μm)	V_{oc} (V)	J_{sc} (mA/cm^2)	FF	η	Reference
Si	1.12	200	0.74	41.8	82.7	25.6	63
GaAs	1.43	1	1.122	29.68	86.5	28.8	64
InP	1.35	3	0.878	29.5	85.4	22.1	65

Abbreviation: FF, fill factor.

Table 2. The efficiencies and cell performances for the most-efficient reported inorganic and hybrid heterojunction junction solar cells with full data

	E_g (eV)	L (μm)	V_{oc} (V)	J_{sc} (mA/cm^2)	FF	η	Reference
GaInP	1.81	1	1.455	16.04	89.3	20.8	66
CdTe	1.45	3	0.872	29.47	79.5	20.4	67
CIGS	1.21	2	0.752	35.3	77.2	20.5	68
MAPbI$_3$	1.42	0.3	1.07	21.5	67.0	15.4	69

Abbreviations: CIGS, CuIn$_x$Ga$_{(1-x)}$Se$_2$; FF, fill factor; MAPbI$_3$, (CH$_3$NH$_3$)PbI$_3$.

Table 3. The efficiencies and cell performances for some of the most-efficient reported organic solar cells with full data

	E_g (eV)	V_{oc} (V)	J_{sc} (mA/cm^2)	FF	η	Reference
SQ	1.47	0.86	13.6	52	6.1	70
DTS	1.40	0.78	14.4	59.3	6.7	71
CuPc	1.46	0.54	15	61	5.4	72
ZnPc	1.39	0.66	12	64	5.0	73
DBP	1.63	1.93	13.2	66	8.1	74
P3HT	1.77	0.87	11.35	75	7.4	75
PTB7	1.51	0.75	17.46	70	9.2	57

Abbreviations: CuPc, copper phthalocyanine; DTS, (5,5-bis(4-(7-hexylthiophen-2-yl)thiophen-2-yl)-[1,2,5]thiadiazolo[3,4–c]pyridine-3,3-di-2-ethylhexylsilylene-2,2'-bithiophene; FF, fill factor; PTB7, (poly[[4,8-bis[(2-ethylhexyl)oxy]benzo[1,2-b:4,5-b0]dithiophene-2,6-diyl][3-fluoro-2-[(2-ethylhexyl)carbonyl]thieno[3,4–b]thiophenediyl]]); P3HT, (poly(3-hexylthiophene-2,5-diyl); SQ, (2,4-bis[4-(N,N-diisobutylamino)-2,6-dihydroxyphenyl]squaraine); ZnPc, zinc phthalocyanine.

bis[(2-ethylhexyl)oxy]benzo[1,2-b:4,5-b0]dithiophene-2,6-diyl][3-fluoro-2-[(2-ethylhexyl)carbonyl]thieno[3,4–b]thiophenediyl]]), Non-excitonics:Si, GaAs, InP, GaInP, CdTe, CuIn$_x$Ga$_{(1-x)}$Se$_2$ (CIGS) and (CH$_3$NH$_3$)PbI$_3$ (MAPbI$_3$).

Before presenting the descriptor model, it is useful to discuss the solar photon flux density (ϕ_{ph}) and to introduce simple approximations for the maximum obtainable current density (J_{ph}). The reference density is tabulated by the American Society for Testing and Materials standard (ASTM G173-03) for AM0, AM1.5g and AM1.5d.[22] For flat panels, AM1.5g is more appropriate and it will be used in this paper. If all the photons above a given E_g are absorbed (i.e., the reflection is neglected and the thickness of the absorbing material is large enough) and each photon was to generate one exciton, the maximal photogenerated current J_{ph} is

$$J_{ph} = q \int_{Eg}^{\infty} \phi_{ph}(E)\, dE, \tag{2}$$

where q is the electron charge. J_{ph} for the different solar spectra is shown in Figure 1. In the Scharber model, J_{sc} is assumed to be a fraction of J_{ph} associated with AM1.5g spectrum.

As J_{ph} is used routinely in solar cells calculations, it would be useful to approximate J_{ph} as a function of E_g in the target range between 1 and 2 eV. This will be used later to develop the improved model. Using the data shown in Figure 1, three possible expressions for approximating J_{ph} as a function of E_g can be suggested:

$$\tilde{J}_{ph,1} = a_1 \exp\left(-b_1 E_g\right), \tag{3}$$

$$\tilde{J}_{ph,2} = A - \delta J\, E_g + a_2 \exp\left(-b_2 E_g\right) \tag{4}$$

and

$$\tilde{J}_{ph,3} = a_3 \exp\left(-b_3 E_g^{c_3}\right). \tag{5}$$

The parameters resulting in the best fit (Figure 2) are:

- for $\tilde{J}_{ph,1}$, $a_1 = 123.62$ and $b_1 = 1.0219$,
- for $\tilde{J}_{ph,2}$, $a_2 = 0.09097$, $b_2 = -2.14$, $A = 85.02$ and $\delta J = 38.69$,
- for $\tilde{J}_{ph,3}$, $a_3 = 73.531$, $b_3 = 0.440$ and $c_3 = 1.8617$.

For its simplicity and good accuracy, the expression for $\tilde{J}_{ph,3}$ will be used in this work. It will be referred to as $\tilde{J}_{ph}$, and, correspondingly, the numerical index will be dropped also for the fitting parameters a, b and c.

Parameters V_{oc}, J_{sc} and FF are tightly coupled. Thus, to have estimations using only E_g, $\alpha(E)$, and L_d, many approximations are

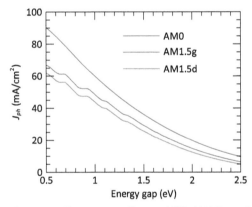

Figure 1. J_{ph} versus E_g corresponding to AM0, AM1.5g and AM1.5d spectra.

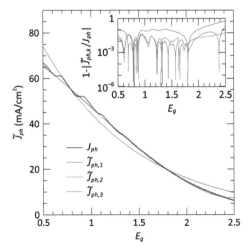

Figure 2. The approximated J_{ph} versus E_g. In the inset, the errors are plotted versus E_g in logarithmic scale.

needed. In this work, we will start by estimating V_{oc} as a function of E_g, where the extracted current is assumed as a fraction of J_{ph}. Then, E_g, $\alpha(E)$, and L_d will be used to estimate J_{sc}. Finally, FF is estimated based on V_{oc}.

The open-circuit voltage (V_{oc})

Theoretically, V_{oc} is the maximum voltage that a solar cell can apply to an external load. It is essentially the difference between electron and hole quasi-Fermi levels resulted from photoexcitation. Typically, it is assumed to be upper bounded by E_g/q, which is standardly defined—unlike the Scharber model—as the difference between highest occupied molecular orbital and lowest unoccupied molecular orbital of a single absorbing material. In the highly unlikely case of extreme charge accumulation, it can exceed the gap. Many relations between V_{oc} and E_g were suggested.[8–10,23–25] Almost all of them are based on the Shockley diode equation (assumed ideal with the identity number set to unity) when the net current vanish. This leads to

$$V_{oc} = \frac{k_B T}{q} \ln\left(\frac{J_{sc}}{J_0} + 1\right), \tag{6}$$

where k_B is the Boltzmann constant, T is the cell temperature and J_0 is the reverse saturation current density. The differences between the suggested models are due to the different assumptions for J_{sc} and J_0. For J_{sc}, in this subsection, it is assumed to be a fraction of J_{ph}. This is acceptable as the scaling constant will be considered by the fitting parameters. As for J_0, many models and empirical equations were suggested.[26–29] Among the best approximations is the Wanlass equation,[28] where the values of J_0 of many of the commonly used semiconductors are fitted to very high accuracy. According to his model,

$$J_0 = \beta(E_g) T^3 \exp\left(-\frac{E_g}{k_B T}\right), \tag{7}$$

where

$$\beta(E_g) = 0.3165 \exp(2.192\, E_g). \tag{8}$$

in mA/cm^2 K^3. Theoretically, β should be constant. However, E_g dependence is introduced empirically as a correction for

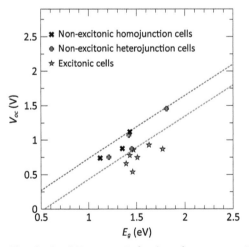

Figure 3. The obtained V_{oc} versus E_g for the reference materials. The black marks are for inorganic homojunction cells (non-excitonic), the red marks are for the inorganic and organometallic heterojunction junction solar cells (non-excitonic), and the green ones are for the organic cells (excitonic). The two dotted lines are for different values of V_L. The blue line is for $V_{L0} = 0.2$ V, which fits non-excitonic cells, and the green line is for $V_{L0} = 0.5$ V for excitonic cells.

homojunction 'solar cells' operation.[28–30] The same form was suggested also for OPVs, which are heterojunction devices, but with slightly smaller value for the prefactor.[31,32] So, by applying a fraction of $\tilde{J}_{ph}$ (Equation 5) and J_0 in equation (6) and by considering the fact that J_0 is very small quantity, we obtain

$$\tilde{V}_{oc} = E_g - \tilde{V}_L, \qquad (9)$$

where at room temperature (at 300 K)

$$\tilde{V}_L = 0.0114 E_g^{1.8617} + 0.057 E_g + V_{L0} \qquad (10)$$

and V_{L0} is used as a fitting parameter (thus the differences in the prefactors of J_0 are accounted for).

In homojunction solar cells, the losses are mainly due to the materials and the excitation; i.e., the losses due to the device design are—in principle—avoidable. For heterojunction devices, the energy offsets between the layers add to the voltage loss. For non-excitonic solar cells, the binding energy of exciton is small and hence it can be dissociated thermally or by a potential gradient. As a result, the heterojunction offset can be made small. On the other hand, in excitonic cells, the binding energy is large and the band offset is used to dissociate excitons. Thus, a considerable additional loss in the voltage is unavoidable. Therefore and as aforementioned, it is essential to make distinction between the two classes of excitonic and non-excitonic cells. The original Scharber model considers the

reduction due to band offset. However, this is routinely ignored in materials screening as it adds extra constraint on the acceptor.

Figure 3 maps the obtained V_{oc} to E_g for the reference materials. Clearly for non-excitonic cells, the blue line ($V_{L0} = 0.2$ V) line provides a good estimation for the upper limit. On the other hand, the maximum experimentally measured V_{oc} values for excitonic cells are much lower. This is mainly due to the sizable, yet needed heterojunction band offset. Thus, larger V_{L0} is indispensable. Here this parameter is set to the lowest reported voltage loss in organic cells. As can be seen in Table 3, it is 0.61 V for SQ-based solar cell leading to $V_{L0} \approx 0.5$ V.

The short-circuit current (J_{sc})

As stated in the introduction, in the Scharber model, J_{sc} is assumed to amount to a constant fraction of J_{ph}. Usually, 0.65 is used as the scaling parameter. So,

$$\tilde{J}_{sc,Sch} = 0.65 J_{ph}. \qquad (11)$$

The two most crucial deficiencies of the Scharber model, namely, the assumption of a homogeneous absorption spectrum (above the bandgap) and that the transport is very efficient such that L_d is much larger than the absorption length, can be addressed by explicitly considering the spectral inhomogeneity (using $a(E)$) and by introducing a proper characterisation of L_d. This shall result in improved predictions while not over-complicating the descriptor model.

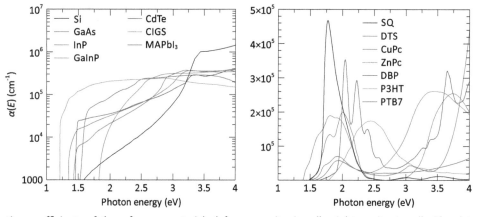

Figure 4. The absorption coefficients of the reference materials; left: non-excitonic cells, right: excitonic cells. The data are extracted from various sources.[49–56]

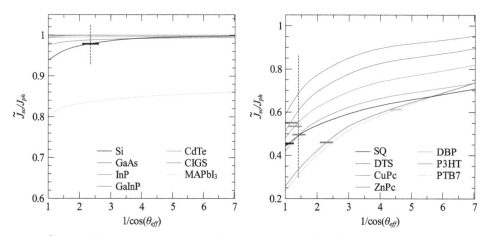

Figure 5. The ratio between $\tilde{J}_{sc}$ and J_{ph} for the reference materials at different 1/cos (θ); left: non-excitonic cells, right: excitonic cells. The small marks correspond to the maximum reported ratios. For the materials without shown marks, the ratios match the maximum possible values without scattering. This is mainly due to their strong absorptions and very efficient transports.

The absorption spectrum $a(E)$ can be computed by means of electronic structure calculations, often based on semiclassical approaches. These calculations also provide numerical values for E_g. So, there is no additional atomistic calculation needed for $a(E)$. However, the calculation of $a(E)$ from these common inputs can be computationally expensive. This fact shall be considered during the design of high-throughput screening.

Commonly, $a(E)$ is calculated by semiclassical approach where the electrons are treated quantum mechanically through the electronic structure and the field is treated classically. The details vary based on the used method for electronic structure calculations. From electronic structure calculations, the complex dielectric function $(\varepsilon(E) = \varepsilon_1(E) + i\varepsilon_2(E))$ can be calculated. $\varepsilon_2(E)$ is calculated by considering all the possible transitions from occupied to unoccupied states. For each transition, its contribution into $\varepsilon_2(E)$ is proportional to the square of the matrix element. Then, $\varepsilon_1(E)$ is calculated from $\varepsilon_2(E)$ using the Kramers–Kronig transformation. Finally, both the refractive index $n(E)$ and $a(E)$ are calculated from the relation $\sqrt{\varepsilon_1(E) + i\varepsilon_2(E)} = n(E) + ia(E)\hbar c/qE$, where $\hbar$ is the Planck constant and c is the speed of light.

As aforementioned, the transport is commonly characterised by L_d. Calculating L_d requires very time-consuming computing, which will complicate the materials screening process. To avoid that, each material is given a value of L_d, which is characteristic for the family of compounds it belongs to.

Many parameters govern L_d. Some of them are related to the intrinsic properties of materials and many others are due to the fabrication quality. For non-excitonic cells, the minority carrier diffusion is the main process and it is limited mainly by material growth quality. For defect-free indirect bandgap materials, lifetimes are in milliseconds and the mobilities are high, which give rise of few hundred microns to L_d.[33,34] However, for direct gap materials, the lifetime is significantly reduced because of the band-to-band recombination. Thus, L_d is reduced to a range between few microns and few tens of microns.[35-37] As for organometallic materials, L_d is estimated to be $>1\,\mu m$ for methylammonium lead iodide.[38]

On the other hand, the main limiting factor in excitonic solar cells is the exciton diffusion length.[39,40] The exciton cannot dissociate at the same location at which it was generated. Rather, it has to travel by hopping to the nearest interface to dissociate. Thus, the transport is limited by exciton diffusion rather than the free carrier diffusion. Exciton diffusion length in organic solar cell materials is normally $<0.1\,\mu m$.[19-21] In this work, we assume the following values for L_d as characteristic for the following material families:

- for indirect-gap semiconductors, $L_d \approx 200\,\mu m$,
- for direct-gap semiconductors, $L_d \approx 10\,\mu m$,
- for organometallic semiconductors, $L_d \approx 0.6\,\mu m$,
- for excitonic cells, $L_d \approx 0.1\,\mu m$.

Conceptually, J_{sc} is the difference between photogenerated and recombination currents, i.e.,

$$J_{sc} = J_g - J_r. \tag{12}$$

In an ideal situation, it is assumed that the thickness of the absorber layer is so large that all photons above E_g are absorbed. Practically, the carrier collection and hence the absorber layer thickness are limited by L_d. Therefore, the maximum possible photogenerated current is[41-44]

$$J_g = q \int_{E_g}^{\infty} \int_{-\pi/2}^{\pi/2} \phi_{ph}(E)\, P(\theta, \theta_{inc}, E) \left[1 - e^{-a(E)L_d/\cos(\theta)}\right] d\theta\, dE, \tag{13}$$

where $P(\theta, \theta_{inc}, E)$ is an angular distribution function that accounts for the scattering of the light at angle θ in the absorbing layer depending on the incidence angle θ_{inc} and photon energy. The scattering results in increasing—positively—the optical path of the light in the absorbing layer by a factor of $1/\cos(\theta)$. From a device-performance perspective, it is important to have L_d much larger than the absorption length $(L_a \propto 1/a(E))$. If $L_d \gg L_a$, the second term in the square bracket gets diminished and J_g increases. Otherwise, J_g is reduced considerably.

Obviously, $P(\theta, \theta_{inc}, E)$ depends on many factors mainly related to the films morphology and microstructure and the structure of interfaces. For the modelling of the distribution function, many different distributions were suggested.[42,43,45-48] In this work, to keep the model simple, we can combine the scattering effects in a

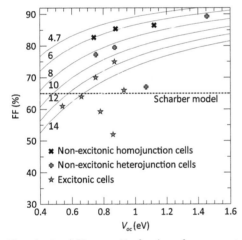

Figure 6. The obtained FF versus V_{oc} for the reference materials. The black marks are for inorganic homojunction cells (non-excitonic), the red marks are for the inorganic and hybrid heterojunction solar cells (non-excitonic) and the green ones are for the organic cells (excitonic). The solid lines are for six different values of a, whereas the dotted black line is the value suggested by Scharber model (Table 4).

Table 4. A summary of the original Scharber model and proposed model for both excitonic and non-excitonic solar cells. For the original Scharber model, ΔV is the band offset

	The original model	The proposed model for excitonic cells	The proposed model for non-excitonic cells
V_{oc}	$E_g - 0.3 - \Delta V$	$E_g - 0.5 - 0.0114E_g^{1.8617} - 0.057E_g$	$E_g - 0.2 - 0.0114E_g^{1.8617} - 0.057E_g$
J_{sc}	$0.65 J_{ph}(E_g)$	$\tilde{J}_g(E_g, a(E), L_d, \pi/4)$	$\tilde{J}_g(E_g, a(E), L_d, \pi/2.75)$
FF	0.65	$V_{oc}/(V_{oc} + 12k_BT_c)$	$V_{oc}/(V_{oc} + 6k_BT_c)$

Abbreviation: FF, fill factor.

single effective angle θ_{eff}. So, J_g becomes

$$\tilde{J}_g(E, \alpha(E), L_d, \theta_{\text{eff}}) = q \int_{E_g}^{\infty} \phi_{\text{ph}}(E) \left[1 - e^{-\alpha(E)L_d / \cos(\theta_{\text{eff}})} \right] dE$$

(14)

The way θ_{eff} is determined will be shown at the end of this subsection.

As for the recombination, there are many mechanisms contributing to it. This is accounted for empirically by L_d and it can be adjusted further by a proper fitting of θ_{eff}. So, J_{sc} can be approximated by a similar form of equation (14), but with a slightly different θ_{eff}, that will be determined based on the actual performances and absorption spectra of the reference materials.

Figure 4 shows the absorption spectra of the reference materials, which are extracted from various sources.[49–56] For the known non-excitonic cells, it is evident that due to the extended absorption spectrum above E_g, $\alpha(E)$ is smooth, whereas the organic materials show a strongly fluctuating bands. For example, the absorption of SQ is strong only between 1.5 and 2.3 eV. Thus, a large portion of solar radiation is not absorbed due to the fact that the device thickness is small.

To determine θ_{eff}, the ratios $\tilde{J}_{\text{sc}}/\tilde{J}_{\text{ph}}$ are calculated and plotted against $1/\cos(\theta_{\text{eff}})$ (Figure 5). For non-excitonic cells, the effect of θ_{eff} for most materials is negligible due to their strong absorption and due to the fact that the growth quality of the studied materials is high and hence the assumed L_d is large. The exceptions are for Si due to its weak absorption and for MAPbI$_3$ due to its relatively short L_d. For Si, $\theta_{\text{eff}} \approx \pi/2.75$ is needed to match the obtained J_{sc}. So, this value will be used for non-excitonic cells. For excitonic cells, $\theta_{\text{eff}} = \pi/4$ is a good approximation for most of the studied materials. The exception is for PTB7, where the difference between the reported J_{sc} and the calculated value from absorption spectrum is high. This can be due to an extremely efficient light trapping used to make the cell.[57] However, to match most of the reported maximum values, $\theta_{\text{eff}} = \pi/4$ is suitable and will be used for excitonic cells.

The fill factor

The third performance parameter is the FF. Practically, many physical mechanisms contribute to it and consequently many models have been suggested to estimate it.[58–61] Generally, the suggested models are based on the relationship between current and voltage; but with different assumptions on the causes and values of shunt and series resistances. FF is usually represented as a function of V_{oc}, which depends as shown above on E_g.

One of the simplest—yet reasonably accurate–forms suggested by Green[61] for conventional inorganic semiconductors is

$$FF = \frac{V_{\text{oc}}}{V_{\text{oc}} + ak_BT}$$

(15)

Originally, he suggested $a = 4.7$. However, this factor can be adjusted for other solar cell technologies. On the basis of the best reported cells, $a = 6$ and $a = 12$ fit better the upper limits of the measured FF for non-excitonic and excitonic solar cells, respectively, as shown in Figure 6, where T is the room temperature.

MODEL IMPLEMENTATION AND EVALUATION

The parameterized expressions used for the three performance factors (V_{oc}, J_{sc} and FF) in the original Scharber model and the presented descriptor are summarised in Table 4.

In the first analysis, we compare the predictions of the two models for excitonic materials with the available experimental data (Table 3). For the original Scharber model, three reasonable values[12,14,62] for ΔV are assumed in the analysis; namely, 0.2, 0.3 and 0.4 V. The results are shown in Figure 7. As can be observed,

the improved model outperforms considerably the original Scharber model in the estimations of J_{sc}, except for PTB7. For V_{oc}, the presented model provides good estimation for most of the studied materials. As for the original Scharber model, this depends obviously on ΔV. For η, the presented model outperforms in most cases the original one. The original model performs

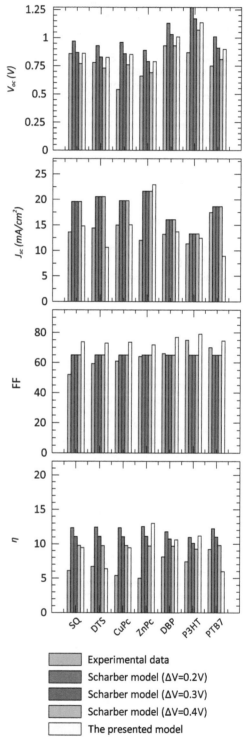

Figure 7. Comparison for organic (excitonic) solar cell performances of the relevant reference materials as estimated by the presented and the Scharber (for three different values for ΔV) models, and as experimentally published for the best reported cells.

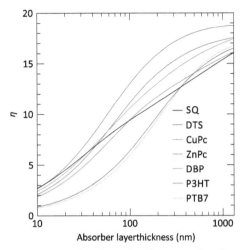

Figure 8. The estimated solar cell efficiencies for the reference organic (excitonic) materials as a function of the absorbing layer thickness.

generally better only in the estimation of *FF*. However, *FF* depends extremely on the device design and optimisation, unlike other parameters that are mostly materials dependent. The improved model suggests that the obtained *FF* values are smaller than the predictions. Thus, there is a reasonable room for improvement through device optimisation.

To illustrate the importance of considering L_d, the presented method is used to estimate conversion efficiencies of the array of organic materials studied as a function of thicknesss as shown in Figure 8. Clearly, material potential ranking varies with the thickness. For example, for very thin films ($L < 50$ nm), SQ is predicted to show better efficiency when compared with all other materials, except ZnPc. Also, PTB7- and DTS-based solar cells would result in better efficiencies when compared with CuPc and SQ only for relatively thicker films.

In the next analysis, the improved model is applied to the reference non-excitonic materials. As aforementioned, the original Scharber model was designed for OPVs. It can be adjusted to also work for inorganic cells in a similar way as our proposed model, i.e., by working with two sets of parameters. Figure 9 shows a comparison between the reported experimental efficiencies and the estimated ones by the proposed model for non-excitonic cells. Again, the model provides very good estimations. For the well-optimised devices such as Si and GaAs, the presented model suggests that the room of improvement is limited. However, it indicates that there is a possibility to considerably improve the performance of MAPbI3, CIGS, CdTe and InP.

The last analysis is for the effect the absorber layer thickness on the expected efficiencies for the non-excitonic solar cell materials. The results are shown in Figure 10. The expected efficiencies for all the studied materials saturate after few micrometre except for Si solar cells. It takes very thick layer to reach a reasonable efficiency. As known, this is due to its weak absorption.[4,5,47,63]

CONCLUSION

A descriptor model for solar cell efficiencies estimation is developed. Relative to the original Scharber model, the developed model presented here revisits the three main performance factors (V_{oc}, J_{sc} and *FF*; Equation 1). For the short-circuit current (J_{sc}), the model takes in full account the details of the absorption spectrum $\alpha(E)$ to evaluate the photogenerated current (J_g), and uses new and more elaborate parametrization for the other components contributing to this quantity, i.e., the scattering distribution and

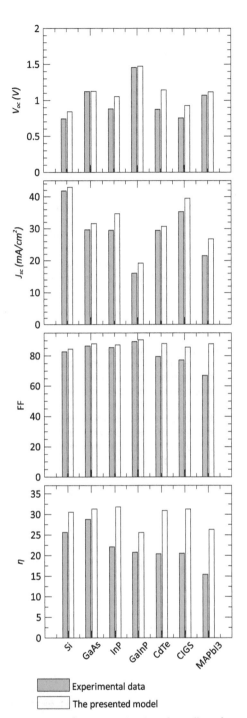

Figure 9. Comparison for non-excitonic solar cell performances of the relevant reference materials as estimated by the presented and as experimental published for the best reported cells.

the diffusion length, which characterize the transport and the recombination. The open-current voltage V_{oc} is expressed in terms of a power series of the energy gap fitted against available experimental data. The FF is estimated using adjusted empirical model originally suggested by M. Green.[61] Using two different sets of parameters, the model can be used for both excitonic and non-excitonic materials.

The analysis of the new model shows that its much better performance arises from the improved predictions for the open-circuit voltage and the short-circuit current (V_{oc} and J_{sc}). On the other hand, the estimation of the original Scharber model for the

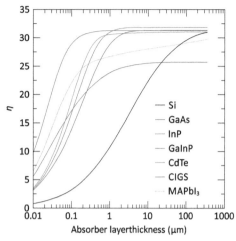

Figure 10. The estimated non-excitonic solar cell efficiencies for the relevant reference materials as a function of the absorbing layer thickness.

FF is slightly better than our model when compared with the experimentally reported values. However, as FF depends extremely on the device design and optimisation, the larger values estimated by the proposed model indicate that the performance of the reference cells can be improved by proper device optimisation.

We expect the proposed descriptor model to allow for more accurate assessments of the performance of light harvesting materials. Even though the results of material screening efforts based on this model are still missing, the model was already shown to be useful to study the change of performance of a given material with the variation of some device parameter such as the layer thickness.

COMPETING INTERESTS

The authors declare no conflict of interest.

REFERENCES

1 Azimi, H., Hou, Y. & Brabec, C. J. Towards low-cost, environmentally friendly printed chalcopyrite and kesterite solar cells. *Energy Environ. Sci.* **7**, 1829–1849 (2014).

2 Habas, S. E., Platt, H. A., van Hest, M. F. & Ginley, D. S. Low-cost inorganic solar cells: from ink to printed device. *Chem. Rev.* **110**, 6571–6594 (2010).

3 Wadia, C., Alivisatos, A. P. & Kammen, D. M. Materials availability expands the opportunity for large-scale photovoltaics deployment. *Energy Environ. Sci.* **43**, 2072–2077 (2009).

4 Alharbi, F. *et al.* Abundant non-toxic materials for thin film solar cells: alternative to conventional materials. *Renew. Energy* **36**, 2753–2758 (2011).

5 Hossain, M. & Alharbi, F. Recent advances in alternative material photovoltaics. *Mater. Technol.* **28**, 88–97 (2013).

6 Dimroth, F. *et al.* Wafer bonded four-junction gainp/gaas//gainasp/gainas concentrator solar cells with 44.7% efficiency. *Prog. Photovoltaics Res. Appl.* **22**, 277–282 (2014).

7 Snaith, H. J. Perovskites: the emergence of a new era for low-cost, high-efficiency solar cells. *J. Phys. Chem. Lett.* **4**, 3623–3630 (2013).

8 Nelson, J., Kirkpatrick, J. & Ravirajan, P. Factors limiting the efficiency of molecular photovoltaic devices. *Phys. Rev. B* **69**, 035337 (2004).

9 Markvart, T. The thermodynamics of optical étendue. *J. Opt. A* **10**, 015008 (2008).

10 Kirk, A. A discussion of fundamental solar photovoltaic cell physics. *Physica B* **423**, 58–59 (2013).

11 Alharbi, F. H. & Kais, S. Theoretical limits of photovoltaics efficiency and possible improvements by intuitive approaches learned from photosynthesis and quantum coherence. *Renew. Sustainable Energy Rev.* **43**, 1073–1089 (2015).

12 Hachmann, J. *et al.* The harvard clean energy project: large-scale computational screening and design of organic photovoltaics on the world community grid. *J. Phys. Chem. Lett.* **2**, 2241–2251 (2011).

13 Olivares-Amaya, R. *et al.* Accelerated computational discovery of high-performance materials for organic photovoltaics by means of cheminformatics. *Energy Environ. Sci.* **4**, 4849–4861 (2011).

14 Hachmann, J. *et al.* Lead candidates for high-performance organic photovoltaics from high-throughput quantum chemistry-the harvard clean energy project. *Energy Environ. Sci.* **7**, 698–704 (2014).

15 Landis, D. D. *et al.* The computational materials repository. *Comput. Sci. Eng.* **14**, 51–57 (2012).

16 Castelli, I. E. *et al.* Computational screening of perovskite metal oxides for optimal solar light capture. *Energy Environ. Sci.* **5**, 5814–5819 (2012).

17 Curtarolo, S. *et al.* The high-throughput highway to computational materials design. *Nat. Mater.* **12**, 191–201 (2013).

18 Scharber, M. C. *et al.* Design rules for donors in bulk-heterojunction solar cells towards 10% energy-conversion efficiency. *Adv. Mater.* **18**, 789–794 (2006).

19 Hoppe, H. & Sariciftci, N. S. Organic solar cells: An overview. *J. Mater. Res.* **19**, 1924–1945 (2004).

20 Forrest, S. R. The limits to organic photovoltaic cell efficiency. *MRS Bull.* **30**, 28–32 (2005).

21 VajjalaáKesava, S. *et al.* Direct measurements of exciton diffusion length limitations on organic solar cell performance. *Chem. Commun.* **48**, 5859–5861 (2012).

22 ASTM G173-03: Standard tables for reference solar spectral irradiances: direct normal and hemispherical on 37° tilted surface. Technical Report. Available at. http://www.astm.org/Standards/G173.htm (2008).

23 Abrams, Z. R., Gharghi, M., Niv, A., Gladden, C. & Zhang, X. Theoretical efficiency of 3rd generation solar cells: comparison between carrier multiplication and down-conversion. *Sol. Energy Mater. Sol. Cells* **99**, 308–315 (2012).

24 Würfel, P., Brown, A., Humphrey, T. & Green, M. Particle conservation in the hot-carrier solar cell. *Prog. Photovoltaics Res. Appl.* **13**, 277–285 (2005).

25 Baruch, P., De Vos, A., Landsberg, P. & Parrott, J. On some thermodynamic aspects of photovoltaic solar energy conversion. *Sol. Energy Mater. Sol. Cells* **36**, 201–222 (1995).

26 Fan, J., Tsaur, B. & Palm, B. In *Proc. 16th IEEE Photovoltaic Specialists Conference* 692–701 (1982).

27 Nell, M. E. & Barnett, A. M. The spectral pn junction model for tandem solar-cell design. *IEEE Trans. Electron Devices* **34**, 257–266 (1987).

28 Wanlass, M. *et al.* Practical considerations in tandem cell modeling. *Sol. Cells* **27**, 191–204 (1989).

29 Coutts, T. A review of progress in thermophotovoltaic generation of electricity. *Renew. Sustainable Energy Rev.* **3**, 77–184 (1999).

30 Coutts, T. J., Emery, K. A. & Scott Ward, J. Modeled performance of polycrystalline thin-film tandem solar cells. *Prog. Photovoltaics Res. Appl.* **10**, 195–203 (2002).

31 Potscavage, W. J. Jr, Yoo, S. & Kippelen, B. Origin of the open-circuit voltage in multilayer heterojunction organic solar cells. *Appl. Phys. Lett.* **93**, 193308 (2008).

32 Katz, E. *et al.* Temperature dependence for the photovoltaic device parameters of polymer-fullerene solar cells under operating conditions. *J. Appl. Phys.* **90**, 5343–5350 (2001).

33 Stutenbaeumer, U. & Lewetegn, E. Comparison of minority carrier diffusion length measurements in silicon solar cells by the photo-induced open-circuit voltage decay (ocvd) with different excitation sources. *Renew. Energy* **20**, 65–74 (2000).

34 Fortini, A., Lande, R., Madelon, R. & Bauduin, P. Carrier concentration and diffusion length measurements by 8-mm-microwave magnetophotoreflectivity in germanium and silicon. *J. Appl. Phys.* **45**, 3380–3384 (1974).

35 Leon, R. In *Proc. 19th IEEE Photovoltaic Specialists Conference* **vol. 1**, 808–812 (1987).

36 Mataré, H. & Wolff, G. Concentration enhancement of current density and diffusion length in iii-v ternary compound solar cells. *Appl. Phys. A* **17**, 335–342 (1978).

37 Gautron, J. & Lemasson, P. Photoelectrochemical determination of minority carrier diffusion length in ii-vi compounds. *J. Cryst. Growth* **59**, 332–337 (1982).

38 Stranks, S. D. *et al.* Electron-hole diffusion lengths exceeding 1 micrometer in an organometal trihalide perovskite absorber. *Science* **342**, 341–344 (2013).

39 Gregg, B. A. Excitonic solar cells. *J. Phys. Chem. B* **107**, 4688–4698 (2003).

40 Gregg, B. A. The photoconversion mechanism of excitonic solar cells. *MRS Bull.* **30**, 20–22 (2005).

41 Poruba, A. *et al.* Optical absorption and light scattering in microcrystalline silicon thin films and solar cells. *J. Appl. Phys.* **88**, 148–160 (2000).

42 Krc, J., Zeman, M., Smole, F. & Topic, M. Optical modelling of thin-film silicon solar cells deposited on textured substrates. *Thin Solid Films* **451**, 298–302 (2004).

43 Krc, J., Smole, F. & Topic, M. Analysis of light scattering in amorphous si: H solar cells by a one-dimensional semi-coherent optical model. *Prog. Photovoltaics Res. Appl.* **11**, 15–26 (2003).

44 Müller, J., Rech, B., Springer, J. & Vanecek, M. Tco and light trapping in silicon thin film solar cells. *Sol. Energy* **77**, 917–930 (2004).

45 Battaglia, C. *et al.* Light trapping in solar cells: can periodic beat random? *ACS Nano* **6**, 2790–2797 (2012).

46 Callahan, D. M., Munday, J. N. & Atwater, H. A. Solar cell light trapping beyond the ray optic limit. *Nano Lett.* **12**, 214–218 (2012).

47 Cho, J.-S. *et al.* Effect of nanotextured back reflectors on light trapping in flexible silicon thin-film solar cells. *Sol. Energy Mater. Sol. Cells* **102**, 50–57 (2012).

48 Kowalczewski, P., Liscidini, M. & Andreani, L. C. Light trapping in thin-film solar cells with randomly rough and hybrid textures. *Opt. Express* **21**, A808–A820 (2013).

49 Sze, S. M. & Ng, K. K. *Physics of Semiconductor Devices* (John Wiley & Sons, 2006).

50 Weber, M. J. *Handbook of Optical Materials* vol. 19 (CRC press, 2002).

51 Theodoropoulou, S., Papadimitriou, D., Anestou, K., Cobet, C. & Esser, N. Optical properties of cuin$_{1-x}$ga$_x$se2 quaternary alloys for solar-energy conversion. *Semicond. Sci. Technol.* **24**, 015014 (2009).

52 Gamliel, S. & Etgar, L. Organo-metal perovskite based solar cells: sensitized versus planar architecture. *RSC Adv.* **4**, 29012–29021 (2014).

53 Wei, G. *et al.* Functionalized squaraine donors for nanocrystalline organic photovoltaics. *ACS Nano* **6**, 972–978 (2011).

54 Duché, D. *et al.* Optical performance and color investigations of hybrid solar cells based on p3ht: Zno, pcpdtbt: Zno, ptb7: Zno and dts (ptth$_2$)$_2$: Zno. *Sol. Energy Mater. Sol. Cells* **126**, 197–204 (2014).

55 Fujishima, D. *et al.* Organic thin-film solar cell employing a novel electron-donor material. *Sol. Energy Mater. Sol. Cells* **93**, 1029–1032 (2009).

56 Senthilarasu, S. *et al.* Characterization of zinc phthalocyanine (znpc) for photovoltaic applications. *Appl. Phys. A* **77**, 383–389 (2003).

57 He, Z. *et al.* Enhanced power-conversion efficiency in polymer solar cells using an inverted device structure. *Nat. Photon.* **6**, 591–595 (2012).

58 Qi, B. & Wang, J. Fill factor in organic solar cells. *Phys. Chem. Chem. Phys.* **15**, 8972–8982 (2013).

59 Taretto, K., Soldera, M. & Troviano, M. Accurate explicit equations for the fill factor of real solar cells applications to thin-film solar cells. *Prog. Photovoltaics Res. Appl.* **21**, 1489–1498 (2013).

60 Gupta, D., Mukhopadhyay, S. & Narayan, K. Fill factor in organic solar cells. *Sol. Energy Mater. Sol. Cells* **94**, 1309–1313 (2010).

61 Green, M. A. Accuracy of analytical expressions for solar cell fill factors. *Sol. Cells* **7**, 337–340 (1982).

62 Dou, L. *et al.* 25th anniversary article: a decade of organic/polymeric photovoltaic research. *Adv. Mater.* **25**, 6642–6671 (2013).

63 Masuko, K. *et al.* Achievement of more than 25% conversion efficiency with crystalline silicon heterojunction solar cell. *IEEE J. Photovoltaics* **4**, 1433–1435 (2014).

64 Kayes, B. M. *et al.* in *Proc. 37th IEEE Photovoltaic Specialists Conference* 4–8 (IEEE, 2011).

65 Keavney, C., Haven, V. & Vernon, S. in *Proc. 21st IEEE Photovoltaic Specialists Conference* 141–144 (IEEE, 1990).

66 Geisz, J., Steiner, M., Garca, I., Kurtz, S. & Friedman, D. Enhanced external radiative efficiency for 20.8% efficient single-junction gainp solar cells. *Appl. Phys. Lett.* **103**, 041118 (2013).

67 Huber, W. H. & Duggal, A. R. in *Thin Films for Solar and Energy Technology* Vol. **9177** (SPIE Proc, 2015).

68 Osborne, M. Hanergys solibro has 20.5% CIGS solar cell verified by NREL. Available at. http://www.pv-tech.org/news/hanergys_solibro_has_20.5_cigs_solar_cell_verified_by_NREL (2014).

69 Liu, M., Johnston, M. B. & Snaith, H. J. Efficient planar heterojunction perovskite solar cells by vapour deposition. *Nature* **501**, 395–398 (2013).

70 Chen, G. *et al.* Co-evaporated bulk heterojunction solar cells with $>$ 6.0% efficiency. *Adv. Mater.* **24**, 2768–2773 (2012).

71 Sun, Y. *et al.* Solution-processed small-molecule solar cells with 6.7% efficiency. *Nat. Mater.* **11**, 44–48 (2012).

72 Xue, J., Rand, B. P., Uchida, S. & Forrest, S. R. A hybrid planar-mixed molecular heterojunction photovoltaic cell. *Adv. Mater.* **17**, 66–71 (2005).

73 Fleetham, T. B. *et al.* Efficient zinc phthalocyanine/c60 heterojunction photovoltaic devices employing tetracene anode interfacial layers. *ACS Appl. Mater. Interfaces* **6**, 7254–7259 (2014).

74 Xiao, X., Bergemann, K. J., Zimmerman, J. D., Lee, K. & Forrest, S. R. Small-molecule planar-mixed heterojunction photovoltaic cells with fullerene-based electron filtering buffers. *Adv. Energy Mater.* **4**, 1301557 (2014).

75 Guo, X. *et al.* High efficiency polymer solar cells based on poly (3-hexylthiophene)/indene-c 70 bisadduct with solvent additive. *Energy Environ. Sci.* **5**, 7943–7949 (2012).

3

Computational understanding of Li-ion batteries

Alexander Urban[1], Dong-Hwa Seo[2] and Gerbrand Ceder[1,3]

Over the last two decades, computational methods have made tremendous advances, and today many key properties of lithium-ion batteries can be accurately predicted by first principles calculations. For this reason, computations have become a cornerstone of battery-related research by providing insight into fundamental processes that are not otherwise accessible, such as ionic diffusion mechanisms and electronic structure effects, as well as a quantitative comparison with experimental results. The aim of this review is to provide an overview of state-of-the-art *ab initio* approaches for the modelling of battery materials. We consider techniques for the computation of equilibrium cell voltages, 0-Kelvin and finite-temperature voltage profiles, ionic mobility and thermal and electrolyte stability. The strengths and weaknesses of different electronic structure methods, such as DFT+U and hybrid functionals, are discussed in the context of voltage and phase diagram predictions, and we review the merits of lattice models for the evaluation of finite-temperature thermodynamics and kinetics. With such a complete set of methods at hand, first principles calculations of ordered, crystalline solids, i.e., of most electrode materials and solid electrolytes, have become reliable and quantitative. However, the description of molecular materials and disordered or amorphous phases remains an important challenge. We highlight recent exciting progress in this area, especially regarding the modelling of organic electrolytes and solid–electrolyte interfaces.

INTRODUCTION

During the last two decades, lithium-ion battery technology has made possible impressive advances in mobile consumer electronics and electric vehicles.[1–4] Electrochemical technology for grid-level energy storage additionally bears the potential to enable the transition away from fossil energy towards renewable energy sources.[5,6] To satisfy the increasing demand for low-weight and low-volume batteries with high-energy storage capacity, researchers have been exploring options to increase the specific energy and energy density of lithium-ion batteries.[7] At the same time, the ability to quickly charge and discharge a battery, i.e., the rate capability, is not only critical for the usability of portable devices, such as smart phones, but is also an important factor to render electric vehicles competitive (imagine taking gas would take hours instead of minutes). Despite the need for better performing lithium batteries, the recent battery fires on commercial airplanes are a reminder that safety issues must not be neglected when developing new materials.

Lithium batteries are collections of electrochemical cells, each composed of two electrodes that are separated by an electrolyte. The battery functions by shuttling lithium ions between the electrodes, which typically are intercalation materials, through the electrolyte. The driving force for this process is the difference of the lithium chemical potential in the two electrodes. During discharge, the cathode material (low lithium chemical potential) is electrochemically reduced by intercalating lithium ions from the electrolyte and taking up electrons from an external circuit. Simultaneously, the anode material (high lithium chemical potential) is oxidised. The resulting electric current through the

external circuit can be used to perform work, i.e., to run an electronic device. The above process is essentially reversed when the battery is recharged by applying an external potential. To sustain the cell reactions with minimal overpotential, electrode materials have to be good electronic and lithium ionic conductors. The electrolyte, on the other hand, needs to conduct lithium ions but has to be electronically insulating to prevent short circuiting. Today, most commercial batteries employ carbon-based anode materials, though research is in progress to investigate alternatives.[8–10] Typical cathode materials are based on rocksalt-type lithium transition-metal oxides that provide high-energy densities, such as $LiCoO_2$, or polyanionic materials with high-rate capability, such as $LiFePO_4$. Commercial electrolytes are organic solutions of lithium salts,[11] but solid lithium ion conductors are under scientific investigation.[12,13]

In the past, the development of new materials took place exclusively in the experimental laboratories, often by trial and error, and largely depending on the researchers' experience and intuition. With today's computational capabilities, however, computer simulations have become an integral part of materials design. Atomistic simulations based on first principles, i.e., simulations that are directly based on physical laws, are especially invaluable, as they provide insight into processes on the atomic and electronic scales without requiring any input from experiment, thereby aiding in the interpretation of experimental observations. For applications in materials science, the electronic density-functional theory (DFT) developed by Kohn and coworkers[14–16] has been particularly successful.

[1]Department of Materials Science and Engineering, University of California, Berkeley, CA, USA; [2]Department of Materials Science and Engineering, Massachusetts Institute of Technology, Cambridge, MA, USA and [3]Materials Science Division, Lawrence Berkeley National Laboratory, Berkeley, CA, USA.
Correspondence: G Ceder (gceder@berkeley.edu)

In recent years, computer-assisted design of entirely new materials has made tremendous progress, not least owing to automated high-throughput calculations that facilitate the rapid screening of thousands of target materials.[17] Computational investigations of specific battery materials have recently been reviewed by Meng and Arroyo-de Dompablo[18] and by Islam and Fisher.[19] Here we will instead focus on the present state of first-principles-based methods for the simulation of lithium-ion battery properties. In the following sections, we will review computational approaches to key properties of lithium-ion batteries, namely the calculation of equilibrium voltages and voltage profiles, ionic mobilities and thermal as well as electrochemical stability. Past research efforts in the field have been predominantly geared towards the discovery and optimisation of cathode materials, and thus most examples from the literature cited in the following are for applications to cathode materials. We stress, however, that the techniques are directly transferable to other battery components, such as anode materials,[20–23] solid electrolytes,[12,13,24] and electrode surface coatings.[25]

EQUILIBRIUM VOLTAGE

One way to increase the specific energy and energy density of a battery is by increasing the cell voltage as much as possible without exceeding the stability window of the electrolyte. In the following section 'Equilibrium cell voltage from first principles' we will discuss the general first-principles framework for the computation of equilibrium cell voltages. The accuracy of computational voltage predictions and strategies for its improvement are considered in section 'Self-interaction and the accuracy of first-principles voltages'.

Equilibrium cell voltage from first principles

The equilibrium lithium intercalation voltage is determined by the difference in lithium chemical potential, μ_{Li}, between cathode and anode

$$V = -\frac{\mu_{Li}^{cathode} - \mu_{Li}^{anode}}{z\,F}, \qquad (1)$$

where z is the charge that is transferred, and F is the Faraday constant. The lithium chemical potential is the change of the free energy of the electrode material with lithium content.[26,27] Integrating Equation (1) over a finite amount of reaction gives the average voltage as function of the free energy change of the combined anode/cathode reaction (Nernst equation)

$$\overline{V} = -\frac{\Delta G_r}{z\,F}. \qquad (2)$$

At low temperatures, the entropic contributions to ΔG_r are small, and the reaction free energy can be approximated by the internal energy, $\Delta G_r \approx \Delta E_r$.

Within this approximation, the equilibrium voltage of a lithium transition-metal oxide intercalation cathode with composition $LiMO_2$ and a lithium metal anode with the cell reaction

$$Li_{x_1}MO_2 \xrightarrow{yields} Li_{x_2}MO_2 + (x_1 - x_2)Li, \qquad (3)$$

can thus be computed as

$$\overline{V}(x_1, x_2) \approx -\frac{E(Li_{x_1}MO_2) - E(Li_{x_2}MO_2) - (x_1 - x_2)E(Li)}{(x_1 - x_2)F}$$

$$\text{with} \quad x_1 > x_2, \qquad (4)$$

where the internal energies of the lithiated and delithiated phases, $E(Li_{x_1}MO_2)$ and $E(Li_{x_2}MO_2)$, and of metallic (body-centered cubic) lithium, $E(Li)$, can be obtained from first principles. Obviously, this approximation is not limited to intercalation electrodes, but can also be applied to conversion or displacement reactions.

At this point it is worth to step back and look at the first principles approximation to the equilibrium voltage, Equation (4). All that is required to compute the voltage are three independent first principles calculations for $Li_{x_1}MO_2$, $Li_{x_2}MO_2$, and Li, and the energy of BCC lithium is independent of the cathode material and hence only needs to be computed once. This means, the average intercalation voltage of, e.g., layered $LiCoO_2$ can be estimated simply based on the results of DFT calculations of $LiCoO_2$ and the delithiated CoO_2, calculations that can be done within minutes on a current desktop computer. Note that if phases with intermediate lithium concentrations are known, Equation (4) can be used to compute a piece-wise approximation to the voltage curve. However, the real challenge lies in the determination of the relevant thermodynamically stable phases $Li_{x_1}MO_2$ and $Li_{x_2}MO_2$ and their respective crystal structures, a topic that will be addressed in section 'Voltage profiles'.

Self-interaction and the accuracy of first-principles voltages

Most early DFT calculations for battery materials were based either on the local density approximation (LDA)[15] or the generalised gradient approximation (GGA).[28] Aydinol et al.[27,29,30] (for the case of LDA) and Deiss et al.[31] (GGA) demonstrated that the approach outlined above reproduces experimental voltage trends. However, a systematic underestimation of experimentally measured voltages of layered lithium transition-metal oxides by up to 1.0 V was found. Despite this large systematic error, DFT calculations were instrumental for the understanding of the fundamental electronic structure origin of redox levels. To this extent, early LDA calculations revealed that the average intercalation voltage of layered $LiMO_2$ (M = first-row transition metals) increases with the atomic number of M, as a result of the increasingly covalent character of the M-O bond in these materials.[27,32] The same general trend was later also confirmed for polyoxianionic cathode materials.[33,34] In addition, LDA predicted a decrease in the intercalation voltage with the electronegativity of the anion, $V(LiMO_2) > V(LiMS_2) > V(LiMSe_2)$.[27,35] The impact of the host structure on the voltage is generally found to be small in comparison to the voltage variations upon cation doping and exchange of the transition-metal species.[27,35]

The reason for the failure of DFT to predict the voltages of lithium transition-metal oxides quantitatively correct can be traced back to a general problem of (semi-)local DFT. The mean-field approximation to the electrostatic interaction of the electrons in DFT introduces a spurious self-interaction, i.e., the interaction of each electron with itself that is not fully canceled out by LDA and GGA functionals. This self-interaction error results in an artificial delocalisation of the electrons that leads to significant errors in systems with strongly correlated electrons.[36] The true electron density of many transition-metal oxides, for example, exhibits strong features of the individual transition-metal atoms with localised d electrons at the metal centers that are not captured by conventional DFT calculations. In fact, GGA incorrectly predicts metallic ground states for many insulating transition-metal oxides that are relevant for lithium ion batteries.[36] Applications that solely rely on relative energies, for example the energetic comparison of different polymorphs, may benefit from error cancellation which effectively reduces the self-interaction error. However, modelling redox reactions requires computing the energy difference of materials with very different electronic configurations, e.g., transition-metal oxides in their oxidised and reduced states, and thus no such error cancellation can be hoped for.

A practical approach to correct the self-interaction error is offered by the DFT+U method,[37–39] which essentially introduces a penalty term for partial occupations, favouring the disproportionation into fully occupied and empty states.[36] The method lends ideas from the Hubbard model Hamiltonian that was developed

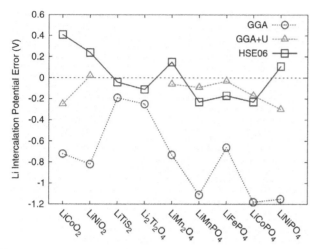

Figure 1. Comparison of the errors in intercalation voltages calculated with GGA (red circles), GGA+U (green triangles), and the HSE hybrid functional (blue squares).[50] The Hubbard U value for Ti is equal to zero, so that GGA = GGA+U.

to investigate the electronic structure of strongly correlated materials. The magnitude of the DFT+U correction is controlled by a system and implementation specific *Hubbard U* parameter that can either be determined self-consistently from linear response theory[36,40] or by fit to reference band structures or formation energies.[41,42] For a given Hubbard U, DFT+U calculations are no more computationally demanding than conventional DFT calculations.

Zhou *et al.*[41] showed that DFT+U calculations using self-consistent U parameters, significantly improve the accuracy of predicted voltages over pure GGA for transition-metal oxides and polyanionic materials, reducing the deviation from the experimental voltages to about ± 0.1 V. The authors report that self-consistent U parameters determined for either the oxidised or the reduced cathode materials result in comparable accuracy. In addition, DFT+U also correctly describes charge localisation and disproportionation, which is not captured by pure GGA/LDA calculations.[43] Yahia *et al.* assessed the impact of different U values on the relative phase stability of different $LiMSO_4F$ (M = Fe, Mn) polymorphs and found only a small dependence, demonstrating the transferability of the U parameter.[44]

Another avenue to reduce the self-interaction error of LDA/GGA is to exactly evaluate the exchange energy, i.e., the non-classical contribution to the electron–electron interaction, based on the expressions from the Hartree–Fock (HF) method.[45,46] Exchange-correlation functionals that implement this approach are called hybrid functionals, as they essentially are a blend of DFT and HF. Instead of requiring additional parameters for selected atomic orbitals (as in DFT+U calculations), the only adjustable parameter in hybrid functionals is the amount of exact exchange, which is only weakly system dependent.[46,47] Note that the long-ranged nature of the electrostatic potential makes it computationally demanding to evaluate the exact exchange energy in periodic calculations. Range-limited hybrid functionals, such as the Heyd-Scuseria-Ernzerhof (HSE) functional,[48,49] reduce the computational effort to some extent by considering only short-ranged contributions. Nevertheless, hybrid-functional calculations scale less favourable with the system size as conventional DFT calculations and are therefore computationally significantly more demanding.

Chevrier *et al.*[50] showed that hybrid functional DFT calculations using the HSE functional predict redox potentials with a similar accuracy as GGA+U, albeit without the need for additional parameters. Even with HSE significant discrepancies can exist between predicted and measured voltage for systems with

complex electronic structure, such as Li_xCoO_2, which undergoes a metal–insulator transition when delithiated.[51,52] Seo *et al.*[47] found that system-specific adjustment of the amount of exact exchange against reference band gaps from spectroscopy can further improve the accuracy of equilibrium voltages and voltage profiles computed with hybrid functional DFT. A comparison of equilibrium voltages of lithium transition-metal oxides and phosphates computed with GGA, GGA+U and HSE06 are shown in Figure 1.

VOLTAGE PROFILES

Many lithium-cathode materials exhibit stable phases at intermediate lithium concentrations, such as lithium-vacancy orderings in intercalation materials or different atomic orderings in alloys. If the structures of all phases are known, then the equilibrium voltage between the intermediate phases can be computed using Equation (4) of the previous section.

An early example is the calculation of the 0-K voltage profile of the Li–Sn alloy by Courtney *et al.*[53] who considered six intermediate Li_xSn phases known from experiment, demonstrating excellent agreement with experimental voltage curves. However, experimental information about intermediate phases is not always available. In particular, when designing novel materials, their phase diagrams are not known *a priori*. A purely first-principles approach for the prediction of stable phases is thus desirable.[54]

0-K intercalation voltage curves

The relevant quantity to compare the stability of different phases at 0 K is the formation energy with respect to stable reference materials. For the example of an intercalation voltage curve for a lithium transition-metal oxide with formula unit $LiMO_2$, the formation energy of any structure with intermediate lithium content can be expressed as

$$E_f(Li_xMO_2) = E(Li_xMO_2) - x\,E(LiMO_2) - (1-x)\,E(MO_2), \quad (5)$$

where E is the internal (DFT) energy and the fully lithiated $LiMO_2$ and delithiated MO_2 phases are the relative energy reference. The formation energies of all Li_xMO_2 phases that are (at 0 K) thermodynamically stable compared with the reference phases lie on the lower convex hull of E_f versus composition x.[55,56]

Once the convex hull construction is available, a piecewise voltage profile can be obtained by evaluating the equilibrium voltage, Equation (4), between phases of adjacent lithium concentrations. Figure 2a shows an example of a formation energy hull construction and the corresponding 0 K voltage curve for sodium intercalation in Na_xMnO_2, a system with a particularly large number of stable intermediate phases.[57] Although technically not a lithium battery system, Figure 2a perfectly illustrates the relationship between the formation-energy hull and the 0-K voltage profile, and it includes the measured voltage profile for reference. As seen in the figure, the computational voltage profile (red solid lines) traces the experimental profile (black dotted lines) well, indicating that phases that are stable at 0 K also dominate at operation conditions.

Given a set of trial atomic structures, the convex hull construction can be used to identify the stable phases among those structures, but the trial structures need to be obtained from somewhere in the first place. In the case of conversion materials, likely candidates for intermediate structures can sometimes be guessed based on the structures of known phases and chemical intuition.[58] However, for intercalation systems, all intermediate phases belong to the same host structure, i.e., lithium ions and vacancies occupy a common sublattice. Although this general problem of finding the distribution of a species over available lattice sites can be solved by the cluster expansion technique, as has been demonstrated for Li_xCoO_2 (refs 55,59) and Li_xNiO_2,[60]

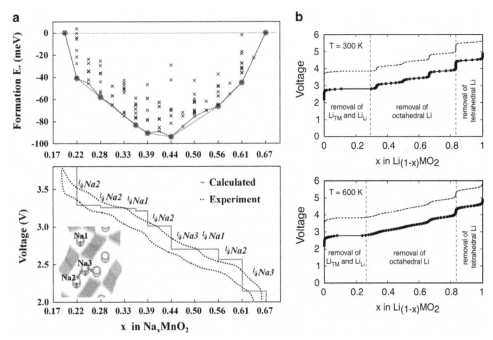

Figure 2. (**a**) Formation energy convex hull construction (top) and voltage profile (bottom) for sodium intercalation in Na_xMnO_2.[57] Top: the formation energies of all considered phases are shown as black crosses, and stable phases are highlighted with filled red circles. Bottom: The computational voltage profile (solid red lines) is overlaid with the experimental voltage curves (black dotted lines) for comparison. (Copyright: American Chemical Society) (**b**) Computational voltage profiles for lithium intercalation in $Li_x(Ni_{1/2}Mn_{1/2})O_2$ at 300 K (top) and 600 K (bottom) as obtained from Monte-Carlo simulations using a cluster expansion Hamiltonian.[81]

one usually defers to a simpler enumeration of likely arrangements using a technique that was originally developed by Hart et al.[61–63] for the enumeration of atomic orderings in alloys. However, the low-vacancy and low-lithium-concentration regimes require large simulations cells, and for intermediate concentrations the number of possible lithium/vacancy orderings even in small cells may become exceedingly large due to the combinatorial explosion, preventing the first principles energy evaluation of all configurations. Intuitively, one could argue that the most stable phases should be those lithium/vacancy orderings that distribute both species homogeneously over the available sites, as such arrangements minimise the electrostatic repulsion between positively charged lithium ions. Following this line of reasoning, Hautier et al. proposed to rank the enumerated configurations by their electrostatic energy, assuming fixed atomic charges, before considering only the N lowest energy configurations for first principles calculations.[64] This approach is generally useful to determine atomic orderings for crystal structures with fractional site occupancies: For example, Kim et al. employed the enumeration technique in conjunction with filtering by electrostatic energy to identify stable $LiMn_{0.5}Fe_{0.375}Mg_{0.125}BO_3$ phases in which Mn, Fe and Mg share a common sublattice.[65]

We note in passing that the Python Materials Genomics package provides a convenient interface for the enumeration of atomic configurations and allows filtering by electrostatic energy.[66]

Finite-temperature voltage profiles

Much of the discrepancies between the experimental voltage curve and the piecewise computational approximation in Figure 2a arises from temperature effects that were neglected in the previous section. Indeed, some of the sodium/vacancy orderings (shown as black crosses in Figure 2a) that are not ground states are within a small energy above the convex hull, so that they may become accessible at finite temperatures.

Including temperature effects in computational voltage profiles requires knowledge of the internal energy and the entropy of the thermalised system to minimise the temperature-dependent free energy of the system. A standard technique for the sampling of the configurational space at some given conditions is the Metropolis Monte-Carlo (MC) method, which stochastically samples the system's states with their correct (Boltzmann) probabilities in a set statistical ensemble.[67,68] However, for the convergence of thermodynamic ensemble averages, MC simulations typically require on the order of millions of energy evaluations and atomic configurations that are too large to be calculated easily by DFT.

The cluster expansion[69–71] previously mentioned is an elegant approach to map accurate DFT energy models onto a simpler Hamiltonian, which captures the dependence of the energy on the Li/vacancy distribution, and can be evaluated fast enough to be used in MC simulations. In the case of lithium/vacancy orderings that only differ in the arrangement of lithium ions and vacancies on their respective sublattice, the largest contribution to entropy can be expected to be due to the configurational degrees of freedom. Although the magnitude of the vibrational entropy may be significant, its variations across phases of the same chemical species are typically small so that it does not much affect the relative phase stability.[72] In addition, electronic contributions to the configurational entropy due to the ordering of transition-metal ions of the same chemical species but in different oxidation states (electron-hole ordering) may become significant in the presence of highly localised electrons.[73,74] Fortunately, the atomic interactions in crystalline solids can be well discretised to lattice models, which allow rapid energy evaluations of configurations of several thousand atoms. In the cluster expansion method, sometimes also referred to as generalised Ising model or lattice gas model,[69–71] the total energy of an atomic configuration $s = \{\sigma_i\}$, i.e., a particular lithium/vacancy ordering, is expressed as an expansion over energy contributions by clusters $\{a\}$ of lattice sites

(i, j, k), $\varphi_\alpha = \sigma_i \sigma_j \ldots \sigma_k$

$$E_s = E(\{\sigma_i\}) = J_0 + \sum_{\alpha}^{\text{clusters}} J_\alpha \, \varphi_\alpha, \tag{6}$$

where J_0 is a constant shift and the expansion coefficients J_α are the effective cluster interactions (ECIs). For most relevant systems, the ECIs decay rapidly with the number of sites within the cluster (n-body interaction) and with their distance from each other (interaction range), so that the expansion (Equation (6)) quickly converges and may be truncated in good approximation. Several standard methods are available to determine the non-zero ECIs $\{J_\alpha\}$ via fit to a small number of first-principles reference energies.[75–77]

The cluster expansion method and related lattice-based models have been applied to various battery materials. Early examples from the literature are the simulation of the room-temperature voltage profile of the Li–Al alloy by Reimers and Dahn,[78] and the investigation of lithium-vacancy orderings in Li_xCoO_2 by Van der Ven et al.[55,59] Wolverton and Zunger also applied first-principles-based cluster expansion to cation (Li/Co) and lithium-vacancy ordering in Li_xCoO_2, reproducing the experimentally observed layered ground state and reporting the temperature-dependence of the voltage profile.[79,80] Van der Ven and Ceder used MC simulations based on a cluster expansion Hamiltonian to compute the temperature dependence of the lithium intercalation voltage profile of $Li(Ni_{1/2}Mn_{1/2})O_2$ (Figure 2b).[81] As seen in Figure 2b, temperature effects mainly smooth out the voltage steps that are present in the 0-K approximation, resulting in a continuous voltage slope instead, so that the difference between finite-temperature and 0-K voltage profiles are often small. As such, the 0-K approximation is usually sufficient to gain qualitative insight into the charge mechanism and to estimate the energy density. More recently, Lee and Persson[82] employed a cluster expansion model to determine the structural phases that give rise to the different features in the voltage profile of $Li_xNi_{0.5}Mn_{1.5}O_4$. Yu et al.[83] calculated the phase diagram of spinel $Li_xCu_yTiS_2$, a material that undergoes lithium displacement reactions, using MC simulations based on a cluster expansion Hamiltonian as part of a multiscale simulation effort.

IONIC MOBILITY

Apart from the energy density, the rate capability is another key factor for the design of new battery materials. The time that is required for a full charge is, for example in the case of a cell phone or electric vehicle, critical for the usability of the device. The discharge rate, on the other hand, determines the amount of power that can be delivered by the battery.

One rate-critical process in lithium intercalation batteries is the extraction of lithium atoms from and their reinsertion into the host structures of the electrode materials. The intercalation rate can either be limited by electric conductivity or ionic conductivity. In this section, we shall focus on the latter, i.e., on the mobility of lithium ions within the intercalation host. Note that the same considerations also apply for ionic diffusion through solid electrolytes.

On the microscopic scale, the chemical diffusivity can be expressed in terms of the Einstein relation[84,85]

$$D = \Theta \frac{1}{2dt} \lim_{t \to \infty} \frac{\left\langle \frac{1}{N} \vec{R}(t)^2 \right\rangle}{t} \quad \text{with} \quad \vec{R}(t) = \sum_{i=1}^{N} \vec{r}_i(t), \tag{7}$$

where d is the dimensionality of the diffusion, e.g., $d = 2$ for lithium diffusion in layered oxides, $R(t)$ is the total displacement of all N diffusing particles during the time period t, $\vec{r}_i(t)$ are the individual particle displacements, and $\langle \ldots \rangle$ indicates the statistical average. The thermodynamic factor Θ accounts for the

fluctuations of the chemical potential of the diffusing species, μ, with the concentration and can be expressed as

$$\Theta = \frac{c}{k_B T} \frac{\partial \mu}{\partial c} = \frac{\partial \mu / (k_B T)}{\partial \ln c}. \tag{8}$$

Note that the thermodynamic factor is related to the negative slope of the voltage profile, $\partial \mu / \partial c = -xF \, \partial V / \partial c$, as evident by comparison of Equation (8) with the expression of the voltage, Equation (1), in terms of the lithium chemical potential.

In the following section 'Direct simulation of the diffusion dynamics', we consider the evaluation of the diffusivity by direct ab initio molecular dynamics simulation of the ionic diffusion. However, in many cases transition-state theory based on 0-K diffusion barriers provides sufficient insight into lithium migration (section 'Diffusivity based on 0-K migration barriers'). In section 'Conceptual insight into lithium diffusion', we discuss how the understanding of lithium diffusion on the atomic scale can be used to devise design criteria to optimise macroscopic lithium transport.

Direct simulation of the diffusion dynamics

The migration of a lithium ion from one site to another is an activated process with a free energy barrier. One way to gain insight into lithium ion diffusion and its microscopic diffusion mechanisms is by direct ab initio molecular dynamics (AIMD) simulations. In AIMD simulations, the atomic forces from first principles (i.e., quantum mechanics) methods are used to propagate the atoms in the system according to the laws of classical mechanics. For a general introduction to the subject we refer the reader to standard text books.[86,87]

AIMD simulations can most readily address the limit of self diffusion, i.e., the case where no concentration gradient is present, for which the thermodynamic factor, Θ of Equation (8), is equal to one, as the explicit evaluation of Θ in AIMD simulations is not straightforward. In the dilute limit, the calculation of the diffusivity further simplifies, as the time-correlations between individual particle positions can be neglected, $\vec{r}_i \vec{r}_j \approx 0$. The ensemble average of the total displacement in Equation (7) can, hence, be replaced by the atomic mean squared displacement (MSD), r^2, so that the diffusion coefficient for $\Theta \approx 1$ becomes

$$D \approx D^* = \frac{1}{2dt} \lim_{t \to \infty} \frac{\overline{\vec{r}(t)^2}}{t} \quad \text{with} \quad \overline{\vec{r}(t)^2} = \frac{1}{N} \sum_{i=1}^{N} \left\langle \vec{r}_i(t)^2 \right\rangle. \tag{9}$$

This approximation is beneficial, as averaging over all N atoms improves the sampling statistics, and hence the ensemble average converges more rapidly with the simulation time. Note that D^* is also called the tracer diffusion coefficient. As pointed out by Alder et al.,[88] the equivalent long-time limit of the change of the MSD with time converges faster

$$D^* = \frac{1}{2dt} \lim_{t \to \infty} \frac{\partial \overline{\vec{r}(t)^2}}{\partial t}, \tag{10}$$

and hence this expression is usually used in practice. Equation (10) provides an efficient approach to obtain the diffusivity from AIMD simulations at constant temperature and volume, i.e., simulations in the canonical (NVT) statistical ensemble. One simply has to compute the MSD and in a plot of the MSD against the simulation time the slope is $2d\,D$.

Note that for systems with highly correlated diffusion, the tracer diffusion coefficient may not be a good approximation, and the diffusivity should be directly evaluated according to Equation (7), i.e., based on the ensemble average of the total displacement, requiring longer MD trajectories to achieve convergence. An in-depth discussion of the different approximations to the diffusivity and their validity and relationship can be found in reference.[89]

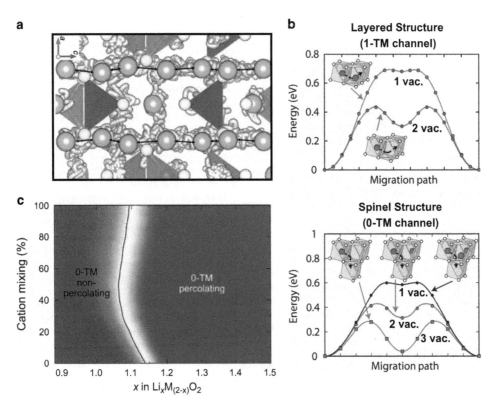

Figure 3. (a) Lithium ion positions (small pink spheres) during an AIMD simulation of lithium diffusion through $Li_{10}GeP_2S_{12}$.[12] The initial lithium sites are indicated by large green spheres, sulfur atoms are yellow, and PS_4 and GeS_4 tetrahedra are shown in purple. (Copyright: *American Chemical Society*) (b) Minimum energy paths obtained from NEB calculations for lithium diffusion through layered and spinel $LiTiS_2$ at different local lithium concentrations (1, 2, and 3 lithium vacancies).[56] Lithium diffusion in the layered structure occurs via 1-TM channels whereas 0-TM channels are present in the spinel structure. (Copyright: American Chemical Society) (c) Map of the lithium percolation probability in cation-mixed $Li_xM_{2-x}O_2$ (adapted from ref. 113). The layered structure corresponds to a cation mixing of 0%, and 100% is the disordered rocksalt structure. Lithium concentrations that result in percolating diffusion channels are shown in blue, whereas non-percolating regions are shaded red.

The direct AIMD simulation of the lithium diffusivity at operation temperature is generally challenging because of the low lithium diffusivities. For the case of lithium diffusion through typical electrode materials, D is of the order of 10^{-10} to 10^{-6} cm^2/s at 400 K,[90] i.e., the mean displacement an atom experiences in one picosecond (at least 500 AIMD steps) is 0.001 to 0.1 Å. AIMD simulations of several nanoseconds would be required to observe ionic migration, yet longer trajectories to converge the value of the diffusivity. However, if the diffusion mechanism is independent of the temperature, then the diffusivity is often found to follow the Arrhenius law

$$D(T) \approx D_0 \, e^{-\frac{E_a}{k_B T}}, \qquad (11)$$

where E_a is the activation energy of the diffusion. In practice, the diffusivity can therefore be evaluated at elevated temperatures (500–1500 K) at which shorter AIMD trajectories suffice, and values at lower temperatures can be obtained by extrapolation of log D.

For comparison with experimental results, it is useful to relate the lithium ion diffusivity to the *ionic conductivity* σ via the Nernst–Einstein relation

$$\sigma(T) = \frac{N_{Li} \, e^2}{V \, k_B \, T} \, D(T), \qquad (12)$$

where N_{Li} is the number of lithium ions, V is the volume of the simulation cell, e is the electronic charge, and T is the temperature.

In the context of lithium ion batteries, Yang and Tse[91] reported AIMD simulations of lithium diffusion in $LiFePO_4$, identifying a diffusion mechanism that involves the creation of Li – Fe anti-sites. Mo *et al.* used AIMD simulations to estimate the lithium diffusivity in $Li_{10}GeP_2S_{12}$, a super ionic conductor material and prospective

solid electrolyte, and report a dependence of the diffusivity on the lattice direction, originating from differences in the diffusion pathways.[12] An AIMD trajectory from this work is shown in Figure 3a. The same authors employed AIMD simulations to identify the sodium diffusion pathways in P2-Na_xCoO_2 and related materials.[92] Hao and Wolverton investigated lithium transport in the amorphous electrode coating materials Al_2O_3 and AlF_3 using AIMD simulations.[93] Xiao *et al.* used AIMD simulations to assess lithium diffusion in the transition-metal layer of lithium rich Li_2MnO_3, observing a diffusion mechanism perpendicular to the layer.[94]

Diffusivity based on 0 K migration barriers

Although AIMD simulations in principle provide a parameter-free route to calculate ionic diffusivities, they are computationally quite demanding. Each step of an AIMD simulation basically is a separate DFT calculation, and the convergence of the diffusivity typically takes on the order of tens of thousands of AIMD steps.[12] Therefore, brute-force AIMD simulations should be the last resort, in cases where no further information about the lithium diffusion mechanism in the system is available, or when the lithium diffusion mechanism is too complex to capture with a simple hopping mechanism.

Consider the microscopic mechanism of lithium hopping, which is at the origin of diffusion. Each such individual hopping event requires a Gibbs free energy of activation, $\Delta G^{\ddagger}$, that is given by the free energy difference of the initial state (i.e., the lithium ion in its original site) and the energetically highest state that has to be overcome during the diffusion, the transition state. According to transition-state theory,[95,96] the rate k at which the hopping process occurs can be expressed as

$$k(T) = \nu^*(T)\, e^{-\frac{\Delta G^{\ddagger}(T)}{k_B T}}, \tag{13}$$

in which ν^* is a temperature-dependent effective attempt frequency. When the *hopping* distance between adjacent sites, a, is known, the chemical diffusivity in the dilute carrier limit can be approximately obtained from the rate as $D(T) = a^2\, k(T)$.[97] However, for systems that follow the Arrhenius law, Equation (10), the pre-exponential coefficient ν^* and the activation energy can be assumed to be independent of the temperature. By neglecting the change of entropy during the diffusion, $\Delta G^{\ddagger}$ can be approximated by a 0-K activation energy $\Delta E^{\ddagger} = E^{\ddagger} - E_i$, where $E^{\ddagger}$ and E_i are the energies of the transition state and the initial state, respectively. Typical values for the prefactor ν^* in Equation (13) are 10^{11} to $10^{13}\,\text{s}^{-1}$.[98]

On the basis of the knowledge of the microscopic diffusion rates k of Equation (13), the actual ionic diffusivity, Equation (7), can be approximated as $D \approx g \cdot a^2 \cdot k$, where g is a geometric factor and a is the hopping distance between two adjacent sites.[99] The geometric factor is usually close to 1 and therefore often neglected,[56] however, an explicit calculation of the diffusivity is feasible with kinetic Monte-Carlo (kMC) simulations on lattice models,[85,100,101] such as the ones discussed in section 'Finite temperature voltage profiles'. By combining the energies of transition states with the cluster-expanded energies of lattice configurations, a complete study of lithium diffusion including composition and ordering dependence can be undertaken.[85] Lattice-based Monte-Carlo simulations have the additional advantage over AIMD simulations that the thermodynamic factor, Equation (8), can be directly evaluated in grand-canonical simulations.[98]

An efficient algorithm for the computation of transition-state energies is the nudged elastic band (NEB) method.[102,103] The NEB algorithm requires the initial and final states of the diffusion as input, from which it generates a number of intermediate states, the images, by linear interpolation. The minimum energy path (MEP) connecting initial and final states is then determined by concurrent minimisation of the atomic forces in all images subject to a harmonic coupling between neighbouring images. See Figure 3b for example MEPs for lithium diffusion in LiTiS$_2$. In our experience, the NEB method is very robust and reliably converges the MEP, as long as the electronic structure of the system does not significantly change during the migration. For practical purposes, this means that GGA+U calculations (section 'Self-interaction and the accuracy of first-principles voltages') tend to be more problematic, as electrons might be localised at different atomic centers along the diffusion path, which may result in the simultaneous diffusion of a polaron that gives rise to an additional charge-transfer barrier.[104] A common strategy to decouple the ionic diffusion barrier from the effects of electron-hole interactions is to turn to plain GGA calculations in which electrons are more delocalised (see also section 'Self-interaction and the accuracy of first-principles voltages').[105]

NEB calculations have clarified the diffusion mechanism in several important cathode materials. As quantitative data and mechanistic information is difficult to obtain experimentally, lithium diffusion is an area where computation has been particularly useful. In early work, Van der Ven and Ceder[90,106] employed NEB calculations to understand the lithium diffusion mechanism in layered LiCoO$_2$, identifying a high-barrier diffusion path through an oxygen–oxygen bond and a low-barrier di-vacancy pathway via a tetrahedrally coordinated activated state. This di-vacancy mechanism controls lithium diffusion at all practical concentrations and has become generally accepted as the reason why layered cathodes have good lithium mobility. The authors subsequently carried out kMC simulations at different lithium concentrations, predicting maximal lithium diffusivity in

partially delithiated phases in agreement with experiment. Morgan et al.[97] estimated the effect of cation substitution on the lithium diffusivity in olivine LiMPO$_4$ (M = Mn, Fe, Co, Ni) from NEB diffusion barriers and TST, finding a one-dimensional diffusion mechanism and generally low activation barriers on the order of 100–300 meV, which set the stage for understanding the size-dependence behaviour of lithium diffusion in LiFePO$_4$ (ref. 107) and led to the evaluation of the very high rate LiFePO$_4$.[108] Van der Ven et al.[98] employed MC and kMC simulations to investigate lithium diffusion in Li$_x$TiS$_2$ accounting for the concentration-dependent thermodynamic factor, Θ. Θ is found to vary four orders of magnitude between the dilute lithium and the dilute vacancy limit due to lithium–lithium interaction. Kang and coworkers employed AIMD simulations at high temperature to identify the lithium diffusion pathways in Al-doped LiGe$_2$(PO$_4$)$_3$, a solid electrolyte, and subsequently used NEB calculations to converge diffusion barriers and to estimate the lithium diffusivity.[109] The simulations predict that Al doping introduces an alternative diffusion mechanism that enhances the lithium diffusivity compared with the undoped material. Du et al.[110] computed the lithium migration barrier in the crystallographic c direction of Li$_{10}$GeP$_2$S$_{12}$ with NEB.[12] The result of 230 meV is slightly larger than the value of 170 meV obtained from AIMD by Mo et al. (see previous section).[12,110] Finally, a recent review of lattice-based simulations of lithium diffusion in intercalation materials can be found in ref. 56

Conceptual insight into lithium diffusion

The previous two sections dealt with different approaches for the computational estimation of lithium mobility by simulating lithium diffusion. Although these computational tools are invaluable to understand new materials, sometimes useful concepts can be identified for entire materials classes.

Conceptually, lithium diffusion is best understood for rocksalt-type lithium transition-metal oxides. In fully lithiated LiMO$_2$ phases, each cation (lithium or transition metal (TM)) is octahedrally coordinated by six oxygen atoms, and lithium diffusion from one octahedral site to another octahedral site takes place via a tetrahedral activated state (o–t–o migration).[106] Figure 3b depicts the MEPs for lithium o–t–o migration through the layered and spinel structures at different local lithium concentrations, and the local minimum halfway corresponds to the activated state.[56] The lower migration barrier in the layered structure (red curve in the top panel of Figure 3b) corresponds to the di-vacancy mechanism mentioned above, i.e., the additional lithium site adjacent to the activated tetrahedral state is vacant.[90,106] As shown for the specific case of LiTiS$_2$ in Figure 3b (top), the di-vacancy mechanism reduces the migration barrier by around 300 meV or almost 50%, which makes it the dominant mechanism in this material. Vacancies lower the migration barrier as the energy of the activated state is mainly determined by the electrostatic repulsion between the activated lithium ion and its neighbouring cations.[111] In the layered structure, every diffusion channel passes along a (TM) ion, i.e., each activated tetrahedral lithium atom has one neighbouring site that is occupied by a TM (1-TM diffusion channels). As a consequence of the electrostatic repulsion between the migrating lithium and the TM, the distance to this neighbouring TM ion (as controlled by the lattice parameters) and the valence of the TM species can be tuned to optimise lithium mobility, as demonstrated by Kang et al.[112] for nickel based layered LiMO$_2$.

The spinel structure also exhibits diffusion channels without an adjacent TM ion, and as a consequence a tri-vacancy diffusion mechanism without any adjacent cations becomes available upon lithium extraction (0-TM channels). The migration barrier associated with 0-TM channels is yet lower (green line in the bottom panel of Figure 3b) and is mostly independent of the TM

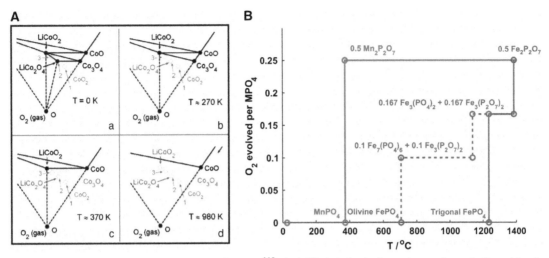

Figure 4. (**a**) Temperature evolution of the Li-Co-O₂ phase diagram.[118] Black filled circles indicate thermodynamically stable phases, and blue empty circles belong to metastable and unstable phases. (Copyright: American Chemical Society) (**b**) Oxygen evolution from MnPO₄ and FePO₄ as function of the temperature.[119]

species in the material. Lee *et al.*[113] showed that 0-TM channels are also present in cation-disordered materials, but that an excess of lithium is required so that they form a percolating network throughout the material and can be utilised for fast macroscopic lithium transport. The lithium concentrations that enable percolating 0-TM channels for different degrees of cation mixing in layered LiMO₂ are mapped as the blue area in Figure 3c. As shown in the figure, the percolation threshold is around 10% excess lithium in the disordered structure (100% cation mixing) and has a minimum of about 6% in partially cation-mixed layered materials (50% cation mixing). Urban *et al.* subsequently generalised this percolation model to other rocksalt-type lithium transition-metal oxides with arbitrary cation order, identifying partially cation-disordered spinel structures as a class of fast lithium ion conductors.[114]

THERMAL AND ELECTROCHEMICAL STABILITY

Although the motivation for the development of new battery materials is foremost to improve the performance of lithium-ion batteries, the safety of the technology must be maintained. In their charged state, many lithium battery materials tend to be thermally unstable. In particular, oxide-based cathode materials, such as LiMO₂ (M = Mn, Co, Ni), experience, as they become more oxidised, a thermodynamic driving force for electrochemical reduction by release of oxygen gas

$$Li_xM_yO_{z+z'} \rightarrow Li_xM_yO_z + \frac{z'}{2}O_2, \qquad (14)$$

a reaction that is potentially exothermic and may thus lead to thermal runaway and ignition of the electrolyte. Oxygen release is of particular importance at particle surfaces, where it may result in the formation of other surface phases or at high temperature lead to combustion of the electrolyte.[115] Decomposition will also occur when the cell voltage exceeds the stability window of the electrolyte, resulting in electrochemical oxidation of the electrolyte molecules. Computational handles to estimate the stability of the cathode and the electrolyte under operation conditions are therefore an essential part of *in silico* design of new battery components.

Thermal stability of cathode materials
The kinetics of cathode decomposition and reaction with the electrolyte is complex and outside of the capabilities of today's

ab initio methods, though interesting work to understand the reaction paths that lead to anode/electrolyte passivation have been explored with *ab initio* approaches.[116] Hence, focus has been on understanding better the driving force for thermal decomposition to other phases. The first step in assessing the thermal stability is the computational construction of a phase diagram that captures the more reduced phases well. As discussed in the section '0-K intercalation voltage curves' for the case of voltage profiles, a 0-K phase diagram can then be obtained by constructing the lower convex hull of the formation energies of all relevant phases. Not unlike voltage calculations, the construction of phase diagrams requires accurate free-energy differences of electronically different phases. Hence, it is clear that the spurious DFT self-interaction (section 'Self-interaction and the accuracy of first-principles voltages') will introduce significant errors in phase diagrams of transition-metal oxides due to the artificial delocalisation of the transition-metal *d* electrons. Another large error in the formation energies of oxides arises from an artificial stabilisation of the O₂ molecule in GGA and LDA calculations. These systematic errors in local DFT need to be addressed as thermal decomposition originates from the competition of the charged state with more reducing phases. Wang *et al.* analysed these two error contributions and proposed a correction scheme based on (i) a constant energy shift of the DFT O₂ energy to account for overbinding, and (ii) a Hubbard U correction (section 'Self-interaction and the accuracy of first-principles voltages') using U values fitted to experimental oxide formation energies.[42] Following this approach it was possible to reduce the error in the formation energies of transition-metal oxides from up to 1.0 eV with GGA to less than 0.1 eV in most cases.[42]

For solid phases, 0 K phase diagrams are generally a good first approximation for the true, finite temperature phase diagrams. For example in the case of LiFePO₄, Ong *et al.* found that the 0 K Li–Fe–P–O phase diagram well predicts the experimentally known stable phases.[117] However, temperature dependence has to be reintroduced to estimate thermal stability. In particular, the oxygen release in reaction (14) results in a significant increase in entropy that has to be accounted for, for instance by expressing the Gibbs free energy of the oxygen release reaction (14) as[118]

$$\Delta G_r \approx E(Li_xM_yO_z) + \frac{z'}{2}E^*(O_2) - E(Li_xM_yO_{z+z'}) - \frac{z'}{2}TS_{O_2}^{p_0}(T), \qquad (15)$$

where the entropy for oxygen gas at a given temperature, $S_{O_2}^{p_0}(T)$, is obtained from experimental thermochemical data for a reference oxygen partial pressure p_0, and the energy of the O₂ molecule is corrected as proposed by Wang. In this approximation, temperature dependence is exclusively due to the O_2 entropy. The thermodynamic decomposition temperature can be estimated by solving for the temperature where ΔG_r vanishes. Note, that there is also a kinetic barrier connected with the oxygen release, which may increase the effective decomposition temperature in particular for systems with very low thermodynamic temperature.

Wang et al. employed this approach to compute the temperature dependent phase diagrams for Li–Mn–O_2, Li–Co–O_2, and Li–Ni–O_2 (the Li–Co–O_2 phase diagrams are shown in Figure 4a),[118] finding that delithiated layered phases are generally metastable and that phase separation into the spinel structure LiM_2O_4 and either the layered $LiMO_2$ or oxygen-deficient structures is thermodynamically preferred. The reaction enthalpies computed according to Equation (15) were generally within 10–20 meV/formula unit of the experimentally measured value for the case of Li–Ni–O_2.

The approximation of the reaction free energy in Equation (15) does not acknowledge whether the reaction conditions are oxidising or reducing, as controlled by the gaseous environment and the temperature. To describe phase equilibria with respect to the ambient conditions in a system that is open to oxygen uptake or release, the oxygen grand potential has to be considered[117]

$$\varphi = G - \mu_{O_2} N_{O_2}, \tag{16}$$

where N_{O_2} is the number of oxygen molecules and the oxygen chemical potential, μ_{O_2}, depends on the oxygen partial pressure, p_{O_2}, and the temperature:

$$\mu_{O_2}(T, p_{O_2}) = \mu_{O_2}(T, p_0) + k_B T \ln \frac{p_{O_2}}{p_0} \approx H_{O_2}^{p_0}(T) - T\left[S_{O_2}^{p_0}(T) - k_B \ln \frac{p_{O_2}}{p_0}\right]. \tag{17}$$

The oxygen enthalpy, $H_{O_2}^{p_0}$, can be approximated using the correction scheme by Wang et al.[42] from above, and the standard oxygen entropy, $S_{O_2}^{p_0}$, can be obtained from thermochemical tables[117].

Ong et al. computed the Li–Fe–P–O and Li–Mn–P–O phase diagrams open to oxygen, i.e., based on the grand potential (16), to assess the condition for oxygen release in delithiated $LiFePO_4$ and $LiMnPO_4$.[119] Their computations predict, in agreement with

experiment, that the manganese phosphate material reduces at lower temperature than the iron-based material, and the predicted oxygen gas evolution versus temperature is shown in Figure 4b. We note, however, that oxygen evolution at the particle surface may be kinetically limited, as pointed out by Mo et al.[120] for the example of lithium peroxide.

Electrochemical stability of electrolytes

Electrolytes in lithium-ion batteries are typically solutions of lithium salts in organic solvents containing additional additives, for example, to enhance the solubility or to create stabilising passivation layers on the electrodes. For the electrolyte to be stable at operation conditions, none of its components may participate in electrochemical reactions with the electrodes. In other words, the cell voltage must at no time leave the voltage range over which the electrolyte is stable, otherwise either reduction or oxidation of the electrolyte molecules would occur.

The redox potentials of the electrolyte molecules are determined by the energy that is required to release and take up electrons, i.e., by the ionisation potential and the electron affinity, both of which can be directly evaluated by first principles calculations of isolated charged molecules.[121] However, the relevant energies are not those of molecules in the gas phase, but instead the solvation energy of all involved species needs to be accounted for. The Gibbs free energies of reduction and oxidation of the solvated molecule M that determine the redox potentials are

reduction : $\Delta G_{red}^s = G[M^{n-}(s)] - G[M(s)] - nG[e^-(s)]$
oxidation : $\Delta G_{ox}^s = G[M^{n+}(s)] + nG[e^-(s)] - G[M(s)],$

where '(s)' indicates species in solution. Zhang et al.[122] proposed an indirect method to compute the electrolyte redox potentials based on thermodynamic cycles. A schematic of an essentially identical approach is shown in Figure 5a. In Zhang's method, the solvation energy of the molecular species is approximated by a continuum solvent model, and the overall reaction free energy in solution is expressed in terms of the free energies in the gas phase and the solvation energies (see equations in Figure 5a). The reference energy of the Li/Li$^+$ redox couple can be calculated based on a similar thermodynamic cycle,[122,123] or can be obtained

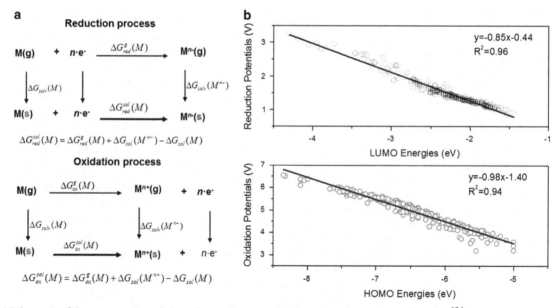

Figure 5. (a) Schematic of the computation of electrolyte redox potentials based on thermodynamic cycles.[124] Species in the gas phase and in solution are indicated by '(g)' and '(s)', respectively. (Copyright: *American Chemical Society*) (b) Correlation of the electrolyte reduction potential with the LUMO energy (top) and the oxidation potential with the HOMO energy (bottom).[124]

from measurement versus the standard hydrogen electrode.[124,125] Note that the free energy of electrons in the gas phase, $G[e^-(g)]$, does not occur in the final expression of the voltage, as it is canceled out when the electrolyte potential is referenced to the Li/Li^+ redox couple.

Shao et al.[123] employed Zhang's approach to the calculation of the electrochemical windows of sulfone-based electrolytes with explicit evaluation of the Li/Li^+ redox potential in solution, comparing the accuracy of different first-principles methods and continuum solvation models. Regarding the solvation models, the authors report that the polarised continuum model (PCM)[126] results in the smallest errors compared with experiment. The most accurate electronic structure method for the prediction of the electrolyte potentials was found to be Møller–Plesset perturbation theory (MP2) with errors in the range of 0.1–0.5 V, whereas DFT (GGA and hybrid functionals) resulted in errors of up to 1.5 V.[123]

The validity of approximating solvation effects using continuum models was further assessed by Ong et al.[127], who compared the electrochemical stability windows of ionic liquids predicted by a molecule-based estimate using PCM with the results of direct classical molecular dynamics simulations of the liquid phase followed by DFT. The authors find that while the stability is generally overestimated by the PCM approach, the identified trends are nonetheless in agreement with explicit solvent calculations.

Wang et al.[128] applied Zhang's method to calculate the oxidation potentials of redox shuttle additives. The authors find a strong correlation of the oxidation potential with the energy of the highest occupied molecular orbital (HOMO). Cheng et al. used Zhang's method for the high-throughput screening of around 1,400 organic molecules, referencing the free energy differences to the experimental voltage of the Li/Li^+ redox couple (1.24 V).[124] The comparison of the results for this large number of molecules also confirmed that the energies of the HOMO and lowest unoccupied molecular orbital (LUMO) correlate well with the ionisation potential and the electron affinity (Figure 5b),[121,124] thus eliminating the need to calculate the energy of the reduced and oxidised species in a rapid screening approach.

Borodin et al. investigated the effect of nearby anions on the oxidation stability of solvent molecules and found that hydrogen or fluorine abstraction may significantly reduce the oxidation potentials.[125] In this work, the electrolyte potentials were also references against the experimental lithium redox voltage, although a slightly different value of 1.37 V was used. A similar proximity effect on electrochemical stability was found by Rajput et al.[129] in a study of Mg-electrolytes. They found that the reduced TFSI (1-ethyl-3-methylimidazolium) anion undergoes significant bond weakening when paired with a Mg cation in the solution.

The decomposition of electrolyte molecules upon reduction at the anode surface is a key process during the formation of the solid-electrolyte interface. Wang et al.[116] investigated the reductive decomposition and polymerisation of ethylene carbonate (EC) with first principles calculations, also employing the PCM solvent and evaluating the redox potentials with the method by Zhang. The calculated reaction mechanisms show that a reduction reaction involving multiple EC molecules (2–4) has a lower reduction potential than the single-molecule reaction, indicating that the accurate calculation of redox potentials may require some explicit solvent molecules. The decomposition products predicted by Wang et al. are in good agreement with experimental observations (see references within ref. 116) including the release of ethylene upon reaction.

Leung studied the reductive decomposition of EC on the surface of a graphite anode with direct AIMD simulations (see also section 'Direct simulation of the diffusion dynamics').[130] On the basis of an AIMD trajectory over 7 ps, the author identifies two reduction pathways resulting in the release of either ethylene or carbon monoxide, both of which had previously been observed in experiments. In contrast to the work by Wang et al. the AIMD simulations by Leung did not impose the reaction mechanism in the computation.

CHALLENGES AND PERSPECTIVES

The overview provided in the previous sections shows that many key properties of lithium batteries, such as the voltage, rate capability and thermal stability, can be reliably addressed by first-principles calculations. Over the last two decades, the accuracy of ab initio methods (thanks to DFT+U and hybrid functionals) and the robustness of the computational methodologies have matured to the point that simulations can often be completely automated, enabling, for example, high-throughput processing of large structural databases.[66,131–134] Notwithstanding these advances, every DFT calculation requires as input a structure model in form of atomic positions, and any computation becomes meaningless if the researcher has no clear conception of the relevant materials phases or makes invalid assumptions about the atomic structure. Today, computational battery research is therefore less limited by technical challenges than by our insufficient understanding of the active phases and relevant processes in some materials.

This challenge has become more pressing since the advent of lithium excess cathode materials which often exhibit (partial) cation disorder and undergo non-coherent phase transitions upon charge and discharge.[113,135] Some lithium excess materials have furthermore been argued to gradually change their stoichiometry due to loss of oxygen gas.[136] Although thermodynamically stable phases can generally be discovered with the methodology of section 'Thermal stability of cathode materials', it is still challenging to predict metastable phases that may be present under operation conditions, such as nanostructures,[137] polymorphs[138] or disordered phases[113] (kMC simulations discussed in section 'Diffusivity based on 0 K migration barriers' are one option). However, without a clear understanding of the structural evolution of a material, it is virtually impossible to make quantitative computational predictions, and it is extremely challenging to identify the material's performance limiting attributes. The difficulty to predict the formation of metastable phases also hampers the computational design of entirely new materials: Presently, we are simply not very good in predicting whether a hypothetical material is actually synthesisable.

Most of the first-principles work on battery materials so far has been focused on crystalline solids, owing to the great interest in crystalline cathode materials but also to the challenges involved in first principles modelling of unordered and molecular materials. However, as touched on in the section 'Electrochemical stability of electrolytes', liquid molecular electrolytes are a crucial element for lithium-ion battery safety and rate capability, and the non-crystalline solid-electrolyte interface has an important role for the battery performance.[7] The simulation of such non-ideal (non-periodic) structures typically requires large length scales that are beyond the means of present first principles methods, but may be accessible by coarse-grained or empirically parametrised models, such as classical force-field-based molecular dynamics simulations[139] or phase-field models.[140] Systematic first-principles computational work in the area of molecular electrolytes is only emerging, but recent initiatives, such as the Electrolyte Genome Project,[141] promise to deliver an increased momentum in the next years.

ACKNOWLEDGEMENTS

This work was supported primarily by the U.S. Department of Energy (DOE) under Contract No. DE-FG02-96ER45571.

COMPETING INTERESTS

The authors declare no conflict of interest.

REFERENCES

1. Whittingham, M. S. Electrical energy storage and intercalation chemistry. *Science* **192**, 1126–1127 (1976).
2. Bruce, P. G. Energy storage beyond the horizon: Rechargeable lithium batteries. *Solid State Ionics* **179**, 752–760 (2008).
3. Goodenough, J. B. & Park, K.-S. The Li-ion rechargeable battery: a perspective. *J. Am. Chem. Soc.* **135**, 1167–1176 (2013).
4. Thackeray, M. M., Wolverton, C. & Isaacs, E. D. Electrical energy storage for transportation—approaching the limits of, and going beyond, lithium-ion batteries. *Energy Environ. Sci.* **5**, 7854 (2012).
5. Whittingham, M. S. Materials challenges facing electrical energy storage. *MRS Bull.* **33**, 411–419 (2008).
6. Dunn, B., Kamath, H. & Tarascon, J.-M. Electrical energy storage for the grid: a battery of choices. *Science* **334**, 928–935 (2011).
7. Goodenough, J. B. & Kim, Y. Challenges for rechargeable Li batteries. *Chem. Mater.* **22**, 587–603 (2010).
8. Zhu, G.-N., Wang, Y.-G. & Xia, Y.-Y. Ti-based compounds as anode materials for Li-ion batteries. *Energy Environ. Sci.* **5**, 6652 (2012).
9. McDowell, M. T., Lee, S. W., Nix, W. D. & Cui, Y. 25th anniversary article: understanding the lithiation of silicon and other alloying anodes for lithium-ion batteries. *Adv. Mater.* **25**, 4966–4985 (2013).
10. Oh, M. H. *et al.* Galvanic replacement reactions in metal oxide nanocrystals. *Science* **340**, 964–968 (2013).
11. Xu, K. Electrolytes and Interphases in Li-Ion Batteries and Beyond. *Chem. Rev.* **114**, 11503–11618 (2014).
12. Mo, Y., Ong, S. P. & Ceder, G. First principles study of the Li$_{10}$GeP$_2$S$_{12}$ lithium super ionic conductor material. *Chem. Mater.* **24**, 15–17 (2012).
13. Wang, Y. *et al.* Design principles for solid-state lithium superionic conductors. *Nat. Mater.* **14**, 1026–1031 (2015).
14. Hohenberg, P. & Kohn, W. Inhomogeneous electron gas. *Phys. Rev.* **136**, B864–B871 (1964).
15. Kohn, W. & Sham, L. J. Self-Consistent Equations Including Exchange and Correlation Effects. *Phys. Rev.* **140**, A1133–A1138 (1965).
16. Koch, W. & Holthausen, M. C. *A Chemist's Guide to Density Functional Theory* (Wiley-VCH Verlag GmbH, 2001).
17. Curtarolo, S. *et al.* The high-throughput highway to computational materials design. *Nat. Mater.* **12**, 191–201 (2013).
18. Meng, Y. S. & Arroyo-de Dompablo, M. E. Recent advances in first principles computational research of cathode materials for lithium-ion batteries. *Acc. Chem. Res.* **46**, 1171–1180 (2013).
19. Islam, M. S. & Fisher, C. A. J. Lithium and sodium battery cathode materials: computational insights into voltage, diffusion and nanostructural properties. *Chem. Soc. Rev.* **43**, 185–204 (2014).
20. Chevrier, V. L. & Dahn, J. R. First principles studies of disordered lithiated silicon. *J. Electrochem. Soc.* **157**, A392–A398 (2010).
21. Persson, K. *et al.* Lithium diffusion in graphitic carbon. *J. Phys. Chem. Lett.* **1**, 1176–1180 (2010).
22. Chan, M. K. Y., Wolverton, C. & Greeley, J. P. First principles simulations of the electrochemical lithiation and delithiation of faceted crystalline silicon. *J. Am. Chem. Soc.* **134**, 14362–14374 (2012).
23. Kirklin, S., Meredig, B. & Wolverton, C. High-throughput computational screening of new Li-ion battery anode materials. *Adv. Energy Mater.* **3**, 252–262 (2013).
24. Richards, W. D., Miara, L. J., Wang, Y., Kim, J. C. & Ceder, G. Interface stability in solid-state batteries. *Chem. Mater.* **28**, 266–273 (2015).
25. Aykol, M., Kirklin, S. & Wolverton, C. Thermodynamic aspects of cathode coatings for lithium-ion batteries. *Adv. Energy Mater.* **4**, 1400690 (2014).
26. McKinnon, W. Insertion electrodes I: Atomic and electronic structure of the hosts and their insertion compounds. in *Solid State Electrochemistry* 163–198 (ed. Bruce, P. G.) (Cambridge University Press, Cambridge, UK, 1994).
27. Aydinol, M. K., Kohan, A. F., Ceder, G., Cho, K. & Joannopoulos, J. Ab initio study of lithium intercalation in metal oxides and metal dichalcogenides. *Phys. Rev. B* **56**, 1354–1365 (1997).
28. Langreth, D. C. & Mehl, M. J. Beyond the local-density approximation in calculations of ground-state electronic properties. *Phys. Rev. B* **28**, 1809–1834 (1983).
29. Aydinol, M. K., Kohan, A. F. & Ceder, G. *Ab initio* calculation of the intercalation voltage of lithium-transition-metal oxide electrodes for rechargeable batteries. *J. Power Sources* **68**, 664–668 (1997).
30. Aydinol, M. K. & Ceder, G. First-principles prediction of insertion potentials in Li-Mn oxides for secondary Li batteries. *J. Electrochem. Soc.* **144**, 3832 (1997).
31. Deiss, E., Wokaun, A., Barras, J. L., Daul, C. & Dufek, P. Average voltage, energy density, and specific energy of lithium-ion batteries. *J. Electrochem. Soc.* **144**, 3877 (1997).
32. Benco, L., Barras, J.-L., Atanasov, M., Daul, C. A. & Deiss, E. First-principles prediction of voltages of lithiated oxides for lithium-ion batteries. *Solid State Ionics* **112**, 255–259 (1998).
33. Arroyo-de Dompablo, M. E., Armand, M., Tarascon, J. M. & Amador, U. On-demand design of polyoxianionic cathode materials based on electronegativity correlations: an exploration of the Li$_2$MSiO$_4$ system (M = Fe, Mn, Co, Ni). *Electrochem. Commun.* **8**, 1292–1298 (2006).
34. Arroyo-de Dompablo, M. E., Rozier, P., Morcrette, M. & Tarascon, J.-M. Electrochemical Data Transferability within Li$_y$VOXO$_4$ (X = Si, Ge$_{0.5}$Si$_{0.5}$, Ge, Si$_{0.5}$As$_{0.5}$, Si$_{0.5}$P$_{0.5}$, As, P) Polyoxyanionic Compounds. *Chem. Mater.* **19**, 2411–2422 (2007).
35. Ceder, G. Predicting properties from scratch. *Science* **280**, 1099 (1998).
36. Cococcioni, M. de Gironcoli S. Linear response approach to the calculation of the effective interaction parameters in the LDA+U method. *Phys. Rev. B* **71**, 035105 (2005).
37. Anisimov, V. I., Zaanen, J. & Andersen, O. K. Band theory and Mott insulators: Hubbard U instead of Stoner I. *Phys. Rev. B* **44**, 943–954 (1991).
38. Anisimov, V. I., Aryasetiawan, F. & Lichtenstein, A. I. First-principles calculations of the electronic structure and spectra of strongly correlated systems: the LDA+U method. *J. Phys. Condens. Matter* **9**, 767 (1997).
39. Dudarev, S. L., Botton, G. A., Savrasov, S. Y., Humphreys, C. J. & Sutton, A. P. Electron-energy-loss spectra and the structural stability of nickel oxide: An LSDA+U study. *Phys. Rev. B* **57**, 1505–1509 (1998).
40. Kulik, H. J., Cococcioni, M., Scherlis, D. A. & Marzari, N. Density functional theory in transition-metal chemistry: a self-consistent hubbard U approach. *Phys. Rev. Lett.* **97**, 103001 (2006).
41. Zhou, F., Cococcioni, M., Marianetti, C. A., Morgan, D. & Ceder, G. First-principles prediction of redox potentials in transition-metal compounds with LDA+U. *Phys. Rev. B* **70**, 235121 (2004).
42. Wang, L., Maxisch, T. & Ceder, G. Oxidation energies of transition metal oxides within the GGA+U framework. *Phys. Rev. B* **73**, 195107 (2006).
43. Zhou, F., Kang, K., Maxisch, T., Ceder, G. & Morgan, D. The electronic structure and band gap of LiFePO$_4$ and LiMnPO$_4$. *Solid State Commun.* **132**, 181–186 (2004).
44. Ben Yahia, M. *et al.* Origin of the 3.6 V to 3.9 V voltage increase in the LiFeSO$_4$F cathodes for Li-ion batteries. *Energy Environ. Sci.* **5**, 9584–9594 (2012).
45. Becke, A. D. A new mixing of Hartree-Fock and local density-functional theories. *J. Chem. Phys.* **98**, 1372 (1993).
46. Becke, A. D. Density-functional thermochemistry. III. The role of exact exchange. *J. Chem. Phys.* **98**, 5648 (1993).
47. Seo, D.-H., Urban, A. & Ceder, G. Calibrating transition metal energy levels and oxygen bands in first principles calculations: accurate prediction of redox potentials and charge transfer in lithium transition metal oxides. *Phys. Rev. B* **92**, 115118 (2015).
48. Heyd, J., Scuseria, G. E. & Ernzerhof, M. Hybrid functionals based on a screened Coulomb potential. *J. Chem. Phys.* **118**, 8207–8215 (2003).
49. Heyd, J., Scuseria, G. E. & Ernzerhof, M. Erratum: 'Hybrid functionals based on a screened Coulomb potential' [J. Chem. Phys.118, 8207 (2003)]. *J. Chem. Phys.* **124**, 219906 (2006).
50. Chevrier, V. L., Ong, S. P., Armiento, R., Chan, M. K. Y. & Ceder, G. Hybrid density functional calculations of redox potentials and formation energies of transition metal compounds. *Phys. Rev. B* **82**, 075122 (2010).
51. Delmas, C. *et al.* Lithium batteries: a new tool in solid state chemistry. *Int. J. Inorg. Mater.* **1**, 11–19 (1999).
52. Marianetti, C. A., Kotliar, G. & Ceder, G. A first-order Mott transition in Li$_x$CoO$_2$. *Nat. Mater.* **3**, 627–631 (2004).
53. Courtney, I. A., Tse, J. S., Mao, O., Hafner, J. & Dahn, J. R. Ab initio calculation of the lithium-tin voltage profile. *Phys. Rev. B* **58**, 15583–15588 (1998).
54. Ceder, G. & Van der Ven, A. Phase diagrams of lithium transition metal oxides: investigations from first principles. *Electrochim. Acta* **45**, 131–150 (1999).
55. Van der Ven, A., Aydinol, M. K., Ceder, G., Kresse, G. & Hafner, J. First-principles investigation of phase stability in Li$_x$CoO$_2$. *Phys. Rev. B* **58**, 2975–2987 (1998).
56. Van der Ven, A., Bhattacharya, J. & Belak, A. A. Understanding Li diffusion in Li-intercalation compounds. *Acc. Chem. Res.* **46**, 1216–1225 (2013).
57. Kim, H. *et al.* Ab Initio Study of the sodium intercalation and intermediate phases in Na$_{0.44}$MnO$_2$ for sodium-ion battery. *Chem. Mater.* **24**, 1205–1211 (2012).
58. Boyanov, S. *et al.* FeP: another attractive anode for the Li-ion battery enlisting a reversible two-step insertion/conversion process. *Chem. Mater.* **18**, 3531–3538 (2006).

59. Van der Ven, A., Aydinol, M. K. & Ceder, G. First-principles evidence for stage ordering in Li$_x$CoO$_2$. *J. Electrochem. Soc.* **145**, 2149–2155 (1998).

60. Arroyo-de Dompablo, M. E., Van der Ven, A. & Ceder, G. First-principles calculations of lithium ordering and phase stability on Li$_x$NiO$_2$. *Phys. Rev. B* **66**, 064112 (2002).

61. Hart, G. L. W. & Forcade, R. W. Algorithm for generating derivative structures. *Phys. Rev. B* **77**, 224115 (2008).

62. Hart, G. L. W. & Forcade, R. W. Generating derivative structures from multi-lattices: Algorithm and application to hcp alloys. *Phys. Rev. B* **80**, 014120 (2009).

63. Hart, G. L. W., Nelson, L. J. & Forcade, R. W. Generating derivative structures at a fixed concentration. *Comput. Mater. Sci.* **59**, 101–107 (2012).

64. Hautier, G., Fischer, C. C., Jain, A., Mueller, T. & Ceder, G. Finding nature's missing ternary oxide compounds using machine learning and Density functional theory. *Chem. Mater.* **22**, 3762–3767 (2010).

65. Kim, J. C., Seo, D.-H. & Ceder, G. Theoretical capacity achieved in a LiMn$_{0.5}$Fe$_{0.4}$Mg$_{0.1}$BO$_3$ cathode by using topological disorder. *Energy Environ. Sci.* **8**, 1790–1798 (2015).

66. Ong, S. P. *et al.* Python materials genomics (pymatgen): a robust, open-source python library for materials analysis. *Comput. Mater. Sci.* **68**, 314–319 (2013).

67. Metropolis, N., Rosenbluth, A. W., Rosenbluth, M. N., Teller, A. H. & Teller, E. Equation of state calculations by fast computing machines. *J. Chem. Phys.* **21**, 1087–1092 (1953).

68. Binder, K. & Heermann, D. W. *Monte Carlo Simulation in Statistical Physics* Vol. 0 (Springer, 2010).

69. Sanchez, J. M., Ducastelle, F. & Gratias, D. Generalized cluster description of multicomponent systems. *Phys A* **128**, 334–350 (1984).

70. Fontaine, D. D. Cluster approach to order-disorder transformations in alloys. *Solid State Phys.* **47**, 33–176 (1994).

71. Li, W., Reimers, J. N. & Dahn, J. R. Lattice-gas-model approach to understanding the structures of lithium transition-metal oxides LiMO$_2$. *Phys. Rev. B* **49**, 826–831 (1994).

72. van de Walle, A. & Ceder, G. The effect of lattice vibrations on substitutional alloy thermodynamics. *Rev. Mod. Phys.* **74**, 11–45 (2002).

73. Zhou, F., Maxisch, T. & Ceder, G. Configurational electronic entropy and the phase diagram of mixed-valence oxides: the case of Li$_x$FePO$_4$. *Phys. Rev. Lett.* **97**, 155704 (2006).

74. Schleger, P., Hardy, W. N. & Casalta, H. Model for the high-temperature oxygen-ordering thermodynamics in YBa$_2$Cu$_3$O$_{6+x}$: Inclusion of electron spin and charge degrees of freedom. *Phys. Rev. B* **49**, 514–523 (1994).

75. van de Walle, A., Asta, M. & Ceder, G. The alloy theoretic automated toolkit: a user guide. *Calphad* **26**, 539–553 (2002).

76. Lerch, D., Wieckhorst, O., Hart, G. L. W., Forcade, R. W. & Müller, S. UNCLE: a code for constructing cluster expansions for arbitrary lattices with minimal user-input. *Model. Simul. Mater. Sci. Eng.* **17**, 055003 (2009).

77. Nelson, L. J., Hart, G. L. W., Zhou, F. & Ozoliņš, V. Compressive sensing as a paradigm for building physics models. *Phys. Rev. B* **87**, 035125 (2013).

78. Reimers, J. N. & Dahn, J. R. Application of *ab initio* methods for calculations of voltage as a function of composition in electrochemical cells. *Phys. Rev. B* **47**, 2995–3000 (1993).

79. Wolverton, C. & Zunger, A. First-principles prediction of vacancy order-disorder and intercalation battery voltages in Li$_x$CoO$_2$. *Phys. Rev. Lett.* **81**, 606–609 (1998).

80. Wolverton, C. & Zunger, A. Cation and vacancy ordering in Li$_x$CoO$_2$. *Phys. Rev. B* **57**, 2242–2252 (1998).

81. Van der Ven, A. & Ceder, G. Ordering in Li$_x$(Ni$_{0.5}$Mn$_{0.5}$)O$_2$ and its relation to charge capacity and electrochemical behavior in rechargeable lithium batteries. *Electrochem. Commun.* **6**, 1045–1050 (2004).

82. Lee, E. & Persson, K. A. Revealing the coupled cation interactions behind the electrochemical profile of Li$_x$Ni$_{0.5}$Mn$_{1.5}$O$_4$. *Energy Environ. Sci.* **5**, 6047 (2012).

83. Yu, H.-C. *et al.* Designing the next generation high capacity battery electrodes. *Energy Environ. Sci.* **7**, 1760 (2014).

84. Heitjans, P. & Kärger, J. (eds). *Diffusion in Condensed Matter: Methods, Materials, Models* (Springer: Berlin, Germany, 2005).

85. Van der Ven, A., Ceder, G., Asta, M. & Tepesch, P. D. First-principles theory of ionic diffusion with nondilute carriers. *Phys. Rev. B* **64**, 184307 (2001).

86. Frenkel, D. & Smit, B. *Understanding Molecular Simulation: From Algorithms to Applications* (Academic Press, 2002).

87. Marx, D. & Hutter, J. *Ab Initio Molecular Dynamics: Basic Theory and Advanced Methods* (Cambridge Univ. Press, 2009).

88. Alder, B. J., Gass, D. M. & Wainright, T. E. Studies in molecular dynamics. VIII. the transport coefficients of a hard-sphere fluid. *J. Chem. Phys.* **53**, 3813 (1970).

89. Van der Ven, A., Yu, H.-C., Ceder, G. & Thornton, K. Vacancy mediated substitutional diffusion in binary crystalline solids. *Prog. Mater. Sci.* **55**, 61–105 (2010).

90. Van der Ven, A. & Ceder, G. Lithium diffusion mechanisms in layered intercalation compounds. *J. Power Sources* **97**, 529–531 (2001).

91. Yang, J. & Tse, J. S. Li ion diffusion mechanisms in LiFePO$_4$: an *ab initio* molecular dynamics study. *J. Phys. Chem. A* **115**, 13045–13049 (2011).

92. Mo, Y., Ong, S. P. & Ceder, G. Insights into diffusion mechanisms in P2 layered oxide materials by first-principles calculations. *Chem. Mater.* **26**, 5208–5214 (2014).

93. Hao, S. & Wolverton, C. Lithium transport in amorphous Al$_2$O$_3$ and AlF$_3$ for discovery of battery coatings. *J. Phys. Chem. C* **117**, 8009–8013 (2013).

94. Xiao, R., Li, H. & Chen, L. Density Functional Investigation on Li$_2$MnO$_3$. *Chem. Mater.* **24**, 4242–4251 (2012).

95. Marcelin, R. Contribution a l'etude de la cinetique physico-chimique. *Ann. Phys.* **3**, 120–231 (1915).

96. Vineyard, G. H. Frequency factors and isotope effects in solid state rate processes. *J. Phys. Chem. Solids* **3**, 121–127 (1957).

97. Morgan, D., Van der Ven, A. & Ceder, G. Li conductivity in Li$_x$MPO$_4$ (M = Mn, Fe, Co, Ni) olivine materials. *Electrochem. Solid State Lett.* **7**, A30 (2004).

98. Van der Ven, A., Thomas, J. C., Xu, Q., Swoboda, B. & Morgan, D. Nondilute diffusion from first principles: Li diffusion in Li$_x$TiS$_2$. *Phys. Rev. B* **78**, 104306 (2008).

99. Kutner, R. Chemical diffusion in the lattice gas of non-interacting particles. *Phys. Lett. A* **81**, 239–240 (1981).

100. Bulnes, F. M., Pereyra, V. D. & Riccardo, J. L. Collective surface diffusion: n-fold way kinetic Monte Carlo simulation. *Phys. Rev. E* **58**, 86–92 (1998).

101. Voter, A. F. *Introduction to the Kinetic Monte Carlo Method, in Radiation Effects in Solids* (Springer, NATO Publishing Unit, 2005).

102. Henkelman, G., Uberuaga, B. P. & Jónsson, H. A climbing image nudged elastic band method for finding saddle points and minimum energy paths. *J. Chem. Phys.* **113**, 9901–9904 (2000).

103. Jónsson, H., Mills, G., Jacobsen, K. W. (eds Ciccotti G., Berne B. J. & Coker D. F.) Ch. *Nudged Elastic Band Method for Finding Minimum Energy Paths of Transitions* 385–404 (World Scientific, 1998).

104. Asari, Y., Suwa, Y. & Hamada, T. Formation and diffusion of vacancy-polaron complex in olivine-type LiMnPO$_4$ and LiFePO$_4$. *Phys. Rev. B* **84**, 134113 (2011).

105. Ong, S. P. *et al.* Voltage, stability and diffusion barrier differences between sodium-ion and lithium-ion intercalation materials. *Energy Environ. Sci.* **4**, 3680–3688 (2011).

106. Van der Ven, A. & Ceder, G. Lithium diffusion in layered Li$_x$CoO$_2$. *Electrochem. Solid State Lett.* **3**, 301–304 (2000).

107. Malik, R., Burch, D., Bazant, M. & Ceder, G. Particle size dependence of the ionic diffusivity. *Nano Lett.* **10**, 4123–4127 (2010).

108. Kang, B. & Ceder, G. Battery materials for ultrafast charging and discharging. *Nature* **458**, 190–193 (2009).

109. Kang, J., Chung, H., Doh, C., Kang, B. & Han, B. Integrated study of first principles calculations and experimental measurements for Li-ionic conductivity in Al-doped solid-state LiGe$_2$(PO$_4$)$_3$ electrolyte. *J. Power Sources* **293**, 11–16 (2015).

110. Du, F., Ren, X., Yang, J., Liu, J. & Zhang, W. Structures, thermodynamics, and Li$^+$ mobility of Li$_{10}$GeP$_2$S$_{12}$: a first-principles analysis. *J. Phys. Chem. C* **118**, 10590–10595 (2014).

111. Kang, K. & Ceder, G. Factors that affect Li mobility in layered lithium transition metal oxides. *Phys. Rev. B* **74**, 094105 (2006).

112. Kang, K., Meng, Y. S., Bréger, J., Grey, C. P. & Ceder, G. Electrodes with high power and high capacity for rechargeable lithium batteries. *Science* **311**, 977–980 (2006).

113. Lee, J. *et al.* Unlocking the potential of cation-disordered oxides for recharge-able lithium batteries. *Science* **343**, 519–522 (2014).

114. Urban, A., Lee, J. & Ceder, G. The configurational Space of rocksalt-type oxides for high-capacity lithium battery electrodes. *Adv. Energy Mater.* **4**, 1400478 (2014).

115. Zheng, J. *et al.* Structural and chemical evolution of Li- and Mn-rich layered cathode material. *Chem. Mater.* **27**, 1381–1390 (2015).

116. Wang, Y., Nakamura, S., Ue, M. & Balbuena, P. B. Theoretical studies to understand surface chemistry on carbon anodes for lithium-ion batteries: reduction mechanisms of ethylene carbonate. *J. Am. Chem. Soc.* **123**, 11708–11718 (2001).

117. Ong, S. P., Wang, L., Kang, B. & Ceder, G. Li-Fe-P-O$_2$ phase diagram from first principles calculations. *Chem. Mater.* **20**, 1798–1807 (2008).

118. Wang, L., Maxisch, T. & Ceder, G. A first-principles approach to studying the thermal stability of oxide cathode materials. *Chem. Mater.* **19**, 543–552 (2007).

119. Ong, S. P., Jain, A., Hautier, G., Kang, B. & Ceder, G. Thermal stabilities of delithiated olivine MPO$_4$ (M = Fe, Mn) cathodes investigated using first principles calculations. *Electrochem. Commun.* **12**, 427–430 (2010).

120. Mo, Y., Ong, S. P. & Ceder, G. First-principles study of the oxygen evolution reaction of lithium peroxide in the lithium-air battery. *Phys. Rev. B* **84**, 205446 (2011).

121. Parker, V. D. Energetics of electrode reactions. II. The relationship between redox potentials, ionization potentials, electron affinities, and solvation energies of aromatic hydrocarbons. *J. Am. Chem. Soc.* **98**, 98–103 (1976).

122. Zhang, X., Pugh, J. K. & Ross, P. N. Computation of thermodynamic oxidation potentials of organic solvents using density functional theory. *J. Electrochem. Soc.* **148**, E183 (2001).

123. Shao, N., Sun, X.-G., Dai, S. & Jiang, D. Electrochemical windows of sulfone-based electrolytes for high-voltage Li-ion batteries. *J. Phys. Chem. B* **115**, 12120–12125 (2011).

124. Cheng, L. *et al.* Accelerating electrolyte discovery for energy storage with high-throughput screening. *J. Phys. Chem. Lett.* **6**, 283–291 (2015).

125. Borodin, O., Behl, W. & Jow, T. R. Oxidative stability and initial decomposition reactions of carbonate, sulfone, and alkyl phosphate-based electrolytes. *J. Phys. Chem. C* **117**, 8661–8682 (2013).

126. Miertuš, S., Scrocco, E. & Tomasi, J. Electrostatic interaction of a solute with a continuum. a direct utilizaion of *ab initio* molecular potentials for the prevision of solvent effects. *Chem. Phys.* **55**, 117–129 (1981).

127. Ong, S. P., Andreussi, O., Wu, Y., Marzari, N. & Ceder, G. Electrochemical windows of room-temperature ionic liquids from molecular dynamics and density functional theory calculations. *Chem. Mater.* **23**, 2979–2986 (2011).

128. Wang, R., Buhrmester, C. & Dahn, J. Calculations of oxidation potentials of redox shuttle additives for Li-ion cells. *J. Electrochem. Soc.* **153**, A445–A449 (2006).

129. Rajput, N. N., Qu, X., Sa, N., Burrell, A. K. & Persson, K. A. The Coupling between stability and ion pair formation in magnesium electrolytes from first-principles quantum mechanics and classical molecular dynamics. *J. Am. Chem. Soc.* **137**, 3411–3420 (2015).

130. Leung, K. Electronic structure modeling of electrochemical reactions at electrode/electrolyte interfaces in lithium ion batteries. *J. Phys. Chem. C* **117**, 1539–1547 (2013).

131. Pizzi, G., Cepellotti, A., Sabatini, R., Marzari, N. & Kozinsky, B. AiiDA: automated interactive infrastructure and database for computational science. *Comput. Mater. Sci.* **111**, 218–230 (2016).

132. Curtarolo, S. *et al.* AFLOW: an automatic framework for high-throughput materials discovery. *Comput. Mater. Sci.* **58**, 218–226 (2012).

133. Kirklin, S. *et al.* The Open Quantum Materials Database (OQMD): assessing the accuracy of DFT formation energies. *NPJ Comput. Mater.* **1**, 15010 (2015).

134. Saal, J., Kirklin, S., Aykol, M., Meredig, B. & Wolverton, C. Materials design and discovery with high-throughput density functional theory: The Open Quantum Materials Database (OQMD). *JOM* **65**, 1501–1509 (2013).

135. Tran, N. *et al.* Mechanisms associated with the 'plateau' observed at high voltage for the overlithiated $Li_{1.12}(Ni_{0.425}Mn_{0.425}Co_{0.15})_{0.88}O_2$ System. *Chem. Mater.* **20**, 4815–4825 (2008).

136. Armstrong, A. R. *et al.* Demonstrating oxygen loss and associated structural reorganization in the lithium battery cathode $Li[Ni_{0.2}Li_{0.2}Mn_{0.6}]O_2$. *J. Am. Chem. Soc.* **128**, 8694–8698 (2006).

137. Huang, J. Y. *et al.* In situ observation of the electrochemical lithiation of a single SnO_2 nanowire electrode. *Science* **330**, 1515–1520 (2010).

138. Seo, D.-H., Kim, H., Park, I., Hong, J. & Kang, K. Polymorphism and phase transformations of $Li_{2-x}FeSiO_4$ ($0 \leq x \leq 2$) from first principles. *Phys. Rev. B* **84**, 220106 (2011).

139. Islam, M. S., Driscoll, D. J., Fisher, C. A. J. & Slater, P. R. Atomic-scale investigation of defects, dopants, and lithium transport in the lifepo₄ olivine-type battery material. *Chem. Mater.* **17**, 5085–5092 (2005).

140. Han, B. C., Van der Ven, A., Morgan, D. & Ceder, G. Electrochemical modeling of intercalation processes with phase field models. *Electrochim. Acta* **49**, 4691–4699 (2004).

141. Qu, X. *et al.* The Electrolyte Genome project: A big data approach in battery materials discovery. *Comput. Mater. Sci.* **103**, 56–67 (2015).

4

Theory and experimental evidence of phonon domains and their roles in pre-martensitic phenomena

Yongmei M Jin[1], Yu U Wang[1] and Yang Ren[2]

Pre-martensitic phenomena, also called martensite precursor effects, have been known for decades while yet remain outstanding issues. This paper addresses pre-martensitic phenomena from new theoretical and experimental perspectives. A statistical mechanics-based Grüneisen-type phonon theory is developed. On the basis of deformation-dependent incompletely softened low-energy phonons, the theory predicts a lattice instability and pre-martensitic transition into elastic-phonon domains via 'phonon spinodal decomposition.' The phase transition lifts phonon degeneracy in cubic crystal and has a nature of phonon pseudo-Jahn–Teller lattice instability. The theory and notion of phonon domains consistently explain the ubiquitous pre-martensitic anomalies as natural consequences of incomplete phonon softening. The phonon domains are characterised by broken dynamic symmetry of lattice vibrations and deform through internal phonon relaxation in response to stress (a particular case of Le Chatelier's principle), leading to previously unexplored new domain phenomenon. Experimental evidence of phonon domains is obtained by *in situ* three-dimensional phonon diffuse scattering and Bragg reflection using high-energy synchrotron X-ray single-crystal diffraction, which observes exotic domain phenomenon fundamentally different from usual ferroelastic domain switching phenomenon. In light of the theory and experimental evidence of phonon domains and their roles in pre-martensitic phenomena, currently existing alternative opinions on martensitic precursor phenomena are revisited.

INTRODUCTION

Pre-martensitic phenomena, also called martensite precursor effects, are long standing critical issues of martensitic phase transformation that have not been fully understood. Martensitic transformation is a typical solid-state displacive (diffusionless) phase transition that breaks crystal symmetry by development of spontaneous anisotropic lattice strain on cooling.[1–4] Before martensitic transformation, the high-symmetry parent phase (cubic austenite) usually undergoes incomplete phonon softening in a wide temperature range 10–100 K above the martensite start temperature,[4–6] which is accompanied by various anomalies that are unexpected in cubic phase. These precursor anomalies include diffuse scattering (streaks and satellites) in diffraction, cross-hatched nanoscale striation image contrast (tweed patterns) in transmission electron microscopy (TEM), and anomalous thermal, acoustic, elastic properties (e.g., anisotropic thermal expansion, increased acoustic attenuation, frequency-dependent elastic moduli, etc.).[4,7–14] Such pre-martensitic phenomena cannot be well explained from the existing phase transition theories.[15,16] At the heart of the martensitic precursor problem is that these anomalies cannot be critical fluctuations before martensitic transformation because of the first-order nature of the martensitic transformation (a great majority of martensitic transformations are first-order phase transitions where certain phonon modes only exhibit incomplete softening thus soft-mode theory is inapplicable[17]), the high temperature range far beyond the vicinity of the transformation (10–100 K above martensite start temperature), and the exceedingly long lifetime (much longer than TEM imaging time of tweed patterns).[4,15] In particular, there

exists a theoretical gap between the well-established experimental fact of incomplete phonon softening and the associated ubiquitous while puzzling martensitic precursor anomalies. To fill this gap and consistently explain these anomalies, a fundamental physical theory beyond a phenomenological model is needed, which may also predict previously unexplored new phenomena that can be tested by new experiments.

This paper addresses martensitic precursor phenomena from new theoretical and experimental perspectives. In the first part of this paper, a statistical mechanics-based phonon theory[18] is discussed. The theory predicts a lattice instability and pre-martensitic transition into elastic-phonon domains via 'phonon spinodal decomposition,' and reveals the nature of this phase transition as phonon pseudo-Jahn–Teller lattice instability. The instability condition and the behaviours of phonon domains are discussed. Exotic domain phenomenon is predicted, which is fundamentally different from the usual ferroelastic domain switching phenomenon. In the second part of this paper, recent experimental evidence of phonon domains[18] is presented. The theoretical predictions discussed in the first part of this paper are tested by specially designed experiments of *in situ* three-dimensional (3D) phonon diffuse scattering and Bragg reflection using high-energy synchrotron X-ray single-crystal diffraction. In the third part of this paper, based on the theory and experimental evidence of phonon domains, the roles of phonon domains in pre-martensitic phenomena are discussed. The notion of phonon domains provides consistent explanations to the martensitic precursor 'anomalies' without resort to extrinsic defects, on the same physical footing of thermal expansion, both being intrinsic

[1]Department of Materials Science and Engineering, Michigan Technological University, Houghton, MI, USA and [2]X-Ray Science Division, Advanced Photon Source, Argonne National Laboratory, Argonne, IL, USA.
Correspondence: YM Jin or YU Wang (ymjin@mtu.edu or wangyu@mtu.edu)

properties and natural consequences of anharmonic crystal lattices. The currently existing alternative explanations of the martensitic precursor anomalies (in particular, tweed patterns) are based on static defects (composition fluctuation, chemical disorder and point defects) that cause strain glass state in analogy to spin glass.[15,19-24] These alternative opinions are revisited and examined against the new theoretical viewpoints and experimental results in light of the phonon domains and their roles in pre-martensitic phenomena.

STATISTICAL MECHANICS-BASED PHONON THEORY OF PRE-MARTENSITIC TRANSITION

A statistical mechanics-based phonon theory[18] is recently developed to gain fundamental understanding of the effects of deformation-dependent low-energy phonons on the behaviours and properties of austenite before martensitic transformation. The theory is based on the experimentally established fact of incomplete phonon softening, which generally occurs in the cubic austenite in a wide temperature range 10-100 K above the martensite start temperature (M_S).[4-6] The incomplete phonon softening phenomenon serves as the premise of the theory to derive consistent explanations to other pre-martensitic phenomena. Although the origin of incomplete phonon softening (also regarded as an anomaly—phonon anomaly[25-28]) is beyond the scope of the theory, it is believed to have an electronic origin.[25,28] Since the same general phenomenon occurs in both nonmagnetic (e.g., Ni–Al[25,26]) and magnetic (e.g., Ni–Mn–Ga[27,28]) systems, and both below and above Curie temperature (T_C) in magnetic systems (e.g., $Ni_{52.0}Mn_{24.4}Ga_{23.6}$ with $M_S = 308 K < T_C = 361 K$ and $Ni_{53.0}Mn_{26.4}Ga_{20.6}$ with $M_S = 479 K > T_C = 368 K$), magnetism seems unlikely to have a general role (although magnetoelastic coupling[29] may have a certain role in magnetic materials), because a general mechanism identified in nonmagnetic systems would also work in magnetic systems. Figure 1 illustrates the incomplete phonon softening phenomenon as measured by inelastic neutron scattering[27,30-32] and high-energy synchrotron X-ray diffuse scattering, which is the premise for the following theoretical formulation.

Phonon-strain coupling and strain-dependent phonon frequencies
Following Born's dynamical approach to crystal lattices,[33] an effective potential function Φ of a cubic crystal is considered. Given the drastically distinct time scales of relatively slow lattice deformation and relatively fast thermal vibrations, Φ is expressed in terms of lattice strain ε_{ij} (quasi-static configurational coordinates) and atomic displacements u_α^r (dynamic vibrational coordinates) around average positions $\mathbf{r}$ of the lattice as defined by ε_{ij}. Taking into account the equilibrium condition of undeformed lattice and the invariance relations associated with lattice periodicity and rigid-body translation,[33] $\Phi(\varepsilon_{ij}, u_\alpha^r)$ is expanded into Taylor series:

$$\Phi = \Phi_0 + \frac{1}{2}VC^0_{ijkl}\varepsilon_{ij}\varepsilon_{kl} + \frac{1}{2}\sum_{\mathbf{r},\mathbf{r}'}K^0_{\alpha\beta}(\mathbf{r}-\mathbf{r}')u_\alpha^r u_\beta^{r'}$$
$$+ \frac{1}{2}\sum_{\mathbf{r},\mathbf{r}'}H_{\alpha\beta ij}(\mathbf{r}-\mathbf{r}')\varepsilon_{ij}u_\alpha^r u_\beta^{r'} \quad (1)$$

where $\Phi_0 = \Phi(\varepsilon_{ij} = 0, u_\alpha^r = 0)$ is a constant and can be omitted, $C^0_{ijkl} = V^{-1}\partial^2\Phi/\partial\varepsilon_{ij}\partial\varepsilon_{kl}$ is adiabatic elastic modulus tensor, V is crystal volume, $K^0_{\alpha\beta}(\mathbf{r}-\mathbf{r}') = \partial^2\Phi/\partial u_\alpha^r\partial u_\beta^{r'}$ is Born-von Kármán force constant matrix of the undeformed crystal, and $H_{\alpha\beta ij}(\mathbf{r}-\mathbf{r}') = \partial^3\Phi/\partial u_\alpha^r\partial u_\beta^{r'}\partial\varepsilon_{ij}$, where all partial derivatives are evaluated at $\varepsilon_{ij} = 0$ and $u_\alpha^r = 0$, and summation convention over repeated indices is implied. The Taylor expansion is truncated

after quadratic terms except the first nonvanishing coupling term between ε_{ij} and u_α^r. Equation (1) can be rewritten in a form of perturbation:

$$\Phi = \frac{1}{2}VC^0_{ijkl}\varepsilon_{ij}\varepsilon_{kl} + \frac{1}{2}\sum_{\mathbf{r},\mathbf{r}'}\left[K^0_{\alpha\beta}(\mathbf{r}-\mathbf{r}') + \Delta K_{\alpha\beta}(\mathbf{r}-\mathbf{r}',\varepsilon_{ij})\right]u_\alpha^r u_\beta^{r'} \quad (2)$$

where $\Delta K_{\alpha\beta}(\mathbf{r}-\mathbf{r}',\varepsilon_{ij}) = H_{\alpha\beta ij}(\mathbf{r}-\mathbf{r}')\varepsilon_{ij}$ is the perturbation to $K^0_{\alpha\beta}(\mathbf{r}-\mathbf{r}')$ due to phonon-strain coupling. Equation (2) formulates a simple lattice dynamics problem of anharmonic crystal that contains the leading term of phonon-strain coupling, which leads to strain dependence of phonon frequencies that can be solved using perturbation method.

The first term in Equation (2) characterises the elastic energy of lattice deformation. The second term characterises the dynamics of lattice vibrations that takes into account the coupling between the vibrations and strain (anharmonicity). The eigen-frequencies $\omega_{\mathbf{k},p}$ of normal mode vibrations (phonons) are determined by the secular equation $|\tilde{D}_{ij}(\mathbf{k},\boldsymbol{\varepsilon}) - \omega_{\mathbf{k},p}^2\delta_{ij}| = 0$, where $\tilde{D}_{ij}(\mathbf{k},\boldsymbol{\varepsilon}) = \tilde{D}^0_{ij}(\mathbf{k}) + \Delta\tilde{D}_{ij}(\mathbf{k},\boldsymbol{\varepsilon})$ is the Fourier coefficients of the dynamical matrix $D_{ij}(\boldsymbol{\rho},\boldsymbol{\varepsilon}) = m^{-1}K_{ij}(\boldsymbol{\rho},\boldsymbol{\varepsilon})$, $K_{ij}(\boldsymbol{\rho},\boldsymbol{\varepsilon}) = K^0_{ij}(\boldsymbol{\rho}) + \Delta K_{ij}(\boldsymbol{\rho},\boldsymbol{\varepsilon})$, m is atomic mass, and the phonon modes are labelled by wave vector $\mathbf{k}$ and polarisation vector $\mathbf{e}(\mathbf{k},p)$. The eigen solutions of the secular equation $|\tilde{D}^0_{ij}(\mathbf{k}) - \omega^2_{\mathbf{k},p}\delta_{ij}| = 0$ give the angular frequencies $\omega^2_0(\mathbf{k},p)$ and polarisations $\mathbf{e}(\mathbf{k},p)$ of the three phonon modes $(p = 1, 2, 3)$ with wave vector $\mathbf{k}$ in the undeformed crystal. Incorporation of the perturbation term $\Delta\tilde{D}_{ij}(\mathbf{k},\boldsymbol{\varepsilon})$ into the secular equation gives new eigen solutions where, to the first-order approximation, the polarisation vectors $\mathbf{e}(\mathbf{k},p)$ remain the same while the eigen-frequencies become $\omega^2(\mathbf{k},p) = \omega^2_0(\mathbf{k},p) + \Delta\omega^2(\mathbf{k},p)$. The frequency change is given by

$$\Delta\omega^2(\mathbf{k},p) = \Delta\tilde{D}_{ij}(\mathbf{k},\boldsymbol{\varepsilon})e_i(\mathbf{k},p)e_j(\mathbf{k},p)$$
$$= \Psi_{ijmn}(\mathbf{k})e_i(\mathbf{k},p)e_j(\mathbf{k},p)\varepsilon_{mn} \quad (3)$$

where $\Psi_{ijmn}(\mathbf{k}) = m^{-1}\tilde{H}_{ijmn}(\mathbf{k})$ is the fourth-rank phonon-strain coupling coefficient tensor. Expanding $\Psi_{ijmn}(\mathbf{k})$ into Taylor series, taking into account the invariance with respect to $\mathbf{k} \leftrightarrow -\mathbf{k}$, and approximating the anisotropy in quadratic form yield $\Psi_{ijmn}(\mathbf{k}) = [\Psi^0_{ijmn}\delta_{rs} + \frac{1}{2}(\partial^2\Psi_{ijmn}/\partial k_r\partial k_s)k^2]\kappa_r\kappa_s = \Gamma_{rsijmn}(k)\kappa_r\kappa_s$, where $\boldsymbol{\kappa} = \mathbf{k}/k$ is a unit vector along the wave vector $\mathbf{k}$. In terms of the sixth-rank phonon-strain coupling coefficient tensor $\Gamma_{rsijmn}(k)$, the strain-dependent phonon frequency change becomes

$$\Delta\omega^2(\mathbf{k},p) = \Gamma_{rsijmn}(k)\kappa_r\kappa_s e_i(\mathbf{k},p)e_j(\mathbf{k},p)\varepsilon_{mn} \quad (4)$$

The sixth-rank tensor $\Gamma_{rsijmn}(k)$ complies with the crystal symmetry. Because of the three symmetry invariances with respect to $(rs) \leftrightarrow (sr)$, $(ij) \leftrightarrow (ji)$ and $(mn) \leftrightarrow (nm)$ as required in Equation (4), Γ_{rsijmn} has 216 independent components in its most general form, and a single-index Voigt notation $\Gamma_{\alpha\beta\gamma}$ can be used for convenience with each index running from 1 to 6 (instead of the double-index tensor notation Γ_{rsijmn} with each index running from 1 to 3). For cubic symmetry, there are only 12 independent components, namely, Γ_{111}, Γ_{122}, Γ_{212}, Γ_{221}, Γ_{123}, Γ_{144}, Γ_{414}, Γ_{441}, Γ_{244}, Γ_{424}, Γ_{442} and Γ_{456}. (It is worth noting that, when all three indices in single-index Voigt notation (or three pairs of symmetric indices in double-index tensor notation) are interchangeable, such as in the case of the third-order elastic modulus tensor $C_{\alpha\beta\gamma}$, the number of independent components of the sixth-rank tensor is further reduced to 6 for cubic symmetry, namely, C_{111}, C_{122}, C_{123}, C_{144}, C_{244} and C_{456}.)

Deformation-dependent low-energy phonons of incompletely softened transverse acoustic $[\zeta\zeta 0]$-TA$_2$ modes
As shown later in Equations (11a and 11b), , the effects of phonon-strain coupling on the behaviours and properties of austenitic

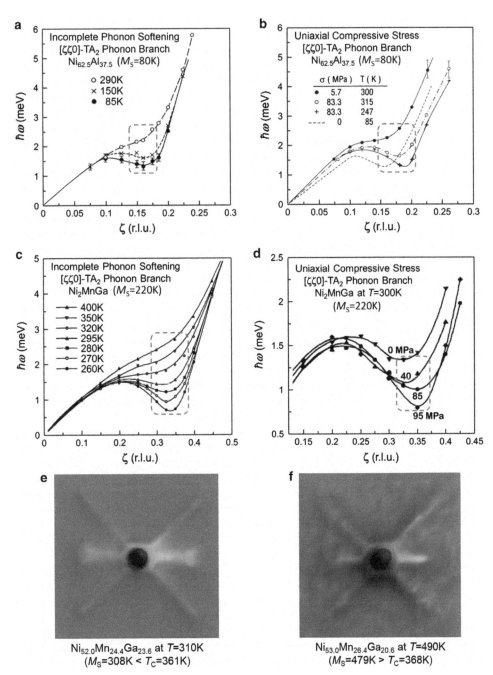

Figure 1. Incomplete phonon softening phenomenon and deformation-dependent low-energy phonons of transverse acoustic $[\zeta\zeta0]$-TA_2 branch in (**a, b**) nonmagnetic $Ni_{62.5}Al_{37.5}$ (refs 30,31) and (**c, d**) magnetic Ni_2MnGa (refs 27,32) before martensitic transformations as measured by inelastic neutron scattering (reproduced with permission—refs 27,30–32). (**a, c**) Dependence on temperature. (**b, d**) Dependence on uniaxial compressive stress. Red dashed lines highlight specific phonon modes at the negative dip of dispersion curve that possess low energy and exhibit strongest dependence on stress. (**e, f**) Diffuse scattering from incompletely softened phonons of $[\zeta\zeta0]$-TA_2 branch before martensitic transformations as measured around (800) Bragg reflection peak using high-energy synchrotron X-ray single-crystal diffraction in (**e**) $Ni_{52.0}Mn_{24.4}Ga_{23.6}$ with $M_S = 308\,K < T_C = 361\,K$ and (**f**) $Ni_{53.0}Mn_{26.4}Ga_{20.6}$ with $M_S = 479\,K > T_C = 368\,K$.

crystals before martensitic transformation depend both on the phonon-strain coupling coefficients and on the phonon energies. It is the phonon modes with low energy and strong strain dependence that have the most significant role. To be specific, transverse acoustic $[\zeta\zeta0]$-TA_2 phonons are considered, which are relevant to most martensitic systems.[5] As illustrated in Figure 1, $[\zeta\zeta0]$-TA_2 phonon branch undergoes incomplete softening in austenite phases.[27,30–32] In particular, the phonon modes corresponding to the negative dip in the dispersion curve at specific short wavelength, as highlighted by red dashed lines, possess low energy and exhibit the strongest dependence on stress (thus

strain). For $[\zeta\zeta0]$-TA_2 phonons of given wavenumber k in cubic crystal, there are 12 degenerate phonon modes, namely, M1–M12 as listed in Table 1. Substituting the wave vector direction κ and polarisation vector $\mathbf{e}$ of each phonon mode into Equation (4) gives the frequency change as a function of strain, as also listed in Table 1. It shows that the phonon frequencies depend on the strain only through 3 independent constants, which are combinations of the 12 independent components of the sixth-rank phonon-strain coupling coefficient tensor Γ_{rsijmn}, namely:

$$a = (\Gamma_{111} + \Gamma_{122} + \Gamma_{123} + \Gamma_{212} + 2\Gamma_{221} - 2\Gamma_{441} - 4\Gamma_{442})/6 \quad (5a)$$

Table 1. Twelve $[\zeta\zeta 0]$-TA$_2$ phonon modes (M1–M12) of specific wavenumber k

Mode: $\sqrt{2}\mathbf{\kappa}\|\sqrt{2}\mathbf{e}$ & $\Delta\omega^2 = \Gamma_{rsijmn}\kappa_r\kappa_s e_i e_j \varepsilon_{mn}$ (note that $\mathbf{\Gamma}$ is a function of k)

$a = (\Gamma_{111}+\Gamma_{122}+\Gamma_{123}+\Gamma_{212}+2\Gamma_{221}-2\Gamma_{441}-4\Gamma_{442})/6$

$b = \Gamma_{424}-\Gamma_{244}$

$c = -(\Gamma_{111}+\Gamma_{122}-2\Gamma_{123}+\Gamma_{212}-\Gamma_{221}+4\Gamma_{441}-4\Gamma_{442})/12$

M1 and M2: $\pm [011]\|[0\bar{1}1]$	M3 and M4: $\pm [0\bar{1}1]\|[01\bar{1}]$
$\Delta\omega^2 = a\varepsilon_V + c(2\varepsilon_{11}-\varepsilon_{22}-\varepsilon_{33})$ $+2b\varepsilon_{23}$	$\Delta\omega^2 = a\varepsilon_V + c(2\varepsilon_{11}-\varepsilon_{22}-\varepsilon_{33}) - 2b\varepsilon_{23}$
M5 and M6: $\pm [101]\|[10\bar{1}]$	M7 and M8: $\pm [10\bar{1}]\|[\bar{1}0\bar{1}]$
$\Delta\omega^2 = a\varepsilon_V + c(2\varepsilon_{22}-\varepsilon_{33}-\varepsilon_{11})$ $+2b\varepsilon_{31}$	$\Delta\omega^2 = a\varepsilon_V + c(2\varepsilon_{22}-\varepsilon_{33}-\varepsilon_{11}) - 2b\varepsilon_{31}$
M9 and M10: $\pm [110]\|[\bar{1}10]$	M11 and M12: $\pm [\bar{1}10]\|[\bar{1}\bar{1}0]$
$\Delta\omega^2 = a\varepsilon_V + c(2\varepsilon_{33}-\varepsilon_{11}-\varepsilon_{22})$ $+2b\varepsilon_{12}$	$\Delta\omega^2 = a\varepsilon_V + c(2\varepsilon_{33}-\varepsilon_{11}-\varepsilon_{22}) - 2b\varepsilon_{12}$

$$b = \Gamma_{424}-\Gamma_{244} \tag{5b}$$

$$c = -\left(\Gamma_{111}+\Gamma_{122}-2\Gamma_{123}+\Gamma_{212}-\Gamma_{221}+4\Gamma_{441}-4\Gamma_{442}\right)/12 \tag{5c}$$

It is readily seen from Table 1 that the symmetry-preserving volume strain $\varepsilon_V = \varepsilon_{11}+\varepsilon_{22}+\varepsilon_{33}$ contributes equally ($a\varepsilon_V$) to all 12 modes thus, as expected, does not lift the phonon degeneracy. On the other hand, a symmetry-breaking strain changes the phonon frequencies differently for different modes, thus lifts the phonon degeneracy and leads to phonon energy level splitting among the 12 modes. The isotropic volume strain is responsible for thermal expansion as in Grüneisen model.[34] The anisotropic deviatoric strain is responsible for the lifting of phonon degeneracy and the breaking of cubic lattice symmetry. Consider two examples of symmetry-breaking lattice deformation strain. Rhombohedral (trigonal) strain with trigonal axis along [111] is:

$$[\varepsilon_{ij}] = \begin{bmatrix} 0 & \gamma/2 & \gamma/2 \\ \gamma/2 & 0 & \gamma/2 \\ \gamma/2 & \gamma/2 & 0 \end{bmatrix} \tag{6}$$

where isotropic volume strain has been separated out. Substituting Equation (6) into $\Delta\omega^2$ of the 12 phonon modes listed in Table 1 gives:

$$\left.\begin{array}{l} \omega_{\perp}^2 = \omega_0^2 + b\gamma \\ \omega_{\|}^2 = \omega_0^2 - b\gamma \end{array}\right\} \quad\Rightarrow$$

$$\begin{cases} \omega_1 = \omega_2 = \omega_5 = \omega_6 = \omega_9 = \omega_{10} = \omega_{\perp} \approx \omega_0 + \frac{b\gamma}{2\omega_0} \\ \omega_3 = \omega_4 = \omega_7 = \omega_8 = \omega_{11} = \omega_{12} = \omega_{\|} \approx \omega_0 - \frac{b\gamma}{2\omega_0} \end{cases} \tag{7}$$

Tetragonal strain with tetragonal axis along [001] and isotropic volume strain separated out is:

$$[\varepsilon_{ij}] = \begin{bmatrix} \eta/3 & 0 & 0 \\ 0 & \eta/3 & 0 \\ 0 & 0 & -2\eta/3 \end{bmatrix} \tag{8}$$

which gives:

$$\left.\begin{array}{l} \omega_{\perp}^2 = \omega_0^2 + c\eta \\ \omega_{\|}^2 = \omega_0^2 - 2c\eta \end{array}\right\} \quad\Rightarrow$$

$$\begin{cases} \omega_1 = \omega_2 = \omega_3 = \omega_4 = \omega_5 = \omega_6 = \omega_7 = \omega_8 = \omega_{\perp} \approx \omega_0 + \frac{c\eta}{2\omega_0} \\ \qquad\quad \omega_9 = \omega_{10} = \omega_{11} = \omega_{12} = \omega_{\|} \approx \omega_0 - \frac{c\eta}{\omega_0} \end{cases} \tag{9}$$

Grüneisen-type phonon theory of pre-martensitic transition

A Grüneisen-type phonon theory is formulated in terms of the free energies of lattice deformation and quasi-harmonic phonons of a cubic crystal.[18,35] For a crystal characterised by the effective potential function $\Phi(\varepsilon_{ij}, u_a^r)$ in Equation (2), the total free energy at temperature T is a sum of lattice free energy $F_L = E_L - TS_L$ and phonon free energy $F_P = E_P - TS_P$. The entropy S_L describes the configurational entropy of a crystal lattice, whose change is negligibly small during diffusionless processes (such as elastic deformation and displacive transformation), thus can be regarded as a constant and omitted in isothermal analysis. Therefore, the lattice free energy is essentially the elastic energy $E_L = \frac{1}{2}VC_{ijkl}^0\varepsilon_{ij}\varepsilon_{kl}$. The phonon energy E_P and entropy S_P are evaluated from Planck distribution of independent harmonic oscillators, which gives the free energy of each phonon mode $F_{P(\mathbf{k},p)} = \frac{1}{2}\hbar\omega_{\mathbf{k},p} + k_BT \ln[1 - \exp(-\hbar\omega_{\mathbf{k},p}/k_BT)]$, where $\hbar$ and k_B are reduced Planck constant and Boltzmann constant, respectively. The total free energy density (normalised by the crystal volume V) of lattice deformation and all phonon modes $\{\mathbf{k},p\}$ is:

$$F = \frac{1}{2}C_{ijkl}^0\varepsilon_{ij}\varepsilon_{kl} + \frac{k_BT}{V}\sum_{\mathbf{k},p} f(x_{\mathbf{k},p}) \tag{10}$$

where $x_{\mathbf{k},p} = \hbar\omega_{\mathbf{k},p}/k_BT$ is dimensionless phonon energy and $f(x) = x/2 + \ln(1 - e^{-x})$. As shown in Equation (4), the phonon frequencies are function of strain due to phonon-strain coupling, thus $x_{\mathbf{k},p} = x_{\mathbf{k},p}(\varepsilon_{ij})$. The effects of the deformation-dependent phonons on the lattice stability is studied by analysing the derivatives of free energy with respect to strain:

$$\frac{\partial F}{\partial\varepsilon_{ij}} = C_{ijkl}^0\varepsilon_{kl} + \frac{\hbar}{V}\sum_{\mathbf{k},p} f'(x_{\mathbf{k},p})\frac{\partial\omega_{\mathbf{k},p}}{\partial\varepsilon_{ij}} \tag{11a}$$

$$\frac{\partial^2 F}{\partial\varepsilon_{ij}\partial\varepsilon_{kl}} = C_{ijkl}^0 + \frac{\hbar^2}{Vk_BT}\sum_{\mathbf{k},p} f''(x_{\mathbf{k},p})\frac{\partial\omega_{\mathbf{k},p}}{\partial\varepsilon_{ij}}\frac{\partial\omega_{\mathbf{k},p}}{\partial\varepsilon_{kl}} \tag{11b}$$

where linear dependence of phonon frequencies on strain is approximated as demonstrated in Equations (7) and (9). Equations (11a and 11b) show that both the phonon energies $\hbar\omega_{\mathbf{k},p}$ and the phonon-strain coupling coefficients have important roles respectively through $f(x_{\mathbf{k},p})$ and $\partial\omega_{\mathbf{k},p}/\partial\varepsilon_{ij}$.

As discussed in the section 'Deformation-dependent low-energy phonons of incompletely softened transverse acoustic $[\zeta\zeta 0]$-TA2 modes' and illustrated in Figure 1, the phonon modes corresponding to the negative dip in the dispersion curve at specific short wavelength possess both low energy and strong strain dependence thus have the dominant roles, which are the most significant terms in the sum in Equations (11a and 11b), i.e., $\sum_{\mathbf{k},p}$ practically sums only over a small fraction of the first Brillouin zone. For simplicity without loss in conceptual generality, an Einstein-type model of density of states is adopted to focus on N' modes of such phonons ($N' << 3N$, the total number of phonon modes), which are assumed to possess the same frequency and phonon-strain coupling coefficients in cubic crystal. To be specific, the transverse acoustic $[\zeta\zeta 0]$-TA$_2$ phonons listed in Table 1 are considered. The lattice stability is analysed with respect to two examples of lattice deformation. With respect to rhombohedral (trigonal) strain in Equation (6), the elastic energy becomes $E_L = \frac{1}{6}\left(C_{11}^0 + 2C_{12}^0\right)\varepsilon_V^2 + \frac{3}{2}C_{44}^0\gamma^2$. According to whether the wave vector $\mathbf{k}$ is perpendicular to the trigonal [111] axis or not, the 12 $[\zeta\zeta 0]$-TA$_2$ phonon modes of each given wavenumber k fall into two groups, each group consisting of six modes with same frequency dependence on the strain, as shown in Equation (7). In this case, Equations (11a and 11b) yield

$$\frac{\partial F}{\partial\varepsilon_V} = \frac{1}{3}\left(C_{11}^0 + 2C_{12}^0\right)\varepsilon_V + \frac{3\hbar a\chi}{2\Omega\omega_0}f'(x_0) \tag{12}$$

$$\frac{\partial^2 F}{\partial\gamma^2} = 3C_{44}^0 + \frac{3\hbar^2 b^2\chi}{4\Omega k_BT\omega_0^2}f''(x_0) \tag{13}$$

where $\Omega = V/N$ is primitive cell volume, $\chi = N'/3N$ and $x_0 = \hbar\omega_0/k_B T$. With respect to tetragonal strain in Equation (8), the elastic energy becomes $E_L = \frac{1}{6}(C_{11}^0 + 2C_{12}^0)\varepsilon_V^2 + \frac{1}{3}(C_{11}^0 - C_{12}^0)\eta^2$. According to whether the wave vector **k** is perpendicular to the tetragonal [001] axis or not, the 12 [$\zeta\zeta$0]-TA$_2$ phonon modes of each given wavenumber k also fall into 2 groups, each group consisting of 8 and 4 modes respectively with same frequency dependence on the strain, as shown in Equation (9). In this case, Equations (11a and 11b) yield the same $\partial F/\partial\varepsilon_V$ in Equation (12) as expected and

$$\frac{\partial^2 F}{\partial\eta^2} = \frac{2}{3}(C_{11}^0 - C_{12}^0) + \frac{3\hbar^2 c^2\chi}{2\Omega k_B T\omega_0^2}f''(x_0) \qquad (14)$$

The isotropic volume strain ε_V characterises the symmetry-preserving thermal expansion phenomenon (as in Grüneisen model[34]) and is determined by $(\partial F/\partial\varepsilon_V)_{\gamma,\eta=0} = 0$, which yields $\varepsilon_V^0 = -9\hbar a\chi f'(x_0)/2\Omega\omega_0(C_{11}^0 + 2C_{12}^0)$—note that $\varepsilon_V^0 = \varepsilon_V^0[\omega_0(T), T]$ is temperature dependent. It is worth noting that $-a/2\omega^2$ corresponds to the Grüneisen parameter of an individual phonon mode, where the phonon-strain coupling constant a is defined in Equation (5a). In fact, the sixth-rank phonon-strain coupling coefficient tensor Γ_{rsijmn} extends the Grüneisen parameter from volume change of thermal expansion to anisotropic strain of symmetry-lifting displacive transformations. In particular, the phonon-strain coupling constants b and c defined in Equations (5b and 5c) respectively characterise the coupling of phonons with rhombohedral (trigonal) and tetragonal lattice deformation, as shown in Equations (7), (9) and (15a and 15b).

For the rhombohedral (trigonal) strain γ and tetragonal strain η characterising the symmetry-breaking anisotropic lattice deformation, $(\partial F/\partial\gamma)_{\gamma=0}\equiv0$ and $(\partial F/\partial\eta)_{\eta=0}\equiv0$ for cubic lattice. The lattice stability of cubic crystal is analysed through the second derivatives. The stability is lost against rhombohedral (trigonal) or tetragonal strain under the following respective conditions:

$$\frac{\partial^2 F}{\partial\gamma^2} = 3\left[C_{44}^0 + \frac{\hbar^2 b^2\chi}{4\Omega k_B T\omega_0'^2}f''(x_0')\right] < 0 \qquad (15a)$$

$$\frac{\partial^2 F}{\partial\eta^2} = \frac{2}{3}\left[(C_{11}^0 - C_{12}^0) + \frac{9\hbar^2 c^2\chi}{4\Omega k_B T\omega_0'^2}f''(x_0')\right] < 0 \qquad (15b)$$

where ω_0' and x_0' incorporate the effect of thermal expansion, i.e., $\omega_0' = \omega_0 + a\varepsilon_V^0/2\omega_0$. Equations (15a and 15b) predict a pre-martensitic transition, on which the cubic symmetry of the crystal is spontaneously broken by anisotropic strain. The critical condition of the pre-martensitic transition is given by $\partial^2 F/\partial\gamma^2 = 0$ and $\partial^2 F/\partial\eta^2 = 0$ in the two exemplary cases, and is analysed via non-positive definiteness of $\partial^2 F/\partial\varepsilon_{ij}\partial\varepsilon_{kl}$ in general. Figure 2a plots $f(x)$, $f'(x)$ and $f''(x)$. With decreasing $x = \hbar\omega/k_B T$, $f''(x)$ rapidly approaches $-\infty$ and, as a result, realises the instability condition in Equations (15a and 15b). The decreasing energy of incompletely softened phonons with decreasing temperature, as illustrated in Figure 1, just satisfies the requirement for a pre-martensitic transition.

'Phonon spinodal decomposition' and formation of phonon domains

Equations (15a and 15b) describe a spinodal instability, which is related to the negative curvature (the second derivative) of the free energy, in analogy to spinodal decomposition.[36,37] In particular, the Grüneisen-type phonon theory predicts the pre-martensitic transition to occur via 'spinodal decomposition' of phonon populations into phonon domains. The symmetry-breaking anisotropic strain changes the phonon frequencies differently for different modes, as given in Equation (4) and exemplified in Equations (7) and (9). Consequently, the thermal equilibrium occupation number of each phonon mode,

$\bar{n}_0 = [\exp(\hbar\omega_0/k_B T) - 1]^{-1}$, changes according to its $\Delta\omega$ by an amount

$$\Delta\bar{n} = -\frac{\Delta\omega\hbar\omega_0}{\omega_0 k_B T}\exp\left(\frac{\hbar\omega_0}{k_B T}\right)\left[\exp\left(\frac{\hbar\omega_0}{k_B T}\right) - 1\right]^{-2} \qquad (16)$$

As illustrated in Figure 2b for the case of rhombohedral (trigonal) deformation (tetragonal case is similar[35]), because $f''(x) < 0$, the total phonon free energy decreases as a result of the changes in phonon frequencies (energies) and populations. Thus, the phonon populations have an intrinsic tendency to 'decompose' into phonon domains. Each phonon domain is a spatial region of dominant phonon mode with excess population, whose strength is characterised by $\Delta\bar{n}/\bar{n}_0$. The tendency of 'phonon spinodal decomposition' is, however, resisted by the elastic energy. Under the condition specified in Equations (15a and 15b), the decrease in phonon free energy outweighs the increase in elastic energy, leading to pre-martensitic transition.

The pre-martensitic transition breaks the cubic crystal symmetry and leads to formation of elastic domains. Elastic domains are characterised by the time-averaged static symmetry of crystal lattice, whose strength is described by the spontaneous aniso-tropic strain (e.g., γ and η). The spontaneous strain of pre-martensitic phase is small as compared with the transformation strain of martensitic phase, and the pre-martensitic phase remains quasi-cubic. More importantly, the elastic domains are also phonon domains of redistributed phonon populations due to 'phonon spinodal decomposition.' Phonon domains are charac-terised by the broken dynamic symmetry of lattice vibrations,[18,38] whose strength is described by the excess population of the dominant phonon mode with lowered phonon energy due to phonon-strain coupling. As illustrated in Figure 5d for a [$\zeta\zeta$0]-TA$_2$ mode of wave vector **k** along [110] axis, the presence of such a dominant phonon mode with excess population in a lattice region (domain) produces anisotropic dynamic effects in the domain. Although the time-averaged atomic positions and thus the static symmetry of the local lattice are retained, the dynamic symmetry of the lattice vibration is broken due to the dominant phonon mode: the directions of wave vector **k** along [110], polarisation **u** along [1$\bar{1}$0] and perpendicular **n** along [001] are inequivalent, and, in particular, the **n** direction along [001] is distinct from [100] and [010], which defines a dynamic tetragonal **n**-axis. Consequently, the phonon domain exhibits anisotropic behaviours due to the broken dynamic symmetry of lattice vibrations. Furthermore, as discussed in the section 'Internal phonon relaxation in phonon domains and exotic domain phenomenon,' phonon domains can deform through internal phonon relaxation in response to applied stress. Such anisotropic phonon domains of broken dynamic symmetry would manifest themselves in phonon diffuse scatter-ing under in situ stress (as observed by in situ synchrotron X-ray diffraction experiments discussed in the section 'Stress-dependent phonon diffuse scattering and Bragg reflection,' where Bragg reflection and phonon diffuse scattering together provide complementary information of crystal lattice and lattice vibra-tions). Therefore, the properties and behaviours of the elastic-phonon domains are primarily characterised by the broken dynamic symmetry of lattice vibrations rather than the broken static symmetry of crystal lattice, in contrast to the conventional ferroelastic domains.

Phonon pseudo-Jahn–Teller lattice instability

Since the pre-martensitic transition via 'phonon spinodal decom-position' into phonon domains lifts the degeneracy of the phonons in cubic crystal, its nature can be interpreted as a phonon pseudo-Jahn–Teller lattice instability. It is caused by the interactions between fast lattice vibrations and slow lattice deformation (i.e., phonon-strain coupling), in analogy to Jahn–Teller instability that is caused by electron–phonon coupling

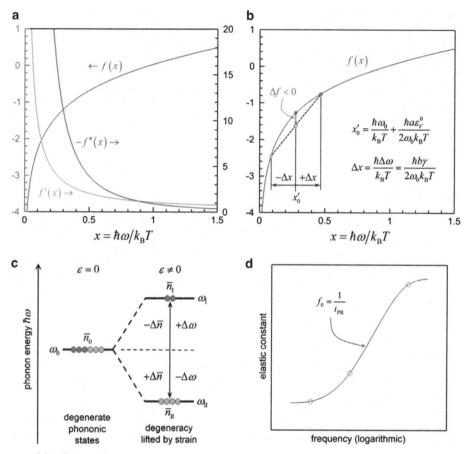

Figure 2. (a) Dependences of $f(x)$, $f'(x)$ and $f''(x)$ on $x = \hbar\omega/k_BT$. (b) Illustration of pre-martensitic transition via 'phonon spinodal decomposition' associated with lattice deformation. (c) Schematic of phonon energy level splitting and phonon population redistribution due to lifting of phonon degeneracy by lattice strain. (d) Schematic of frequency dependence of elastic constant.

(the interactions between fast electronic and slow phononic processes). During the pre-martensitic transition, the crystal lattice spontaneously deforms to break the cubic symmetry and lift the phonon degeneracy, and in so doing lower the total phonon energy $E_P = \sum_{\mathbf{k},p}(1/2 + \bar{n}_{\mathbf{k},p})\hbar\omega_{\mathbf{k},p}$, where $\bar{n}_{\mathbf{k},p} = [\exp(\hbar\omega_{\mathbf{k},p}/k_BT) - 1]^{-1}$. For cubic lattice, the above considered 12 equivalent $[\zeta\zeta 0]$-TA$_2$ phonon modes of given wavenumber k are degenerate with the same frequency ω_0 and equilibrium occupation number $\bar{n}_0$ for each mode. Such a degenerate phononic state is unstable due to phonon-strain coupling. For example, under rhombohedral (trigonal) lattice strain, the 12 phonon modes fall into 2 groups (6 in each) that undergo phonon energy level splitting: $\omega_I = \omega_0 + \Delta\omega$ and $\omega_{II} = \omega_0 - \Delta\omega$, as shown in Equation (7) and illustrated in Figure 2c. Consequently, the phonon populations redistribute: $\bar{n}_I = \bar{n}_0 - \Delta\bar{n}$ and $\bar{n}_{II} = \bar{n}_0 + \Delta\bar{n}$—note here $\Delta\bar{n}$ represents absolute value to explicitly emphasise the opposite sign to the frequency change magnitude $\Delta\omega$. Lifting phonon degeneracy lowers the total phonon energy: $\Delta E_P = 6\hbar\omega_I\bar{n}_I + 6\hbar\omega_{II}\bar{n}_{II} - 12\hbar\omega_0\bar{n}_0 = -12\hbar\Delta\omega\Delta\bar{n} < 0$. Similar results can be demonstrated for tetragonal lattice strain.[18] The phonon energy decrease $\Delta E_P \propto -\Delta\omega\Delta\bar{n}$ is a general feature. Because $\Delta\omega \propto \varepsilon$ as shown in Equations (7) and (9) and $\Delta\bar{n} \propto \Delta\omega$ as shown in Equation (16), the phonon energy decrease is quadratically proportional to strain, $\Delta E_P \propto -\varepsilon^2$, which competes with the elastic energy. With the effects of temperature and entropy taken into account, a statistical mechanics picture of strain-induced phonon energy level splitting and phonon population redistribution is equivalent to the thermodynamics picture of 'phonon spinodal decomposition.'

The phonon-strain pseudo-Jahn–Teller lattice instability exhibits some features distinct from the electron–phonon Jahn–Teller

instability. Unlike electrons being fermions thus degenerate orbitals must be partially filled to produce Jahn–Teller instability, phonons are boson quasiparticles whose occupation numbers are free to change, thus phonon-strain coupling is a general mechanism of lattice instability. Moreover, the energy decrease in phonon pseudo-Jahn–Teller effect is quadratically proportional to strain (i.e., $\Delta E_P \propto -\Delta\omega\Delta\bar{n} \propto -\varepsilon^2$), while in Jahn–Teller effect the energy decrease is linearly proportional to strain because the filling of lower-energy orbitals in ground state does not depend on the strain. Furthermore, the strength of the phonon pseudo-Jahn–Teller effect as measured by ΔE_P depends not only on the phonon-strain coupling coefficients but also, more importantly, on the low phonon energy $\hbar\omega_0$. As the phonon energy decreases during incomplete phonon softening, the strength increases rapidly and eventually leads to the phonon pseudo-Jahn–Teller lattice instability and pre-martensitic transition.

Internal phonon relaxation in phonon domains and exotic domain phenomenon

Because the behaviours of the elastic-phonon domains are primarily characterised by the broken dynamic symmetry of lattice vibrations associated with the excess population of the dominant phonon mode, and the phonon populations redistribute in response to deformation strain due to phonon-strain coupling, phonon population redistribution provides an extra degree of freedom for internal relaxation of the domains in response to applied stress. Such internal phonon relaxation is a particular case of Le Chatelier's principle.[34] According to Le Chatelier's principle, an external interaction with a system to disturb its internal equilibrium brings about internal processes to

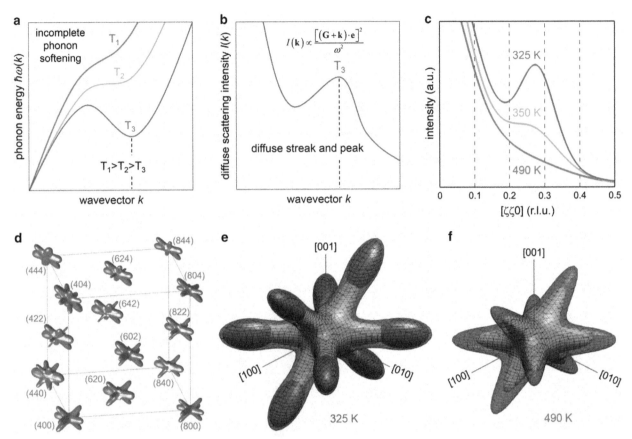

Figure 3. (**a**) Schematic of phonon dispersion curve at temperatures $T_1 > T_2 > T_3 > M_S$; a negative dip develops at T_3. (**b**) Schematic of intensity distribution of phonon diffuse scattering at T_3; a peak develops at the same wave vector of the dispersion dip. (**c**) Experimental measurement of temperature-dependent phonon diffuse scattering intensity distribution; a peak develops at 325 K. (**d**) Experimental measurement of 3D phonon diffuse scattering from $< \zeta\zeta 0 >$-TA$_2$ modes around different Bragg reflections at 330 K without applied stress; extinction rule is determined by $(\mathbf{G} \cdot \mathbf{e})^2$. (**e, f**) Experimental measurement of 3D phonon diffuse scattering around (800) Bragg reflection at 325 K with satellite peaks and at 490 K without satellite peaks.

restore the equilibrium and reduce the effects of this interaction. In our case here, an externally imposed deformation disturbs the internal equilibrium of a pre-martensitic crystal and produces stress, which brings about internal relaxation processes via phonon population redistributions; on re-establishing the phonon populations into new thermal equilibrium, the stress response is reduced for the given externally imposed strain, leading to lower elastic modulus. In particular, it allows the pre-martensitic crystal to respond to applied stress via domain deformation rather than domain switching, in drastic contrast to the usual ferroelastic domain switching behaviour in martensitic phase. Thus, exotic domain phenomenon in pre-martensitic phase under external stress is expected from the perspective of phonon domains, where the exoticness is characterised by the following distinct features: domain deformation via internal phonon relaxation and phonon-strain coupling without domain switching, reversibility and anhysteresis on unloading through thermal equilibration of phonon populations, and elasticity with low stiffness. As internal phonon relaxation takes time, the relaxation time of phonon population redistribution also leads to frequency dependence of the properties of pre-martensitic crystal. In particular, higher values of elastic constants are expected when measured by high-frequency technique (e.g., ultrasonic wave velocity) than by low-frequency technique (e.g., quasi-static stress-strain curve), as illustrated in Figure 2d, where a transition frequency f_0 corresponds to the reciprocal of phonon relaxation time t_{PR}. Such frequency dependence offers a means to determine the characteristic time t_{PR} of internal phonon relaxation in the domains.

EXPERIMENTAL EVIDENCE OF PHONON DOMAINS IN PRE-MARTENSITIC AUSTENITE

Above theoretical predictions of phonon domains and their behaviours in austenite and pre-martensite phases are critically tested by specially designed experiments of *in situ* 3D phonon diffuse scattering and Bragg reflection using high-energy synchrotron X-ray single-crystal diffraction,[18,38,39] which provide experimental evidence of phonon domains. Phonon diffuse scattering and Bragg reflection together provide complementary information of crystal lattice and lattice vibrations. High-flux high-energy synchrotron X-ray single-crystal diffraction is essential for such experimental tests[39]: high energy provides deep penetration to probe bulk specimen, high flux enables short exposure time to measure weak diffuse scattering intensity that is critical for 3D reciprocal space mapping, single crystal allows loading along specific crystallographic axis, and 3D single-crystal diffraction allows measurement of diffuse scattering intensities from individual phonon modes—2D measurement using rocking crystal method causes intensity overlapping problem where scattering intensities from multiple phonon modes (such as 12 $< \zeta\zeta 0 >$-TA$_2$ branches) are all projected onto one image plate, as demonstrated in Figures 1e,f and 7. The diffuse scattering intensity from a phonon mode of wave vector $\mathbf{k}$, polarisation vector $\mathbf{e}$ and frequency ω around reciprocal lattice vector $\mathbf{G}$ (corresponding to Bragg reflection) is[40]:

$$I_{\mathbf{G}}(\mathbf{k}) \propto \frac{[(\mathbf{G} + \mathbf{k}) \cdot \mathbf{e}]^2}{\omega^2} \qquad (17)$$

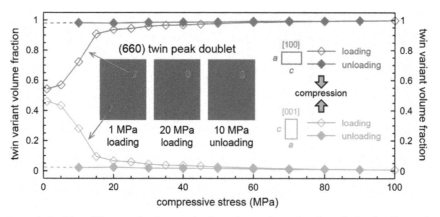

Figure 4. Evolutions of twin peak doublet of Bragg reflections and twin variant volume fractions during loading-unloading cycle in tetragonal martensite, showing usual ferroelastic domain switching (detwinning) and hysteresis.

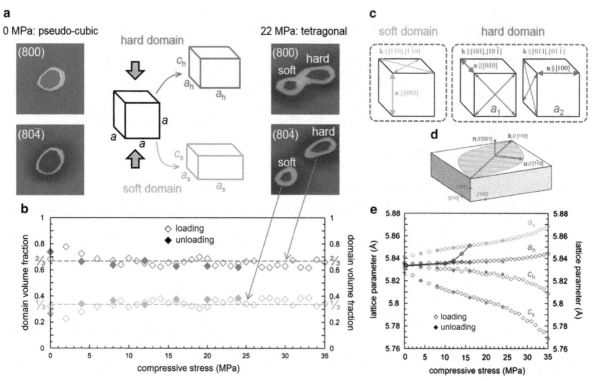

Figure 5. (**a**) Evolution of Bragg reflection peaks and (**b**) conservation of domain variant volume fractions during loading-unloading cycle in pre-martensite. (**c**) 12 [$\zeta\zeta$0]-TA$_2$ phonon modes fall into two groups according to their tetragonal axes **n** with respect to [001] stress axis. (**d**) Schematic of dynamic tetragonal axis **n** of phonon domain of dominant [$\zeta\zeta$0]-TA$_2$ phonon mode. (**e**) Stress-dependent lattice parameters of hard (red, blue) and soft (green, pink) phonon domains as determined from Bragg peaks in (**a**). Lattice parameters (black) are also determined from the centres of 3D diffuse scattering in Figure 6a–c.

In light of phonon dispersion curves directly measured by inelastic neutron scattering as illustrated in Figure 1a–d, Equation (17) explicitly correlates the phenomena of phonon diffuse scattering and incomplete phonon softening (X-ray scattering has been used to determine phonon dispersions[41]). In particular, as illustrated in Figure 3a,b, when a negative dip develops at specific wave vector in the phonon dispersion curve during incomplete phonon softening, the intensity distribution of phonon diffuse scattering from this softened branch accordingly develops a peak at the same wave vector, as highlighted by dashed line. Thus, the diffuse streaks and satellites observed around Bragg reflections manifest the behaviours of the incompletely softened phonon modes. Together with Bragg reflection, phonon diffuse scattering is employed to study phonon domains in single crystals of

Ni$_{49.90}$Mn$_{28.75}$Ga$_{21.35}$ ($M_S = 323$ K, $T_C = 373$ K, prepared by Adaptamat).

Temperature-dependent phonon diffuse scattering

The experimentally measured phonon diffuse scattering intensity distributions along wave vector **k** = [$\zeta\zeta$0] (in reciprocal lattice unit, r.l.u.) around **G** = [800] at three representative temperatures are shown in Figure 3c. In agreement with the evolution of phonon dispersion curve during incomplete phonon softening as exemplified in Figure 1a,c and with Equation (17), the diffuse streak becomes stronger on cooling, and diffuse satellite peak develops at low temperature (still above M_S) manifesting a negative dip developed in the phonon dispersion curve. Shifting in the peak position to higher wave vector value with decreasing temperature

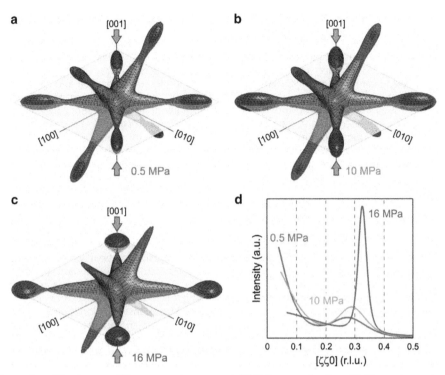

Figure 6. (**a–c**) Stress-dependent 3D phonon diffuse scattering around (800) Bragg reflection. (**d**) Stress dependence of diffuse scattering intensity distribution.

is also in agreement with the evolution of the negative dip position in the phonon dispersion curve. Figure 3e,f shows the experimentally measured 3D phonon diffuse scattering around (800) Bragg reflection, where satellite peaks are observed at 325 K while not at 490 K. Figure 3d shows the experimentally measured 3D phonon diffuse scattering around different Bragg reflections at 330 K, where each diffuse rod originates from one of the $<\zeta\zeta0>$-TA$_2$ phonon branches. As $\mathbf{k}\cdot\mathbf{e}=0$ for transverse acoustic phonons, the phonon diffuse scattering exhibits systematic extinction around different Bragg reflections following the rule determined by $(\mathbf{G}\cdot\mathbf{e})^2$ according to Equation (17). The observed phenomenon of temperature-dependent phonon diffuse scattering shown in Figure 3 agrees with the phenomenon of incomplete phonon softening.

Stress-dependent phonon diffuse scattering and Bragg reflection
To contrast the exotic domain phenomenon predicted in pre-martensite to the usual ferroelastic domain switching phenomenon in martensite, Ni$_{52.0}$Mn$_{24.4}$Ga$_{23.6}$ at 150 K is first studied, which transforms into nonmodulated tetragonal martensite below 220 K. Figure 4 shows the evolution of Bragg reflection peak doublet from the twin-related martensite domain variants under loading-unloading cycle. The change in the peak intensities manifests the domain switching/detwinning process and allows determination of the twin variant volume fraction. The evolution of twin variant volume fraction as a function of stress exhibits large hysteresis on unloading, which is one important characteristic of usual ferroelastic domain switching phenomenon.

Figure 5a,b shows the evolution of Bragg reflection peaks and conservation of domain variant volume fractions during loading-unloading cycle in pre-martensite of Ni$_{49.90}$Mn$_{28.75}$Ga$_{21.35}$ at 350 K (above $M_S=323$ K). Uniaxial stress is applied along single crystal [001] axis. Exotic domain phenomenon is observed from Bragg reflection, where the most unusual behaviour is that the domains deform into two (soft and hard) tetragonal lattices but do not switch. In particular, the volume fractions of soft and hard domain variants remain 1/3 and 2/3, respectively, during entire

loading-unloading cycle. As illustrated in Figure 5d, the presence of dominant phonon modes in a domain breaks the dynamic symmetry of the lattice and makes the **n** direction a tetragonal axis. As shown in Equation (9) and illustrated in Figure 5c, according to whether their tetragonal axes **n** are parallel to the [001] stress axis or not (or equivalently, their wave vectors **k** perpendicular to the [001] stress axis or not), the 12 [$\zeta\zeta0$]-TA$_2$ phonon modes of each given wavenumber k fall into 2 groups, each group consisting of 4 and 8 modes, respectively. As illustrated in Figure 1b,d, the [$\zeta\zeta0$]-TA$_2$ phonon modes at the dispersion dip soften further in response to [001] stress, such internal phonon relaxation makes the 4 domains softer than the other 8 domains, leading to Bragg peak splitting under [001] stress shown in Figure 5a, and the ⅓ and ⅔ domain variant volume fractions manifest the 4:8 volume ratio of the soft and hard phonon domains. The values of lattice parameters a_1 and a_2 are slightly different under stress due to the anisotropy of the domains as illustrated in Figure 5c, resulting in anisotropic shape of the Bragg peak of the hard tetragonal lattice shown in Figure 5a. Figure 5e shows the lattice parameters as function of stress, which exhibit reversibility free of hysteresis, indicating elastic deformation behaviours of both soft and hard domains. The lattice parameters $a_{s[100]}$ and $a_{s[001]}$ respectively along crystal [100] and [001] axes correspond to the elastically deformed soft domains with **k** along $\pm[110]$ or $\pm[1\bar{1}0]$, while $a_{h[100]}$ and $a_{h[001]}$ correspond to the hard domains with **k** along $\pm[011]$, $\pm[01\bar{1}]$, $\pm[101]$, or $\pm[10\bar{1}]$. Although the hard phonon domains exhibit Young's modulus $E=14$ GPa and normal Poisson's ratio $\nu=0.33$, the soft domains exhibit significantly lower Young's modulus $E=4.8$ GPa and high Poisson's ratio $\nu=0.47$. The observed large, reversible, anhysteretic elastic strain response with low elastic stiffness agrees with the predictions from internal phonon relaxation and phonon-strain coupling.

The 3D phonon diffuse scattering around (800) Bragg reflection as function of stress shown in Figure 6a–c reveals more information on the phonon domains. The in-plane diffuse scattering corresponding to **k** perpendicular to [001]-stress (**k** along $\pm[110]$ or $\pm[1\bar{1}0]$ within the yellow plane) grows in intensity with increasing [001] stress, while the out-of-plane diffuse scattering is depressed by the stress (note that only 4 diffuse rods with **k** along $\pm[101]$ or $\pm[10\bar{1}]$ are present while the other 4 with **k** along $\pm[011]$ or $\pm[01\bar{1}]$ are absent due to extinction). The enhanced in-plane diffuse scattering manifests the softening of the 4 phonon modes in the soft domains with tetragonal axis **n** along [001] stress axis. The in-plane diffuse satellites not only grow in intensity with increasing stress, but also concentrate into narrower and stronger peaks, and their incommensurate positions gradually shift towards $\zeta = \frac{1}{3}$, as shown in Figure 6d. These observations are in agreement with the evolution of phonon dispersion curve under stress as exemplified in Figure 1b,d. To confirm that each of the two tetragonal domains produces own phonon diffuse scattering, the lattice parameter a is also determined respectively from the centres of in-plane and out-of-plane diffuse scatterings around (800) Bragg peak, as also plotted in black symbols and lines in Figure 5e. The out-of-plane diffuse scattering gives the same a as $a_{h[100]}$, indicating its association with the hard domains. The in-plane diffuse scattering gives a that is close to $a_{h[100]}$ at low stress and approaches $a_{s[100]}$ with increasing stress, indicating its association with the soft domains. The transition from $a_{h[100]}$ at low stress to $a_{s[100]}$ with increasing stress is caused by the fact that the in-plane diffuse scattering is increasingly dominated by the soft phonon domains with increasing stress and the out-of-plane diffuse scattering is dominated by the hard phonon domains, while at low stress their diffuse scatterings overlap. Such exotic domain behaviours observed by combined 3D phonon diffuse scattering and Bragg reflection agree with the theoretical predictions and provide experimental evidence of phonon domains in pre-martensitic phase.

THE ROLES OF PHONON DOMAINS IN PRE-MARTENSITIC PHENOMENA

The notion of phonon domains formed by pre-martensitic transition via 'phonon spinodal decomposition' provides consistent explanations to the martensitic precursor 'anomalies' without resort to extrinsic defects. These 'anomalies' are shown to be natural consequences of incomplete phonon softening and are explained through the key roles played by phonon domains. In particular, the very existence of phonon domains makes the pre-martensitic austenite no longer homogeneous, which in and of itself explains various martensite precursor effects. As has been previously pointed out[16]: 'The explanation of most of precursor effects is based on the existence of hetero-phase fluctuation in the parent phase. However, such fluctuation requires that the system climb over large energy barrier. It has been questioned whether such fluctuation is really possible.' The pre-martensitic transition produces heterogeneous structures in the pre-martensite phase consisting of phonon domains. It is worth noting that such pre-martensite is a thermodynamically equilibrium new phase rather than hetero-phase fluctuation in the parent phase. The currently existing alternative explanations of the pre-martensitic phenomena (in particular, tweed patterns) are based on static defects (composition fluctuation, chemical disorder and point defects) that induce martensite embryos and cause strain glass state in analogy to spin glass.[15,19–24] In light of the theory and experimental evidence of phonon domains and their roles in pre-martensitic phenomena,[18] these alternative opinions are revisited.

'Phonon spinodal decomposition' versus phonon condensation

The pre-martensitic anomalies have been previously attributed to $[\zeta\zeta 0]$-TA$_2$ phonon mode condensation into static displacement waves, resulting in lattice modulations.[42] According to the

Grüneisen-type phonon theory,[18] the pre-martensitic transition occurs via 'phonon spinodal decomposition' into phonon domains. With decreasing temperature, the phonons increasingly soften, and the lower phonon energy $x = \hbar\omega/k_BT$ makes $f''(x)$ rapidly approach $-\infty$, which eventually fulfills the instability condition in Equations (15a and 15b) before the phonon energy softens to zero. It is important that such instability condition does not require phonon frequency to vanish, that is, soft mode is not required. In particular, the pre-martensitic transition takes place at incompletely softened phonon energy $\sim 1\,\text{meV}$, which corresponds to finite phonon frequency $\sim 1\,\text{THz}$. At such finite frequency, phonons 'decompose' into domains but do not condense into static displacement waves or lattice modulations. Thus, the pre-martensitic transition is fundamentally different from a soft-mode transition. Moreover, Equations (15a and 15b) imply that, even in the case of a soft mode whose frequency softens to zero on cooling, as long as there is phonon-strain coupling as characterised by nonzero coupling constants, the lattice instability will be reached and the pre-martensitic transition will occur before a soft-mode transition.

It is worth noting that the effects of phonon-strain coupling on the behaviours and properties of austenitic crystals before martensitic transformation depend not only on the phonon-strain coupling coefficients Γ_{rsijmn} through $\partial\omega_{\mathbf{k},p}/\partial\varepsilon_{ij}$ but also, more strongly, on the phonon energies $\hbar\omega_{\mathbf{k},p}$ through $f(x_{\mathbf{k},p})$, as shown in Equations (11)–(15). The strong effect of incompletely softened phonon energy can be illustrated through $f''(x)$ from the experimental data shown in Figure 1c: $\hbar\omega \approx 1.4\,\text{meV}$ at 295 K gives $f''(x) \approx -330$, which is enhanced to $f''(x) \approx -1,024$ for further softened phonon energy $\hbar\omega \approx 0.7\,\text{meV}$ at 260 K. The phonon-strain coupling coefficients have an electronic origin and, in principle, can be determined through $\omega_{\mathbf{k},p} = \omega_{\mathbf{k},p}(\varepsilon_{ij})$ by first-principles density functional theory computations at finite temperature, e.g., using self-consistent *ab initio* lattice dynamics approach.[43] Such studies would identify the mechanisms of phonon-strain coupling and guide a search for materials with high phonon-strain coupling coefficients. Although it could be challenging to experimentally measure the phonon-strain coupling coefficients, the magnitude of the coefficients can be estimated from the stress-dependent phonon dispersion. For example, from the phonon dispersion curves in Ni$_2$MnGa under [001] uniaxial compressive stress at 300 K shown in Figure 1d, it is estimated that $d(\hbar\omega)/d\sigma \approx 5\,\text{meV/GPa}$ for the incompletely softened low-energy phonon modes at the dispersion dip (the modes that have the dominant roles in pre-martensitic transition). According to Equation (9) for tetragonal strain, $d\omega/d\varepsilon \approx c/\omega$, where the phonon-strain coupling constant c is defined in Equation (5c). Using $d\omega/d\varepsilon = \hbar^{-1}[d(\hbar\omega)/d\sigma](d\sigma/d\varepsilon)$, $d\sigma/d\varepsilon \approx E \sim 10\,\text{GPa}$ and $\hbar\omega \sim 1\,\text{meV}$, it gives $c \sim 1 \times 10^{26}\,\text{s}^{-2}$. It would be meaningful to show that the lattice instability condition with respect to tetragonal strain shown in Equation (15b) is satisfied in pre-martensitic Ni–Mn–Ga crystals: using $T = 300\,\text{K}$, $\Omega \approx 25\,\text{Å}^3$ and $\chi \sim 0.01$, Equation (15b) becomes $C' < 5\,\text{GPa}$, where $C' = (C_{11}^0 - C_{12}^0)/2$. The estimated value of C' agrees in order of magnitude with the experimental values from ultrasonic measurements.[44] This condition indeed coincides with the general phenomenon of significantly reduced C' before martensitic transformation,[5] which could be the reason that such phonon-driven effects appear as pre-martensitic phenomena or martensite precursor effects.

Phonon diffuse scattering (streaks and satellites) versus superlattice reflection

The diffuse scattering streaks have been previously attributed to strain diffuse scattering from static displacement field associated with martensite embryos, and the diffuse satellite spots have been regarded as superlattice reflection peaks thus interpreted as evidence of static displacement waves (periodic lattice

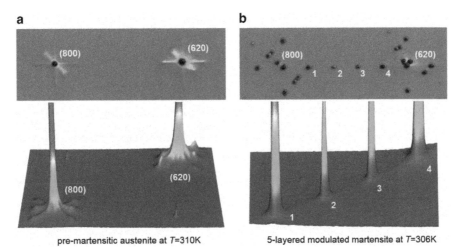

Figure 7. Diffuse scattering and Bragg reflections from $Ni_{52.0}Mn_{24.4}Ga_{23.6}$ ($M_S = 308$ K) as measured by high-energy synchrotron X-ray diffraction using single crystal rocking method. (**a**) Diffuse scattering from incompletely softened phonons of $[\zeta\zeta0]$-TA_2 branch in pre-martensite at 2 K above M_S. (**b**) Superlattice reflections from lattice modulations in 5-layered modulated martensite at 2 K below M_S. Lower row shows 3D visualisations of the 2D intensities in upper row.

modulations).[4,6–10,15,16,19–24,45] As shown in Equation (17), the diffuse scattering is directly related to the incomplete phonon softening and, in particular, manifests the shape of the phonon dispersion curve. When a phonon branch softens, it produces diffuse scattering streaks. When the phonon branch further softens on cooling, the diffuse streaks become stronger. When the phonon dispersion curve develops a negative dip at specific wave vector, the diffuse scattering produces satellite peaks accordingly around fundamental Bragg reflections. When the negative dip deepens on cooling, the diffuse satellite spots grow stronger. It is worth noting that the positions of the diffuse satellites correspond to the same wave vector of the negative dip in the phonon dispersion curve. In particular, the wave vector of both the negative dip and the diffuse satellites exhibits incommensurability to the cubic lattice and an evolution towards commensurability on cooling and/or stress loading. In the picture of static lattice modulations, the nature of the incommensurate lattice structure and its evolution in pre-martensitic phase have been puzzling.[16] In light of phonon diffuse scattering, the incommensurate diffuse satellites correspond to the wave vector of the negative dip in the incompletely softened phonon dispersion curve, rather than the wave vector of lattice modulation waves.

It is also worth noting that, even from the viewpoint of superlattice reflections, the intensity profiles of the diffuse satellite peaks are highly unusual as compared with the normal superlattice reflection peaks. Figure 7 compares the satellite peak intensity profiles from pre-martensite phase and from modulated martensite phase. Modulated martensite usually produces arrays of well-defined strong satellite peaks from superlattice reflections with very weak or little diffuse streaks between them. On the other hand, pre-martensite produces weak diffuse peaks only around fundamental Bragg peaks, while there are no satellite arrays, and significant streaking intensity connects each diffuse peak to the fundamental Bragg peak. These features of diffuse scattering streaks and satellites in pre-martensite phase are difficult to explain by lattice modulations. The absence of satellite arrays implies that the lattice modulations assume perfect sine waveforms thus higher-order superlattice reflections have zero intensities, which is highly unusual for normal modulated phases. Moreover, to produce the diffuse streaks connecting to Bragg peaks, a spectrum of sinusoidal lattice modulations with increasing wavelengths is needed, which is also highly unusual for normal long-period structures. In light of phonon diffuse

scattering, these features are manifestations of the shape of the incompletely softened phonon dispersion curve.

Tweed patterns versus strain contrast

The tweed patterns observed by TEM in pre-martensitic austenite have been previously attributed to static lattice displacements caused by point defects or martensite embryos.[4,6,9,10,15,16,19–24,45] It is worth noting that, unlike twin that defines a crystallographic lattice configuration, 'tweed' is a term commonly used to describe a group of TEM images with qualitatively similar contrast features, i.e., cross-hatched nanoscale striations. A tradition to seek static structural origin of tweed patterns in martensitic precursor follows an early systematic TEM study of diffraction contrast from static shear strains associated with coherent precipitates (i.e., G.P. zones formed by atomic diffusion) in Cu-Be alloys.[46] Owing to high elastic anisotropy, the elastic strains caused by tetragonal coherency of {100} disc-shaped G.P. zones are resolved into $\{110\}\langle1\bar{1}0\rangle$ shears distributed in the matrix, which produce diffuse streaking, strain contrast and tweed patterns. The static displacements of $\{110\}\langle1\bar{1}0\rangle$ shears have the same qualitative nature as the atomic displacements of $[\zeta\zeta0]$-TA_2 phonons, both producing diffuse streaking along the same directions and exhibit the same systematic extinction rule. However, such strain diffuse scattering does not produce satellite peaks, while phonon diffuse scattering produces satellite peaks when dispersion curve develops a negative dip. Nevertheless, the diffuse scattering in both cases can be used to produce diffraction contrast in TEM to form tweed patterns. Thus, the tweeds of pre-martensite phase and coherent precipitation must be distinguished.

It is worth noting that only the phonons localised in the phonon domains can produce tweed patterns in TEM imaging. Travelling waves carry instantaneous image contrast to different regions while propagating through the specimen, thus averaging out the image contrast. In contrast, localised lattice vibrations within the phonon domains consistently produce diffraction contrast in the same lattice regions, producing tweed patterns. The tweed image contrast depends on the strength of diffuse scattering, which in turn depends on the degree of phonon softening. With decreasing temperature, the phonons soften further and the diffuse scattering becomes stronger, leading to increased diffraction image contrast of the tweed patterns.

Although the lattice strains caused and/or stabilised by extrinsic defects (composition heterogeneities, chemical disorders, point defects) may produce strain contrast and tweed patterns, they do not produce satellite peaks unless the defects order into super-lattice structures characterised by the same wave vectors of the satellite peaks. Such defect arrangement is unlikely. Moreover, to explain the diffuse scattering satellites and their reversible evolutions with temperature and stress shown in Figures 3 and 6, implausible *ad hoc* assumptions have to be made about ordered defect distribution and its reversible evolution and rapid response to stress, making defect-based explanations questionable. Furthermore, defects cannot explain the exotic domain phenomenon shown in Figure 5. In light of phonon diffuse scattering and phonon domains, these observations are naturally explained.

The pre-martensitic phase with phonon domains is a thermodynamically equilibrium state, which only depends on temperature and stress but not thermal-mechanical history, thus the observations shown in Figures 3, 5 and 6 are reproducible without exhibiting annealing effect. In addition, the tweed patterns are uniform throughout the crystal. In contrast, annealing effect is expected for defect-induced strain glass state, which is a thermodynamically nonequilibrium frozen state with random defects and uncorrelated local strain clusters in untransformed parent phase matrix.[23] A nonequilibrium frozen strain glass state will evolve towards equilibrium on annealing, leading to history dependence. The experiments shown in Figures 3, 5 and 6 do not show annealing effect or history dependence in the single crystal samples used in the *in situ* diffraction.

Pre-martensitic transition versus martensitic transformation

There is a fundamental difference that distinguishes the 'phonon spinodal decomposition' pre-martensitic transition from a conventional martensitic transformation. The phonon dispersion in the cubic austenite phase and the pre-martensite phase exhibits a continuity in the dependence on temperature across the pre-martensitic transition. In contrast, there is a drastic discontinuous change in the phonon dispersion during martensitic transformation. The continuity of phonon dispersion as a function of temperature before martensitic transformation is illustrated in Figure 1a,c. The discontinuity of phonon dispersion across martensitic transformation is demonstrated in Figure 7. Although the incompletely softened phonons produce strong diffuse scattering in pre-martensite phase, the phonons become hard in martensite phase thus the diffuse scattering is very weak. The hard phonons in martensite phase imply that such 'phonon spinodal decomposition' and phonon-strain coupling do not have a role in martensite because these effects require low-energy softened phonons. It is worth noting that 'phonon spinodal decomposition' occurs at pre-martensitic transition and is characterised by the same phonon dispersion, while martensitic transformation is accompanied by drastic phonon spectrum reconstruction (indeed, displacive transformations are generally accompanied by phonon spectrum reconstruction resulting in different phonon contributions to the free energy of the phases). The nature of pre-martensitic transition has been previously treated as another martensitic transformation preceding the conventional one.[47] In light of the phonon theory, the following new picture emerges: due to incomplete phonon softening in austenite phase, pre-martensitic transition naturally precedes martensitic transformation, where the former leads to elastic-phonon domains that are primarily characterised by broken dynamic symmetry of lattice vibrations while the latter leads to elastic domains that are primarily characterised by broken static symmetry of crystal lattice.

Thermal, acoustic and elastic anomalies

The anomalous thermal, acoustic and elastic properties are hard to understand for cubic austenite phase, which is expected to have a homogeneous structure without domains. The pre-martensitic transition produces heterogeneous structures in the pre-martensite phase consisting of elastic-phonon domains, which

explains these anomalous effects. In particular, the anisotropic thermal expansion[11,12] manifests the anisotropic behaviour of the elastic-phonon domains under the unavoidable mounting stress for dilatometer measurement. As shown in Figure 5, the elastic-phonon domains are sensitive to stress, and small stress can induce significant strain due to the low elastic stiffness resulting from internal phonon relaxation. The increased acoustic attenuation[13] also results from the heterogeneous structures consisting of multiple elastic-phonon domains and the phonon relaxation processes. As discussed in the section 'Internal phonon relaxation in phonon domains and exotic domain phenomenon,' since the internal phonon relaxation takes time, it leads to frequency dependence of the elastic constants, where higher values are expected when measured at higher frequencies. Such frequency dependence has been observed in experiments. For example, in pre-martensitic Ni–Mn–Ga alloys, ultrasonic wave velocity measurement reports elastic modulus ~ 150 GPa,[44] while three-point bending measurement at 1 Hz reports elastic modulus ~ 20 GPa.[48] Figure 2d demonstrates three data points measured in In–Tl respectively at ~ 10 Hz, ~ 1 KHz and ~ 1 MHz.[14] A systematic measurement of such kind would allow determination of the phonon relaxation time t_{PR} in the elastic-phonon domains through the transition frequency f_0. It is worth noting that these explanations are derived from the statistical mechanics-based phonon theory[18] rather than a phenomenological model and do not resort to extrinsic defects.

Fluctuation stabilisation and phonon localisation

The Grüneisen-type phonon-strain coupling mechanism and the notion of phonon domains discussed above reveal additional aspects of the pre-martensitic phenomena, which are beyond the scope of this paper but worth brief discussions before concluding this paper. As mentioned in the 'Introduction' section, at the heart of the pre-martensitic problem, it has been thought that critical fluctuations before martensitic transformation could not have a significant role in the martensite precursor effects because of the first-order nature of the martensitic transformation and the high temperature range far above the vicinity of the transformation. Such an opinion must be changed in light of the 'phonon spinodal decomposition' pre-martensitic transition caused by phonon-strain coupling. The Grüneisen-type phonon-strain coupling mechanism enhances the fluctuations before the pre-martensitic transition and stabilises the fluctuations in the pre-martensite phase, both at temperatures above the martensite start temperature, M_S. In particular, at temperature right above the pre-martensitic transition temperature (T_{PM}), although the instability condition in Equations (15a and 15b) is not reached yet, the lattice stability has been greatly weakened by the phonon-strain coupling, thus heterogeneous fluctuations are significantly enhanced, which appear and disappear in the lattice thus the cubic symmetry is retained. At temperature below T_{PM} but still above M_S, these large heterogeneous fluctuations are stabilised by the phonon-strain coupling to form elastic-phonon domains, where the local cubic symmetry is broken. In this sense, the Grüneisen-type phonon-strain coupling is responsible for the entire pre-martensitic phenomena, where strongly developed heterogeneous fluctuations occur in a wide temperature range above M_S. Therefore, the pre-martensitic transition and its associated anomalous behaviours can be further investigated from the perspectives of phonon-strain coupling-enabled fluctuations and critical phenomena.[34,49]

The stabilisation of heterogeneous fluctuations into elastic-phonon domains can be treated as phonon localisations due to phonon-strain coupling. It is worth noting that phonon localisation due to different mechanisms has been intensively studied theoretically and experimentally[50–53] where, unlike the pre-martensitic phenomena discussed here, strain is not essential for phonon localisation. In perfect systems, intrinsic localised modes (also known as discrete breathers) form in discrete nonlinear systems, which are stable and do not involve extrinsic defects[50,51] (note that perfect anharmonic crystals just possess the two critical components of intrinsic localised modes: nonlinearity and discreteness). In imperfect systems, Anderson localisation occurs in disordered systems, where interference effects of coherent scattering from multiple random scattering centres produce localised regions of trapped standing waves.[52,53] In particular, Anderson localisation of ferroelectric transverse optical phonons has been employed to explain polar nanoregions in relaxor ferroelectrics, where nanoregions of standing ferroelectric phonons develop from preexisting resonance modes of Anderson localisation[53] (note that the Grüneisen-type phonon-strain coupling mechanism discussed in this paper involves ferroelastic transverse acoustic phonons). It would be intriguing to address the elastic-phonon domains from the perspectives of phonon localisation and standing waves.[50,52]

Finally, the notion of phonon domains addresses the pre-martensitic phenomena as intrinsic properties of perfect anharmonic crystal lattices caused by phonon-strain coupling without resort to extrinsic defects. In our phonon theory, both the phonon-strain coupling and the incomplete phonon softening are treated as intrinsic properties of ideal austenitic crystals. First-principles studies employing supercells without extrinsic defects confirm such phonon behaviours as intrinsic properties.[25,28,43] Although extrinsic defects are omnipresent in real materials and their effects undoubtedly exist, they are believed to have secondary roles through their interactions with the phonon domains. Understanding the effects of extrinsic defects in pre-martensitic transition is an important topic and deserves further investigations.

CONCLUSION

Martensitic precursor phenomena have been known for decades, while the puzzling anomalies are unexpected for cubic austenite phases above martensite start temperatures. They have been the topics of research workshops, conference symposia, review articles and book chapters[4,6,15,16,54,55] (a large body of literature can be found in the references therein and in the proceedings of the conferences and workshops). This paper addresses martensitic precursor phenomena from new theoretical and experimental perspectives, and revisits and examines the currently existing alternative opinions in light of theory and experimental evidence of phonon domains and their roles in pre-martensitic phenomena.

Recent new theory and critical experiments are reviewed. A statistical mechanics-based phonon theory is developed by employing Born's dynamical approach to anharmonic crystal lattices. On the basis of a simple Taylor expansion of the effective potential function of anharmonic cubic crystal that contains the leading term of phonon-strain coupling, strain-dependent phonon frequencies are determined using perturbation method. A Grüneisen-type phonon theory is formulated in terms of the free energy of lattice deformation and quasi-harmonic phonons of the crystal, which predicts a loss of lattice stability and pre-martensitic transition into elastic-phonon domains via 'phonon spinodal decomposition.' The transition lifts the degeneracy of the phonons in cubic crystal thus has the nature of phonon pseudo-Jahn–Teller lattice instability. The predicted instability condition depends both on the phonon-strain coupling coefficients and, more strongly, on the phonon energies, thus the low energies of incompletely softened phonons in austenitic crystals before martensitic transformation just satisfy the requirement. The behaviours of the elastic-phonon domains are primarily characterised by the broken dynamic symmetry of lattice vibrations rather than the broken static symmetry of crystal lattice, in contrast to the conventional ferroelastic domains. As a particular case of Le Chatelier's principle, phonon population redistribution in the elastic-phonon domains provides extra degree of internal relaxation in response to applied stress, which predicts exotic domain phenomenon in pre-martensitic austenite that is fundamentally different from the usual ferroelastic domain switching phenomenon in martensitic phase. The theoretical predictions are tested by specially designed experiments of in situ 3D phonon diffuse scattering and Bragg reflection using high-energy synchrotron X-ray single-crystal diffraction, which provide experimental evidence of phonon domains.

The roles of phonon domains in pre-martensitic phenomena are discussed, which provide consistent explanations to the martensitic precursor 'anomalies' on the same physical footing of thermal expansion, both being intrinsic properties and natural consequences of anharmonic crystal lattices. The incompletely softened phonons produce diffuse streaks and satellites in diffraction, and the dominant phonon modes localised in individual phonon domains produce diffraction contrast in TEM imaging of tweed patterns (not from the strain contrast of static displacements). The dependences of the diffuse satellite intensity and tweed image contrast on temperature are determined by the dependence of incomplete phonon softening on temperature as directly manifested in phonon dispersion curves. Other pre-martensitic 'anomalies' are also explained by the behaviours of phonon domains, e.g., the anisotropic thermal expansion manifests the anisotropic properties of the elastic-phonon domains, the increased acoustic attenuation results from the heterogeneous structures consisting of multiple elastic-phonon domains, and the frequency dependence of the elastic moduli is attributed to the internal relaxation process via the phonon population redistributions in response to external stress, which offers a means to determine the characteristic time of phonon relaxation. These explanations are derived from the statistical mechanics-based phonon theory rather than a phenomenological model and do not resort to extrinsic defects.

The theory is developed from the premise of incomplete phonon softening that is generally observed in both nonmagnetic and magnetic systems and both above and below Curie temperature. Although the origin of incomplete phonon softening is beyond the scope of the theory, it poses a question to future fundamental research. The theory adopts an Einstein-type model of density of states to focus on the deformation-dependent incompletely softened low-energy phonon modes that have dominant roles in pre-martensitic phenomena. Although such treatment can be improved by including all phonon modes if the phonon dispersion and phonon-strain coupling can be determined for individual modes, this simplification allows an illustration of the basic features determined only by symmetry and a few material parameters without loss of conceptual generality. This phonon theory fills a gap between the well-established experimental fact of incomplete phonon softening and the ubiquitous while puzzling martensitic precursor 'anomalies.' The theory can be extended to describe martensitic transformations. The current theory only considers the incompletely softened phonon modes that are characteristic of the cubic austenite and the pre-martensite phases. To describe martensitic transformation, it needs to incorporate the effective lattice potential, deformation and phonons of the martensite phase. Such further development would potentially lead to a unified phonon theory capable of describing the sequence of pre-martensitic, martensitic and inter-martensitic phase transformations.

ACKNOWLEDGEMENTS

We are very grateful to Professor Armen G. Khachaturyan and Professor Long-Qing Chen for valuable discussions. Supports from NSF DMR-1409317 (YMJ) and DMR-1506936 (YUW) are acknowledged. Use of Advanced Photon Source at Argonne National Laboratory was supported by DOE DE-AC02-06CH11357.

COMPETING INTERESTS

The authors declare no conflict of interest.

REFERENCES

1 Nishiyama, Z. *Martensitic Transformation* (Academic Press, 1978).
2 Roitburd, A. L. Martensitic transformation as a typical phase transformation in solids. *Solid State Phys.* **33**, 317–390 (1978).
3 Khachaturyan, A. G. *Theory of Structural Transformations in Solids* (John Wiley & Sons, 1983).
4 Salje, E. K. H. *Phase Transitions in Ferroelastic and Co-elastic Crystals* (Cambridge University Press, 1990).
5 Nakanishi, N. Elastic constants as they relate to lattice properties and martensite formation. *Prog. Mater. Sci.* **24**, 143–265 (1980).
6 Planes, A. & Mañosa, L. Vibrational properties of shape-memory alloys. *Solid State Phys.* **55**, 159–267 (2001).
7 Shapiro, S. M., Noda, Y., Fujii, Y. & Yamada, Y. X-ray investigation of the pre-martensitic phase in $Ni_{46.8}Ti_{50}Fe_{3.2}$. *Phys. Rev. B* **30**, 4314–4321 (1984).
8 Salamon, M. B., Meichle, M. E. & Wayman, C. M. Premartensitic phases of $Ti_{50}Ni_{47}Fe_3$. *Phys. Rev. B* **31**, 7306–7315 (1985).
9 Oshima, R., Sugiyama, M. & Fujita, F. E. Tweed structures associated with fcc-fct transformations in Fe-Pd alloys. *Metall. Trans. A* **19**, 803–810 (1988).
10 Tanner, L. E., Schryvers, D. & Shapiro, S. M. Electron microscopy and neutron scattering studies of premartensitic behavior in ordered Ni-Al β_2 phase. *Mater. Sci. Eng. A* **127**, 205–213 (1990).
11 Liu, M., Finlayson, T. R., Smith, T. F. & Tanner, L. E. Martensite precursor observations using thermal expansion: Ni-Al. *Mater. Sci. Eng. A* **157**, 225–232 (1992).
12 Liu, M., Finlayson, T. R. & Smith, T. F. Thermal expansion of V_3Si with controlled martensite-phase morphology. *Phys. Rev. B* **52**, 530–535 (1995).
13 Finlayson, T. R. Pretransformation phenomena as revealed by elastic waves. *Metall. Trans. A* **19**, 185–191 (1988).
14 Wuttig, M. Unpublished experimental data in In-Tl (private communication).
15 Krumhansl, J. A. Multiscale science: materials in the 21st century. Section 3 'Precursors: the next frontier'. *Mater. Sci. Forum* **327–328**, 1–8 (2000).
16 Otsuka, K. & Ren, X. Physical metallurgy of Ti-Ni-based shape memory alloys. Section "5. Precursor effects to martensitic transformations," in particular Section "5.9. Unsolved issues". *Prog. Mater. Sci.* **50**, 511–678 (2005).
17 Krumhansl, J. A. & Gooding, R. J. Structural phase transitions with little phonon softening and first-order character. *Phys. Rev. B* **39**, 3047–3053 (1989).
18 Wang, Y. U & Jin, Y. M. Martensitic transformation precursors: phonon theory and critical experiments. In: Militzer, M *et al.* (eds) *Proc Int Conf Solid-Solid Phase Transformations in Inorganic Materials 2015 (PTM 2015)*, Whistler, BC, Canada, 2015: 467–474. Full-version preprint at arXiv:1503.00027 (2015).
19 Semenovskaya, S. & Khachaturyan, A. G. Kinetics of strain-related morphology transformation in $YBa_2Cu_3O_{7-\delta}$. *Phys. Rev. Lett.* **67**, 2223–2226 (1991).
20 Semenovskaya, S., Zhu, Y., Suenaga, M. & Khachaturyan, A. G. Twin and tweed microstructures in $YBa_2Cu_3O_{7-\delta}$ doped by trivalent cations. *Phys. Rev. B* **47**, 12182–12189 (1993).
21 Kartha, S., Castán, T., Krumhansl, J. A. & Sethna, J. P. Spin-glass nature of tweed precursors in martensitic transformations. *Phys. Rev. Lett.* **67**, 3630–3633 (1991).
22 Kartha, S., Krumhansl, J. A., Sethna, J. P. & Wickham, L. K. Disorder-driven pretransitional tweed pattern in martensitic transformations. *Phys. Rev. B* **52**, 803–822 (1995).
23 Sarkar, S., Ren, X. & Otsuka, K. Evidence for strain glass in the ferroelastic-martensitic system $Ti_{50-x}Ni_{50+x}$. *Phys. Rev. Lett.* **95**, 205702 (2005).
24 Wang, D., Wang, Y., Zhang, Z. & Ren, X. Modeling abnormal strain states in ferroelastic systems: the role of point defects. *Phys. Rev. Lett.* **105**, 205702 (2010).
25 Zhao, G. L. & Harmon, B. N. Phonon anomalies in β-phase Ni_xAl_{1-x} alloys. *Phys. Rev. B* **45**, 2818–2824 (1992).
26 Chou, H. & Shapiro, S. M. Observation of predicted phonon anomalies in β-phase $Ni_{50}Al_{50}$. *Phys. Rev. B* **48**, 16088–16090 (1993).
27 Zheludev, A. *et al.* Phonon anomaly, central peak, and microstructures in Ni_2MnGa. *Phys. Rev. B* **51**, 11310–11314 (1995).
28 Bungaro, C., Rabe, K. M. & Corso, A. D. First-principles study of lattice instabilities in ferromagnetic Ni_2MnGa. *Phys. Rev. B* **68**, 134104 (2003).
29 Uijttewaal, M. A., Hickel, T., Neugebauer, J., Gruner, M. E. & Entel, P. Understanding the phase transitions of the Ni_2MnGa magnetic shape memory system from first principles. *Phys. Rev. Lett.* **102**, 035702 (2009).
30 Shapiro, S. M., Yang, B. X., Shirane, G., Noda, Y. & Tanner, L. E. Neutron scattering study of the martensitic transformation in a Ni-Al β-phase alloy. *Phys. Rev. Lett.* **62**, 1298–1301 (1989).
31 Shapiro, S. M., Svensson, E. C., Vettier, C. & Hennion, B. Uniaxial-stress dependence of the phonon behavior in the premartensitic phase of $Ni_{62.5}Al_{37.5}$. *Phys. Rev. B* **48**, 13223–13229 (1993).
32 Zheludev, A. & Shapiro, S. M. Uniaxial stress dependence of the $[\zeta\zeta 0]$-TA_2 anomalous phonon branch in Ni_2MnGa. *Solid State Commun.* **98**, 35–39 (1996).
33 Born, M. & Huang, K. *Dynamical Theory of Crystal Lattices* (Oxford University Press, 1954).
34 Landau, L. D. & Lifshitz, E. M. *Statistical Physics* (Pergamon Press, 1980).
35 Jin, Y. M. & Wang, Y. U. Phonon theory of martensitic transformation precursors. Preprint at arXiv:1412.3725 (2014).
36 Cahn, J. W. On spinodal decomposition. *Acta Metall.* **9**, 795–801 (1961).
37 Grimvall, G., Magyari-Köpe, B., Ozoliņš, V. & Persson, K. A. Lattice instabilities in metallic elements. *Rev. Mod. Phys.* **84**, 945–986 (2012).
38 Jin, Y. M. *et al.* Broken dynamic symmetry and phase transition precursor. Preprint at arXiv:1302.5479 (2013).
39 Cheng, T. L. *et al.* In-situ three-dimensional reciprocal-space mapping of diffuse scattering intensity distribution and data analysis for precursor phenomenon in shape-memory alloy. *JOM* **64**, 167–173 (2012).
40 Warren, B. E. *X-Ray Diffraction* (Addison-Wesley Publishing, 1969).
41 Holt, M. *et al.* Determination of phonon dispersions from x-ray transmission scattering: the example of silicon. *Phys. Rev. Lett.* **83**, 3317–3319 (1999).
42 Mañosa, L., Gonzàlez-Comas, A., Obradó, E. & Planes, A. Anomalies related to the TA_2-phonon-mode condensation in the Heusler Ni_2MnGa alloy. *Phys. Rev. B* **55**, 11068–11071 (1997).
43 Souvatzis, P., Eriksson, O., Katsnelson, M. I. & Rudin, S. P. Entropy driven stabilization of energetically unstable crystal structures explained from first principles theory. *Phys. Rev. Lett.* **100**, 095901 (2008).
44 Stipcich, M. *et al.* Elastic constants of Ni-Mn-Ga magnetic shape memory alloys. *Phys. Rev. B* **70**, 054115 (2004).
45 Yamada, Y., Noda, Y., Takimoto, M. & Furukawa, K. 'Modulated lattice relaxation' and incommensurability of lattice waves in β-based premartensitic phase. *J. Phys. Soc. Jpn* **54**, 2940–2947 (1985).
46 Tanner, L. E. Diffraction contrast from elastic shear strains due to coherent phases. *Philos. Mag.* **14**, 111–130 (1966).
47 Planes, A., Obradó, E., Gonzàlez-Comas, A. & Mañosa, L. Premartensitic transition driven by magnetoelastic interaction in bcc ferromagnetic Ni_2MnGa. *Phys. Rev. Lett.* **79**, 3926–3929 (1997).
48 Chernenko, V. A., Pons, J., Seguí, C. & Cesari, E. Premartensitic phenomena and other phase transformations in Ni-Mn-Ga alloys studied by dynamical mechanical analysis and electron diffraction. *Acta Mater.* **50**, 53–60 (2002).
49 Patashinskiĭ, A. Z. & Pokrovskiĭ, V. L. *Fluctuation Theory of Phase Transitions* (Pergamon Press, 1979).
50 Campbell, D. K., Flach, S. & Kivshar, Y. S. Localizing energy through nonlinearity and discreteness. *Phys. Today* **57**, 43–49 (2004).
51 Manley, M. E., Abernathy, D. L., Agladze, N. I. & Sievers, A. J. Symmetry-breaking dynamical pattern and localization observed in the equilibrium vibrational spectrum of NaI. *Sci. Rep.* **1**, 4 (2011).
52 Lagendijk, A., van Tiggelen, B. & Wiersma, D. S. Fifty years of Anderson localization. *Phys. Today* **62**, 24–29 (2009).
53 Manley, M. E. *et al.* Phonon localization drives polar nanoregions in a relaxor ferroelectric. *Nat. Commun.* **5**, 3683 (2014).
54 Tanner, L. E. & Soffa, W. A. Pretransformation behavior related to displacive transformations in alloys: foreword. *Metall. Trans. A* **19**, 760 (1988).
55 Tanner, L. E. & Wuttig, M. Workshop on first-order displacive phase transformations: review and recommendations. *Mater. Sci. Eng. A* **127**, 137–144 (1990).

The Open Quantum Materials Database (OQMD): assessing the accuracy of DFT formation energies

Scott Kirklin[1], James E Saal[1], Bryce Meredig[1], Alex Thompson[1], Jeff W Doak[1], Muratahan Aykol[1], Stephan Rühl[2] and Chris Wolverton[1]

The Open Quantum Materials Database (OQMD) is a high-throughput database currently consisting of nearly 300,000 density functional theory (DFT) total energy calculations of compounds from the Inorganic Crystal Structure Database (ICSD) and decorations of commonly occurring crystal structures. To maximise the impact of these data, the entire database is being made available, without restrictions, at www.oqmd.org/download. In this paper, we outline the structure and contents of the database, and then use it to evaluate the accuracy of the calculations therein by comparing DFT predictions with experimental measurements for the stability of all elemental ground-state structures and 1,670 experimental formation energies of compounds. This represents the largest comparison between DFT and experimental formation energies to date. The apparent mean absolute error between experimental measurements and our calculations is 0.096 eV/atom. In order to estimate how much error to attribute to the DFT calculations, we also examine deviation between different experimental measurements themselves where multiple sources are available, and find a surprisingly large mean absolute error of 0.082 eV/atom. Hence, we suggest that a significant fraction of the error between DFT and experimental formation energies may be attributed to experimental uncertainties. Finally, we evaluate the stability of compounds in the OQMD (including compounds obtained from the ICSD as well as hypothetical structures), which allows us to predict the existence of ~ 3,200 new compounds that have not been experimentally characterised and uncover trends in material discovery, based on historical data available within the ICSD.

INTRODUCTION

The development of new materials is critical to continued technological advancement, a fact that has spurred the creation of the Materials Genome Initiative.[1] One component required to build material innovation infrastructure and accelerate material development is the creation of large sets of shared and comprehensive data. In the 1960s, the development of density functional theory (DFT)[2,3] created a theoretical framework for accurately predicting the electronic-scale properties of a crystalline solid from first principles. However, it was many years before the first practical DFT algorithms were constructed and calculations performed,[4–7] and even then it was an impressive and noteworthy accomplishment to describe the electronic structure of a single compound. Since then, computational resources have advanced to the point where it is now feasible to predict the properties of many thousands of compounds in an efficient, high-throughput manner.[8–16] We extend the promise of high-throughput DFT to its logical extreme, calculating in a consistent and accurate manner the properties of a significant fraction of known crystalline solids. Using experimentally measured crystal structures obtained from our partnership with the Inorganic Crystal Structure Database (ICSD),[17,18] we have created a new database of DFT-relaxed structures and total energies, which is called the Open Quantum Materials Database (OQMD). The OQMD has already been used to perform several high-throughput DFT analyses for a variety of material applications.[14,19–23]

There are other efforts in the high-throughput calculation of compounds from large crystal structure databases, including the Materials Project,[11,24] the Computational Materials Repository[16] and AFLOWLIB.[25] We intend the current data set to be freely available, in its entirety, to the scientific community without any conditions or limitations. It is currently available for download at www.oqmd.org/download. We envision three important benefits from making the entire data set available. First, the availability of such a large data set of DFT data enables new and creative uses of these results by others in the field who lack the resources to create their own database. This outcome is strongly in line with the goals of the Materials Genome Initiative. Second, this data set may serve as a nucleus from which external—and unaffiliated—projects can grow. By providing calculated electronic structures for a large fraction of known materials, along with a utility for performing new calculations, new projects can begin more quickly. Third, multiple calculations of the same data set (e.g. the ICSD structures) enable confirmation of the accuracy of the calculations across all databases. Minor differences in the approach, i.e., the use of a slightly different set of potentials or a different choice for GGA+U parameters, makes it possible to see whether a particular choice gives systematically better results.

This paper is composed of two main sections. First, we provide the details of the construction of the OQMD—a description of the calculation settings and chemical-potential fitting approach—and review the current state of the database, which contains the DFT-predicted 0-K relaxed ground-state structures and total energies for every calculable ICSD compound with 34 atoms or less, a total of over 32,559 compounds. In addition, the database contains 259,511 hypothetical compounds described in the

[1]Department of Materials Science and Engineering, Northwestern University, Evanston, IL, USA and [2]FIZ Karlsruhe—Leibniz Institute for Information Infrastructure, Eggenstein-Leopoldshafen, Germany.
Correspondence: C Wolverton (c-wolverton@northwestern.edu)

Table 1. GGA+U $U - J$ values and their corresponding fitted chemical-potential corrections employed in OQMD calculations for oxides of the listed elements

Element	$U - J$ (eV)	Correction (eV/atom)
V	3.1	2.675
Cr	3.5	2.818
Mn	3.8	1.987
Fe	4.0	2.200
Co	3.3	1.987
Ni	6.4	2.530
Cu	4.0	1.381
Th	4.0	0.999
U	4.0	2.614
Np	4.0	2.705
Pu	4.0	2.177

Abbreviation: OQMD, Open Quantum Materials Database.

section 'Structures in the OQMD', based on decorations of commonly occurring crystal structures. The database also contains a growing number of additional structures (currently ~ 5,000) calculated for ongoing material discovery projects such as for structural alloys and energy materials.[14] These numbers give OQMD a total size of 297,099 DFT calculations to date. Second, we employ the OQMD to investigate a fundamental question of DFT: how accurately can we use DFT to reproduce experimentally known elemental ground states and compound-formation energies? We use these comparisons as a basis for establishing confidence in the database, before using the database to calculate the stability of every compound in the database. With this large library of DFT calculations of hypothetical compounds included in the OQMD, we are able to make predictions of compositions where new, previously unknown compounds are likely to exist; in this work we identify 3,231 such compositions in which one of our hypothetical structures is predicted to be stable. Finally, we use the breadth of our database to examine historical trends in material discovery.

RESULTS AND DISCUSSION

The OQMD Methodology

A critical component of any high-throughput DFT database is the infrastructure to create and access the contained knowledge, e.g., pymatgen,[26] ASE[27] and AFLOW.[15] We have developed an infrastructure for such high-throughput DFT calculations and database management, dubbed qmpy. qmpy is written in python, and it uses the django web framework as an interface to a MySQL database. In the same 'Open' spirit as the database itself, qmpy is freely available for download as well. It is our goal to develop tools that any research group can use to catalogue, access and analyse large sets of calculations. To this end, qmpy is designed with a decentralised model—any user can download and use it to build a database (e.g., PostgreSQL, MySQL, sqlite or Oracle), and have simple, programmatic access to their calculations. The package has a built-in web interface, and, because we utilise a django backend, it is very easy to customise the web interface depending on the specific needs of any user. Details of qmpy and its analysis algorithms can be found at www.oqmd.org/static/docs.

The calculation of many thousands of compounds within DFT in a reasonable timeframe demands that optimal efficiency within a constrained standard of convergence be found. Furthermore, the comparison of calculation results across many different types of materials (e.g., metals, semiconductors and oxides) requires that all the calculations be performed at a consistent level of theory, which is acceptable for all classes of materials, e.g., consistent plane-wave cutoff, smearing schemes and k-point densities.

To that end, extensive testing on a sample of ICSD structures has resulted in the calculation flow described in Materials and Methods, which ensures converged results in an efficient manner for a variety of material classes. Furthermore, the settings are consistent across all the calculations, ensuring that results between different compounds are directly comparable (e.g., predictions of energetic stability). Using DFT and DFT+U calculations (with the parameters listed in Table 1) and the scheme described in the Materials and Methods section, at the time of the writing of this paper we have calculated 32,559 compounds from the ICSD and 259,511 hypothetical compounds based on decorations of prototype structures.

Structures in the OQMD

The structures in the OQMD come from two sources. The first source, and the origin of the majority of our lowest-energy structures, is the ICSD. For the ICSD structures, we start with a list of 148,279 entries. Of those, 64,412 structures contain atomic positions with partial occupancy, which are substantially more difficult to treat in a high-throughput manner and thus are not included in the current study. A further 32,202 are found to be duplicates of other entries and have been discarded. Structural uniqueness is determined with respect to the lattice and the internal coordinates. Two lattices are compared by finding the reduced primitive cells of each structure and comparing all lattice parameters. Internal coordinates of different structures are compared by testing all rotations allowed by each lattice, and searching for a rotation + translation that maps atoms of the same species onto one another within a given tolerance. Here, any two structures in which all atoms can be mapped to within 0.2 Å of an identical atom are considered identical. Of the remaining structures, 13,934 have incomplete entries in the database, missing either atomic coordinates or spacegroup information. The removal of these structures leaves a pool of 44,506 unique, eligible structures to be calculated. Having started with the structures with the fewest number of atoms, the database currently consists of calculations for all ICSD entries consisting of less than 34 atoms and passing the above filters, a total of 32,559 structures.

The differences between what we identify as calculable, unique structures and the total set of structures in the ICSD must be understood in the context of the challenges of creating a repository of measured crystal structures. There is no general way to judge the quality of a crystal structure. Moreover, it is not simple to determine which of the several measurements/refinements is the best for a given structure, because there are many ways to manipulate the typical criteria used (r-values, goodness-of-fit and low estimated standard deviation). These challenges are compounded by the fact that every measured crystal structure is an average in time (i.e., the time over which data are collected) and space (size of the crystal). DFT energetics can help resolve some of these difficulties by providing another criterion for determining the physicalness of a refined structure, as well as differentiating between structures that give similar fits to experimental data.[28]

In addition to ICSD structures, we have also calculated decorations of many simple prototype structures over a wide range of compositions. We define a prototype structure to be a crystal structure commonly observed in nature for a variety of chemical compositions, e.g., A1 FCC and L1$_2$. Table 2 gives a complete listing of the prototype structures that we have calculated. For every elemental prototype, we have calculated every element for which Vienna Ab-initio Simulation Package (VASP) includes projected augmented wave-Perdew, Burke and Ernzerhof (PAW-PBE) potentials (89 elements). For all of the binary prototypes, we have calculated each structure with every combination of elements excluding the noble gases (84 elements).

Table 2. Prototype structures calculated in the OQMD

Designation	Number of atoms	SG	Formula	Description	Example
Unary					
A1	1	$Fm\bar{3}m$	A	FCC	Cu
A2	1	$Im\bar{3}m$	A	BCC	W
A3	2	$P6_3/mmc$	A	HCP	Mg
A3'	4	$P6_3/mmc$	A	Distorted HCP	a-La
A4	2	$Fd\bar{3}m$	A	Diamond	C
A5	2	$I4_1/amd$	A		β-Sn
A6	1	$I4/mmm$	A		In
A7	2	$R\bar{3}m$	A		a-As
A8	3	$P3_121$	A		γ-Se
A9	4	$P6_3/mmc$	A	Graphite	C
A10	1	$R\bar{3}m$	A	Simple rhombohedral	a-Hg
A11	4	Cmca	A	BCO	a-Ga
A12	29	$I\bar{4}3m$	A		a-Mn
A13	20	$P4_132$	A		β-Mn
A15	8	$Pm\bar{3}m$	A	W_3O	β-W
A17	4	Cmca	A	Black phosphorus	P
A20	2	Cmcm	A		a-U
A_a	1	$I4/mmm$	A	BCT	a-Pa
A_h	1	$Pm\bar{3}m$	A	Simple cubic	a-Po
C19	3	$R\bar{3}m$	A		a-Sm
Binary					
B2	2	$Pm\bar{3}m$	AB	BCC superstructure	CsCl
$D0_3$	4	$Fm\bar{3}m$	A_3B	BCC superstructure	$AlFe_3$
$L1_0$	2	P4/mmm	AB	FCC superstructure along [100]	AuCu
$L1_1$	2	$R\bar{3}m$	AB	FCC superstructure along [111]	CuPt
$L1_2$	4	$Pm\bar{3}m$	A_3B	FCC superstructure with maximised unlike bonds	Cu_3Au
B_h	2	P6m2	AB	HCP superstructure along [0001]	WC
B19	4	Pmma	AB	HCP superstructure	AuCd
$D0_{19}$	8	$P6_3/mmc$	A_3B	HCP superstructure with maximised unlike bonds	Ni_3Sn
$E2_1$	5	$Pm\bar{3}m$	ABO_3	Perovskite	$CaTiO_3$
defect-$E2_1$	9	P4/mmm	$A_2B_2O_5$	Defect-perovskite	
$D5_1$	10	$R\bar{3}c$	A_2B_3	Corundum	Al_2O_3
C4	6	$P4_2/mnm$	AB_2	Rutile	TiO_2
B1	2	$Fm\bar{3}m$	AB	Rocksalt	NaCl
B3	2	$F\bar{4}3m$	AB	Zincblende	ZnS
B4	4	$P6_3mc$	AB	Wurtzite	ZnS
$D0_{22}$	4	$I4/mmm$	A_3B	FCC superstructure along [012]	Al_3Ti
Ternary					
$L2_1$	4	$Fm\bar{3}m$	A_2BC	Heusler BCC superstructure	Cu_2MnAl

Abbreviation: OQMD, Open Quantum Materials Database.
For each structure the number of atoms in the primitive cell, general formula, a simple description and an example compound are given.

This results in 3,486 compounds for structures with symmetrically equivalent sites and 6,972 compounds for structures with symmetrically distinct sites (e.g., AB_2 or A_3B structures). Perovskite and defect-pervoskite prototypes are only calculated for oxides, and thus are treated as a binary, e.g., ABO_3. We also calculated one ternary prototype, the Heusler or $L2_1$ structure. We have calculated 186,596 compounds in the Heusler structure. A relatively small number of additional hypothetical compounds (~5,000 to date) computed for projects that utilised OQMD are also included in the database.

We included these prototype structures for two purposes. First, by including all compositions, we have a more complete picture of the energetic landscape of phase space. It is important to understand the energy landscape over a comprehensive range of compositions and structures, as, in order to reliably assess the stability of any individual compound, its energy must be compared with the energy of all possible competing phases and combinations of phases. These hypothetical compounds based on common prototype structures are useful because they ensure that our stability calculations are reasonable, even at compositions where limited experimental data are available. Furthermore, at compositions where prototype compounds are predicted to be stable, it is likely that new compounds are waiting to be discovered. Although in general the prototype compounds that are predicted to be stable are not likely to be the true stable ground-state structures, they indicate a region of composition space where *some* new stable compounds must exist, but which has not been found (or at least not in the OQMD set of ICSD structures). Thus, these ground states represent predictions of new ordered compounds that should be validated experimentally. Second, internal interest in particular crystal structures has motivated the calculation of some prototypes for specific applications—a demand that is easily accomplished using the qmpy framework.

Elemental Ground-State Prediction

We begin by examining the phase stability of all 89 elements included in OQMD in a variety of structure types. We determine the lowest DFT energy structure for every element with a VASP PAW-PBE potential (89 elements) and compare them with the experimentally observed low-temperature structures.[29,30] In addition to the elemental structures in the ICSD, we also

Table 3. Elemental ground-state structures and chemical potentials predicted by DFT at 0 K

Potential	DFT ground state μ_i (eV/atom)					Exp. LT			Exp. RT		
	ID	SG	Fit-none	Fit-partial	Fit-all	ID	SG	ΔE (eV/atom)	ID	SG	ΔE (eV/atom)
H		I4/mmm	−3.327	−3.394	−3.434					Gas	
He	A5	I41/amd	−0.004	−0.004	−0.004	A3	$P6_3/mmc$	0.006		Gas	
LLsv	C19	$R\bar{3}m$	−1.907	−1.907	−1.731				A2	$Im\bar{3}m$	0.003
Be	A3	$P6_3/mmc$	−3.755	−3.755	−3.653						
B		$R\bar{3}m$	−6.678	−6.678	−6.656						
C	A9	$P6_3/mmc$	−9.217	−9.217	−9.044						
N		$Pa\bar{3}$	−8.235	−8.122	−8.195					Gas	
O		C2/m	−4.844	−4.485	−4.523					Gas	
F	A11	Cmca	−1.666	−1.429	−1.443		C2/c	0.006		Gas	
Ne	A1	$Fm\bar{3}m$	−0.029	−0.029	−0.029					Gas	
Na_pv	A3	$P6_3/mmc$	−1.303	−1.212	−1.196				A1	$Fm\bar{3}m$	0.003
Mg	A3	$P6_3/mmc$	−1.542	−1.542	−1.417						
Al	A1	$Fm\bar{3}m$	−3.746	−3.746	−3.660						
Si	A4	$Fd\bar{3}m$	−5.425	−5.425	−5.386						
P		$P\bar{1}$	−5.405	−5.161	−5.175	A17	Cmca	0.031			
S		P21	−4.114	−3.839	−3.868						
Cl		Cmca	−1.820	−1.465	−1.479					Gas	
Ar	A3	$P6_3/mmc$	−0.006	−0.006	−0.006	A1	$Fm\bar{3}m$	0.009		Gas	
K_sv	A7	$R\bar{3}m$	−1.097	−1.097	−0.987	A2	$Im\bar{3}m$	0.000			
Ca_pv	A1	$Fm\bar{3}m$	−1.978	−1.978	−1.780						
Sc_sv	A3	$P6_3/mmc$	−6.328	−6.328	−6.344						
Ti		P6/mmm	−7.776	−7.712	−7.702				A3	$P6_3/mmc$	0.014
V	A2	$Im\bar{3}m$	−8.941	−8.941	−8.898						
Cr	A2	$Im\bar{3}m$	−9.508	−9.508	−9.463						
Mn	A12	$I\bar{4}3m$	−9.027	−9.027	−8.898						
Fe	A2	$Im\bar{4}3m$	−8.308	−8.308	−8.499						
Co	A3	$P6_3/mmc$	−7.090	−7.090	−7.078						
Ni	A1	$Fm\bar{3}m$	−5.567	−5.567	−5.587						
Cu	A1	$Fm\bar{3}m$	−3.716	−3.716	−3.710						
Zn	A3	$P6_3/mmc$	−1.266	−1.266	−1.157						
Ga_d	A11	Cmca	−3.032	−3.032	−2.902						
Ge_d	A4	$Fd\bar{3}m$	−4.624	−4.624	−4.522						
As	A7	$R\bar{3}m$	−4.652	−4.652	−4.593						
Se	A8	$P3_121$	−3.481	−3.481	−3.374						
Br		Cmca	−1.606	−1.317	−1.333					Liquid	
Kr	A3	$P6_3/mmc$	−0.004	−0.004	−0.004	A1	$Fm\bar{3}m$	0.003		Gas	
Rb_sv	C19	$R\bar{3}m$	−0.963	−0.963	−0.881	A2	$Im\bar{3}m$	0.001			
Sr_sv	A1	$Fm\bar{3}m$	−1.683	−1.683	−1.549						
Y_sv	A3	$P6_3/mmc$	−6.464	−6.464	−6.449						
Zr_sv	A3	$P6_3/mmc$	−8.547	−8.547	−8.438						
Nb_pv	A2	$Im\bar{3}m$	−10.094	−10.094	−10.017						
Mo_pv	A2	$Im\bar{3}m$	−10.848	−10.848	−10.921						
Tc_pv	A3	$P6_3/mmc$	−10.361	−10.361	−10.457						
Ru	A3	$P6_3/mmc$	−9.202	−9.202	−9.210						
Rh	A1	$Fm\bar{3}m$	−7.269	−7.269	−7.319						
Pd	A1	$Fm\bar{3}m$	−5.177	−5.177	−5.197						
Ag	A3′	$P6_3/mmc$	−2.822	−2.822	−2.907	A1	$Fm\bar{3}m$	0.000			
Cd	A3	$P6_3/mmc$	−0.900	−0.900	−0.861						
In_d	A6	I4/mmm	−2.720	−2.720	−2.609						
Sn_d	A4	$Fd\bar{3}m$	−4.007	−3.895	−3.938				A5	I41/amd	0.042
Sb	A7	$R\bar{3}m$	−4.118	−4.118	−4.155						
Te	A8	$P3_121$	−3.142	−3.142	−3.027						
I		Cmca	−1.509	−1.344	−1.365						
Xe	A6	I4/mmm	0.003	0.003	−0.640	A1	$Fm\bar{3}m$	0.004		Gas	
Cs_sv	A3	$P6_3/mmc$	−0.855	−0.855	−0.744	A2	$Im\bar{3}m$	0.002			
Ba_sv	A2	$Im\bar{3}m$	−1.924	−1.924	−1.479						
La	A3′	$P6_3/mmc$	−4.935	−4.935	−4.959						
Ce_3	A3′	$P6_3/mmc$	−4.777	−4.777	−4.564	A1	$Fm\bar{3}m$	0.006	A3′	$P6_3/mmc$	0.000
Pr_3	A3′	$P6_3/mmc$	−4.775	−4.775	−4.627						
Nd_3	A3′	$P6_3/mmc$	−4.763	−4.763	−4.697						
Pm_3	A3′	$P6_3/mmc$	−4.745	−4.745	−4.716						
Sm_3	A3′	$P6_3/mmc$	−4.715	−4.715	−4.606				C19	$R\bar{3}m$	0.004
Eu_2	A2	$Im\bar{3}m$	−1.888	−1.888	−1.708						
Gd_3	A3′	$P6_3/mmc$	−4.655	−4.655	−4.718	A3	$P6_3/mmc$	0.017			
Tb_3	C19	$R\bar{3}m$	−4.629	−4.629	−4.731	A20	Cmcm	0.017	A3	$P6_3/mmc$	0.012
Dy_3	C19	$R\bar{3}m$	−4.602	−4.602	−4.660	A20	Cmcm	0.013	A3	$P6_3/mmc$	0.008
Ho_3	C19	$R\bar{3}m$	−4.577	−4.577	−4.573	A3	$P6_3/mmc$	0.003			
Er_3	A3	$P6_3/mmc$	−4.563	−4.563	−4.582						

Table 3. (Continued)

Potential	DFT ground state μ_i (eV/atom)					Exp. LT			Exp. RT		
	ID	SG	Fit-none	Fit-partial	Fit-all	ID	SG	ΔE (eV/atom)	ID	SG	ΔE (eV/atom)
Tm_3	A3	P6$_3$/mmc	−4.475	−4.475	−4.451						
Yb_2	A1	Fm$\bar{3}$m	−1.513	−1.513	−1.125	A3	P6$_3$/mmc	0.007	A1	Fm$\bar{3}$m	0.000
Lu_3	A3	P6$_3$/mmc	−4.524	−4.524	−4.549						
Hf_pv	A3	P6$_3$/mmc	−9.955	−9.955	−9.902						
Ta_pv	A2	Im$\bar{3}$m	−11.853	−11.853	−11.941						
W_pv	A2	Im$\bar{3}$m	−12.960	−12.960	−13.130						
Re	A3	P6$_3$/mmc	−12.423	−12.423	−12.378						
Os_pv	A3	P6$_3$/mmc	−11.226	−11.226	−11.374						
Ir	A1	Fm$\bar{3}$m	−8.855	−8.855	−8.953						
Pt	A1	Fm$\bar{3}$m	−6.056	−6.056	−6.162						
Au	A1	Fm$\bar{3}$m	−3.267	−3.267	−3.283						
Hg	A12	I$\bar{4}$3m	−0.298	−0.376	−0.374	A$_a$	I4/mmm	0.074		Liquid	
Tl_d	A3	P6$_3$/mmc	−2.359	−2.359	−2.480						
Pb_d	A1	Fm$\bar{3}$m	−3.704	−3.704	−3.951						
Bi_d	A7	R$\bar{3}$m	−4.039	−4.039	−4.199						
Ac	A3'	P6$_3$/mmc	−4.106	−4.106	−4.106	A1	Fm$\bar{3}$m	0.012			
Th	A1	Fm$\bar{3}$m	−7.413	−7.413	−7.237						
Pa	A1	Fm$\bar{3}$m	−9.496	−9.496	−9.497	A$_a$	I4/mmm	0.017			
U	A20	Cmcm	−11.292	−11.292	−11.032						
Np		Pnma	−12.940	−12.940	−12.797						
Pu		P2$_1$/m	−14.298	−14.298	−13.950						

Abbreviations: DFT, density functional theory; Exp. LT, experimentally observed lowest temperature; Exp. RT, experimentally observed room temperature PAW-PBE, projected augmented wave-Perdew, Burke and Ernzerhof; VASP, Vienna *Ab-initio* Simulation Package.
The 'Potential' column corresponds to VASP PAW-PBE potential names for the elements used in the current work. The ID and SG columns list the Strukturbericht designation and the space group of the crystal structures, respectively. The 'fit-none' chemical potentials are the DFT ground-state total energies of each element, whereas the 'fit-partial' and 'fit-all' correspond to the chemical-potential correction schemes described in the section 'Formation Energy Calculation'. The 'Exp. LT' and/or 'Exp. RT' ground states[29,30] are given if they differ from the DFT-predicted 0-K ground state (RT only provided if it differs from LT), along with the difference between the DFT-predicted total energy of the Exp. LT or RT structure and DFT ground-state energy (energy difference only provided if the DFT ground state differs from Exp. LT or Exp. RT).

calculate each element in 20 elemental ground-state structures, listed in Table 2. For each element, therefore, we have the energetics of a large number and geometric variety of (⩾20) crystal structure types, and we refer to the structure with the lowest energy as the DFT-predicted $T = 0$ K ground state. Previous efforts at a comprehensive comparison of experimentally observed elemental ground-state structures and DFT-predicted elemental ground-state structures have been limited to HCP, FCC and BCC structures.[31] Hence, our present study provides a more complete, systematic investigation of the ability of DFT (at the settings of the OQMD) to predict the correct 0-K elemental ground-state crystal structures.

Table 3 shows the DFT-predicted ground-state structure and the experimentally observed ground-state structure[29,30] for all 89 elements, for which a potential is available in VASP. During DFT calculations of the 89 elements in the wide spectrum of crystal structure types tested, it is possible for an element to relax to a higher-symmetry parent structure starting from a lower-symmetry prototype-based crystal structure. Therefore, for all elements in Table 3, we analysed the OQMD-relaxed structures of elements that are almost degenerate with the ground-state structure (taken as within ~2 meV/atom) using the structure comparison approach outlined in the section 'Structures in the OQMD' We found that only a small set of structures (Al, Ni, Rh, Ir and Th in A6, K, Sr, Al and Th in A7, Sr in A8 and Pa in A10 structures) relaxed to the higher-symmetry A1 (FCC) structure, and only Mg and Er in the A20 structure relaxed to the higher-symmetry A3 (HCP) structure.

Of the 89 elements in Table 3, the OQMD prediction agrees with the experimentally observed low-temperature structure for 82 elements to within 12 meV/atom, and 77 elements to within 5 meV/atom. For the 12 elements with a discrepancy between the OQMD-predicted ground state and the experimentally observed

ground state greater than 5 meV/atom, we look for possible sources of error. The elements for which we fail to correctly predict the ground-state structure are He, F, P, Ar, Ce, Gd, Tb, Dy, Yb, Hg, Ac and Pa. There are three distinct groups among these elements: first are noble gases and molecular solids, i.e., He, Ar, F and P, second are solids of elements with f-electrons, i.e., Ce, Gd, Tb, Dy, Yb, Ac and Pa and third is Hg. For most of these elements, we found no change in the predicted ground states with calculations at ~30–100% higher plane-wave basis-set cutoffs. The exceptions are He, F and Ar for which the higher-cutoff calculations can reproduce the experimental ground states. However, for such noble gas and molecular solids, it is expected that van der Waals interactions contribute a significant portion of the total energy and therefore this change in stabilities cannot be substantiated. For phosphorous, which has the second largest discrepancy between the OQMD-predicted ground state and the experimental ground state, the experimental ground state consists of layers of covalently bonded atoms that interact primarily through van der Waals forces, and increasing the cutoff has no effect on the predicted relative stability of P allotropes. For all elements in the first group above, the ground-state structure cannot be determined reliably with (semi)-local exchange-correlation functionals used in DFT because of the lack of van der Waals interactions, and can likely be corrected by their inclusion.[32]

The second group—elements that contain f-electrons—presents its own set of known challenges for DFT.[33–38] There are two sources of error for these elements. For Ce, Gd, Tb, Dy and Yb we use the 'frozen' potentials, meaning that the f-electrons are frozen in the core and not treated explicitly as valence. These potentials may introduce errors because freezing the f-electrons into the core neglects interactions involving these electrons. For example, for Ce, we found that the experimental low-

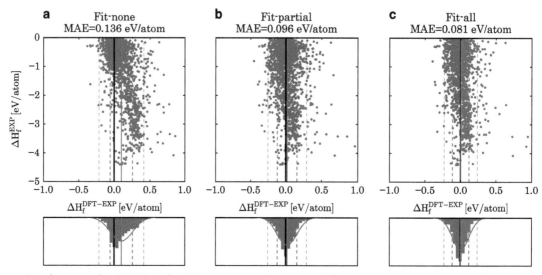

Figure 1. Comparison between the OQMD and 1,670 experimentally measured formation energies for three different sets of elemental chemical potentials. (**a**) Fit-none reference states: the DFT energy of the OQMD ground-state structure is taken as the chemical potential for each element. (**b**) Fit-partial reference states: chemical potentials for 13 elements where the DFT ground-state structures are known to poorly represent the STP elemental chemical potentials are fit to experimental formation energies (for all other elements, the reference state is the DFT energy). (**c**) Fit-all reference states: the chemical potentials for all elements are fit to experimental formation energies. The solid red line in each plot corresponds to the average error between DFT and experiment. The dashed red lines indicate the first and second s.d.'s. The curves in the lower plots correspond to normal distributions computed from the mean and s.d. of each data set.

temperature structure (a-Ce) can be reproduced as the ground state using the Ce potential with f-electrons in the valence. However, in the OQMD, frozen potentials are used because the high degree of correlation of f-electrons and related self-interaction error when they are included as valence is hard to accurately treat with DFT. Pa and Ac are examples where frozen potentials are not available, and therefore the 'free' f-electrons lead to errors. The issues with f-electrons in DFT, as well as the trade-offs that come with using frozen potentials, are thoroughly discussed elsewhere.[33–37] Lastly, it is already known that the relative stabilities of allotropes of Hg cannot be reproduced with local density approximation or generalized gradient approximation (GGA),[39] and relativistic effects such as spin–orbit coupling (excluded from our DFT calculations in the OQMD) are essential for accurate treatment of Hg.[40]

Formation Energies of Compounds

One of the most useful quantities that we have calculated for each compound is its formation energy—the energy required to form (positive formation energy), or given off by forming (negative formation energy), a compound from its constituent elements. Compound-formation energies are required to predict compound stability, generate phase diagrams, calculate reaction enthalpies and voltages and determine many other material properties. Because this quantity is so ubiquitous, it is important to determine the trustworthiness of our predictions. Although previous large-scale investigations of DFT's accuracy in predicting formation energies have been performed, these investigations have either been limited in the scope of material type considered[41–44] or in the quantity of structures assessed.[45] The database of formation energies for 297,099 structures in the OQMD allows for the most comprehensive assessment of the ability of DFT to predict formation energies for solids to date.

As DFT calculations are performed at 0 K and experimental formation energies are typically measured at room temperature, we must consider how much we can expect the formation energy to change between 0 and 300 K. The largest source of differences between 0 and 300 K formation energies is the existence of phase transformations in this temperature range. These phase transformations can take the form of solid–liquid, solid–gas or

solid–solid transformations and lead to significant changes in energetics. In addition to the elements that are gaseous or liquid at room temperature, at least five elements are known to exhibit a solid–solid transformation below 300 K: Ce, Na, Li, Ti and Sn.[29] As shown in Table 3, the energy differences between the 0-K- and room-temperature structures of Ce, Na and Li are less than 7 meV/atom. For Ti and Sn, however, the energy differences are 14 and 42 meV/atom, respectively, which will introduce systematic errors in the OQMD-predicted formation energies when comparing with experimental formation energies. For this reason, the chemical potentials of Ti and Sn have been fit to experimental data, as discussed in the section 'Formation Energy Calculation'.

In the following sections we make extensive use of a large collection of experimental formation energies. These experimental formation energies come from two sources: the SGTE Solid SUBstance (SSUB) database,[46] from which we obtain 1,702 compound-formation energies, and the thermodynamic database at the Thermal Processing Technology Center at the Illinois Institute of Technology (IIT),[47] from which we obtain 994 compound-formation energies. The SSUB contains many oxides (680), nitrides (75), hydrides (102) and halides (369), and relatively few intermetallics (272). In contrast, the IIT database is exclusively intermetallic compounds. By combining these databases we have a total of 2,712 experimental formation energies to compare with. For Th, Pa and Np oxides, which do not appear in either of these databases, we also include formation energies reported in a review of actinide thermodynamics,[48] which includes measured formation energies for ThO_2, Np_2O_5 and NpO_2 and estimated formation energies for PaO_2 and NpO_3 from trends among similar actinide oxides.

Because the different experimental data sources specialise in different types of materials, they have relatively limited overlap, but they do have some compositions in common. As a result, after controlling for double counting of compositions within and between databases, we still have 2,233 distinct compositions with experimental formation energies. The number of comparisons is further reduced because some of the compositions for which we have experimental formation energies have no known corresponding crystal structure. One might wish to ascertain whether, for a given stoichiometry, a given experimental formation energy

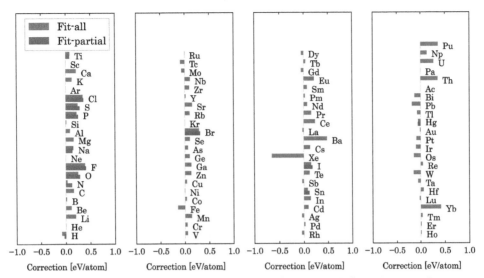

Figure 2. Corrections to chemical potentials ($\mu_{fit} - \mu_{DFT}$) as determined by fitting OQMD formation energies to experimental formation energies. The corrections in blue were obtained by fitting only the chemical potentials of elements whose STP phase differs significantly from the 0-K phase, while the corrections in red were obtained by fitting the chemical potentials of all elements simultaneously.

corresponds precisely to the phase with crystal structure reported in the ICSD. However, this comparison is not possible because the thermodynamic databases often do not have any explicit crystal structure information. We therefore assume that the existence of a corresponding crystal structure in the ICSD is sufficient to include a composition in the formation energy comparison, and, if more than one crystal structure is available, we always compare with the lowest-energy OQMD structure. As a result, we have a total of 1,670 formation energies to compare between the OQMD and experiment—the largest comparison of this kind that has been performed to date.

Formation Energy Calculation. In general, the formation energy for a compound is given by,

$$\Delta H_f = E_{tot} - \sum_i \mu_i x_i \qquad (1)$$

where E_{tot} is the DFT total energy of the compound, μ_i is the chemical potential of element i and x_i is the quantity of element i in the compound. The standard convention is to take the chemical potential of each species to be the DFT total energy of the elemental ground state. With this choice, the computed formation energy is valid only for 0 K. However, as the available experimental formation energies are typically measured at room temperature or above, it is useful to understand how well equation 1 approximates the standard temperature and pressure (STP) formation energy. In order to answer this question, we calculate the formation energy of all OQMD-calculated compounds and compare the results with experimentally measured formation energies.

Using the elemental DFT total energies as chemical potentials, we find that the OQMD-calculated formation energies have an average error of 0.105 eV/atom with respect to experimentally measured formation energies and a mean absolute error (MAE) of 0.136 eV/atom. The 1,670 formation energies used in this comparison include a wide range of compositions, including oxides, semiconductors and intermetallics. This error is plotted against experimental formation energy in Figure 1a. There is a clear systematic error: compounds with more-negative formation energies underestimate formation energies compared with those having less-negative formation energies. This trend is easily understood; highly stable compounds, such as oxides and halides, frequently contain elements whose 0-K ground states differ from the STP stable phases, e.g., gases. In order to improve these

formation energies, we fit chemical potentials to experimental formation energies.

The use of experimental data to fit elemental reference energies has been shown to reduce systematic error in DFT formation energies.[11,44,45,49,50] In this work we perform simultaneous least squares fitting in the manner of the Fitted Elemental Reference Energies method,[45] which was developed for finding chemical potentials for GGA+U oxides; however, in our work we extend this method to a wider set of elements involving a much wider set of compounds. In our approach we apply two distinct corrections; therefore, we perform two independent fits. The first fit is to calculate chemical potentials for elements with ground states that are not applicable to STP, and the second fit is to find corrections to chemical potentials for elements to which we apply GGA+U (Table 1). In the first step, we correct all the elemental chemical potentials simultaneously, fitting to the experimental formation energies (binary, ternary, quaternary and so on) of every applicable compound that does not contain a GGA+U element, and in the second step we determine values for the chemical potentials of GGA+U elements (described below).

To evaluate the efficacy of chemical-potential fitting, we define three fit sets that we will compare throughout this section and the next. The first fit set is empty, which we will label 'fit-none,' and corresponds to completely uncorrected DFT. The second fit set we will label 'fit-partial,' which refers to the elements for which we can rationally argue that the DFT energy of the $T = 0$ K ground state is not an accurate reference state for STP formation energies. We have identified five groups of elements to include in this fit set. These groups are room-temperature diatomic gases (H, N, O, F and Cl), room-temperature liquids (Br and Hg), molecular solids (P, S and I), several elements with structural phase transformations between 0 and 298 K (Na, Ti and Sn) and elements employing GGA+U for oxide-formation energies (Table 1). The third fit set, which we will label 'fit-all,' has the chemical potential for every element fit to experiment; none of the chemical potentials come from DFT.

For the case of GGA+U elements, the chemical potentials of the elements in structures with the U correction (e.g., Fe in Fe_3O_4) and without the U correction (e.g., Fe in BCC Fe) are different.[11] To address this, we fit corrections for the GGA+U elements (Table 1) using the method of ref. 11 *after* all other corrections described in this section; no compounds for which GGA+U was applied were used in the fitting of chemical potentials for non-GGA+U species. For GGA+U compounds, all corrected chemical potentials (e.g., O_2)

Table 4. Comparison of errors between experimental and OQMD-predicted formation energies for a variety of material classes, under different sets of elemental chemical potentials

Category	$\langle\Delta H_f^{EXP}\rangle$	Number of comps.	Fit-none		Fit-partial		Fit-all	
			Err	MAE	Err	MAE	Err	MAE
Magnetism								
Magnetic	−1.231	327	0.135	0.151	0.052	0.102	0.023	0.097
Non-magnetic	−1.234	1343	0.098	0.132	0.011	0.093	−0.005	0.076
Bandgap								
Metallic ($E_g = 0$)	−0.706	921	0.056	0.098	0.031	0.086	0.005	0.077
Semi-conductor ($0 < E_g < 2$)	−1.104	249	0.137	0.155	0.017	0.115	0.007	0.093
Wide bandgap ($E_g > 2$)	−2.175	491	0.180	0.195	0.003	0.101	−0.006	0.083
Number of components								
Binary	−0.935	1376	0.089	0.124	0.025	0.097	0.002	0.082
Ternary	−2.187	259	0.179	0.191	0.002	0.094	0.002	0.082
Quaternary	−2.439	33	0.207	0.214	0.019	0.085	0.020	0.071
Bonding type								
I–VII (ionic)	−2.225	18	0.224	0.224	0.090	0.090	0.052	0.057
III–V (covalent)	−0.577	15	0.144	0.146	0.095	0.123	0.056	0.090
Intermetallic binary (metallic)	−0.434	278	0.006	0.071	−0.002	0.069	−0.013	0.069
Binary compounds that contain a/an:[a]								
Alkali metal	−1.047	87	0.120	0.126	0.052	0.081	0.017	0.056
Alkali earth metal	−1.256	92	0.161	0.191	0.098	0.128	0.009	0.092
Transition metal	−0.727	855	0.051	0.094	0.002	0.079	−0.006	0.074
f-block metal	−1.106	414	0.098	0.128	0.060	0.098	0.010	0.074
Semi-metal[b]	−0.573	317	0.046	0.093	0.027	0.093	−0.009	0.082
Post-transition metal[c]	−0.651	233	0.078	0.101	0.028	0.093	−0.005	0.075
Halide	−1.717	258	0.199	0.214	0.015	0.135	0.014	0.122
Compounds w/o corrections[d]	−0.514	780	0.031	0.081	0.031	0.081	−0.009	0.066
Overall	−1.232	1670	0.105	0.136	0.020	0.096	0.002	0.081

Abbreviations: comps, compounds; Err, average error; MAE, mean absolute error; OQMD, Open Quantum Materials Database; w/o, with out.
The definitions of the different fits are given in the section 'Formation Energy Calculation'. Average experimental formation energies ($\langle\Delta H_f^{EXP}\rangle$), Err and MAE are given in eV/atom.
[a]At least one of the components in the binary compound belongs to a given subcategory of elements, and the second component can be any element.
[b]We take semimetal elements to be B, Si, Ge, As, Sb and Te.
[c]We take post-TM elements to be Al, Ga, In, Sn, Tl, Pb and Bi.
[d]Includes all compounds that do not contain any elements that require GGA+U or chemical-potential corrections.

were applied first, and then the GGA+U correction was determined.

The difference between the fitted chemical potential (for both fit-partial and fit-all sets) and DFT ground-state energy for each element is shown in Figure 2, and the values of the chemical potentials for each element in each fit set are provided in Table 3. The fit-partial corrections are consistently larger than the fit-all corrections for the same elements. However, compared with the magnitude of the corrections, the difference between fit-partial and fit-all corrections is quite small in all cases as observed in Figure 2. Adding more fitting parameters in fit-all does not substantially alter the fitted chemical potentials from fit-partial, suggesting that the corrections in fit-partial capture the majority of the error associated with those particularly problematic elemental chemical potentials.

In the fit-all set, some elements appear to have surprisingly large corrections. These are elements that have relatively few compounds to fit to—the actinides and some lanthanides have comparatively large corrections. The small pool of formation energies may lead to overcorrection, and less-predictive calculated formation energies. *For this reason, the formation energies that are available through www.oqmd.org/download are all computed using the fit-partial correction scheme.* In the section 'Comparison of OQMD to Experiment', we next provide a detailed analysis of the agreement between the OQMD and experimental

formation energies, and how the agreement with experiment depends on the fit set being used.

Comparison of OQMD to Experiment. To understand the limits of applicability of our large database of calculated thermochemical data, it must be comprehensively validated against a known reliable source. In the case of the OQMD, we attempt to validate the predicted energies of formation by comparing with as many experimental formation energies as possible. In this section, we compare OQMD values for formation energies to experiment in several ways. First, we will look at the statistics of the agreement between OQMD and experiment and investigate the outliers in the data set. For this analysis we will also consider the effects of different fit sets, described in the section 'Formation Energy Calculation', and determine which set of chemical potentials we find to be most trustworthy. Then, we explore variation in the error for various material classes. Finally, we consider how to correctly distribute error between DFT and experiment, and how to improve the predictive power of DFT-formation energies.

In Figure 1 we compare the OQMD formation energies to experimental formation energies, and in Table 4 we present detailed statistics for the same data. In Figure 1a–c, we show the difference between OQMD and experiment for the fit-none, fit-partial and fit-all chemical-potential sets, respectively. In the case of fit-none, the average difference is 0.105 eV/atom, with a MAE of

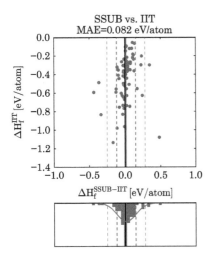

Figure 3. Illustration of the lack of agreement between the IIT[47] and SSUB[46] experimental thermochemical databases. Plots average error ($\Delta H_f^{SSUB-IIT}$) against IIT formation energy (ΔH_f^{IIT}), with the distribution of errors summarised in the histogram at the bottom. The significant range of ΔH_f^{IIT} values demonstrates the surprising degree of disagreement between these experimental formation energy databases. This experiment to experiment comparison for the 75 intermetallic compounds common to both databases gives a MAE of 0.082 eV/atom, whereas the MAE of OQMD formation energies for intermetallic compounds is 0.071 eV/atom (using 'fit-none' chemical potentials).

0.136 eV/atom. Using chemical potentials from the fit-partial set, we find that the average error is reduced to 0.020 eV/atom and the MAE is 0.096 eV/atom. Finally, we find that, by fitting the chemical potentials of all elements, the average error is 0.002 eV/atom and MAE is 0.081 eV/atom, a slight improvement compared with the difference between fit-none and fit-partial chemical potentials.

With such a diverse database of experimental formation energies to compare with, we are able to look for trends in the errors across various material classes as a function of chemical-potential fitting. Table 4 compares the errors between experiment and the OQMD formation energies for a variety of material classes for the fit-none, fit-partial and fit-all chemical-potential sets. Note that for all classes of compounds in the fit-none set, the average error (OQMD–EXP) is positive, which agrees with the expectation that, in the generalised gradient approximation, on average DFT underbinds or underestimates the stability of compounds.

Below, we discuss in more detail the comparison between the OQMD and experimental formation energies for the specific classes of materials in Table 4:

Magnetism: The MAE of magnetic-structure formation energies is larger than that of non-magnetic structures regardless of chemical-potential choice—0.097 vs. 0.076 eV/atom, using the fit-all chemical potentials. Including more complicated magnetic ordering beyond the ferromagnetism assumed here could lead to lower-energy structures, reducing the MAE of magnetic-compound-formation energies.

Bandgap: Without any fitting, the difference in the formation energy of a compound between the OQMD and experiment is largest for wide bandgap insulators, smaller for semiconductors and smallest for metals. As expected, formation energy error magnitudes follow the trend in absolute formation energy magnitudes for these classes of compounds. Wide bandgap compounds are expected to be found in material classes such as nitrides, fluorides or oxides, for which the uncorrected chemical potentials are unreasonable, which results in systematically skewed formation energies. Once corrections to the chemical

potentials are applied, however, the error becomes largely independent of band gap.

Number of components: We find that the average and the MAE between OQMD-calculated and experimental formation energies (using the fit-none chemical-potential set) increase as the number of elements in the compound increases from 0.124 to 0.191 to 0.214 eV/atom for the MAE of binary, ternary and quaternary compounds, respectively. This trend disappears when either the fit-partial or fit-all chemical potentials are used. There are only 33 quaternary compound-formation energies to compare between OQMD and experiment, and all of them are oxides, causing the effect of chemical-potential fitting to be large for this category.

Bonding type: We compare DFT accuracy for three groups of binary compounds: alkali-metal–halide (I–VII), III–V and intermetallic (both elements are metals, including alkali metals, alkaline earth, lanthanides, actinides, transition metals and poor metals). These groups correspond roughly to ionic, covalent and metallic bonding characters, respectively. Without any fitting, intermetallic binary compound-formation energies have the smallest MAE with respect to experimental formation energies, followed by covalent binary compound-formation energies, and finally ionic compound-formation energies have the largest MAE of the three bonding types. In addition, the accuracy of intermetallic formation energies is almost completely unaffected by chemical-potential fitting. In contrast, ionic compounds have formation energies that are systematically more positive than experiment (Err. equals MAE for fit-none and fit-partial chemical potential sets in Table 4), and ionic compound-formation energies were significantly improved by both chemical-potential corrections, with their MAE shrinking from 0.224 to 0.090 to 0.057 eV/atom for fit-none, fit-partial and fit-all, respectively.

Binary compounds that contain a/an:

Alkali metal: Binary compounds containing alkali metals have formation energies with MAEs slightly below the overall average for all chemical-potential sets, and have reductions in MAE from fit-none to fit-partial and from fit-partial to fit-all of 0.045 and 0.025 eV/atom, respectively.

Alkaline earth metal: Alkaline earth binaries have the largest reduction in MAE between fit-partial and fit-all—an improvement of 0.036 eV/atom. Most categories have less than a 0.02-eV/atom improvement over the same range, suggesting that the alkaline earth elements have systematic errors in their DFT reference state energies.

Transition metal: As the largest binary compound data set with 855 compounds, transition metal containing binary compounds have agreement with experiment that is slightly below the overall average, and show little improvement between the fit-partial and fit-all chemical potentials.

f-block element: Binary compounds containing f-block elements show consistent improvement in MAE across all fitting levels and have MAEs comparable to the overall MAE.

Semi-metal: Semi-metal containing binaries have one of the lowest MAEs before fitting and one of the smallest changes in MAE between the fit-none and fit-all chemical-potential sets, suggesting that the largest error component in these compounds cannot be addressed by simply adjusting chemical potentials.

Post-transition metal: These compounds show consistent improvement in accuracy across all levels of fitting, and overall average accuracy.

Halide: Across all fitting levels, binary halides have the largest MAE of all binary compound categories.

Next, we consider several possible sources of error between the OQMD and experiment. First, there are several ways in which the OQMD formation energies can be expected to be improved. One significant approximation made during our calculations is the assumption that all magnetic structures are ferromagnetic. Another potential improvement we could make would be to assign *U*-values for more elements and systems. In this study we applied DFT+U correction only to transition metal oxides and actinide oxides; however, by applying this correction to more compositions—both applying GGA+U corrections to additional cations and applying those corrections in the presence of additional anions—further improvements in formation energies may be achieved. For example, local-environment (anion and oxidation state)-dependent GGA+U calculations have recently been shown to provide improved thermochemical accuracy in transition metal oxides and fluorides.[51] For systems where dispersion interactions are important, such as molecular or layered crystals, more accurate predictions may require van der Waals inclusive methods beyond GGA.[32,52] Lastly, we should note that DFT-predicted bulk formation energies serve as a $T = 0$ K starting point for further thermodynamic analysis. Enthalpic and entropic contributions at finite temperatures such as lattice dynamics, configurational defects and order–disorder transitions can be captured more accurately with relevant statistical, mechanical and DFT methods.

Assessing the Accuracy of Experimental Formation Energies. Improvements to the DFT calculation scheme may lead to some reductions in the discrepancy between the OQMD and experimental formation energies. However, some of the errors can also be attributed to the experimental formation energies themselves. We wish to ascertain the size of this error or uncertainty. With multiple experimental data sources to draw from, we can compare experimental measurements of a given compound with one another. Figure 3 shows the discrepancies between formation energies from different experimental sources for the same compounds. The resulting MAE of the experimental values is surprisingly large, 0.082 eV/atom. This experimental error is calculated based on a comparison between the 75 compounds common to both of the experimental databases used in this study. Note that this comparison is limited by the fact that the IIT database is strictly intermetallic compounds, and therefore all of the energies compared are for intermetallics. Of course we acknowledge that not all experimental data are equally reliable or accurate and that advances in techniques can yield more accurate data. However, these comparisons are for curated databases, and therefore might reasonably be expected to represent a high degree of experimental accuracy.

The OQMD formation energies (using the 'fit-none' chemical potentials, i.e., uncorrected DFT formation energies) have a MAE relative to experiment of 0.071 eV/atom for similar compounds, i.e., intermetallics, which is slightly less than the experimental error with a second experiment. For this same set of 75 compounds, we compare the OQMD formation energy with each of the experimental values. The minimum MAE between experiment and DFT (i.e., comparing DFT with the experimental formation energy closer in energy to the DFT formation energy for each compound) is 0.057 eV/atom, while the maximum MAE between experiment and DFT (i.e., comparing DFT with the experimental formation energy farther away in energy from the DFT formation energy for each compound) is 0.116 eV/atom. From this result it is clear that, where experiments disagree, DFT is often significantly closer to one experiment than the other.

Given this level of disparity in experimental formation energies, it is highly unlikely that *all* of the errors should be attributed to DFT. In fact, without additional information, it is impossible to fairly determine which values are in error. *We conclude from these comparisons that there remains a need for additional experimental thermochemical data.* This is particularly urgent as many computational schemes[45,49,50] rely on these experimental values to obtain elemental correction factors.

To explore the source of disparity between DFT and experiment further, we looked at several of the compounds that have the poorest agreement with experiment, and searched the literature for alternative values for their formation energies. For the cases

Table 5. Comparison of SSUB database[46] and alternative sources for experimental formation energies

Composition	SSUB	ΔH_f (eV/atom) alternative	OQMD
$LiNbO_3$	−5.660	−2.774[a]	−2.739
Eu_2O_3	−3.437	−3.423[b]	−2.574
SiB_6	−0.173	−0.181[c]	0.455
EuF_3	−4.104	−4.195[d]	−3.483
MnSe	−0.889	−0.922[e]	−0.363
InN	−0.500	−0.148[f]	−0.021
AlB_2	−0.521	−0.055[g]	−0.044
PtZr	−1.561	−0.99[d]	−1.086
Ba_2Pb	−1.012	−0.694[h]	−0.540
$PuCl_3$	−2.614	−2.486[i]	−2.176
PrO_2	−3.280	−3.316[j]	−2.846
$CaAl_2$	−0.759	−0.346[k]	−0.335
$BaSiO_3$	−3.354	−3.330[l]	−2.945
CeN	−1.715	−1.763[m]	−1.326

Abbreviations: OQMD, Open Quantum Materials Database; SSUB, SGTE Solid SUBstance.

Comparison made for outlying compounds that show a large disagreement between SSUB data and OQMD-predicted formation energies (using fit-partial chemical potentials).

Differences between SSUB and alternative-source formation energies ranging from 0.008 eV/atom (SiB_6) to 2.885 eV/atom ($LiNbO_3$).

[a]Reference 53.
[b]Reference 54.
[c]Reference 55.
[d]Reference 56.
[e]Reference 57.
[f]Reference 58.
[g]Reference 59.
[h]Reference 60.
[i]Reference 61.
[j]Reference 62.
[k]Reference 63.
[l]Reference 64.
[m]Reference 65.

where we were able to find another value, we show the composition, SSUB[46] formation energy and the literature formation energy in (Table 5) these compounds with very large discrepancies between experimental formation energies—the second value we found in literature is often closer to the OQMD-predicted formation energy. As a result, we conclude that some of the *very significant disagreements* between DFT and experiment are more likely to be due to experimental or transcription errors than to problems in DFT. On the basis of these findings, we believe that for many other compounds with large formation-energy errors, and for which no alternative formation energies could be found, the source of error might also be the experimental measurement, rather than only the DFT calculation.

Comparison of OQMD to Other DFT Databases and the Miedema Model. The Miedema model has historically been widely used to provide estimates of formation enthalpies of solid alloys and intermetallic compounds.[66] The Miedema model is a semi-empirical approach wherein atoms are conceptually treated as space-filling polyhedra. Chemical bonding is treated by considering the overlap of the surface areas of neighbouring atomic polyhedra, weighted by the difference in charge density of each atom at the boundary and the electronegativity difference between the atoms.[66] The model contains several element-dependent parameters that Miedema fit to trends in the formation energies of a range of binary intermetallic compounds, as well as elemental properties (bulk modulus, molar volume and work function), which were adjusted to give the best fit to experimental

Table 6. Comparison of predicted formation energies with experimental values for two DFT databases (the OQMD with 'fit-partial' corrections and Materials Project) and an empirical model for intermetallic compounds (Miedema model)

		Materials Project[a]	Miedema
Number of compounds		1,386	820
ΔH_f^{X-EXP}	Avg Err	0.006	0.033
	MAE	0.133	0.199
$\Delta H_f^{OQMD-EXP}$	Avg Err	0.032	0.029
	MAE	0.108	0.090

Abbreviations: API, application programming interface; Avg Err, average error; DFT, density functional theory; MAE, mean absolute error; OQMD, Open Quantum Materials Database.
Average error (ΔH_f^{X-EXP}) and MAE ($|\Delta H_f^{X-EXP}|$) are given in eV/atom.
[a]Data from the Materials Project used in this comparison were retrieved via the Materials Project API[24,26] on 16/12/2013.

formation energies. A comparison of the accuracy of the Miedema model with the accuracy of the OQMD is important as the Miedema model is still actively employed.[67,68] Table 6 contains a comparison of Miedema model predictions for formation energy to experiment. The MAE between the Miedema model and experiment is 0.199 eV/atom, greater than twice that of the OQMD for the same set of compounds, 0.090 eV/atom. This result indicates that, in addition to the inherent drawbacks of the Miedema model (i.e., applicable only to binary intermetallics), the OQMD is a much more accurate predictor than the Miedema model for formation energies.

The Materials Project[11,24] was one of the first high-throughput databases to be developed. As the Materials Project database uses slightly different calculation parameters from OQMD, and a different chemical potential correction scheme, there is an interesting opportunity to directly compare the results of different DFT databases with one another. Table 6 shows the statistics of the agreement between Materials Project and experimental formation energies. The average error of formation energies for the Materials Project is 0.006 eV/atom, smaller than the average error of the OQMD formation energies, 0.032 eV/atom. However, the MAE for the Materials Project is 0.133 eV/atom, which is larger than that of the OQMD over the same compounds, which is 0.108 eV/atom.

We attribute the difference in MAE with experiment between OQMD and Materials Project to the difference in chemical-potential fitting procedures for the two data sets.[11,24] The chemical-potential fitting used in the OQMD is performed on the same set of compounds on which the computed accuracy is based, which gives the OQMD a 'natural advantage.' Further evidence to support this argument can be found by calculating the mean absolute difference between the OQMD and Materials Project formation energies for all compounds for which we do not fit any chemical potentials. For this set of 563 compounds, the mean absolute difference between the OQMD and Materials Project is 0.028 eV/atom, much smaller than the difference between the OQMD and experiment (0.093 eV/atom for these 563 compounds) or between Materials Project and experiment (0.086 eV/atom for these 563 compounds). As a result, we conclude that in general the two databases contain very similar results, and that different choices for DFT parameters have a much smaller impact on compound-formation energies than do the different approaches to chemical-potential fitting. Finally, as new calculations are continuously added to both OQMD and Materials Project, the analysis above corresponds only to certain *snapshots* of each database; however, we expect this conclusion to be valid as long as the chemical-potential fitting approaches are not significantly revised.

Historical Trends in Material/Compound Discovery

A large thermodynamic database of energetics and phase stability of the type presented here can be used to address many interesting general trends. For instance, we leverage the fact that we have evaluated a significant fraction of known ground-state compounds to answer several questions about trends and patterns in material discovery and stability. Without a large database such as the OQMD it would otherwise be impossible to answer many of these questions.

How many stable compounds are in the database? How many of these are experimentally known versus theoretically predicted? In order to answer these questions, first we determine phase stability. Phase stability is determined by constructing the energy convex hull of a given region of composition space.[69] Once this has been determined using existing computational geometry algorithms,[70] (Kirklin, S. & Wolverton, C. (2015, unpublished)) every phase that lies on the convex hull is stable at $T = 0$ K (i.e., it is lower in energy than any other phase or combination of phases in the database). Of the 297,099 calculated compounds in the OQMD, we find that 19,757 are thermodynamically stable at $T = 0$ K. Of these, 16,526 were from the ICSD with the remaining 3,231 being prototype structures.

All the 3,231 compounds that we predict to be stable, but are not in the ICSD, represent new compounds to be discovered. The prototype compounds were constructed from commonly occurring, simple, crystal structures, and do not represent an exhaustive crystal structure determination for each predicted compound. For this reason, we do not assert that in all cases the predicted compounds are stable in the crystal structure we list. Rather, in these cases, our predicted convex hull is an upper bound to the true ground-state hull. Thus, for all the 3,231 cases, we predict that some new compound(s) are awaiting experimental discovery in these (binary and ternary) systems. A detailed crystal structure search in such systems can be made using evolutionary[71] or minima hopping[72] methods. Furthermore, because the prototype compounds have identified so many holes in our knowledge of ground-state phase stability, we expect that by including the prototype structures in our list of ground-state structures we are providing a better estimate of the energy landscape where we do not have experimentally measured structures.

What is the rate of stable material discovery? New materials and compounds are being discovered all the time, some of which are stable and some of which are not. Utilising the publication data associated with ICSD records, we can study the historical rate of material discovery. In Figure 4a we plot the total rate of material discovery as the number of new compounds reported per year since 1910. Each compound only appears in the year in which it was *first* reported. In Figure 4b, we plot the number of *stable* ICSD compounds discovered per year. The 'material discovery' data in Figure 4a come directly from the ICSD (no DFT is required). However, to classify a compound as 'stable' is not possible from the ICSD alone but requires some measures of energetics, as well as those of competing (combinations of) phases. The latter is possible only with a large material database containing formation energies, such as the OQMD.

We find that the rate of compound discovery is increasing with time—in most years, more compounds are discovered than the year before. In contrast, the rate of *stable* material discovery has been fairly constant since the 1960s. By decomposing the number of discovered materials into binary, ternary, quaternary and pentanary compounds, we observe that the number of stable binary compounds discovered each year has been dwindling since the 1970s, when the number of ternary compounds discovered began to significantly rise. As of the 1990s, the number of stable ternaries has also stagnated, while the number of stable quaternary compounds began to increase.

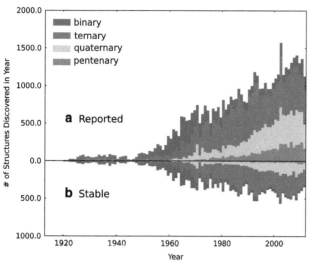

Figure 4. (a) The total number of compound discovery within the ICSD by year. (b) The number of *stable* ($T = 0$ K) compound discovery in the ICSD, where the stability is assesed by the OQMD energies. The year for a structure corresponds to the earliest publication year for ICSD entries at that given structure's composition.

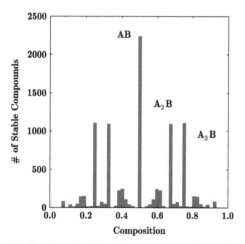

Figure 5. Distribution of stable binary compounds as a function of composition. Note the presence of large peaks at low integer ratios, i.e., 1:1, 2:1 and 3:1.

What compositions are most likely to be stable? On the basis of the same stability data, we can also look at which compositions are most frequently stable. In Figures 5 and 6 we provide histograms showing the frequency at which compositions are stable for binary and ternary compositions, respectively. We find that the most commonly stable binary composition is at 1:1 (AB), followed by 3:1 (A_3B) and 2:1 (A_2B). For ternary compounds, the most common composition is 2:1:1 (A_2BC), which is the composition of the L2₁ (Heusler) prototype. We believe that the preponderance of newly predicted L2₁ compounds is primarily because L2₁ is the only ternary prototype that we have calculated in a wide range of compositions. Many ternary systems that have favourable ordering, but for which the true ground state is unknown, could yield predictions that the L2₁ compound is stable (the upper bound to the convex hull), demonstrating the need for further exploration of a wide range of composition spaces. Following this interpretation, we checked every stable ternary L2₁ prototype in the OQMD and searched for any ternary ICSD compound in that system—that is, if Ca_2GaLi is a stable Heusler, is there any ternary Ca-Ga-Li compound reported in the ICSD? Of the

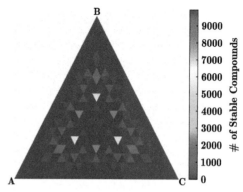

Figure 6. Distribution of stable ternary compounds as a function of composition. Plotted on log scale to account for the extremely high density of phases at A_2BC compositions because of the calculation of over 180,000 decorations of the L2₁ structure.

2,290 stable Heusler prototypes in the OQMD, only 781, or 22%, have *any* ICSD structure in that region of phase space. In the remaining ~ 1,500 ternary systems, we predict the existence of stable compounds waiting to be discovered.

In order to facilitate the discovery of new, stable compounds in the thousands of regions of composition space where we predict stable compounds to exist, we provide a full list of compositions where we predict a prototype to be stable online. We break this list into (i) prototypes that are more stable than an experimentally measured structure and (ii) prototypes that have no experimental structure at that composition. In the first case, finite-temperature effects may cause the formation energy of the experimental structure to lower relative to the formation energy of the prototype. However, in the second case, *some* compounds should be found at the listed composition, although possibly not the prototype. A current list of predicted compounds can be found online at www.oqmd.org/materials/discovery where one can obtain the entire list or filter by composition.

CONCLUSIONS

The OQMD is a high-throughput database of DFT calculations of 32,559 ICSD structures and 259,511 prototypical structures, growing steadily as new structures are added continuously. OQMD is available for download without restrictions at www.oqmd.org/download. Included in the download is a complete framework for performing additional calculations that are commensurate with the database. We use the breadth of the OQMD to compare DFT-calculated structures and formation energies with experiment at an unprecedented scale. We find the following:

Elemental Ground-State Structures. Using the capabilities of qmpy and the OQMD, we find that for 77 out of the 89 elements DFT as implemented in VASP at the settings of the OQMD is able to correctly predict the observed ground-state structure as being lowest in energy out of 20 possible structures, chosen from among the known ground states of all elements. In all cases where DFT finds a lower-energy structure, the observed ground state is nearly degenerate with the lowest-energy structure (by far, the largest errors being phosphorous and mercury, with 0.036 and 0.074 eV/atom error, respectively).

Formation Energies. In order to most accurately determine compound-formation energies, we evaluate the effects of three different choices of chemical potentials: using DFT ground-state reference energies, fitting chemical potentials to experiment for elements where the DFT ground state differs significantly from the room temperature stable phase, and fitting chemical potentials for all elements (labelled fit-none, fit-partial and fit-all). We find that fit-partial exhibits significant improvements over fit-none, with a reduction in MAE against experiment from 0.136 to 0.096 eV/atom. In comparison, fit-all has only marginal gains over fit-partial, with the MAE reducing to 0.081 eV/atom. However, by increasing the number of fit parameters, we also increase the risk of over-fitting, and, as a result, take fit-partial to be the optimal choice of chemical potentials for predicting formation energies.

To put the discrepancy between the OQMD and experiment into appropriate context, we also compared formation energies between the two database experimental formation energies used in this work (the SSUB database,[46] and the thermodynamic database at the Thermal Processing Technology Center at the IIT).[47] We find that the MAE between these two databases is 0.082 eV/atom, a value that is similar to the error between the OQMD and experiment. As a result, it is impossible to assign all of the errors between DFT and experiment solely to DFT, and leads us to conclude that, in order to establish a more accurate evaluation of the accuracy of formation energies based on DFT total energies, more accurate measurements of formation energies should be undertaken in future.

We also compare the OQMD with two other databases of calculated formation energies: the Miedema model and the Materials Project. We compare the Miedema model with experiment for all compositions for which we have (i) an experimental formation energy, (ii) an OQMD formation energy, and (iii) the Miedema model is applicable. Over the resulting pool of 820 compounds, the OQMD has a MAE of 0.090 eV/atom, less than half the error found in the Miedema model, 0.199 eV/atom. We made an identical comparison with the Materials Project, which had a comparison pool of 1,386 compounds. For this set of compounds, we found that the Materials Project had an MAE of 0.133 eV/atom, which is slightly larger than the OQMD error for the same set of compounds, 0.108 eV/atom. By comparing cases in which the formation energy is calculated without any chemical-potential fitting with cases in which the chemical potentials have been fit to experiment, we determine that the majority of the difference between the error in the OQMD and Materials Project is attributed to differences in the chemical-potential-fitting approach.

Historical Trends in Material/Compound Discovery. The OQMD allows us to explore trends, both historical and stoichiometric, in compound discovery and stability. From a historical perspective, the number of reported structures has been increasing linearly with time, while the number of *stable* structures reported annually has remained roughly constant since the 1960s. In addition, the scope of compound discovery has progressed from binary to ternary to quaternary compounds over time. This trend has now been disrupted by recent advances in structure prediction, structure determination and high-throughput structure calculation. In particular, in this study we predict the existence of ~3,200 new compounds. Using our predictions of the compositions at which new phases should be found, experimentalists can now more efficiently discover and characterise new materials.

In this work we demonstrate a few examples of how a large database of DFT calculations can be used to extract information beyond what can be gleaned from many distributed collections containing similar data. We believe that there is much more that can be understood by looking at large-scale material property databases, and in order to facilitate such discovery, we make the entire database available for download, without restriction.

MATERIALS AND METHODS

All DFT calculations are performed with the VASP[73,74] (v5.3.2). The electron exchange and correlation are described with the GGA of PBE,[75] using the

potentials supplied by VASP with the PAW method.[76] PAW-PBE potentials for 89 elements are supplied with VASP, and those employed in the OQMD are listed in Table 3. We follow the VASP guidelines concerning the optimum choice of potentials.[77] Potentials where electrons have been moved from the core and treated as valence are appended with _sv, _pv and _d in Table 3 for s-, p- and d-electrons, respectively. For the 4f-elements, we employ potentials where the valence f-electrons are treated as core electrons (appended with _2 and _3). For all calculations, Γ-centred k-point meshes are constructed to have the same relative ratios of mesh points to reciprocal lattice vector length, and with the number of k-points such that the k-points per reciprocal atom (calculated from $N_{k-points} \times N_{atoms}$) is as close to a target value as possible. The electronic self-consistency (for a given set of ion positions) is converged to within 10^{-4} eV/atom.

Any calculation containing d-block or actinide elements are spin-polarised with a ferromagnetic alignment of spins to capture possible magnetism, with initial magnetic moments of 5 and 7 μ_B for the d-block and actinide elements, respectively. It should be noted that this approach will not capture more complex magnetic ordering, such as antiferromagnetism. For many d-block oxides, a typical difference in total energy between ferromagnetic and antiferromagnetic states is on the order of 10–20 meV/atom.[45] However, given that in this work the range of compounds being calculated is extremely broad, it is likely that in some cases this error will be larger.

For several d- and f-block elements listed in Table 1, the GGA+U approach is implemented to improve the exchange and correlation description of the localised charge density when these elements are in compounds with oxygen. We employed the Dudarev approach to GGA+U,[78] where the only input parameter is $U-J$. For several transition metals, previously determined $U-J$ values were used.[49] For actinide elements in oxides, we apply a $U-J$ parameter of 4 eV. This value was chosen because no reliable values had been reported in the literature when these calculations were begun, and we found that the formation energies and band gaps of compounds containing these elements are relatively insensitive to the exact value of $U-J$; therefore, we elected to use a moderate $U-J$ value of 4 eV for these elements. Recent lines of work have identified $U-J$ values for a few of these elements,[51,79] with $U-J$ values close to those used herein. All $U-J$ values are given in Table 1.

All calculations were completed in a two-step scheme. First, the structures were fully relaxed, followed by a static calculation. In relaxing an ICSD structure, we begin with the given ICSD structure parameters and perform several relaxation runs sequentially, until the volume change within the last relaxation run is less than 10%. The relaxation calculations are performed at a plane-wave basis-set energy cutoff at the energy recommended in the VASP potentials of the elements in the structure, and 6,000 k-points per reciprocal atom. The quasi-Newton scheme is used to optimise the structure to within 10^{-3} eV/atom. In these relaxation steps, Gaussian smearing is applied with a width of 0.2 eV. The final static calculation of the structure is performed at an energy cutoff of 520 eV using tetrahedral k-point integration.[76] The 520-eV cutoff is chosen because it is 25% higher than the highest recommended energy cutoff over all of the potentials used (including Li_sv, for which we use the version with lower recommended cutoff). This constant cutoff for all calculations ensures that all the energies calculated in OQMD are compatible, and can be used to evaluate the formation energies of compounds and $T=0$ ground-state phase diagrams.

ACKNOWLEDGEMENTS

We acknowledge funding support from the following sources: the Center for Electrical Energy Storage: Tailored Interfaces, an Energy Frontier Research Center funded by the U.S. Department of Energy, Office of Science, Office of Basic Energy Sciences (SK); the Revolutionary Materials for Solid State Energy Conversion, an Energy Frontier Research Center funded by the U.S. Department of Energy, Office of Science, Office of Basic Energy Sciences under Award Number DE-SC00010543 (JWD); the Ford-Boeing-Northwestern Alliance (JES); the Department of Defense through the National Defense Science and Engineering Graduate Fellowship Program with further support by DOE under Grant No. DE-FG02-07ER46433 (BM and AT); The Dow Chemical Company (MA) and the National Science Foundation under grant DRL-1348800 (CW). We acknowledge the openness of the Materials Project for making it possible to obtain a large set of energies with which to compare with the OQMD and experiment. We acknowledge a large allocation on Northwestern University's Quest high-performance computing system. Calculations were also performed on resources of the National Energy Research Scientific Computing

Center, which is supported by the Office of Science of the U.S. Department of Energy under Contract No. DE-AC02-05CH11231.

CONTRIBUTIONS

SK developed DFT automation and thermodynamic analysis tools, and had primary writing responsibilities. JWD carried out comparison with Miedema model. JES carried out comparison with Materials Project, evaluated SSUB formation energies at STP. BM and AT selected DFT+U parameterisations and VASP potentials. MA analysed the elemental ground states. BM and SR organised the partnership with the ICSD. MA, JWD, JES and CW extensively edited the manuscript. CW conceptualised, guided all aspects and led the project.

COMPETING INTERESTS

The authors declare no conflict of interest.

REFERENCES

1. National Science and Technology Council. Materials Genome Initiative for Global Competitiveness Tech. Rep. http://www.whitehouse.gov/blog/2011/06/24/materials-genome-initiative-renaissance-american-manufacturing (2011).
2. Hohenberg, P. & Kohn, W. Inhomogeneous electron gas. *Phys. Rev.* **136**, 864–871 (1964).
3. Kohn, W. & Sham, L. J. Self-consistent equations including exchange and correlation effects. *Phys. Rev.* **140**, 1133–1138 (1965).
4. Perdew, J. P. & Zunger, A. Self-Interaction correction to density-functional approximations for many-electron systems. *Phys. Rev. B* **23**, 5048–5079 (1981).
5. Ihm, J., Zunger, A. & Cohen, M. L. Momentum-space formalism for the total energy of solids. *J. Phys. C Solid State Phys.* **12**, 4409–4422 (1979).
6. Ceperley, D. M. & Alder, B. J. Ground state of the electron gas by a stochastic method. *Phys. Rev. Lett.* **45**, 566–569 (1980).
7. Ihm, J., Yin, M. T. & Cohen, M. L. Quantum mechanical force calculations in solids: the phonon spectrum of Si. *Solid State Commun.* **37**, 491–494 (1981).
8. Levy, O., Hart, G. L. W. & Curtarolo, S. Uncovering compounds by synergy of cluster expansion and high-throughput methods. *J. Am. Chem. Soc.* **132**, 4830–4833 (2010).
9. Curtarolo, S. *et al.* The high-throughput highway to computational materials design. *Nat. Mater.* **12**, 191–201 (2013).
10. Hautier, G., Fischer, C., Ehrlacher, V., Jain, A. & Ceder, G. Data mined ionic substitutions for the discovery of new compounds. *Inorg. Chem.* **50**, 656–663 (2010).
11. Jain, A. *et al.* A high-throughput infrastructure for density functional theory calculations. *Comput. Mater. Sci.* **50**, 2295–2310 (2011).
12. Armiento, R., Kozinsky, B., Fornari, M. & Ceder, G. Screening for high-performance piezoelectrics using high-throughput density functional theory. *Phys. Rev. B* **84**, 014103 (2011).
13. Setyawan, W., Gaume, R. M., Lam, S., Feigelson, R. S. & Curtarolo, S. High-throughput combinatorial database of electronic band structures for inorganic scintillator materials. *ACS Comb. Sci.* **13**, 382–390 (2011).
14. Saal, J., Kirklin, S., Aykol, M., Meredig, B. & Wolverton, C. Materials design and discovery with high-throughput density functional theory: the Open Quantum Materials Database (OQMD). *JOM* **65**, 1501–1509 (2013).
15. Curtarolo, S. *et al.* AFLOW: an automatic framework for high-throughput materials discovery. *Comput. Mater. Sci.* **58**, 218–226 (2012).
16. Landis, D. D. *et al.* The computational materials repository. *Comput. Sci. Eng.* **14**, 51–57 (2012).
17. Bergerhoff, G. & Brown, I. D. *Crystallographic Databases. chap. Inorganic* 147–156 (International Union of Crystallography: Chester, 1987).
18. Belsky, A., Hellenbrandt, M., Karen, V. L. & Luksch, P. New developments in the Inorganic Crystal Structure Database (ICSD): accessibility in support of materials research and design. *Acta Crystallogr. B* **58**, 364–369 (2002).
19. Saal, J. E. & Wolverton, C. Thermodynamic stability of Mg-Y-Zn long-period stacking ordered structures. *Scr. Mater.* **67**, 798–801 (2012).
20. Kirklin, S., Meredig, B. & Wolverton, C. High-throughput computational screening of new Li-ion battery anode materials. *Adv. Energy Mater.* **3**, 252–262 (2013).
21. Meredig, B. *et al.* Combinatorial screening for new materials in unconstrained composition space with machine learning. *Phys. Rev. B* **89**, 094104 (2014).
22. Aykol, M., Kirklin, S. & Wolverton, C. Thermodynamic aspects of cathode coatings for lithium-ion batteries. *Adv. Energy Mater.* **4**, 1400690 (2014).
23. Kirklin, S., Chan, M., Trahey, L., Thackeray, M. M. & Wolverton, C. M. High-throughput screening of high-capacity electrodes for hybrid Li-ion/Li-O 2 cells. *Phys. Chem. Chem. Phys.* **16**, 22073–22082 (2014).
24. Jain, A. *et al.* Commentary: The Materials Project: a materials genome approach to accelerating materials innovation. *APL Mater.* **1**, 011002 (2013).
25. Curtarolo, S. *et al.* AFLOWLIB.ORG: a distributed materials properties repository from high-throughput ab initio calculations. *Comput. Mater. Sci.* **58**, 227–235 (2012).
26. Ong, S. P. *et al.* Python Materials Genomics (pymatgen): a robust, open-source python library for materials analysis. *Comput. Mater. Sci.* **68**, 314–319 (2013).
27. Bahn, S. R. & Jacobsen, K. W. An object-oriented scripting interface to a legacy electronic structure code. *Comput. Sci. Eng.* **4**, 56–66 (2002).
28. Meredig, B. & Wolverton, C. A hybrid computational-experimental approach for automated crystal structure solution. *Nat. Mater.* **12**, 123–127 (2013).
29. Young, D. A. *Phase Diagrams of the Elements* (University of California Press: Berkeley and Los Angeles, 1991).
30. King, H. W. in *CRC Handbook of Chemistry and Physics* 95th edn (ed. Haynes W. M.) Ch. 12, 15–18 (CRC Press, Taylor & Francis, 2012).
31. Wang, Y. *et al. Ab initio* lattice stability in comparison with CALPHAD lattice stability. *Calphad* **28**, 79–90 (2004).
32. Klimeš, J. & Michaelides, A. Perspective: advances and challenges in treating van der Waals dispersion forces in density functional theory. *J. Chem. Phys.* **137**, 120901 (2012).
33. Tao, X. *et al.* Phase stability of magnesium-rare earth binary systems from first-principles calculations. *J. Alloys Compd.* **509**, 6899–6907 (2011).
34. Gao, M. C., Rollett, A. D. & Widom, M. First-principles calculation of lattice stability of C15M2R and their hypothetical C15 variants (M = Al, Co, Ni; R = Ca, Ce, Nd, Y). *Calphad* **30**, 341–348 (2006).
35. Mao, Z., Seidman, D. N. & Wolverton, C. First-principles phase stability, magnetic properties and solubility in aluminumare-earth (AlRE) alloys and compounds. *Acta Mater.* **59**, 3659–3666 (2011).
36. Temmerman, W. *et al. Handbook on the Physics and Chemistry of Rare Earths* vol. 39 (Elsevier, 2009); http://www.sciencedirect.com/science/article/pii/S0168127308000019.
37. Zhou, F. & Ozoliņš, V. Obtaining correct orbital ground states in f-electron systems using a nonspherical self-interaction-corrected LDA+U method. *Phys. Rev. B* **80**, 125127 (2009).
38. Duthie, J. C. & Pettifor, D. G. Correlation between d-band occupancy and crystal structure in the rare earths. *Phys. Rev. Lett.* **38**, 564–567 (1977).
39. Biering, S. & Schwerdtfeger, P. High-pressure transitions in bulk mercury: a density functional study. *Theor. Chem. Acc.* **130**, 455–462 (2011).
40. Mishra, V., Gyanchandani, J., Chaturvedi, S. & Sikka, S. Effect of spin-orbit coupling on the ground state structure of mercury. *Solid State Commun.* **186**, 38–41 (2014).
41. Wolverton, C. Crystal structure and stability of complex precipitate phases in Al-Cu-Mg-(Si) and Al-Zn-Mg alloys. *Acta Mater.* **49**, 3129–3142 (2001).
42. Wolverton, C., Ozoliņš, V. & Asta, M. Hydrogen in aluminum: First-principles calculations of structure and thermodynamics. *Phys. Rev. B* **69**, 144109 (2004).
43. Lany, S. Semiconductor thermochemistry in density functional calculations. *Phys. Rev. B* **78**, 245207 (2008).
44. Hautier, G., Ong, S. P., Jain, A., Moore, C. J. & Ceder, G. Accuracy of density functional theory in predicting formation energies of ternary oxides from binary oxides and its implication on phase stability. *Phys. Rev. B* **85**, 155208 (2012).
45. Stevanović, V., Lany, S., Zhang, X. & Zunger, A. Correcting density functional theory for accurate predictions of compound enthalpies of formation: fitted elemental-phase reference energies. *Phys. Rev. B* **85**, 115104 (2012).
46. SGTE. *Thermodynamic Properties of Inorganic Materials* Vol. 19. (Springer-Verlag: Berlin, Heidelberg, 1999).
47. Nash, P. Thermodynamic database https://tptc.iit.edu/index.php/thermo-data base (2013).
48. Konings, R. J. M., Morss, L. R., Fuger, J. in *The Chemistry of the Actinide and Transactinide Elements* 3rd edn (eds Morss L. R. *et al.*) Ch. 19, 2113–2224 (Springer: Dordrecht, 2006); http://www.springerlink.com/index/l17213068108mk22.pdf.
49. Wang, L., Maxisch, T. & Ceder, G. Oxidation energies of transition metal oxides within the GGA+U framework. *Phys. Rev. B* **73**, 195107 (2006).
50. Grindy, S., Meredig, B., Kirklin, S., Saal, J. E. & Wolverton, C. Approaching chemical accuracy with density functional calculations: diatomic energy corrections. *Phys. Rev. B* **87**, 075150 (2013).
51. Aykol, M. & Wolverton, C. Local environment dependent GGA+U method for accurate thermochemistry of transition metal compounds. *Phys. Rev. B* **90**, 115105 (2014).
52. Aykol, M., Kim, S. & Wolverton, C. van der waals interactions in layered lithium cobalt oxides. *J. Phys. Chem. C* **119**, 19053–19058 (2015).
53. Pozdnyakova, I., Navrotsky, A., Shilkina, L. & Reznitchenko, L. Thermodynamic and structural properties of sodium lithium niobate solid solutions. *J. Am. Ceram. Soc.* **85**, 379–384 (2004).
54. CRC. *Handbook of Chemistry and Physics*, 93rd edn 2012–2013 http://www.hbcpnetbase.com/ (2012).
55. Gordienko, S. P. Enthalpies of formation for boron silicides. *Powder Metall. Met. Ceram.* **34**, 660–662 (1995).
56. Kubaschewski, O., Alcock, C. B. & Spencer, P. J. *Materials Thermochemistry* 6th edn (Pergamon Press, 1993).

57. Olin, Å., Noläng, B., Öhman, L.-O., Osadchii, E. & Rosén, E. *Chemical Thermodynamics* (Elsevier Science, 2005).

58. Ranade, M. R., Tessier, F., Navrotsky, A. & Marchand, R. Calorimetric determination of the enthalpy of formation of InN and comparison with AlN and GaN. *J. Mater. Res.* **16**, 2824–2831 (2001).

59. van Setten, M. & Fichtner, M. On the enthalpy of formation of aluminum diboride, AlB2. *J. Alloys Compd.* **477**, L11–L12 (2009).

60. Sommer, F., Borzone, G., Parodi, N. & Ferro, R. Enthalpy of formation of CaPb and BaPb alloys. *Intermetallics* **14**, 287–296 (2006).

61. Lemire, R. J. *Chemical Thermodynamics of Neptunium and Plutonium* Vol. C (Elsevier Science, 2001).

62. Gramsch, S. A. & Morss, L. R. Standard molar enthalpies of formation of PrO2 and SrPrO3: the unusual thermodynamic stability of APrO3 (A = Sr,Ba). *J. Chem. Thermodyn.* **27**, 551–560 (1995).

63. Notin, M., Mejbar, J., Bouhajib, A., Charles, J. & Hertz, J. The thermodynamic properties of calcium intermetallic compounds. *J. Alloys Compd.* **220**, 62–75 (1995).

64. Huntelaar, M. E., Cordfunke, E. H. P. & Ouweltjes, W. The standard molar enthalpies of formation of BaSiO3 (s) and Ba2SiO4 (s). *J. Chem. Thermodyn.* **24**, 1099–1102 (1992).

65. Tessier, F. *et al.* Thermodynamics of formation of binary and ternary nitrides in the system Ce/Mn/N. *Z. Anorg. Allg. Chem.* **627**, 194–200 (2001).

66. de Boer, F. R., Boom, R., Mattens, W. C. M., Miedema, A. R. & Niessen, A. K. *Cohesion in Metals: Transition Metal Alloys* (North-Holland, Amsterdam 1988).

67. Das, N. *et al.* Miedema model based methodology to predict amorphous-forming-composition range in binary and ternary systems. *J. Alloys Compd.* **550**, 483–495 (2013).

68. Ray, P. K., Akinc, M. & Kramer, M. J. Applications of an extended Miedema's model for ternary alloys. *J. Alloys Compd.* **489**, 357–361 (2010).

69. Akbarzadeh, A. R., Ozoliņš, V. & Wolverton, C. First-principles determination of multicomponent hydride phase diagrams: application to the Li-Mg-N-H system. *Adv. Mater.* **19**, 3233–3239 (2007).

70. Barber, C., Dobkin, D. & Huhdanpaa, H. The quickhull algorithm for convex hulls. *ACM Trans. Math. Softw.* **22**, 469–483 (1996).

71. Oganov, A. R. & Glass, C. W. Crystal structure prediction using *ab initio* evolutionary algorithms: principles and applications. *J. Chem. Phys.* **124**, 244704 (2006).

72. Amsler, M. & Goedecker, S. Crystal structure prediction using the minima hopping method. *J. Chem. Phys.* **133**, 224104 (2010).

73. Kresse, G. & Furthmuller, J. Efficiency of *ab-initio* total energy calculations for metals and semiconductors using a plane-wave basis set. *Comput. Mater. Sci.* **6**, 15–50 (1996).

74. Kresse, G. & Furthmüller, J. Efficient iterative schemes for *ab initio* total-energy calculations using a plane-wave basis set. *Phys. Rev. B* **54**, 11169–11186 (1996).

75. Perdew, J. P., Burke, K. & Ernzerhof, M. Generalized gradient approximation made simple. *Phys. Rev. Lett.* **77**, 3865–3868 (1996).

76. Blöchl, P. E. Projector augmented-wave method. *Phys. Rev. B* **50**, 17953–17979 (1994).

77. Kresse, G. & Joubert, D. From ultrasoft pseudopotentials to the projector augmented-wave method. *Phys. Rev. B* **59**, 1758–1775 (1999).

78. Dudarev, S. L., Botton, G. A., Savrasov, S. Y., Humphreys, C. J. & Sutton, A. P. Electron-energy-loss spectra and the structural stability of nickel oxide. *Phys. Rev. B* **57**, 1505–1509 (1998).

79. Dorado, B. & Garcia, P. First-principles DFT+U modeling of actinide-based alloys: application to paramagnetic phases of UO2 and (U,Pu) mixed oxides. *Phys. Rev. B* **87**, 195139 (2013).

The ReaxFF reactive force-field: development, applications and future directions

Thomas P Senftle[1], Sungwook Hong[2], Md Mahbubul Islam[2], Sudhir B Kylasa[3], Yuanxia Zheng[4], Yun Kyung Shin[2], Chad Junkermeier[2], Roman Engel-Herbert[4], Michael J Janik[1], Hasan Metin Aktulga[5], Toon Verstraelen[6], Ananth Grama[3] and Adri CT van Duin[2]

The reactive force-field (ReaxFF) interatomic potential is a powerful computational tool for exploring, developing and optimizing material properties. Methods based on the principles of quantum mechanics (QM), while offering valuable theoretical guidance at the electronic level, are often too computationally intense for simulations that consider the full dynamic evolution of a system. Alternatively, empirical interatomic potentials that are based on classical principles require significantly fewer computational resources, which enables simulations to better describe dynamic processes over longer timeframes and on larger scales. Such methods, however, typically require a predefined connectivity between atoms, precluding simulations that involve reactive events. The ReaxFF method was developed to help bridge this gap. Approaching the gap from the classical side, ReaxFF casts the empirical interatomic potential within a bond-order formalism, thus implicitly describing chemical bonding without expensive QM calculations. This article provides an overview of the development, application, and future directions of the ReaxFF method.

INTRODUCTION

Atomistic-scale computational techniques provide a powerful means for exploring, developing and optimizing promising properties of novel materials. Simulation methods based on quantum mechanics (QM) have grown in popularity over recent decades due to the development of user-friendly software packages making QM level calculations widely accessible. Such availability has proved particularly relevant to material design, where QM frequently serves as a theoretical guide and screening tool. Unfortunately, the computational cost inherent to QM level calculations severely limits simulation scales. This limitation often excludes QM methods from considering the dynamic evolution of a system, thus hampering our theoretical understanding of key factors affecting the overall behaviour of a material. To alleviate this issue, QM structure and energy data are used to train empirical force fields that require significantly fewer computational resources, thereby enabling simulations to better describe dynamic processes. Such empirical methods, including reactive force-field (ReaxFF),[1] trade accuracy for lower computational expense, making it possible to reach simulation scales that are orders of magnitude beyond what is tractable for QM.

Atomistic force-field methods utilise empirically determined interatomic potentials to calculate system energy as a function of atomic positions. Classical approximations are well suited for nonreactive interactions, such as angle-strain represented by harmonic potentials, dispersion represented by van der Waals potentials and Coulombic interactions represented by various polarisation schemes. However, such descriptions are inadequate for modelling changes in atom connectivity (i.e., for modelling chemical reactions as bonds break and form). This motivates the inclusion of connection-dependent terms in the force-field description, yielding a reactive force-field. In ReaxFF, the interatomic potential describes reactive events through a bond-order formalism, where bond order is empirically calculated from interatomic distances. Electronic interactions driving chemical bonding are treated implicitly, allowing the method to simulate reaction chemistry without explicit QM consideration.

The classical treatment of reactive chemistry made available by the ReaxFF methodology has opened the door for numerous studies of phenomena occurring on scales that were previously inaccessible to computational methods. In particular, ReaxFF enables simulations involving reactive events at the interface between solid, liquid, and gas phases, which is made possible because the ReaxFF description of each element is transferable across phases. For example, an oxygen atom is treated with the same mathematical formalism whether that oxygen is in the gas phase as O_2, in the liquid phase within an H_2O molecule, or incorporated in a solid oxide. Such transferability, coupled with a lower computational expense allowing for longer simulation timescales, allows ReaxFF to consider phenomena dependent not only on the reactivity of the involved species, but also on dynamic factors, such as diffusivity and solubility, affecting how species migrate through the system. This allows ReaxFF to model complex processes involving multiple phases in contact with one another.

To demonstrate these capabilities, this article will review the development and application of the ReaxFF method. In 'History of ReaxFF development', we discuss initial development choices shaping the overall ReaxFF formalism, whereas 'Current ReaxFF methodology' and 'Overview of available ReaxFF

[1]Department of Chemical Engineering, Pennsylvania State University, University Park, PA, USA; [2]Department of Mechanical and Nuclear Engineering, Pennsylvania State University, University Park, PA, USA; [3]Department of Computer Science and Engineering, Purdue, West Lafayette, IN, USA; [4]Department of Materials Science and Engineering, Pennsylvania State University, University Park, PA, USA; [5]Department of Computer Science and Engineering, Michigan State University, East Lansing, MI, USA and [6]Center for Molecular Modeling (CMM), Ghent University, Zwijnaarde, Belgium.
Correspondence: ACT van Duin (acv13@psu.edu)

parameterisations and development branches' outline the currently employed method and available parameter sets, respectively. Comparisons with other reactive potentials available in the literature are briefly discussed in 'Comparison to similar methods'. Examples of various ReaxFF applications are provided in 'Applications of ReaxFF', with emphasis placed on demonstrating the breadth of systems that can be modelled with the method. Finally, plans for future extensions and improvements are discussed in 'Future developments and outlook'.

DEVELOPMENT OF THE ReaxFF METHOD

History of ReaxFF development

The current functional form of the ReaxFF potential, best described in the Chenoweth et al.[2] hydrocarbon combustion work (herein referred to as 2008-C/H/O),[2] has demonstrated significant transferability across the periodic table. It is important to note, however, that the 2008 functional form is different from the original 2001 ReaxFF hydrocarbon description,[1] as well as from the 2003 extension to silicon and silica.[3] Although conceptually similar to the current 2008-C/H/O functional form, the 2001 hydrocarbon description employed the same dissociation energy for C–C single, double and triple bonds. This approach was reasonable for hydrocarbons, but could not be extended to treat Si–O single and double bonds. As such, the 2003 Si/O/H extension required separate parameters describing single-, double- and triple-bond dissociation. Furthermore a lone-pair energy term was introduced to handle formation and dissociation of oxygen lone-pairs. The 2003 Si/O functional form was further augmented by a three-body conjugation term introduced to handle $-NO_2$ group chemistry in nitramines, where a triple-bond stabilisation term was added to improve the description of terminal triple bonds. This led to the 2003–2005 ReaxFF description for the RDX high-energy material employed by Strachan et al.[4,5] to study RDX initiation.

Since 2005, the ReaxFF functional form has been stable, although optional additions, such as angular terms to destabilise Mg–Mg–H zero-degree angles[6] or double-well angular terms necessary for describing aqueous transition metal ions,[7] have occasionally been added to the potential. Goddard and co-workers implemented an additional attractive van der Waals term to improve performance for nitramine crystals (ReaxFF-lg).[8] This concept, however, was not made transferable with previous or later ReaxFF parameter sets. The 2005 functional form developed for RDX is the current version of ReaxFF distributed by the van Duin group (commonly referred to as 'standalone ReaxFF'), as well as integrated in the open-source LAMMPS code,[9] supported through Nanohub, (http://www.nanohub.org) and available through the PuReMD (Purdue Reactive Molecular Dynamics) code.[10–12] Apart from these open-source distributions, the ReaxFF method is also integrated in ADF[13] (released by SCM (http://www.scm.com)) and in Materials Studio (released under license by Accelrys (http://www.accelrys.com)). The pre-2005 ReaxFF parameter sets, including the aforementioned 2001-C/H,[1] 2003-Si/O,[3] and 2004-Al/O[14] descriptions, are not supported by any codes curated by the van Duin group or its collaborators. The materials described by the three pre-2005 ReaxFF parameterisations are equally, if not better, described by later parameterisations. The 2008-C/H/O parameter set was trained against the entire 2001-C/H training set, while the 2010 and 2011 Si/O/H parameterisations (Fogarty et al.[15] on the 'aqueous branch' and Neyts et al.[16] on the 'combustion branch') were validated against the full 2003-Si/O/H training set. Finally, the 2008-Al/H ReaxFF description, and later applications to aluminium oxides and aluminosilicates,[17–23] fully contain and extend the 2004-Al/O description. We are aware of other ReaxFF implementations, often developed by individual scientists based on the 2008-C/H/O formalism. Given the

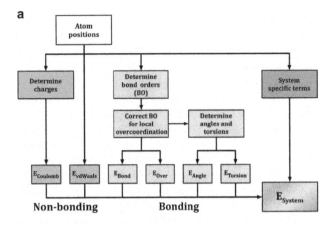

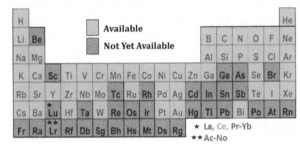

Figure 1. (a) Overview of the ReaxFF total energy components and (b) elements currently described in available parameter sets.

complexity of the ReaxFF functional form, it is advisable to validate ReaxFF implementations against the standalone ReaxFF code prior to applying them in production-scale simulations.

Current ReaxFF methodology

The currently implemented form of the ReaxFF potential is described in detail in a recent article,[24] and therefore here we only provide a brief overview of the method's central concepts. ReaxFF employs a bond-order formalism in conjunction with polarisable charge descriptions to describe both reactive and non-reactive interactions between atoms (Figure 1). This allows ReaxFF to accurately model both covalent and electrostatic interactions for a diverse range of materials. Energy contributions to the ReaxFF potential are summarised by the following:

$$E_{system} = E_{bond} + E_{over} + E_{angle} + E_{tors} + E_{vdWaals} + E_{Coulomb} + E_{Specific}. \qquad (1)$$

E_{bond} is a continuous function of interatomic distance and describes the energy associated with forming bonds between atoms. E_{angle} and E_{tors} are the energies associated with three-body valence angle strain and four-body torsional angle strain. E_{over} is an energy penalty preventing the over coordination of atoms, which is based on atomic valence rules (e.g., a stiff energy penalty is applied if a carbon atom forms more than four bonds). $E_{Coulomb}$ and $E_{vdWaals}$ are electrostatic and dispersive contributions calculated between all atoms, regardless of connectivity and bond-order. $E_{Specific}$ represents system specific terms that are not generally included, unless required to capture properties particular to the system of interest, such as lone-pair, conjugation, hydrogen binding, and C_2 corrections. Full functional forms can be found in the Supplementary Information of the 2008-C/H/O publication.[2]

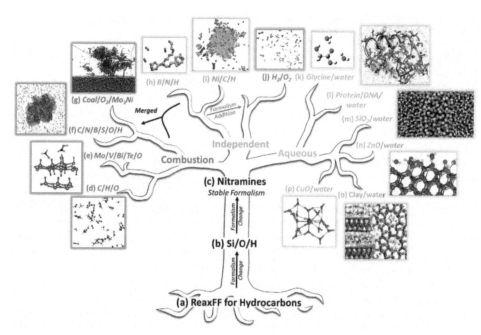

Figure 2. ReaxFF development tree, where parameter sets on a common 'branch' are fully transferable with one another. Parameter sets are available in (**a**) van Duin *et al.*;[1] (**b**) van Duin *et al.*;[3] (**c**) Strachan *et al.*;[4,5] (**d**) Chenoweth *et al.*;[2] (**e**) Goddard *et al.*[29] and Chenoweth *et al.*;[28] (**f**) Castro-Marcano *et al.*[163] and Kamat *et al.*;[13] (**g**) Vasenkov *et al.*;[164] (**h**) Weismiller *et al.*;[32] (**i**) Mueller *et al.*;[33] (**j**) Agrawalla *et al.*;[165] (**k**) Rahaman *et al.*;[116] (**l**) Monti *et al.*;[53] (**m**) Fogarty *et al.*;[15] (**n**) Raymand *et al.*;[50] (**o**) Pitman *et al.*[23] and Manzano *et al.*;[166] and ( ) van Duin *et al.*[7] adapted with permission from the provided references: (**d, e, h–k, o, p**)

As depicted in Figure 1a, the potential is divided into bond-order-dependent and -independent contributions. Bond order is calculated directly from interatomic distance using the empirical formula:

$$BO_{ij} = BO_{ij}^{\sigma} + BO_{ij}^{\pi} + BO_{ij}^{\pi\pi}$$

$$= \exp\left[p_{bo1}\left(\frac{r_{ij}}{r_o^{\sigma}}\right)^{p_{bo2}}\right] + \exp\left[p_{bo3}\left(\frac{r_{ij}}{r_o^{\pi}}\right)^{p_{bo4}}\right]$$

$$+ \exp\left[p_{bo5}\left(\frac{r_{ij}}{r_o^{\pi\pi}}\right)^{p_{bo6}}\right] \tag{2}$$

where BO is the bond order between atoms i and j, r_{ij} is interatomic distance, r_o terms are equilibrium bond lengths, and p_{bo} terms are empirical parameters. Equation (2) is continuous, containing no discontinuities through transitions between σ, π, and $\pi\pi$ bond character. This yields a differentiable potential energy surface, as required for the calculation of interatomic forces. This bond-order formula accommodates long-distance covalent interactions characteristic in transition state structures, allowing the force-field to accurately predict reaction barriers. This covalent range is typically taken to be 5 Angstrom—which is sufficient for most elements to capture even the weakest of covalent interactions—but can be extended beyond this range; this may occasionally be required for elements with very large covalent radii. This long-distance covalent bond feature, however, necessitates the addition of a bond-order correction to remove spurious bond character between non-bonded neighbours, such as neighbouring H atoms in a methane molecule. Terms in the potential that are dependent on bond order, such as bond energy and angle strains, are calculated directly from the corrected bond order. Finally, a charge equilibration scheme is applied at each iteration to calculate partial atomic charges (see 'Charge description improvements'), which are then used to calculate Coulombic interactions.

Note that the non-bonded and bonded terms in ReaxFF are calculated independently—there is no information transfer between the bond-order-based terms and the van der Waals- and Coulomb-related terms. For all materials and molecules, both bond-order-based terms and nonbonded terms are calculated—without exclusions—enabling ReaxFF to be applied to both predominantly covalent and ionic materials without user input.

Overview of available ReaxFF parameterisations and development branches

Each element of the periodic table for which a ReaxFF parameter set has been published is highlighted in Figure 1b. Although a ReaxFF description exists for these elements, one cannot simply use these parameter sets in any combination and expect to obtain satisfactory transferability. As shown in Figure 2, there are currently two major groupings (i.e., the ReaxFF branches) of parameter sets that are intra-transferable with one another: (1) the combustion branch and (2) the aqueous branch. In order to explain the existence of branches within a common functional form, we must consider that ReaxFF does not employ atom typing strategies, in contrast to popular non-reactive force-fields like AMBER[25] and CHARMM.[26] For example, there is only one oxygen type in ReaxFF regardless of the chemical environment in which the oxygen atom finds itself. This is quite helpful because it allows atoms to migrate seamlessly between phases during a simulation. At the same time, however, this results in a significantly more complex force-field development process. The lack of transferability between branches is evident in the performance of the 2008-C/H/O combustion force-field,[2] which accurately describes water as a gas-phase molecule yet fails to describe water as a liquid. During development at that time, describing liquid water was not a particular aim for ReaxFF. Since all intended applications were at temperatures well above the water boiling point, this was not a major development concern. In 2009, efforts were initiated to redevelop ReaxFF for aqueous chemistry, and it became clear that the 2008-C/H/O combustion force field, which at that time was already extended to a significant range of metal oxide (Me = V/Bi/Mo/Nb/Si) materials and catalysts,[27–31] could not be parameterised to treat liquid water without changing general-ReaxFF and atom-specific parameters. As such, the decision was made to initiate a new branch—the aqueous-branch—that employs the same functional form as the 2008-C/H/O description, but with different O/H atom and bond parameters. This resulted in the creation of a number of parameter sets that are not directly transferable with those on the pre-existing combustion branch, thus leading to the new aqueous branch (Figure 2).

Parameter sets on the same development branch can be directly combined, where the required force-field fitting consists only of parameterising bond and angular terms between the newly combined elements. Transferring parameters between branches, however, requires more extensive refitting. In general, combustion branch descriptions are more straightforwardly transferred to the aqueous branch. In addition to the combustion and aqueous branches, there have been several independent branches (e.g., the B/N/H and Ni/C/H sets shown in Figure 2).[32,33] Typically these sets began as non-transferable ReaxFF descriptions constituting independent development branches, but many have later been merged, through extensive refitting, with the combustion (C/B/N/H/O)[34] or aqueous branches (Ni/C/H/O).[35–37] It is worth mentioning that the popular ReaxFF high-energy material description[4,5,38–44] is older than the combustion branch, but was recently merged—without an obvious loss in accuracy—with this branch.[39,45] Notable developments on the aqueous branch include water–liquid and proton/anion transfer extensions to a range of transition metals and metal oxides (Fe/Ni/Cu/Zn/Al/Ti/Ca/Si),[7,15,19,35,46–52] along with C/H/O/N/S/P developments aimed at biomolecules and their interactions with inorganic interfaces.[53–60]

Comparison to similar methods

Although in this article we focus almost exclusively on ReaxFF and its applications, the ReaxFF method is not unique in its aim: to provide a simulation environment for describing the dynamics of chemical reactions at an atomistic scale with significantly fewer computational resources compared with QM. Purely empirical methods essentially abandon—or simplify—QM concepts,[61] providing significantly more freedom in their choice of functional form. Alternatively, some methods stay within the QM domain while applying substantial empirical approximations. Examples of the QM-based approaches include the MOPAC semiempirical[62] and tight-binding DFT[63,64] concepts, which have been developed with significant success for a wide range of systems. On the purely empirical side, restricting our discussion to the more transferable methods, Baskes and co-workers have developed the embedded atom method (EAM[65,66]), which—opposite to ReaxFF—was mainly formulated for metals, yet has since been modified (MEAM[67]) to treat oxides, hydrides and hydrocarbons. Furthermore, the bond-order concept, as initiated by Abell,[68] Tersoff,[69,70] and Brenner,[71] was further developed into the AIREBO method[72] by Stuart, Tutein and Harrison, as well as into the highly transferable COMB method[73–77] by Sinnott, Philpott, and co-workers. We refer readers to recent reviews for more in-depth comparisons of empirical reactive methods,[73,74] and of simulation methods for large-scale molecular dynamics on reactive systems.[78]

In general, empirical methods tend to be faster and scale better than QM-based approaches,[79] although relatively few head-to-head comparisons have appeared in the literature so far. Within the empirical methods, ReaxFF's origin in hydrocarbon chemistry has led to a focus on reproducing energy barriers, while methods developed by the materials community (e.g., EAM and COMB) tend to focus more on elastic properties. Long-range bond-order terms render ReaxFF suitable for transition states, but complicate bond-order descriptions in condensed systems when non-bonded

neighbours may be in close proximity. This does not necessarily exclude the method from reproducing elastic properties (e.g., refs 80,81), although it makes these properties less straightforward to include in force-field parameterisation.

APPLICATIONS OF REAXFF

Here we highlight a few of the many studies that have employed the ReaxFF methodology, with the objective of demonstrating the diverse applicability of ReaxFF rather than providing a comprehensive review of the literature. In 'Heterogeneous catalysis' and 'Atomic layer deposition' we discuss the application of ReaxFF to heterogeneous catalysis and to atomic layer deposition (ALD), respectively, as these applications demonstrate the methodological strength of ReaxFF: modelling reactive chemistry at heterogeneous interfaces. A broader overview of other applications will be provided in 'Other applications'. Finally, in Section 3.4 we discuss the development of computational methodologies that take advantage of the ReaxFF formalism to explore phenomena on time and length scales that are inaccessible to traditional MD simulation methodologies.

Heterogeneous catalysis

Nickel catalysts. Ni-catalysed carbon nanotube (CNT) growth involves dissolution of C atoms into Ni nanoclusters, with CNT formation dependent on the dynamic reformation and transport of carbon through the cluster. Modelling such processes clearly requires some treatment of bond scission and formation, yet the time and length scales inherent to C diffusion in nm-sized clusters is not tractable with QM. For this reason, the ReaxFF force-field has been extensively applied to model catalytic processes involving CNT growth initiated from dissociative hydrocarbon adsorption on Ni surfaces. Mueller et al. developed[82] and employed[33] a Ni/C/H parameter set to investigate the onset of CNT formation via the dissociative adsorption of various hydrocarbons on Ni surfaces, where they found that surface defects likely play an essential role initiating CNT growth. ReaxFF enabled MD simulations to reach beyond the time and length scales available to ab initio methods, which proved essential in modelling defective Ni surfaces and clusters exposing active sites on irregular surface terminations and cluster edges. The Ni/C/H parameter set was further utilised by Neyts et al.[83] to model the effect of Ar+ bombardment on CNT formation. Simulations showed that the energy of impinging Ar+ ions can be tuned to break weak C–C bonds between under-coordinated C atoms, thus healing CNT defects. In similar works,[84–86] a force-biased Monte Carlo (MC) method, which will be further discussed in 'Uniform-acceptance force-biased Monte Carlo', was employed to demonstrate the growth of CNTs with well-defined chirality and the effect of applied electric fields on plasma-assisted CNT growth.

Vanadia catalysts. In addition to catalysis by metal surfaces, ReaxFF is also an effective methodology for modelling the catalytic properties of oxides. Among the first studies in this area, Chenoweth et al. developed and implemented a V/O/C/H description aimed at capturing the interaction of hydrocarbons with vanadium oxide.[27,28] Mixed-metal-oxide (MMO) catalysts typically feature partial, mixed, and irregular metal occupations at various crystallographic sites. This characteristic is difficult to capture with QM models, as prohibitively large unit cells are required if the system is to be treated with periodic boundary conditions. In ref. 28, a combined MC/reactive dynamics (MC/RD) procedure was employed to explore possible MMO structures, in which the MC/RD scheme was used to systematically interchange metal atoms between crystallographic sites to determine the lowest energy structures. Thermodynamically favoured structures were then exposed to hydrocarbons through MD simulations,

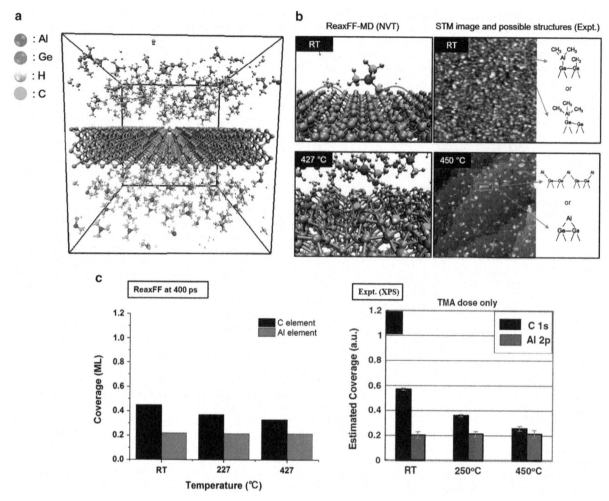

Figure 3. Application of ReaxFF in MD simulations demonstrating TMA nucleation on a bare Ge(100) slab. (**a**) Initial configuration of 200 TMA molecules on a bare Ge(100) surface. (**b**) Comparison of TMA nucleation on a bare Ge(100) surface as modelled by ReaxFF and as observed by STM.[95] (**c**) Estimated coverage of TMA sites obtained by ReaxFF simulations (left panel) and determined experimentally using XPS (right panel).[95] (STM images adapted with permission from ref. 95

demonstrating that optimal catalysts contain several surface channels exposing active metal sites. This study demonstrates the utility of ReaxFF when modelling catalytic processes on oxides, which can be readily combined with metal parameter sets to model cluster–support interactions impacting catalysis on oxide-supported metal catalysts—this extension is an active area of ReaxFF development.

Atomic layer deposition

ReaxFF has recently been applied to study reactive events playing a critical role in the atomic layer deposition process (ALD), which is central to microchip manufacturing. ALD provides high conformity and excellent film thickness control in the ultrathin limit. Although ALD was introduced to the semiconductor industry quite some time ago as a cost-effective, scalable, low-temperature deposition process, a comprehensive and quantitative understanding of its kinetic mechanisms at the atomic scale has not yet been attained. This may be attributed to three main factors: (1) lack of suitable experimental probes to directly image or otherwise characterise the surface state with sufficient temporal and lateral resolution under realistic deposition conditions using spectroscopic or diffraction techniques; (2) challenges in computationally describing ALD relevant processes, such as adsorption, chemisorption, diffusion, steric hindrance, desorption and reaction dynamics; which are all strongly dependent on the 'chemical state' of the

solid surface; and (3) lack of interest in acquiring such detailed knowledge, as it is not entirely necessary in order to utilise ALD for Si-based technology.

With the current drive to develop CMOS technology beyond Si into large-scale, manufacturable processes with high yield enabling high performance, low-power logic by replacing the channel using high mobility, low bandgap semiconductors (Ge or III–V),[87–89] the situation has markedly changed. The integration of a thermodynamically stable high-κ dielectric film that is in contact with the semiconductor and forms an electrically well-behaved interface, allows for aggressive thickness scaling to achieve a large gate capacitance while limiting the gate leakage is a critical roadblock toward realizing this technology.[90–92] Here an atomic scale understanding of the chemical state of the surface, suitable precursors and successful passivation strategies to avoid electrically active traps at the interface is critical. The ReaxFF methodology can help to rationalise and rapidly identify promising approaches, thus shortening the time required to develop this technology.

Recent studies have focused on ALD of Al_2O_3 using trimethyl-aluminum (TMA) and H_2O cycles as a route for developing Ge-based metal oxide semiconductors (CMOS).[89,93,94] To demonstrate the temperature dependence of TMA nucleation on a bare Ge(100) surface, which has been previously reported in the literature,[95] MD simulations were performed at 23 °C, 227 °C, and 427 °C. The analysis of simulation trajectories revealed that TMA molecules non-dissociatively chemisorb on Ge dangling bonds at room temperature, whereas at higher temperature (427 °C) residual methyl groups on TMA molecules either recombine to form methane (and ethane) or bind to neighbouring Ge dangling bonds resulting in Ge-Al-Ge connections on the Ge(100) surface. These results are qualitatively consistent with scanning tunneling microscopy (Figure 3). In addition, the coverage of TMA sites with respect to the number Ge dangling bonds was evaluated as a function of temperature and compared to values obtained with X-ray photoelectron spectroscopy (XPS). Both theoretical and experimental approaches confirm that carbon coverage decreases with elevated temperature due to the dissociation of methyl groups from TMA, while Al coverage remains nearly constant, indicating that TMA sites become saturated under operating conditions.

The recently developed ReaxFF description demonstrates the ability to correctly describe the TMA chemisorption on the Ge surface. This will enable future studies to assess ALD of Al_2O_3, as well as other feasible high-κ candidates and passivation strategies (such as hafnia, zirconia, tantalum, germanium oxynitride[96–100] and bi/tri-layer dielectrics[101–104]) on pristine and oxidised Ge surfaces. As such, ReaxFF can be instrumental in identifying optimal processing conditions to form high quality high-κ dielectrics/non-Si semiconductor interfaces.

Other applications

The utility of ReaxFF lies in its ability to treat reactive processes at the interface between gas, liquid, and solid phases, which is pertinent when dealing with heterogeneous catalysis on metals and oxides (Figures 4a and b, respectively). This capability is not only relevant to catalysis, but rather is essential for modelling any nanoscale phenomena in complex systems. Onofrio et al.[105] employed a modified ReaxFF potential to conduct atomistic-scale simulations of electrode-bridge-electrode systems (Figure 4c). By modifying ReaxFF to enable the direct simulation of an applied potential, the authors were able to model the formation and degradation of metallic filaments bridging electrodes in resistance-switching cells. Tavazza et al.[106] conducted MD simulations of nanoindentation processes on Ni substrates, where ReaxFF was used to more accurately describe attractive interactions between diamond indenter tips and Ni/NiO surfaces. The utility of using ReaxFF for modelling reactive events at the interface between gas and solid phases is demonstrated in the work of Bagri et al.,[107] who investigated defect formation during the reduction of graphene oxide (GO). MD simulations revealed reorganisation patterns in the graphene sheet during reduction (Figure 4d), allowing the authors to identify more efficient pathways for minimising defect formation during GO reduction. This is an important issue, as GO reduction is a promising synthesis route for the rapid production of graphene sheets. Similarly, Srinivasan et al.[108,109] studied hyperthermal oxygen collisions with graphene, which replicates degradation processes affecting the performance of graphene-based heat shields during spacecraft re-entry. Oxidation of metal surfaces was investigated by Fantauzzi et al.[110] and by Senftle et al.,[111] studies in which MD simulations were employed to assess kinetic limitations affecting the initial oxidation of Pt and Pd surfaces, respectively (Figure 4e). ReaxFF has also been employed to model the kinetics of hydrogen adsorption and diffusion in various Pt,[112] Pd,[113] and Fe[114] phases. The effect of solid-gas interactions is further demonstrated in the work of Raju et al.,[52,115] in which the authors investigated oriented attachment mechanisms governing the organisation of TiO_2 nanocrystals (Figure 4f). MD simulations revealed that, in vacuum, TiO_2 crystals tend to form polycrystalline aggregates. In solution, approaching TiO_2 crystals effectively dissociate water, thus creating a network of hydrogen bonds facilitating oriented attachment.

In addition to gas phase chemistry, the development of the 'aqueous branch' parameter set has led to numerous ReaxFF studies involving reactive processes occurring in the liquid phase. Rahaman et al.[54,116] developed a ReaxFF description capable of modelling glycine tautomerisation in water, which was later employed to investigate the interaction between glycine and TiO_2 surfaces.[54] Monti et al.[53] further extended this parameter set to include amino acids and short peptide structures, allowing ReaxFF to simulate the conformational dynamics of biomolecules in solution (Figure 4g). Aqueous proton transfer across graphene was investigated by Achtyl et al.,[51] where ReaxFF helped to establish that proton transfer is enabled by hydroxyl-terminated atomic defects in the graphene sheet (Figure 4h). Hatzell et al.[117] used ReaxFF to determine the impact of strong acid functional groups on graphene electrodes in capacitive mixing devices, in which salinity gradients are used for power generation (Figure 4i). For numerous examples of water–solid interfaces that have been modelled with the ReaxFF potential, we refer the reader to references on the 'aqueous branch' shown in Figure 2.

RECENTLY DEVELOPED METHODS

MD approaches with reactive force-fields are tremendously useful for examining dynamic and kinetic properties of chemical processes on timescales beyond what is accessible with QM. However, numerous chemical processes, and in particular processes requiring thermal activation to overcome significant kinetic barriers, occur on timescales that are still beyond the simulation capabilities of empirical methods. Such limitations have provided the impetus for developing MC and accelerated molecular dynamics (AMD) approaches for the ReaxFF potential. MC methods are most useful when the properties of interest in a system are dominated by thermodynamic driving forces (i.e., when kinetic factors do not limit the deterministic evolution of the system). Conversely, AMD methods are suitable for modelling kinetic processes, as arguments from transition state theory are employed to accelerate system evolution between rare events while maintain system determinism. This section highlights four methods (two MC-based and two AMD-based) that have been recently introduced to ReaxFF: (1) hybrid Grand Canonical MC/Molecular dynamics (GC-MC/MD), (2) uniform-acceptance force-biased MC (UFMC), (3) parallel replica dynamics (PRD), and (4) adaptive accelerated ReaxFF reaction dynamics (aARRDyn).

Grand canonical Monte Carlo/molecular dynamics (GC-MC/MD)

MC methods in the grand canonical ensemble stochastically insert, remove, and displace MC-atoms in the host system. The equilibrated number of MC-atoms, as well as system structure, is dictated by the chemical potential of the MC-atom reservoir (i.e., the free energy per MC-atom in the original source, such as $\frac{1}{2}G_{O2}$ for modelling oxide formation upon exposure to O_2). When applying GC-MC with the ReaxFF potential, we have introduced a force relaxation step (via either low-temperature MD or similar energy minimisation algorithms) that allows the structure of the host system to locally reorganise as MC-steps are executed (i.e., to find the nearest local energy minimum). The introduction of this MD step, motivated by the MC/MD methodology employed by Chenoweth et al.,[28] is essential for reaching an equilibrated structure. Otherwise the majority of MC moves involving the addition of an atom into a bulk region will be rejected because such moves inevitably result in a very high energy. Slow acceptance rates are effectively mitigated by allowing the system to restructure as it accommodates the inserted MC-atoms, mimicking the kinetic restructuring of the system as the new phase is formed.

The GC-MC/MD approach was initially developed to model the extent of oxidation in Pd nanoclusters, as PdO formation

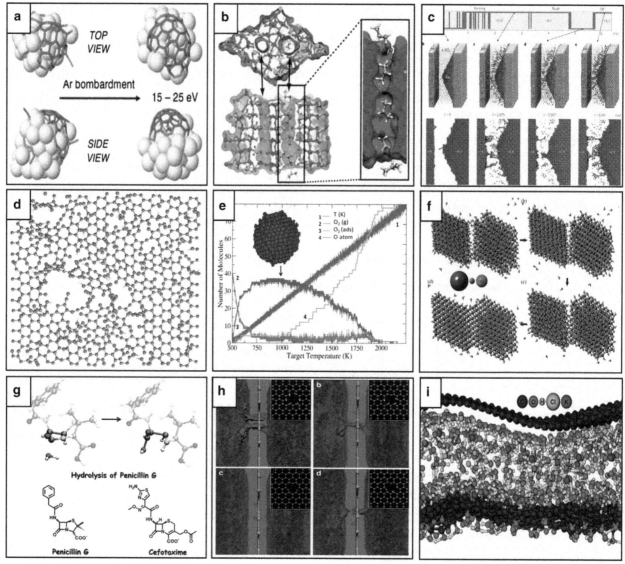

Figure 4. Application of the ReaxFF method to (**a**) Ni-catalysed CNT growth,[83] (**b**) oxidative dehydrogenation over MMO catalysts,[28] (**c**) electrometallisation cells,[105] (**d**) reduction of graphene oxide,[107] (**e**) Pd surface oxidation,[111] (**f**) oriented attachment mechanisms in TiO_2 nanocrystals,[115] (**g**) conformational dynamics of biomolecules,[53] (**h**) proton diffusion membranes[51] and (**i**) capacitive mixing by double layer expansion.[117] (Images adapted with permission from the provided references: **a**, Copyright 2013 American Physical Society; **b**, Copyright 2009 John Wiley and Sons; **c**, **d** and **h**, Copyright 2010 and 2015 Nature Publishing Group; **e**, Copyright 2013 AIP Publishing LLC; **g**, Copyright 2013 Royal Society of Chemistry; **f** and **i**

significantly affects the performance of Pd-based catalysts.[111] To validate the GC-MC/MD approach, we derived an oxidation phase diagram for Pd clusters, where the predicted T, P ranges of PdO to Pd transitions were in agreement with experimental observations (Figure 5a). Coupled with MD simulations capturing the extent of oxygen migration through various Pd surface facets, this study demonstrated the utility of ReaxFF for investigating both kinetic and thermodynamic driving forces affecting phase stability. A similar approach has been applied to investigate the initial stages of oxide formation on Pt surfaces.[110] The method has additionally been applied to investigate hydrogen and carbon uptake by palladium, where it successfully captured phase transitions in bulk, cluster and oxide-supported systems.[113,118–120] It has also been used to produce open-circuit voltage profiles for Li intercalation in graphite[121] and sulfur[122] electrodes during cell discharge.

Applied to catalysis, GC-MC/MD provides an approach for predicting the thermodynamic stability of phases likely to form upon exposure to the reactant mixture, which can be used to establish plausible surface models that better approximate the structure of the catalyst under reaction conditions. To investigate kinetics, the derived surface models can then be directly employed in MD simulations of the catalytic process, or can serve to motivate smaller scale models for more accurate DFT calculations. More importantly, it provides a means to explore the phase space of a system without *a priori* knowledge of the phase diagram in question—making it a powerful low-CPU-cost tool when searching for novel system properties.

Uniform-acceptance force-biased Monte Carlo (UFMC)

Pure MC methods are stochastic in nature, and correspondingly cannot be used to evaluate the deterministic evolution of a system in time. However, stochastic approaches can be incorporated in deterministic MD to accelerate reaction steps past kinetic barriers, decreasing the number of iterations required to model

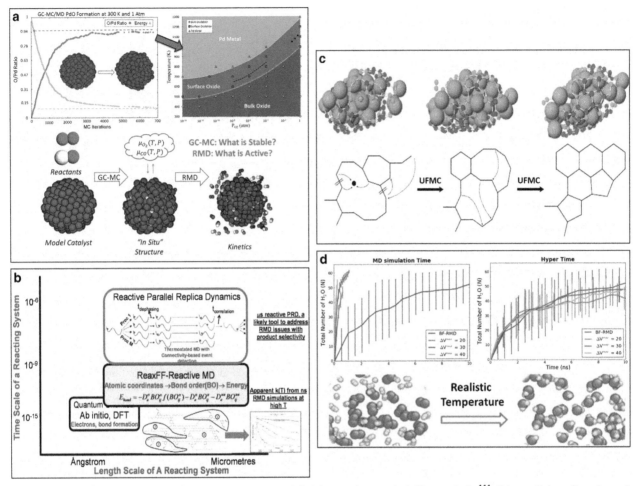

Figure 5. New simulation techniques for the ReaxFF potential: (**a**) grand canonical Monte Carlo,[111] (**b**) parallel replica dynamics,[126] (**c**) uniform-acceptance force-biased Monte Carlo,[85] and (**d**) adaptive accelerated molecular dynamics.[127] (Images adapted with permission from the provided references:

state-to-state transitions in slower processes. Although information regarding the timescale of such processes may be lost, the deterministic nature of the simulation is preserved. This is valuable when one wishes to understand how the system evolves, rather than how fast the system evolves. In UFMC simulations,[123] stochastic spatial displacements (accepted at every iteration) are applied along the force acting on each atom, where the extent and direction of the displacement is controlled by both the magnitude of the force and the overall system temperature. The displacement is restricted along the acting force when large forces are present or the system is at low temperature, whereas displacements are fully random when small forces are acting or the system is at high temperature. Uniform-acceptance prevents the system from becoming restricted to its current potential energy basin, while force-bias ensures a preferential evolution along the lowest-force reaction coordinate.

This strategy, called uniform-acceptance force-biased MC (UFMC), was adapted for ReaxFF by Neyts et al.[84–86,124] to investigate the initial stages of CNT growth over Ni clusters. During simulations of CNT nucleation, UFMC simulation stages were incorporated between MD runs to allow deposited carbon atoms to relax into chiral CNT structures. Such relaxation steps with pure MD would necessitate unphysical carbon deposition rates, as well as intractable simulation timeframes. As described above, this methodology was also successfully employed to assess the effect of applied potentials and Ar$^+$ ion bombardment on CNT nucleation and growth (Figure 5c).

Parallel replica dynamics and adaptive accelerated ReaxFF reaction dynamics

PRD and aARRDyn are two MD strategies designed to increase the sampling rate of infrequent events. PRD employs a temporal parallelisation strategy, in which independent trajectories of multiple system replicas are evaluated simultaneously across many processors. When a state transition occurs in one replica, all remaining replicas are reverted to the new state and are reinitialised with independent starting trajectories. The total elapsed simulation time is summed from contributions of each replica, where such parallelisation is indistinguishable from an analogous single-processor simulation if the transition process follows first-order escape kinetics, as demonstrated in the original PRD publication.[125] Joshi et al.[126] implemented PRD with the ReaxFF potential, where a state-transition event was defined as a change in molecular connectivity (i.e., a chemical reaction). When implemented with 180 replicas to track the thermal pyrolysis of n-heptane, PRD-ReaxFF was able to reach simulation times on the order of 1 µs with a parallel scaling efficiency of 93% (Figure 5b)

aARRDyn, introduced to ReaxFF by Cheng et al.,[127] employs a 'bond boost' algorithm to accelerate dynamic evolution across the potential energy surface. In this implementation, the bond-order formalism of ReaxFF is aptly suited for identifying elongated bonds that are approaching a transition state, where the boost potential is calculated using an envelope function[128] that directs energy toward bonds that are closest to reacting. The authors demonstrated the aARRDyn methodology by modelling H$_2$/O$_2$

ignition in simulations reaching ~10 ns, which successfully reproduced previously determined reaction pathways (Figure 5d). Altogether, PRD and aARRDyn demonstrate the extension of ReaxFF to low-temperature applications, where kinetic processes occur on timescales inaccessible to traditional MD trajectory integration.

FUTURE DEVELOPMENTS AND OUTLOOK

Charge description improvements

Incorporation of ACKS2. A polarisable charge calculation method is essential for a transferable reactive force-field method. ReaxFF typically employs the electronegativity equalisation method (EEM) developed by Mortier et al.[129] Unfortunately, EEM has a few drawbacks,[130–134] the worst being its inability to restrain long-range charge-transfer, even between molecular fragments that are well separated. This issue is especially apparent in low-density gas phase simulations, where EEM-ReaxFF predicts small, but certainly nonzero, charges on isolated molecular species, which significantly affects accommodation coefficients.[135] In simulations of dense systems, unrealistic charge-transfer may also occur (e.g., between two dielectric phases with a different intrinsic electronegativity). To ameliorate this issue, we have recently incorporated atom-condensed Kohn-Sham DFT approximated to second order (ACKS2) in ReaxFF.[136,137] ACKS2 is an extension of EEM that penalises long-range charge transfer with a bond-polarisation energy, in line with the split-charge equilibration (SQE) model.[138] The bond polarisability is a function of interatomic distance, which slightly increases beyond the equilibrium bond length but then quickly decays to zero, effectively enforcing fragment neutrality. Transferring from EEM-ReaxFF to ACKS2-ReaxFF does require reparameterisation, although the EEM-ReaxFF parameters are typically a very good starting point for deriving ACKS2-ReaxFF parameters. As such, redevelopment is relatively straightforward. We believe that ACKS2-ReaxFF is most relevant for obtaining reliable accommodation coefficients, as well as for describing physisorption of neutral and ionic molecules on surfaces. It is also vital for incorporating explicit electronic degrees of freedom into ReaxFF, as described in the next section.

Explicit electron description in ReaxFF (eReaxFF). The treatment of explicit charge and polarisation is essential for force-field descriptions applied to systems such as rechargeable battery interfaces and ferro-/piezoelectric materials. Essentially, such descriptions require a classical treatment for an explicit electron. Recently, potentials including some form of explicit electron description have been introduced, such as the electron force-field[139] and the LEWIS[140] force-field. Nonetheless, these methods have not yet been demonstrated to accurately simulate complex materials and intricate chemistries.

To extend the ReaxFF description to include chemistry dependent on electron diffusion, we have introduced explicit electron-like and hole-like particles that carry negative (−1) and positive (+1) charges, respectively. The electron and hole particles interact with atomic centers through a single Gaussian function.[139] We implemented charge-valence coupling, which allows the electron or hole particle to modify the number of valence electrons in a host atom, thus ensuring the appropriate change in valence when calculating the degree of over or under coordination in the host atom. To demonstrate the capability of the electron-explicit version of ReaxFF (eReaxFF), we trained our force-field to capture the electron affinity (EA) of various hydrocarbon species. Figure 6 summarises the performance of eReaxFF compared with both standard ReaxFF and available literature data.[141,142] eReaxFF qualitatively reproduces the literature data, whereas standard ReaxFF entirely fails to capture the EA of most species considered in this training set. Still, eReaxFF significantly

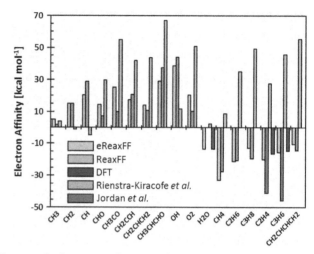

Figure 6. Performance of eReaxFF for calculating electron affinities in radical species compared to standard ReaxFF, QM, and experimental data.[141,142]

underestimates the EA of ethylene and propene compared with experimental data.[142] To further investigate this discrepancy, we performed DFT calculations with the M06-2X functional (aug-cc-pVTZ basis set) implemented in Jaguar 7.5,[143] which shows that DFT also underestimates the EA of these two species. Overall, eReaxFF provides a significant improvement in capturing EAs in comparison with standard ReaxFF. Future development of eReaxFF will focus on the description of interfacial chemistry in batteries and on structural/polarisation behaviour in piezo/ferro-electric materials.

ReaxFF parameter optimisation

Most ReaxFF parameter development is currently performed using a straightforward, single-parameter parabolic search method.[144] This method has a high level of transparency—which, given the complexity of the ReaxFF functional form and the typically large size of its training set, is a significant advantage. However, this method is clearly not the most efficient for sampling the complex force-field error landscape, as it does not implicitly include parameter correlation and has a significant risk of becoming ensnared in local energy minima. As such, various groups have recently focused on developing more sophisticated parameter sampling methods. In particular, MC and genetic algorithm (GA) based approaches have risen in popularity.[145–149] So far, these methods have only been used for isolated force-field development efforts, and as such they still have to demonstrate their transferability and user friendliness. They do hold promise for reducing force-field development time, as well as for making the process more accessible to nonexpert users.

ReaxFF implementations and current efforts for modern architectures

A number of different ReaxFF implementations have been developed over the years. The first-generation ReaxFF implementation of van Duin et al.[1] established the utility of the force field in the context of various applications. This serial, fortran-77, implementation was integrated into the publicly available, open-source LAMMPS code[150] by Thompson et al.[151] as the Reax package[152]—the first publicly available parallel implementation of ReaxFF. Nomura et al. have developed a parallel ReaxFF implementation, which has been used in a number of large-scale simulations, including high-energy materials, metal grain boundary decohesion, water bubbles and surface chemistry.[153–159] This Nomura et al. code is not publicly available.

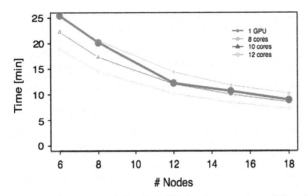

Figure 7. Performance of the PuReMD implementation of ReaxFF. Time required to run 1,000 time steps of a MD simulation on half a million atoms where either GPU or CPU is used for computing the interatomic forces.

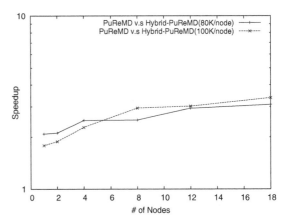

Figure 8. Speedup achieved by Hybrid-PuReMD compared to PuReMD with 20 MPI processors per node.

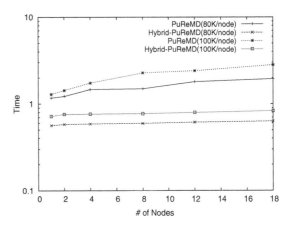

Figure 9. Total time per time-step comparison between Hybrid-PuReMD and PuReMD for weak scaling scenarios.

The PuReMD software package, and its integration into LAMMPS, which is available as the USER-REAXC package, represent the state-of-the-art open-source implementations of ReaxFF. Since PuReMD and USER-REAXC are, to the best of our knowledge, the most widely used open-source codes for ReaxFF simulations, we summarise algorithmic and software design choices, as well as future directions in PuReMD. PuReMD uses novel algorithms and data structures to achieve high performance in force computations, with a small memory footprint. An optimised binning-based neighbour generation method, elimination of the bond-order derivatives list in bonded interactions, lookup tables to accelerate non-bonded interaction computations, and preconditioned fast iterative solvers for the charge equilibration problem are the major algorithmic innovations in PuReMD.[10,12,15] The dynamic nature of the bond, three-body and four-body interactions in a reactive molecular system presents challenges in terms of memory management and data structures for efficiently computing bonded interactions. PuReMD introduces novel data structures to store three-body and four-body interactions in a compact form. Its dynamic memory management system automatically adapts to the needs of the input system over the course of a simulation, significantly reducing the memory footprint and minimizing the effort to setup a simulation. Its parallel formulation, called PuReMD, uses conventional message passing (MPI) to enable the simulation of large molecular systems[12] on scalable parallel platforms.

Recent efforts by us and others have focused on enabling fast ReaxFF simulations on GPUs. Zheng et al.[160] reported the first GPU implementation of ReaxFF, called GMD-Reax. GMD-Reax is reported to be up to six times faster than the USER-REAXC package on small systems, and about 2.5 times faster on typical workloads using a quad core Intel Xeon CPU. However, this performance is significantly predicated on the use of single-precision arithmetic operations (PuReMD codebase is fully double precision) and low-precision transcendental functions, which can potentially lead to significant energy drifts in long NVE simulations. GMD-Reax can run on single GPUs only, and is not publicly available.

Recently, Kylasa et al. have developed a publically available GPU version of the PuReMD code, called PuReMD-GPU[10], which uses CUDA along with an extensive set of optimisations to deliver significant speedups on single GPU systems. For a variety of benchmarks, this code shows up to a 16-fold speedup on an Nvidia C2075 GPU, compared with a single core of an Intel Xeon processor.

Due to device memory limitations, single GPU simulations are typically limited to 30 to 40 thousand atoms. Our more recent development efforts have resulted in the PuReMD-PGPU code, which builds on PuReMD-GPU to enable ReaxFF simulations on multiple GPUs. PuReMD-PGPU has been benchmarked on Texas Advanced Computing Center's Maverick cluster, which has 132 nodes each with two Intel E5–2680 v2 Ivy Bridge processors running at 2.86 GHz and one NVIDIA Tesla K40 GPU. We ran NVT MD calculations of DNA strands suspended in water, having a total of 531,494 atoms. In Figure 7, we present the performance of this simulation on 6, 8, 12, 15 and 18 nodes, where we use either 1 GPU per node to perform the calculations using PuReMD-PGPU or the MPI-based CPU code (PuReMD) with a varying number of cores per node. In terms of speed, we find that using one GPU per node with PuReMD-PGPU is equivalent to using 8 to 10 cores when using PuReMD.

More recently, a Hybrid-PuReMD, an MPI/pthreads/CUDA implementation by Kylasa et al., leverages all the available compute resources on GPU-equipped nodes. Hybrid-PuReMD uses an effective task-parallel implementation by overlapping the execution of bonded interactions on CPU and non-bonded and charge equilibration computations on the GPU. This code has been benchmarked on Michigan State University's High Performance Computing Center GPU cluster, which has 20 nodes each with two 2.5 GHz 10-core Intel Xeon E5–2670v2 processors and two NVIDIA K20 GPUs. Figures 8 and 9 plot the speedup achieved by Hybrid-PuReMD compared with PuReMD (20 MPI processes per node) and total time per time-step for weak scaling scenarios. We notice that for bulk water system with 100 K atoms per node Hybrid-PuReMD effectively delivers a speedup of over 3.3-fold over PuReMD on a per-node basis (i.e., 3.3 times the performance of PuReMD running on all 20 cores of the node)—achieving near optimal resource utilisation.

Other ongoing developments with PuReMD includes performance optimisation for many-core architectures, such as IBM BGQ and Intel Xeon Phi processors. Using a hybrid MPI/OpenMP parallel implementation, Aktulga et al. have achieved speedups of 1.5–4.5× over the current USER-REAXC implementation for PETN crystal benchmarks of sizes ranging from 32 thousand to 16.6 million atoms[161]. This study has been conducted using up to 16,384 nodes (262 K cores) on Mira, an IBM BGQ supercomputer.

While promising, performance results obtained on GPUs and many-core CPUs suggest that there is still room for performance improvements. Our future work will focus on achieving improved performance on GPUs, other accelerators, and many-core CPUs by better managing on-node parallelism, exploiting data locality, and the development of scalable solvers for the charge equilibration problem, which represents a major bottleneck in massively parallel runs.

SUMMARY

In this article, we reviewed the development, application, and future directions of the ReaxFF method. ReaxFF helps to bridge the gap in simulation scale separating QM and classical methods. By employing a bond-order formalism within a classical approach, ReaxFF implicitly describes chemical bonding without expensive QM calculations. We provided an overview of ReaxFF development history, lending insight into the development choices that shape the currently employed ReaxFF formalism. In addition, we discussed numerous applications of the method, which include processes ranging from the combustion of coal to the conformational dynamics of biomolecules. A particular emphasis was placed on studies that use ReaxFF to investigate reactive events at the interphase between solid, liquid and gas phases, thus demonstrating the diverse transferability of the ReaxFF formalism. In addition, we reviewed the recent development and application of MC and accelerated molecular dynamics methods tailored to the bond-length formalism of ReaxFF. Finally, we provided an overview of future research directions we will be taking, as we seek to improve the performance of ReaxFF, as well as to extend the range of chemical phenomena that can be studied with the method.

ACKNOWLEDGEMENTS

TPS, MJJ, and ACTvD acknowledge funding from the National Science Foundation, grant CBET-1032979. ACTvD acknowledges support from the Fluid Interfaces Reactions, Structures and Transport (FIRST) center funded by the US Department of Energy, Office of Energy, Office of Basic Energy Sciences. ACTvD and MMI acknowledge support from a grant from the US Army Research Laboratory through the Collaborative Research Alliance (CRA) for Multi Scale Multidisciplinary Modeling of Electronic Materials (MSME). TV is a post-doctoral fellow of the Fund for Scientific Research-Flanders (FWO) and received additional funding from the Research Board of the Ghent University (BOF) and BELSPO in the frame of IAP/7/05.

COMPETING INTERESTS

The authors declare no conflict of interest.

REFERENCES

1. van Duin, A. C. T., Dasgupta, S., Lorant, F. & Goddard, III W. A. ReaxFF: a reactive force field for hydrocarbons. J. Phys. Chem. A **105**, 9396–9409 (2001).

2. Chenoweth, K., van Duin, A. C. T. & Goddard, W. A. ReaxFF reactive force field for molecular dynamics simulations of hydrocarbon oxidation. J. Phys. Chem. A **112**, 1040–1053 (2008).

3. van Duin, A. C. T. et al. ReaxFFSiO reactive force field for silicon and silicon oxide systems. J. Phys. Chem. A **107**, 3803–3811 (2003).

4. Strachan, A., van Duin, A. C. T., Chakraborty, D., Dasgupta, S. & Goddard, W. A. Shock waves in high-energy materials: The initial chemical events in nitramine RDX. Phys. Rev. Lett. **91**, 098301 (2003).

5. Strachan, A., Kober, E. M., van Duin, A. C. T., Oxgaard, J. & Goddard, W. A. Thermal decomposition of RDX from reactive molecular dynamics. J. Chem. Phys. **122**, 054502 (2005).

6. Cheung, S., Deng, W. Q., van Duin, A. C. T. & Goddard, W. A. ReaxFF(MgH) reactive force field for magnesium hydride systems. J. Phys. Chem. A **109**, 851–859 (2005).

7. van Duin, A. C. T. et al. Development and validation of a ReaxFF reactive force field for Cu cation/water interactions and copper metal/metal oxide/metal hydroxide condensed phases. J. Phys. Chem. A **114**, 9507–9514 (2010).

8. Liu, L., Liu, Y., Zybin, S. V., Sun, H. & Goddard, W. A. ReaxFF-lg: correction of the ReaxFF reactive force field for London dispersion, with applications to the equations of state for energetic materials. J. Phys. Chem. A **115**, 11016–11022 (2011).

9. Plimpton, S. J. & Thompson, A. P. Computational aspects of many-body potentials. MRS Bull. **37**, 513–521 (2012).

10. Kylasa, S. B., Aktulga, H. M. & Grama, A. Y. PuReMD-GPU: A reactive molecular dynamics simulation package for GPUs. J. Comput. Phys. **272**, 343–359 (2014).

11. Aktulga, H. M., Fogarty, J. C., Pandit, S. A. & Grama, A. Y. Parallel reactive molecular dynamics: Numerical methods and algorithmic techniques. Parallel Comput. **38**, 245–259 (2012).

12. Aktulga, H. M., Pandit, S. A., van Duin, A. C. T. & Grama, A. Y. Reactive molecular dynamics: numerical methods and algorithmic techniques. SIAM J. Sci. Comput. **34**, C1–C23 (2012).

13. Kamat, A. M., van Duin, A. C. T. & Yakovlev, A. Molecular dynamics simulations of laser-induced incandescence of soot using an extended reaxff reactive force field. J. Phys. Chem. A **114**, 12561–12572 (2010).

14. Zhang, Q. et al. Adhesion and nonwetting-wetting transition in the Al/α-Al2O3 interface. Phys. Rev. B **69**, 045423 (2004).

15. Fogarty, J. C., Aktulga, H. M., Grama, A. Y., van Duin, A. C. T. & Pandit, S. A. A reactive molecular dynamics simulation of the silica-water interface. J. Chem. Phys. **132**, 174704 (2010).

16. Neyts, E. C., Khalilov, U., Portois, G. & van Duin, A. C. T. Hyperthermal oxygen interacting with silicon surfaces: adsorption, implantation and damage creation. J. Phys. Chem. C **115**, 4818–4823 (2011).

17. Castro-Marcano, F. & van Duin, A. Comparison of thermal and catalytic cracking of hydrocarbon fuel from ReaxFF reactive molecular dynamics simulations. Combust. Flame **160**, 766–775 (2013).

18. Sen, F. G., Alpas, A. T., van Duin, A. C. T. & Qi, Y. Oxidation assisted ductility in aluminum nanowires. Nat. Commun. **5**, 3959 (2014).

19. Russo, M., Li, R., Mench, M. & van Duin, A. C. T. Molecular dynamic simulation of aluminum-water reactions using the ReaxFF reactive force field. Int. J. Hydrogen Energ. **36**, 5828–5835 (2011).

20. Abdolhosseini Qomi, M. J. et al. Combinatorial molecular optimization of cement hydrates. Nat. Commun. **5**, 4960 (2014).

21. Joshi, K., Psofogiannakis, G., Raman, S. & van Duin, A. C. T. Reactive molecular simulations of protonation of water clusters and depletion of acidity in H-ZSM-5 zeolite. Phys. Chem. Chem. Phys. **16**, 18433–18441 (2014).

22. Joshi, K. & van Duin, A. C. T. Molecular dynamics study on the influence of additives on the high temperature structural and acidic properties of ZSM-5 zeolite. Energ. Fuel. **27**, 4481–4488 (2013).

23. Pitman, M. C. & van Duin, A. C. T. Dynamics of confined reactive water in smectite clay-zeolite composites. J. Am. Chem. Soc. **134**, 3042–3053 (2012).

24. Russo, M. F. Jr & van Duin, A. C. T. Atomistic-scale simulations of chemical reactions: Bridging from quantum chemistry to engineering. Nucl. Instrum. Methods Phys. Res. B **269**, 1549–1554 (2011).

25. Case, D. A. et al. The Amber biomolecular simulation programs. J. Comput. Chem. **26**, 1668–1688 (2005).

26. Brooks, B. R. et al. CHARMM: The biomolecular simulation program. J. Comput. Chem. **30**, 1545–1614 (2009).

27. Chenoweth, K., van Duin, A. C. T. & Goddard, W. A. ReaxFF reactive force field for molecular dynamics simulations of hydrocarbon oxidation. J. Phys. Chem. A **112**, 1040–1053 (2008).

28. Chenoweth, K., van Duin, A. C. T. & Goddard, W. A. The ReaxFF Monte Carlo reactive dynamics method for predicting atomistic structures of disordered ceramics: application to the Mo3VOx Catalyst. Angew. Chem. Int. Ed. **48**, 7630–7634 (2009).

29. Goddard, W. A., Chenoweth, K., Pudar, S., van Duin, A. C. T. & Cheng, M. J. Structures, mechanisms, and kinetics of selective ammoxidation and oxidation of propane over multi-metal oxide catalysts. Top. Catal. **50**, 2–18 (2008).

30. Buehler, M. J., van Duin, A. C. T. & Goddard, W. A. Multiparadigm modeling of dynamical crack propagation in silicon using a reactive force field. Phys. Rev. Lett., **96**, 095505 (2006).

31. Buehler, M. J., Tang, H., van Duin, A. C. T. & Goddard, W. A. Threshold crack speed controls dynamical fracture of silicon single crystals. Phys. Rev. Lett. **99**, 165502 (2007).

32. Weismiller, M. R., Duin, A. C. T. v., Lee, J. & Yetter, R. A. ReaxFF reactive force field development and applications for molecular dynamics simulations of ammonia borane dehydrogenation and combustion. *J. Phys. Chem. A* **114**, 5485–5492 (2010).

33. Mueller, J. E., van Duin, A. C. T. & Goddard, W. A. Application of the ReaxFF reactive force field to reactive dynamics of hydrocarbon chemisorption and decomposition. *J. Phys. Chem. C* **114**, 5675–5685 (2010).

34. Paupitz, R., Junkermeier, C. E., van Duin, A. C. T. & Branicio, P. Fullerenes generated from porous structures. *Phys. Chem. Chem. Phys.* **16**, 25515–25522 (2014).

35. Verners, O. & van Duin, A. C. T. Comparative molecular dynamics study of fcc-Ni nanoplate stress corrosion in water. *Surf. Sci.* **633**, 94–101 (2015).

36. Zou, C., Shin, Y. K., Liu, Z. K., Fang, H. Z. & van Duin, A. C. T. Molecular dynamics simulations of the vacancy effect on the nickel self diffusion, oxygen diffusion and oxidation initiation in nickel using the ReaxFF reactive force field. *Acta Mater.* **83**, 102–112 (2015).

37. Fang, H. Z. *et al.* First-Principles Studies on Vacancy-modified Interstitial Diffusion Mechanism of Oxygen in Nickel, Associated with Large-Scale Atomic Simulation Techniques. *J. Appl. Phys.* **115**, 043501 (2014).

38. Han, S.-p., Strachan, A., van Duin, A. C. T. & Goddard, W. A. Thermal decomposition of condense-phase nitromethane from molecular dynamics using reaxff reactive dynamics. *J. Phys. Chem. B* **115**, 6534–6540 (2011).

39. Cherukara, M. J., Wood, M. A., Kober, E. M. & Strachan, A. Ultra-fast chemistry under non-equilibrium conditions and the shock to deflagration transition at the nanoscale. *J. Phys. Chem. C* **119**, 22008–22015 (2015).

40. Guo, F., Zhang, H. & Cheng, X. Molecular dynamic simulations of solid nitromethane under high pressures. *J. Theor. Comput. Chem.* **9**, 315–325 (2010).

41. Zhang, L. Z. *et al.* Carbon cluster formation during thermal decomposition of octahydro-1,3,5,7-tetranitro-1,3,5,7-tetrazocine and 1,3,5-triamino-2,4,6-trinitrobenzene high explosives from ReaxFF reactive molecular dynamics simulations. *J. Phys. Chem. A* **113**, 10619–10640 (2009).

42. Zhang, L. Z., van Duin, A. C. T., Zybin, S. V. & Goddard, W. A. Thermal decomposition of hydrazines from reactive dynamics using the ReaxFF reactive force field. *J. Phys. Chem. B* **113**, 10770–10778 (2009).

43. Rom, N. *et al.* Change of mechanism in the decomposition of hot dense liquid nitromethane as a function of density: MD simulations based on REAXFF. *J. Phys. Chem. A* **115**, 10181–10202 (2011).

44. Chen, N., Lusk, M. T., van Duin, A. C. T. & Goddard, W. A. Mechanical properties of connected carbon nanorings via molecular dynamics simulation. *Phys. Rev. B* **72**, 085416 (2005).

45. Wood, M. A., van Duin, A. C. T. & Strachan, A. Coupled Thermal and Electromagnetic Induced Decomposition in the Molecular Explosive αHMX; A Reactive Molecular Dynamics Study. *J. Phys. Chem. A* **118**, 885–895 (2014).

46. Kim, S.-Y. *et al.* Development of a ReaxFF reactive force field for Titanium dioxide/water systems. *Langmuir* **29**, 7838–7846 (2013).

47. Pitman, M. C. & van Duin, A. C. T. Dynamics of confined reactive water in smectic clay-zeolite composites. *J. Am. Chem. Soc.* **134**, 3042–3053 (2012).

48. Gale, J. D., Ratieri, P. & van Duin, A. C. T. A reactive force field for aqueous-calcium carbonate systems. *Phys. Chem. Chem. Phys.* **13**, 16666–16679 (2011).

49. Aryanpour, M., van Duin, A. C. T. & Kubicki, J. D. Development of a reactive force field for iron-oxyhydroxide systems. *J. Phys. Chem. A* **114**, 6298–6307 (2010).

50. Raymand, D., van Duin, A. C. T., Spångberg, D., Goddard, W. A. III & Hermansson, K. Water adsorption on stepped ZnO surfaces from MD simulation. *Surf. Sci.* **604**, 741–752 (2010).

51. Achtyl, J. L. *et al.* Aqueous proton transfer across single-layer graphene. *Nat. Commun.* **6**, 6539 (2015).

52. Raju, M., Kim, S.-Y., van Duin, A. C. T. & Fichthorn, K. A. ReaxFF reactive force field study of the dissociation of water on titania surfaces. *J. Phys. Chem. C* **117**, 10558–10572 (2013).

53. Monti, S. *et al.* Exploring the conformational and reactive dynamics of biomolecules in solution using an extended version of the glycine reactive force field. *Phys. Chem. Chem. Phys.* **15**, 15062–15077 (2013).

54. Monti, S., van Duin, A. C. T., Kim, S.-Y. & Barone, V. Exploration of the conformational and reactive dynamics of glycine and diglycine on TiO2: computational investigations in the gas phase and in solution. *J. Phys. Chem. C* **116**, 5141–5150 (2012).

55. Yue, D.-C. *et al.* Tribochemistry of phosphoric acid sheared between quartz surfaces: a reactive molecular dynamics study. *J. Phys. Chem. C* **117**, 25604–25614 (2013).

56. Rahaman, O. *et al.* Development of a ReaxFF reactive force field for aqueous chloride and copper chloride. *J. Phys. Chem. A* **114**, 3556–3568 (2010).

57. Zhu, R. *et al.* Characterization of the active site of yeast RNA polymerase II by DFT and ReaxFF calculations. *Theor. Chem. Acc.* **120**, 479–489 (2008).

58. Yusupov, M. *et al.* Atomic scale simulations of plasma species interacting with bacteria cell walls. *New. J. Phys.* **14**, 093043 (2012).

59. Abolfath, R. M., van Duin, A. C. T., Biswas, P. & Brabec, T. Reactive Molecular Dynamics study on the first steps of DNA-damage by free hydroxyl radicals. *J. Phys. Chem. A* **115**, 11045–11049 (2011).

60. Keten, S., Chou, C.-C., van Duin, A. C. T. & Buehler, M. J. Tunable nanomechanics of protein disulfide bond begets weakening in reducing and stabilization in oxidizing chemical microenvironments. *J. Mech. Behav. Biochem. Mater.* **5**, 32–40 (2012).

61. Pettifor, D. G. New Many-Body Potential for the Bond Order. *Phys. Rev. Lett.* **63**, 2480–2483 (1989).

62. Stewart, J. P. MOPAC: A semiempirical molecular orbital program. *J. Comput. Aid. Mol. Des.* **4**, 1–103 (1990).

63. Porezag, D., Frauenheim, T., Köhler, T., Seifert, G. & Kaschner, R. Construction of tight-binding-like potentials on the basis of density-functional theory: Application to carbon. *Phys. Rev. B* **51**, 12947–12957 (1995).

64. Lewis, J. P. *et al.* Advances and applications in the FIREBALLab initio tight-binding molecular-dynamics formalism. *Phys. Stat. Solid. B* **248**, 1989–2007 (2011).

65. Lee, B.-J. & Baskes, M. I. Embedded-atom method: Derivation and application to impurities, surfaces, and other defects in metals. *Phys. Rev. B* **62**, 8564–8567 (2000).

66. Daw, M. S. & Baskes, M. I. Embedded-atom method: derivation and application to impurities, surfaces, and other defects in metals. *Phys. Rev. B* **29**, 6443–6453 (1984).

67. Baskes, M. I. Modified embedded-atom potentials for cubic materials and impurities. *Phys. Rev. B* **46**, 2727–2742 (1992).

68. Abell, G. C. Empirical chemical pseudopotential theory of molecular and metallic bonding. *Phys. Rev. B* **31**, 6184–6196 (1985).

69. Tersoff, J. New empirical approach for the structure and energy of covalent systems. *Phys. Rev. B* **37**, 6991–7000 (1988).

70. Tersoff, J. Empirical interatomic potential for carbon, with applications to amorphous carbon. *Phys. Rev. Lett.* **61**, 2879–2882 (1988).

71. Brenner, D. W. Empirical potential for hydrocarbons for use in simulating the chemical vapor deposition of diamond films. *Phys. Rev. B* **42**, 9458–9471 (1990).

72. Stuart, S. J., Tutein, A. B. & Harrison, J. A. A reactive potential for hydrocarbons with intermolecular interactions. *J. Chem. Phys.* **112**, 6472 (2000).

73. Liang, T. *et al.* Reactive Potentials for Advanced Atomistic Simulations. *Annu. Rev. Mater. Sci.* **43**, 109–129 (2013).

74. Shin, Y. K. *et al.* Variable Charge Many-Body Interatomic Potentials. *MRS Rev.* **37**, 504–512 (2012).

75. Fonseca, A. F. *et al.* Reparameterization of the REBO-CHO potential for graphene oxide molecular dynamics simulations. *Phys. Rev. B* **84**, 075460 (2011).

76. Shan, T.-R. *et al.* Second generation charge optimized many-body (COMB) Potential for Si/SiO2 and amorphous silica. *Phys. Rev. B* **82**, 235302 (2010).

77. Phillpot, S. R. & Sinnott, S. B. Simulating multifunctional structures. *Science* **325**, 1634–1635 (2009).

78. Akimov, A. V. & Prezhdo, O. V. Large-scale computations in chemistry: a bird's eye view of a vibrant field. *Chem. Rev.* **115**, 5797–5890 (2015).

79. Qian, H.-J., van Duin, A. C. T., Morokuma, K. & Irle, S. Reactive molecular dynamics simulation of fullerene combustion synthesis: ReaxFF vs DFTB potentials. *J. Chem. Theor. Comput.* **7**, 2040–2048 (2011).

80. Raymand, D., van Duin, A. C. T., Baudin, M. & Hermansson, K. A reactive force field (ReaxFF) for zinc oxide. *Surf. Sci.* **602**, 1020–1031 (2008).

81. LaBrosse, M. R., Johnson, J. K. & van Duin, A. C. T. Development of a transferable reactive force field for cobalt. *J. Phys. Chem. A* **114**, 5855–5861 (2010).

82. Goddard, W. A., Mueller, J. E. & van Duin, A. C. T. Development and validation of ReaxFF reactive force field for hydrocarbon chemistry catalyzed by nickel. *J. Phys. Chem. C* **114**, 4939–4949 (2010).

83. Neyts, E. C. *et al.* Defect healing and enhanced nucleation of carbon nanotubes by low-energy ion bombardment. *Phys. Rev. Lett.* **110**, 065501 (2013).

84. Neyts, E. C., van Duin, A. C. T. & Bogaerts, A. Insights in the plasma-assisted growth of carbon nanotubes through atomic scale simulations: effect of electric field. *J. Am. Chem. Soc.* **134**, 1256–1260 (2012).

85. Neyts, E. C., Shibuta, Y., van Duin, A. C. T. & Bogaerts, A. Catalyzed growth of carbon nanotube with definable chirality by hybrid molecular dynamics-force biased monte carlo simulations. *ACS Nano* **4**, 6665–6672 (2010).

86. Neyts, E. C., van Duin, A. C. T. & Bogaerts, A. Changing chirality during single-walled carbon nanotube growth: a reactive molecular dynamics/Monte Carlo study. *J. Am. Chem. Soc.* **133**, 17225–17231 (2011).

87. del Alamo, J. A. Nanometre-scale electronics with III-V compound semi-conductors. *Nature* **479**, 317–323 (2011).

88. Frank, M. M. in *2011 Proceedings of the ESSCIRC (ESSCIRC)* 50–58 (IEEE, Helsinki, 2011).

89. Kamata, Y. High-k/Ge MOSFETs for future nanoelectronics. *Mater. Today* **11**, 30–38 (2008).

90. Houssa, M., Chagarov, E. & Kummel, A. Surface defects and passivation of Ge and III-V interfaces. *MRS Bull.* **34**, 504–513 (2009).

91. Wallace, R. M., McIntyre, P. C., Kim, J. & Nishi, Y. Atomic layer deposition of dielectrics on ge and iii-v materials for ultrahigh performance transistors. *MRS Bull.* **34**, 493–503 (2009).

92. Engel-Herbert, R., Hwang, Y. & Stemmer, S. Comparison of methods to quantify interface trap densities at dielectric/III-V semiconductor interfaces. *J. Appl. Phys.* **108**, 124101 (2010).

93. George, S. M. Atomic layer deposition: an overview. *Chem. Rev.* **110**, 111–131 (2009).

94. Swaminathan, S., Sun, Y., Pianetta, P. & McIntyre, P. C. Ultrathin ALD-Al2O3 layers for Ge (001) gate stacks: Local composition evolution and dielectric properties. *J. Appl. Phys.* **110**, 094105 (2011).

95. Lee, J. S. et al. Atomic imaging of nucleation of trimethylaluminum on clean and H2O functionalized Ge(100) surfaces. *J. Chem. Phys.* **135**, 054705 (2011).

96. Fadida, S. et al. Hf-based high-k dielectrics for p-Ge MOS gate stacks. *J. Vac. Sci. Technol. B* **32**, 03D105 (2014).

97. Seo, K.-I. et al. Chemical states and electronic structure of a HfO2/Ge(001) interface. *Appl. Phys. Lett.* **87**, 042902 (2005).

98. Yoshiki, K., Yuuichi, K., Tsunehiro, I. & Akira, N. Direct Comparison of ZrO_2 and HfO_2 on Ge Substrate in Terms of the Realization of Ultrathin High-κ Gate Stacks. *Jpn J. Appl. Phys.* **44**, 2323 (2005).

99. Kutsuki, K., Okamoto, G., Hosoi, T., Shimura, T. & Watanabe, H. Germanium oxynitride gate dielectrics formed by plasma nitridation of ultrathin thermal oxides on Ge(100). *Appl. Phys. Lett.* **95**, 022102 (2009).

100. On Chui, C., Ramanathan, S., Triplett, B. B., McIntyre, P. C. & Saraswat, K. C. Germanium MOS capacitors incorporating ultrathin high-/spl kappa/ gate dielectric. *IEEE Electr. Device Lett.* **23**, 473–475 (2002).

101. Swaminathan, S., Shandalov, M., Oshima, Y. & McIntyre, P. C. Bilayer metal oxide gate insulators for scaled Ge-channel metal-oxide-semiconductor devices. *Appl. Phys. Lett.* **96**, 082904 (2010).

102. Kana, H. et al. Fabrication of Ge Metal-Oxide-Semiconductor Capacitors with High-Quality Interface by Ultrathin SiO_2/GeO_2 Bilayer Passivation and Postmetallization Annealing Effect of Al. *Jpn J. Appl. Phys.* **50**, 04DA10 (2011).

103. Xie, Q. et al. Implementing TiO_2 as gate dielectric for Ge-channel complementary metal-oxide-semiconductor devices by using HfO_2/GeO_2 interlayer. *Appl. Phys. Lett.* **97**, 112905 (2010).

104. Zheng, Y. X. et al. In situ process control of trilayer gate-stacks on p-germanium with 0.85-nm EOT. *IEEE Electr. Device Lett.* **36**, 881–883 (2015).

105. Onofrio, N., Guzman, D. & Strachan, A. Atomic origin of ultrafast resistance switching in nanoscale electrometallization cells. *Nat. Mater.* **14**, 440–446 (2015).

106. Tavazza, F., Senftle, T. P., Zou, C., Becker, C. A. & van Duin, A. C. T. Molecular dynamics investigation of the effects of tip-substrate interactions during nanoindentation. *J. Phys. Chem. C* **119**, 13580–13589 (2015).

107. Bagri, A. et al. Structural evolution during the reduction of chemically derived graphene oxide. *Nat. Chem.* **2**, 581–587 (2010).

108. Srinivasan, S. G., van Duin, A. C. T. & Ganesh, P. Development of a ReaxFF potential for carbon condensed phases and its application to the thermal fragmentation of a large fullerene. *J. Phys. Chem. A* **119**, 571–580 (2015).

109. Goverapet Srinivasan, S. & van Duin, A. C. T. Molecular-dynamics-based study of the collisions of hyperthermal atomic oxygen with graphene using the ReaxFF reactive force field. *J. Phys. Chem. A* **115**, 13269–13280 (2011).

110. Fantauzzi, D. et al. Development of a ReaxFF potential for Pt-O systems describing the energetics and dynamics of Pt-oxide formation. *Phys. Chem. Chem. Phys.* **16**, 23118–23133 (2014).

111. Senftle, T. P., Meyer, R. J., Janik, M. J. & van Duin, A. C. T. Development of a ReaxFF potential for Pd/O and application to palladium oxide formation. *J. Chem. Phys.* **139**, 044109–044115 (2013).

112. Ludwig, J., Vlachos, D. G., van Duin, A. C. T. & Goddard, W. A. Dynamics of the dissociation of hydrogen on stepped platinum surfaces using the reaxff reactive force field. *J. Phys. Chem. B* **110**, 4274–4282 (2006).

113. Senftle, T. P., Janik, M. J. & van Duin, A. C. T. A ReaxFF investigation of hydride formation in palladium nanoclusters via Monte Carlo and molecular dynamics simulations. *J. Phys. Chem. C* **118**, 4967–4981 (2014).

114. Zou, C., Duin, A. T. & Sorescu, D. Theoretical investigation of hydrogen adsorption and dissociation on iron and iron carbide surfaces using the ReaxFF reactive force field method. *Top. Catal.* **55**, 391–401 (2012).

115. Raju, M., van Duin, A. C. T. & Fichthorn, K. A. Mechanisms of oriented attachment of TiO_2 nanocrystals in vacuum and humid environments: reactive molecular dynamics. *Nano Lett.* **14**, 1836–1842 (2014).

116. Rahaman, O., van Duin, A. C. T., Goddard, W. A. & Doren, D. J. Development of a ReaxFF reactive force field for glycine and application to solvent effect and tautomerization. *J. Phys. Chem. B* **115**, 249–261 (2011).

117. Hatzell, M. C. et al. Effect of strong acid functional groups on electrode rise potential in capacitive mixing by double layer expansion. *Environ. Sci. Technol.* **48**, 14041–14048 (2014).

118. Senftle, T. P., van Duin, A. C. T. & Janik, M. J. Determining in situ phases of a nanoparticle catalyst via grand canonical Monte Carlo simulations with the ReaxFF potential. *Catal. Commun.* **52**, 72–77 (2014).

119. Addou, R. et al. Influence of hydroxyls on Pd Atom mobility and clustering on rutile TiO2(011)-2×1. *ACS Nano* **8**, 6321–6333 (2014).

120. Spanjers, C. S. et al. Illuminating surface atoms in nanoclusters by differential X-ray absorption spectroscopy. *Phys. Chem. Chem. Phys.* **16**, 26528–26538 (2014).

121. Raju, M., Ganesh, P., Kent, P. R. C. & van Duin, A. C. T. Reactive force field study of Li/C systems for electrical energy storage. *J. Chem. Theor. Comput.* **11**, 2156–2166 (2015).

122. Islam, M. M. et al. ReaxFF molecular dynamics simulations on lithiated sulfur cathode materials. *Phys. Chem. Chem. Phys.* **17**, 3383–3393 (2015).

123. Dereli, G. Stillinger-Weber type potentials in monte carlo simulation of amorphous silicon. *Mol. Simulat.* **8**, 351–360 (1992).

124. Neyts, E. C. & Bogaerts, A. Numerical study of the size-dependent melting mechanisms of nickel nanoclusters. *J. Phys. Chem. C* **113**, 2771–2776 (2009).

125. Voter, A. F. Parallel replica method for dynamics of infrequent events. *Phys. Rev. B* **57**, R13985–R13988 (1998).

126. Joshi, K. L., Raman, S. & van Duin, A. C. T. Connectivity-based parallel replica dynamics for chemically reactive systems: from femtoseconds to microseconds. *J. Phys. Chem. Lett.* **4**, 3792–3797 (2013).

127. Cheng, T., Jaramillo-Botero, A., Goddard, W. A. & Sun, H. Adaptive accelerated ReaxFF reactive dynamics with validation from simulating hydrogen combustion. *J. Am. Chem. Soc.* **136**, 9434–9442 (2014).

128. Miron, R. A. & Fichthorn, K. A. Accelerated molecular dynamics with the bond-boost method. *J. Chem. Phys.* **119**, 6210–6216 (2003).

129. Mortier, W. J., Ghosh, S. K. & Shankar, S. Electronegativity equalization method for the calculation of atomic charges in molecules. *J. Am. Chem. Soc.* **108**, 4315–4320 (1986).

130. Chen, J. & Martinez, T. J. QTPIE: Charge transfer with polarization current equalization. A fluctuating charge model with correct asymptotics. *Chem. Phys. Lett.* **438**, 315–320 (2007).

131. Lee Warren, G., Davis, J. E. & Patel, S. Origin and control of superlinear polarizability scaling in chemical potential equalization methods. *J. Chem. Phys.* **128**, 144110 (2008).

132. Chelli, R., Procacci, P., Righini, R. & Califano, S. Electrical response in chemical potential equalization schemes. *J. Chem. Phys.* **111**, 8569–8575 (1999).

133. Verstraelen, T. et al. Assessment of atomic charge models for gas-phase computations on polypeptides. *J. Chem. Theor. Comput.* **8**, 661–676 (2012).

134. Verstraelen, T. et al. The significance of parameters in charge equilibration models. *J. Chem. Theor. Comput.* **7**, 1750–1764 (2011).

135. Valentini, P., Schwartzentruber, T. E. & Cozmuta, I. Molecular dynamics simulation of O_2 sticking on Pt(111) using the ab initio based ReaxFF reactive force field. *J. Chem. Phys.* **133**, 084703/084701–084703/084709 (2010).

136. Verstraelen, T., Ayers, P. W., Van Speybroeck, V. & Waroquier, M. ACKS2: Atom-condensed Kohn-Sham DFT approximated to second order. *J. Chem. Phys.* **138**, 074108 (2013).

137. Verstraelen, T., Vandenbrande, S. & Ayers, P. W. Direct computation of parameters for accurate polarizable force fields. *J. Chem. Phys.* **141**, 194114 (2014).

138. Nistor, R. A., Polihronov, J. G., Müser, M. H. & Mosey, N. J. A generalization of the charge equilibration method for nonmetallic materials. *J. Chem. Phys.* **125**, 094108 (2006).

139. Su, J. T. & Goddard, W. A. The dynamics of highly excited electronic systems: applications of the electron force field. *J. Chem. Phys.* **131**, 244501 (2009).

140. Kale, S., Herzfeld, J., Dai, S. & Blank, M. Lewis-inspired representation of dissociable water in clusters and Grotthuss chains. *J. Biol. Phys.* **38**, 49–59 (2012).

141. Rienstra-Kiracofe, J. C., Tschumper, G. S., Schaefer, H. F., Nandi, S. & Ellison, G. B. Atomic and molecular electron affinities: photoelectron experiments and theoretical computations. *Chem. Rev.* **102**, 231–282 (2002).

142. Jordan, K. D. & Burrow, P. D. Studies of the temporary anion states of unsaturated hydrocarbons by electron transmission spectroscopy. *Acc. Chem. Res.* **11**, 341–348 (1978).

143. Bochevarov, A. D. et al. Jaguar: A high-performance quantum chemistry software program with strengths in life and materials sciences. *Int. J. Quantum. Chem.* **113**, 2110–2142 (2013).

144. van Duin, A. C. T., Baas, J. M. A. & van de Graaf, B. Delft molecular mechanics: A new approach to hydrocarbon force fields. *J. Chem. Soc.* **90**, 2881–2895 (1994).

145. Iype, E., Hutter, M., Jansen, A. P. J., Nedea, S. V. & Rindt, C. C. M. Parameterization of a Reactive Force Field using a Monte Carlo Algorithm. *J. Comput. Chem.* **34**, 1143–1154 (2013).

146. Jaramillo-Botero, A., Naserifar, S. & Goddard, W. A. General Multiobjective Force Field Optimization Framework, with Application to Reactive Force Fields for Silicon Carbide. *J. Chem. Theor. Comput.* **10**, 1426–1439 (2014).

147. Larsson, H., Hartke, B. & van Duin, A. Global optimization of parameters in the reactive force field ReaxFF for SiOH. *J. Comput. Chem.* **34**, 2178–2189 (2013).

148. Rice, B. M., Larentzos, J. P., Byrd, E. F. C. & Weingarten, N. S. Parameterizing complex reactive force fields using multiple objective evolutionary strategies (MOES): Part 2: transferability of ReaxFF models to C-H-N-O energetic materials. *J. Theor. Comput. Chem.* **11**, 392–405 (2015).

149. Dittner, M., Muller, J., Aktulga, H. M. & Hartke, B. Efficient global optimization of reactive force-field parameters. *J. Comput. Chem.* **36**, 1550–1561 (2015).

150. Plimpton, S. J. Fast parallel algorithms for short-range molecular dynamics. *J. Comput. Phys.* **117**, 1–19 (1995).

151. Budzien, J., Thompson, A. P. & Zybin, S. V. Reactive molecular dynamics simulations of shock through a single crystal of pentaerythritol tetranitrate. *J. Phys. Chem. B* **113** (2009).

152. Zybin, S. V., Goddard, W. A. III, Xu, P., van Duin, A. C. T. & Thompson, A. P. Physical mechanism of anisotropic sensitivity in pentaerythritol tetranitrate from compressive-shear reaction dynamics simulations. *Appl. Phys. Lett.* **96**, 081918/081911–081918/081913 (2010).

153. Li, Y., Kalia, R. J., Nakano, A., Nomura, K. & Vashishta, P. Multistage reaction pathways in detonating high explosives. *Appl. Phys. Lett.* **105**, 204101–204103 (2014).

154. Nomura, K., Kalia, R. K., Nakano, A., Vashishta, P. & van Duin, A. C. T. Mechanochemistry of nanobubble collapse near silica in water. *Appl. Phys. Lett.* **101**, 073108/073101–073108/073104 (2012).

155. Vedadi, M. *et al.* Structure and dynamics of shock-induced nanobubble collapse in water. *Phys. Rev. Lett.* **105**, 014503/014501–014503/014504 (2010).

156. Chen, H.-P. *et al.* Embrittlement of metal by solute segregation-induced amorphization. *Phys. Rev. Lett.* **104**, 155502/155501–155502/155504 (2010).

157. Nakano, A. *et al.* De novo ultrascale atomistic simulations on high-end parallel supercomputers. *Int. J. High Perform. Comput. Appl.* **22**, 113–128 (2008).

158. Nomura, K. I. *et al.* Dynamic transition in the structure of an energetic crystal during chemical reactions at shock front prior to detonation. *Phys. Rev. Lett.* **99**, 148303 (2007).

159. Nakano, A. *et al.* A divide-and-conquer/cellular-decomposition framework for million-to-billion atom simulations of chemical reactions. *Comput. Mater. Sci.* **38**, 642–652 (2007).

160. Zheng, M., Li, X. & Guo, L. Algorithms of GPU-enabled reactive force field (ReaxFF) molecular dynamic. *J. Mol. Graph. Model.* **41**, 1–11 (2013).

161. Aktulga, H. M., Knight, C., Coffman, P., Shan, T.-R. & Jiang, W. Optimizing the performance of reactive molecular dynamics simulations for multi-core architectures. *IEEE Transac. Parallel Distrib. Syst.* (2015) (in preparation).

162. Liu, B., Lusk, M. T., Ely, J. F., van Duin, A. C. T. & Goddard, W. A. Reactive molecular dynamics force field for the dissociation of light hydrocarbons on Ni(111). *Mol. Simulat.* **34**, 967–972 (2008).

163. Castro-Marcano, F., Kamat, A. M., Russo, M. F. Jr, van Duin, A. C. T. & Mathews, J. P. Combustion of an Illinois No. 6 coal char simulated using an atomistic char representation and the ReaxFF reactive force field. *Combust. Flame* **159**, 1272–1285 (2012).

164. Vasenkov, A. *et al.* Reactive molecular dynamics study of Mo-based alloys under high-pressure, high-temperature conditions. *J. Appl. Phys.* **112**, 013511 (2012).

165. Agrawalla, S. & van Duin, A. C. T. Development and application of a ReaxFF reactive force field for hydrogen combustion. *J. Phys. Chem. A* **115**, 960–972 (2011).

166. Manzano, H. *et al.* Confined water dissociation in microporous defective silicates: mechanism, dipole distribution, and impact on substrate properties. *J. Am. Chem. Soc.* **134**, 2208–2215 (2012).

First-principles calculations of lattice dynamics and thermal properties of polar solids

Yi Wang[1], Shun-Li Shang[1], Huazhi Fang[1], Zi-Kui Liu[1] and Long-Qing Chen[1]

Although the theory of lattice dynamics was established six decades ago, its accurate implementation for polar solids using the direct (or supercell, small displacement, frozen phonon) approach within the framework of density-function-theory-based first-principles calculations had been a challenge until recently. It arises from the fact that the vibration-induced polarization breaks the lattice periodicity, whereas periodic boundary conditions are required by typical first-principles calculations, leading to an artificial macroscopic electric field. The article reviews a mixed-space approach to treating the interactions between lattice vibration and polarization, its applications to accurately predicting the phonon and associated thermal properties, and its implementations in a number of existing phonon codes.

INTRODUCTION

Lattice dynamics is the study of the collective atomic vibrations in a crystal. The concept of phonons was introduced by Tamm[1] in 1930 through an observation of the particle-like energetics of atomic vibrations in a crystal, similar to the wave–particle duality in quantum mechanics. Lattice dynamics has since become an important branch of condensed matter physics and is critical for understanding the thermal properties of crystalline solids at finite temperatures.[2,3] For example, the phonon densities of states are required for evaluating thermodynamic properties of a crystal,[4–6] such as thermal expansion coefficients, heat capacity, entropy and lattice thermal conductivity.[7–9] There exist excellent reference books on lattice dynamics, e.g., by Wallace[10] and by Born and Huang.[2]

With the advances in density functional theory calculations,[11–15] all the input data needed by lattice dynamics can now be obtained by the first-principles approach solely based on the crystal structure and atomic numbers. Currently, there are essentially two implementations in wide use for the first-principles calculations of lattice dynamics: the linear-response approach[16] and the direct approach.[17,18] A general review of first-principles approach to phonon theory was given by Baroni et al.[16] focused on the linear-response approach.[16,19,20]

The linear-response approach directly evaluates the dynamical matrix at a predetermined reference coarse grid in the wave-vector space through the density functional perturbation theory.[16,19,20] Then the backward Fourier transform of the calculated dynamical matrix at the coarse wave-vector grid is employed to extract the interatomic force constants on the corresponding real-space grid. In contrast, the direct approach first calculates the force constants using a predetermined reference supercell of the primitive cell. In the literature, the direct approach[4,16,17,21,22] is also referred as the supercell method, the small-displacement method or the frozen-phonon approach. The features of the linear-response approach and the direct approach are compared in Table 1. A collection of phonon/first-principles codes, including YPHON,[23] ShengBTE,[8] PhonTS,[7] Phonopy,[18] ALAMODE,[24] PHON,[21] ATAT,[4,22] PHONON,[17,25–27] PWSCF/QUANTUM ESPRESSO,[14] ABINIT,[15] CASTEP,[13] CRYSTAL[28] and VASP (the Vienna Ab initio Simulation Package)[11,12] that can be employed to calculate phonon and related properties are briefed in Table 2.

The present review focuses on the theory of lattice dynamics for polar solids. Here a polar solid implies an insulator or a semiconductor composed of cations with positive charges and anions with negative charges. As a matter of fact, the majorities of modern functional materials are made of polar solids, such as the topological crystalline insulator group-IV tellurides,[29] the ferroelectrics and multiferroics,[30] and materials for solar cells.[31] The accurate descriptions of phonon properties have key roles for the understandings and developments of these materials.

For certain optical atomic vibration modes, cations and anions vibrate in opposite directions creating dipole–dipole interactions and hence homogeneous electric fields, which have to be treated with caution in phonon calculations. As an illustration, Figure 1 shows the effects of vibration-induced polarization within super-cells. The supercell described in Figure 1a is commensurate with the wavelength of the lattice vibration of interest, so the averaged electric polarization is zero, and thus no macroscopic electric field is generated. It should be pointed out that there are still internal dipole–dipole interactions within an individual supercell, but they are already accounted for in first-principles calculations of the interatomic force constants.[32,33] In Figure 1b, the supercell is incommensurate with the wavelength of a lattice vibration, and the corresponding phonon produces a nonzero-averaged electric polarization and thus a nonzero artificial macroscopic electric field. Figure 1c shows how the artificial electric polarization varies with the supercell size (or the supercell geometry for the three-dimentional case). In existing literature,[2,34] polar effects were mainly discussed in the long-wavelength limit. It has been shown that the homogeneous field presents significant difficulties in calculating the phonon frequencies of polar materials for the

---

[1]Department of Materials Science and Engineering, The Pennsylvania State University, University Park, Pennsylvania, PA, USA.
Correspondence: Y Wang (yuw3@psu.edu)

Table 1. The linear-response approach versus the direct approach

Features	Linear response approach	Direct approach
Implementation in DFT	Calculate the dynamical matrix in the reciprocal space	Calculate the interatomic interaction force constants in the real space
Advantage	Dynamical matrix can be accurately calculated at any arbitrary $\mathbf{q}$ (wave-vector) points	Straightforward to determine the total energy as a function of atomic displacements
Disadvantage	Extensive programming required and it sometimes has special requirements for the form of pseudopotential in DFT calculations	Phonon frequencies can be accurately calculated only at $\mathbf{q}$ points that are commensurate with supercell geometry
Limitation on $\mathbf{k}$ point mesh	$\mathbf{k}$ point meshes for electronic structure and phonon calculations should be compatible	No need
Additional calculations for polar materials	Separate calculation of dynamical matrix at Γ point, Born effective charges and dielectric constants; additional calculation to separate coulombic contribution from dynamical matrix	Separate calculation of Born effective charges and dielectric constants
For accurate phonon dispersions or density of states for polar solids	Forward Fourier interpolation using the interatomic force constants followed by adding the coulombic contribution back.	Adding coulombic contribution as a constant term to the interatomic force constants followed by forward Fourier interpolation

Abbreviation: DFT, density functional theory.

Table 2. A collection of phonon/first-principles codes

Codes	Abilities	Method to compute polar effects on phonons
PWSCF/QUANTUM ESPRESSO[14]	Electronic structure; phonon	Linear-response approach
ABINIT[15]	Electronic structure; phonon	Linear-response approach
CASTEP[13]	Electronic structure; phonon	Linear-response approach
CRYSTAL[28]	Electronic structure; phonon	Mixed-space approach
VASP[11,12]	Electronic structure; phonon	Not available
YPHON[23]	Phonon	Mixed-space approach
ShengBTE[8]	Phonon; thermal conductivity; thermodynamic properties	Mixed-space approach
PhonTS[7]	Phonon; thermal conductivity; thermodynamic properties	Mixed-space approach
Phonopy[18]	Phonon; thermal conductivity; thermodynamic properties	Mixed-space approach
ALAMODE[24]	Phonon; thermal conductivity; thermodynamic properties; anharmonicity	Mixed-space approach
PHONON[17,25–27]	Phonon; thermodynamic properties	Only accurate at the Γ point
PHON[21]	Phonon; thermodynamic properties	Not available
ATAT[4,22]	Phonon; thermodynamic properties	Not available

direct approach under the Born-von Kármán boundary conditions.[35]

In statistical thermodynamics, a phonon represents a quantized vibrational mode characterized by a frequency at a given reciprocal lattice wave-vector point.[2,10] An accurate thermodynamic calculation[4–6] requires the phonon frequency distribution, i.e., the phonon density of states, calculated over a fine mesh in the wave-vector space. Although in principle the phonon frequencies at any wave-vector points can be calculated by the first-principles approach, it is still computationally too expensive to account for all mesh points in the wave-vector space. To reduce the computational cost, a practical approach is to first calculate either the force constants for a predefined supercell in the real space or the dynamical matrix for a predefined grid in the reciprocal space,[16] and followed by Fourier interpolations to evaluate phonon properties at arbitrary wave-vector points. For polar solids, such a strategy has only been implemented in the linear-response approach until recently when a mixed-space method becomes available.[21,36,37] The mixed-space approach makes it possible to accurately calculate the phonon properties for polar solids within the framework of direct approach for phonon calculations.[9,28] The mixed-space approach has also been extended to lattice thermal conductivity calculations[7–9,38] where the third-order force constants are needed.

By this review, we will show that (i) the Born-von Kármán boundary conditions[39] still apply when the phonon frequencies

are calculated at the exact wave-vector points;[40] (ii) The effects of vibration-induced polarization on phonons at an arbitrary wave-vector point can be understood in the real space; and (iii) The longitudinal optical–transverse optical (LO–TO) phonon splitting[34,40] can be predicted accurately at the exact wave-vector points by the direct approach without explicitly handling the effects of macroscopic electric fields.

The present paper is organized as follows: 'Basic lattice dynamics of polar solids' describes the basics of lattice dynamics and the fundamentals of the mixed-space approach. 'Helmholtz energy and quasiharmonic approximation' outlines the first-principles thermodynamics based on the phonon theory; More discussions of the mixed-space approach are given in 'The mixed-space approach'. 'Computational procedure' summarizes the common procedures in phonon calculations; 'Phonon software packages have implemented the mixed-space approach' briefs the implementation of the mixed-space approach in several software packages. Extensive applications of the mixed-space approach are summarized in 'Recent calculations using the mixed-space approach'. 'Other phonon software packages' briefs a list of other phonon codes implemented differently from the mixed-space approach for polar solids. 'Software packages for both electronic and phonon calculations' introduces a few widely in use first-principles codes for both electronic and phonon calculations. Finally, the last section is the 'Summary'.

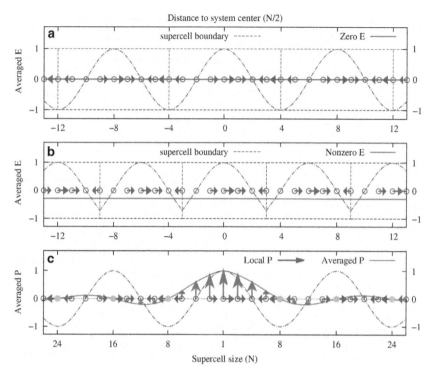

Figure 1. Vibration-induced polarization at a selected wave-vector point $q = 2\pi(1/8,0,0)$. The horizontal (solid blue) arrows indicate the local electric polarization of a primitive unit cell due to the optical vibration. The cosine type (blue dot-dashed) curves are used as a guide to eyes for the periodicity of electric polarization. The open (blue) circles mark the positions of primitive unit cells. The dashed (black) lines in (**a,b**) designate the supercell boundaries. (**a**) The supercell geometry (supercell size $N = 8$) is commensurate with the wave-vector, so the averaged electric polarization is zero, and no macroscopic electric field is generated. (**b**) The supercell (supercell size $N = 6$) is incommensurate with the wave-vector, and therefore, the average electric polarization is nonzero, creating an artificial macroscopic electric field. (**c**) The variation of the averaged electric polarization with supercell size is shown by the vertical (solid red) arrows enveloped by the solid (red) curve. The solid (cyan) circles in **c** mark the supercell sizes by which the averaged polarizations are zero.

BASIC LATTICE DYNAMICS OF POLAR SOLIDS

As phonons represent waves of collective atomic vibrations in a periodic pattern, we can choose to model a system by a repeated parallelepiped, i.e., a supercell. To formulate atomic vibrations within such a supercell, it is convenient to start from equations of motion for the atoms. For a polar solid, the forces can be divided into two additive contributions—analytic and nonanalytic.[20] The terms analytic and nonanalytic can be traced back to the mathematical definition[41] that a function is called analytic if and only if its Taylor series expansion about a reference point converges to the function at each point in some neighbourhood of the reference point, otherwise it is nonanalytic. Under the framework of first-principles calculations, the analytic contribution accounts for all the forces under the restricted periodic boundary conditions under which the averaged electric field is assumed to be zero. The nonanalytic contribution accounts the additional forces owing to a nonzero-averaged electric field.

The classical Newton's second law of motion for describing the atomic vibrations of a polar solid is given by

$$m_j \frac{\partial^2 u_a(t,j;P)}{\partial t^2} = -\frac{\partial E(\mathbf{U})}{\partial u_a(t,j;P)} + e\mathbf{Z}_a(j) \cdot \mathbf{E}. \qquad (1)$$

where m_j represents the atomic mass of the jth atom in the primitive unit cell, t, the time, the a (a = x, y, z) component of the atomic displacement from its equilibrium position of the jth atom in the Pth reference primitive unit cell within a supercell. The first term in the right-hand side of Equation (1) accounts for the analytic force due to the short-range interatomic interaction, where $\mathbf{U}$ represent the whole set of atomic displacements and $E(\mathbf{U})$ the total energy. The second term in the right-hand side of

Equation (1), with $-e$ the electron charge, accounts for the nonanalytic force due to the long-range coulombic interactions shown as the dot product between the Born effective charge (**Z**, a second-rank tensor, i.e., 3×3 matrix) and the averaged electric field (**E**) induced by the atomic vibration. The Born effective charges can be calculated as the change in electric polarization divided by the amount an ion is displaced.[20,42]

The analytic contribution in the Harmonic approximation

Within the harmonic approximation, one can truncate the first term $E(\mathbf{U})$ in the right-hand side of Equation (1) at the second order in Taylor series expansion

$$E(\mathbf{U}) = \frac{1}{2} \sum_{P,Q}^{N} \sum_{j,k}^{N_P} \sum_{\alpha,\beta}^{3} \Phi_{\alpha\beta}^{jk}(P,Q) u_\alpha(t,j;P) u_\beta(t,k;Q), \qquad (2)$$

where N represents the number of primitive unit cells contained in the supercell, N_P the number of atoms in the primitive cell P, and α and β the Cartesian components x, y and z. $\Phi_{\alpha\beta}^{jk}(P,Q)$ is the real-space interatomic force constant matrix representing the interactions between the jth atom within the primitive unit cell P and kth atom within the primitive cell Q under zero macroscopic electric field.

The nonanalytic contribution due to the vibration-induced electric field

The nonanalytic contribution accounts for the LO–TO splitting, i.e., the removal of degeneracy between the LO and TO phonons at the Brillouin zone center.[35] In particular, the LO–TO splitting generally depends upon the direction along which the wave-vector approaches the Brillouin zone center (mainly for non-cubic solids),[43] making the splitting discontinuous.

We now prove that the second term $e\mathbf{Z}_a(j)\cdot\mathbf{E}$ at the right-hand side of Equation (1) can be explicitly formulated at any wave-vector points, going beyond the long-wavelength limit by Cochran and Cowley.[34] On the basis of the Born effective charge and atomic displacement, the averaged polarization[44] can be evaluated by

$$\mathbf{P} = \frac{e}{NV_P}\sum_{k,Q}^{N_P,N}\mathbf{Z}(k)\cdot\mathbf{u}(t,k;Q), \qquad (3)$$

where V_P is the volume of the primitive cell. The vibration-induced polarization is parallel to the direction of the wave-vector $\hat{\mathbf{q}}$ and can only have effects on the atomic vibrations along the direction of the wave-vector $\hat{\mathbf{q}}$. It is noticed that a normal mode with a wave-vector $\mathbf{q}$ is nothing but a collective vibration of some parallel charged crystal planes with normal along $\hat{\mathbf{q}}$. As a result, the induced electric field by the lattice vibrations can be formulated as[45]

$$\mathbf{E} = -\frac{4\pi\mathbf{P}}{\hat{\mathbf{q}}\cdot\boldsymbol{\varepsilon}_\infty\cdot\hat{\mathbf{q}}} = -\frac{4\pi e}{NV_P}\sum_{k,Q}^{N_P,N}\frac{\mathbf{Z}(k)\cdot\mathbf{u}(t,k;Q)}{\hat{\mathbf{q}}\cdot\boldsymbol{\varepsilon}_\infty\cdot\hat{\mathbf{q}}}, \qquad (4)$$

where $\hat{\mathbf{q}}\cdot\boldsymbol{\varepsilon}_\infty\cdot\hat{\mathbf{q}}$ represents a projection of the macroscopic dielectric constant tensor into $\hat{\mathbf{q}}$. As $\mathbf{E}$ is along $\hat{\mathbf{q}}$, we can also project $\mathbf{Z}_a(j)$ into $\hat{\mathbf{q}}$ in evaluating $e\mathbf{Z}_a(j)\cdot\mathbf{E}$ and obtain

$$e\mathbf{Z}_a(j)\cdot\mathbf{E} = -\frac{4\pi e}{NV_P}\sum_{k,Q}^{N_P,N}\frac{[\mathbf{Z}(j)\cdot\hat{\mathbf{q}}]_a[\mathbf{Z}(k)\cdot\hat{\mathbf{q}}]\cdot\mathbf{u}(t,k;Q)}{\hat{\mathbf{q}}\cdot\boldsymbol{\varepsilon}_\infty\cdot\hat{\mathbf{q}}}. \qquad (5)$$

For a normal mode[2,10] in which the atoms oscillate at the same frequency, ω, in a periodic pattern, the displacements $\mathbf{u}(t,j;P)$ can be expressed as

$$u_a(t,k;P) = u_a(j;\mathbf{q})\exp\{i\mathbf{q}\cdot[\mathbf{R}(P)+\mathbf{r}(j)]-i\omega(\mathbf{q})t\}, \qquad (6)$$

where $\mathbf{q}$ is wave-vector, $\mathbf{R}(P)$ the position of the Pth primitive cell in the supercell, and $\mathbf{r}(j)$ the position of the jth atom in the primitive cell.

Then, inserting equation (6) into equation (5), we obtain

$$e\mathbf{Z}_a(j)\cdot\mathbf{E} = -\frac{4\pi e}{NV_P}\sum_{k,Q}^{N_P,N}\sum_{\beta}$$
$$\frac{[\mathbf{Z}(j)\cdot\hat{\mathbf{q}}]_a[\mathbf{Z}(k)\cdot\hat{\mathbf{q}}]_\beta\cdot u_\beta(k;\mathbf{q})\exp\{i\mathbf{q}\cdot[\mathbf{R}(Q)+\mathbf{r}(k)]-i\omega(\mathbf{q})t\}}{\hat{\mathbf{q}}\cdot\boldsymbol{\varepsilon}_\infty\cdot\hat{\mathbf{q}}}. \qquad (7)$$

The dynamical matrix

Substituting Equation (2) into Equation (1) together with utilizing Equation (7), the equation of motion for the atoms can be expressed in the reciprocal space, $\mathbf{q}$,

$$-\omega^2 w_a(j;\mathbf{q}) = -\sum_k^{N_P}\sum_\beta^3 D_{\alpha\beta}^{jk}(\mathbf{q})w_\beta(k;\mathbf{q}), \qquad (8)$$

where we have replaced $u_a(j;\mathbf{q})$ by $w_a(j;\mathbf{q})$ which is defined as

$$w_a(j;\mathbf{q}) = \sqrt{m_j}u_a(j;\mathbf{q}). \qquad (9)$$

$D_{\alpha\beta}^{jk}(\mathbf{q})$ in Equation (8) is called the dynamical matrix,[10] which takes the form

$$D_{\alpha\beta}^{jk}(\mathbf{q}) = \exp\{i\mathbf{q}\cdot[\mathbf{r}(k)-\mathbf{r}(j)]\}\left[DA_{\alpha\beta}^{jk}(\mathbf{q})+DN_{\alpha\beta}^{jk}(\mathbf{q})\right]. \qquad (10)$$

$DA_{\alpha\beta}^{jk}(\mathbf{q})$ is called the reduced dynamical matrix,[10] accounting for the analytic contribution under zero-averaged electric field, whereas $DN_{\alpha\beta}^{jk}(\mathbf{q})$ results entirely from the effects of the vibration-induced macroscopic field. They have the following forms

$$DA_{\alpha\beta}^{jk}(\mathbf{q}) = \sum_Q^N \frac{\Phi_{\alpha\beta}^{jk}(P,Q)}{\sqrt{m_j m_k}}\exp\{i\mathbf{q}\cdot[\mathbf{R}(Q)-\mathbf{R}(P)]\}, \qquad (11)$$

$$DN_{\alpha\beta}^{jk}(\mathbf{q}) = \frac{4\pi e[\mathbf{Z}(j)\cdot\hat{\mathbf{q}}]_a[\mathbf{Z}(k)\cdot\hat{\mathbf{q}}]_\beta}{\sqrt{m_j m_k}V_P\hat{\mathbf{q}}\cdot\boldsymbol{\varepsilon}_\infty\cdot\hat{\mathbf{q}}}\frac{1}{N}\sum_Q^N\exp\{i\mathbf{q}\cdot[\mathbf{R}(Q)-\mathbf{R}(P)]\}. \qquad (12)$$

The choice of the reference P is arbitrary due to the translational invariance by which $\Phi_{\alpha\beta}^{jk}(P,Q)$ depends on P and Q only through the difference $\mathbf{R}(Q)-\mathbf{R}(P)$. Note that the prefactor term in the right-hand side of Equation (12)

$$\Delta_{\alpha\beta}^{jk}(\mathbf{q}) = \frac{4\pi e[\mathbf{Z}(j)\cdot\hat{\mathbf{q}}]_a[\mathbf{Z}(k)\cdot\hat{\mathbf{q}}]_\beta}{\sqrt{m_j m_k}V_P\hat{\mathbf{q}}\cdot\boldsymbol{\varepsilon}_\infty\cdot\hat{\mathbf{q}}} \qquad (13)$$

is generally not continuous at $\mathbf{q}=0$ (except for cubic crystals), which is the main reason why contribution given by Equation (12) is called 'nonanalytic'.

With Equation (8), determining phonon frequencies is reduced to finding the eigenvalues for the secular equation

$$\det\left|D_{\alpha\beta}^{jk}(\mathbf{q})-\omega^2(\mathbf{q})\right| = 0. \qquad (14)$$

There are generally $3N_P$ eigenvalues whose roots $\omega_j(\mathbf{q})$ $(j=1, 2, \ldots, 3N_P)$, are the normal phonon frequencies.

HELMHOLTZ ENERGY AND QUASIHARMONIC APPROXIMATION

Once the phonon frequencies are obtained, all the thermodynamic quantities can be calculated using statistical physics without further approximations. Neglecting the electron–phonon coupling and the thermal electronic contributions, it is a well-demonstrated procedure[5] to decompose the Helmholtz energy F of a system at temperature T into two additive contributions as follows

$$F(V_P,T) = E_c(V) + F_{vib}(V,T), \qquad (15)$$

where E_c is the static total energy per primitive unit cell at 0 K, and F_{vib} is the vibrational contribution to the Helmholtz energy given by[46]

$$F_{vib}(V,T) = k_B T\int_0^\infty \ln[2\sinh\frac{\hbar\omega}{2k_B T}]g(\omega,V)d\omega, \qquad (16)$$

where k_B is the Boltzmann constant and $g(\omega,V)$ is the phonon density of states.

The term 'quasiharmonic approximation' arises from the approach that for a given volume, $F_{vib}(V,T)$ is calculated under the harmonic approximation, and the anharmonic effects are included solely through the volume dependence of the phonon frequency. Once the Helmholtz energy is calculated as a function of volume and temperature, other thermodynamic quantities can be calculated as usual, such as entropy $S=-(\partial F/\partial T)_V$, enthalpy $H=F+TS$ and so on.

THE MIXED-SPACE APPROACH

The procedure presented in the above section represents a combined solution to the phonon problem for polar materials that (i) the long-ranged coulombic interactions are accounted for through Equation (12) in the reciprocal space through $\mathbf{Z}$ and $\boldsymbol{\varepsilon}_\infty$ calculated at $\mathbf{q}=0$; and (ii) short-ranged interatomic interactions are accounted for through Equation (11) in terms of $\Phi_{\alpha\beta}^{jk}(P,Q)$ by a supercell in the real space. We therefore refer our solution as mixed-space approach.

In this section, we discuss how the analytic and the nonanalytic contributions are related to the supercell geometry and the type of wave-vector points for evaluating the normal vibration frequencies of a polar solid. Let us first examine the nonanalytic contribution to the dynamical matrix in Equation (12) from which we can extract the mathematical geometry factor

$$f(\mathbf{q}) = \frac{1}{N} \sum_{Q}^{N} \exp\{i\mathbf{q} \cdot [\mathbf{R}(Q) - \mathbf{R}(P)]\}. \tag{17}$$

The values for $f(\mathbf{q})$ depends on whether $\mathbf{q}$ is one of the exact wave-vector points[10,16,47] in the direct approach.[21,37] The exact wave-vector points satisfy the following condition

$$\mathbf{q}_{ex} \cdot \mathbf{S}_i = 2\pi \times \text{Integer} \tag{18}$$

where $\mathbf{S}_i$ with $i = 1, 2$, and 3 represent the three lattice vectors of the supercell in the direct approach.

(i)   if $\mathbf{q}$ is an exact wave-vector point, i.e., at $\mathbf{q} = \mathbf{q}_{ex}$ (see Figure 1a). This is the case that no corrections are needed in calculating the dynamical matrix, i.e., the interatomic force constants calculated by the direct approach can be used directly in calculating the dynamical matrix. The internal dipole–dipole interactions within the supercell are already accounted at all exact wave-vector points except for the point at $\mathbf{q} = 0$.

(ii) $f(\mathbf{q}) \neq 0$ if $\mathbf{q}$ is not an exact wave-vector point (see Figure 1b). In this case, phonon properties are determined based on the interatomic force constants obtained with a supercell that is incommensurate with the wavelength corresponding to the wave-vector $\mathbf{q}$. As a result, an artificial electric field is introduced by the incomplete supercell and must be accounted for as a correction to those interatomic force constants calculated under zero macroscopic electric field.

(iii) $f(0) = 1$. This is the limiting case as $\mathbf{q} \to 0$ of case (ii) in which all the local polarizations within the supercell would be equal to each other. The long-wavelength limit $\mathbf{q} \to 0$ of Equation (12) recovers the results by Cochran and Cowley.[34]

COMPUTATIONAL PROCEDURE

An actual first-principles phonon calculation of polar solids can be summarized as follows:[13,37]

(i)   Calculate the interatomic force constants $\Phi_{\alpha\beta}^{jk}(P, Q)$ in the right-hand side of Equation (11) in the real space,[4,48] based on a designated supercell; or calculate the dynamical matrix $D_{\alpha\beta}^{jk}(\mathbf{q})$ in Equation (8) in the wave-vector space based on a designated wave-vector grid.[16]

(ii) Calculate the dielectric constant and Born effective charge tensors used in Equation (13) based on the primitive cell by employing either the linear-response approach[42] or the Berry phase expressions of electric polarization.[44]

(iii) Combine the dielectric properties from step (ii) with $\Phi_{\alpha\beta}^{jk}(P, Q)$ in step (i) to interpolate the phonon frequencies at any wave-vector points.[13,37]

It should be reiterated that in step (i), an implicit condition is that the averaged electric field or the macroscopic electric field is zero. The enforcement of the zero macroscopic electric-field condition is due to the periodic condition adopted in most computer codes for the total electronic energy calculations. For step (iii), a tedious procedure exists in several major computer software packages employing the linear-response approach[20,49] by separating the dipole–dipole interaction from the short-range interactions. In comparison, except for the codes based on the mixed-space approach,[36,37] other computational implementations[25–27] of step (iii) in the direct approach are inaccurate.

PHONON SOFTWARE PACKAGES HAVE IMPLEMENTED THE MIXED-SPACE APPROACH

The mixed-space approach has been adopted in a number of software packages, including YPHON,[23] ShengBTE,[8] CRYSTAL,[28] PhonTS,[7] Phonopy[18] and ALAMODE.[24] In these codes, a generalized force constant $\Psi_{\alpha\beta}^{jk}(P, Q)$ was introduced as

$$\Psi_{\alpha\beta}^{jk}(P, Q) = \Phi_{\alpha\beta}^{jk}(P, Q) + \phi_{\alpha\beta}^{jk}(\hat{\mathbf{q}}), \tag{19}$$

where

$$\phi_{\alpha\beta}^{jk}(\hat{\mathbf{q}}) = \frac{1}{N} \frac{4\pi e [\mathbf{Z}(j) \cdot \hat{\mathbf{q}}]_\alpha [\mathbf{Z}(k) \cdot \hat{\mathbf{q}}]_\beta}{V_P \hat{\mathbf{q}} \cdot \hat{\boldsymbol{\varepsilon}}_\infty \cdot \hat{\mathbf{q}}}. \tag{20}$$

Therefore, the evaluation of the dynamical matrix in Equation (14) becomes

$$D_{\alpha\beta}^{jk}(\mathbf{q}) = \exp\{i\mathbf{q} \cdot [\mathbf{r}(k) - \mathbf{r}(j)]\} \sum_{Q}^{N} \frac{\Psi_{\alpha\beta}^{jk}(P, Q)}{\sqrt{m_j m_k}} \exp\{i\mathbf{q} \cdot [\mathbf{R}(Q) - \mathbf{R}(P)]\}. \tag{21}$$

This greatly simplifies the computational procedure since one only needs to add a constant term to the calculated force constants by the direct approach. In this respect, the mixed-space approach is a generalization of the approach for the specific case of GaAs.[50] It should be pointed out that the mixed-space approach is significantly different from previous implementations accounting for the presence of a macroscopic field[51] in the linear-response approach,[14,20] where rather tedious and expensive mathematical calculations are involved in order to decompose the calculated interatomic force constants into the short-range contributions and the long-range one from the polar effects.

YPHON[23] is an open-source code (c++) for the calculations of phonon dispersions and phonon density of states.

ShengBTE is a software package for computing the lattice thermal conductivity of crystalline materials and nanowires with diffusive boundary conditions by Li et al.[8] Both the linear-response approach and the mixed-space approach were implemented in ShengBTE.

The CRYSTAL package performs ab initio calculations of the ground state energy, energy gradient, electronic wave function, and properties of periodic systems. Hartree–Fock or Kohn–Sham Hamiltonians (that adopt an exchange-correlation potential following the postulates of the density functional theory) can be used. Like ShengBTE, PhonTS developed by Chernatynskiy and Phillpot[7] is a code mainly for thermal conductivity calculations. The mixed-space approach has been implemented in PhonTS exampled by the calculation of phonon lifetime and thermal conductivity of UO_2.[52,53]

Phonopy is a Python code developed by Togo[18] for phonon, thermodynamic properties and thermal conductivity calculations. The mixed-space approach has been implemented in Phonopy for the phonon properties, replacing that by Parliński et al.[25–27] In particular, a formulation for the evaluation of the third-order force constants is also derived by Togo et al.,[9] extending the mixed-space approach.

ALAMODE (Anharmonic LAttice MODEl)[24] is designed for estimating harmonic and anharmonic properties of lattice vibrations (phonons) in solids.

RECENT CALCULATIONS USING THE MIXED-SPACE APPROACH
Phonon and thermodynamic properties

Using the mixed-space approach, phonon and associated properties have been studied for a variety of polar (and non-polar) solids. Most of these calculations are based on the direct approach using the output data from first-principles codes such as VASP[11,12] as input. Examples are firstly shown for several energy conversion and storage materials.

Polar crystal Li$_2$S with the anti-fluorite structure and band gap of 4.4 eV is a major compound in Li-S batteries.[54] Phonons of Li$_2$S have been calculated[55] using the mixed-space approach. In addition to a nearly perfect agreement between experimental and calculated dispersions of Li$_2$S, a large LO–TO splitting was found for the T$_{1u}$ infrared (IR) mode of Li$_2$S (~30% and >120 cm^{-1}).

Chalcogenide Cu$_2$ZnSn(S,Se)$_4$ (labelled as CZTSSe) with the kesterite structure (space group $P\bar{4}2c$) and band gap 1.0 ~ 1.5 eV is a photovoltaic absorber material that has helped achieving significant recent improvement in photovoltaic device cell efficiency (12.6%).[56] Phonon frequencies at the Γ point and phonon density of state of CZTS were calculated using the mixed-space approach, a 64-atom supercell, and a combined PBEsol[57] and HSE06. A gap band in the phonon density of state (from 170 ~ 250 cm^{-1}) was predicted for CZTS, and especially a negative thermal expansion was suggested at low temperatures (e.g., < 50 K) in terms of phonon density of states.

LiMPO$_4$ (M = Mn, Fe, Co, and Ni) compounds with the olivine structure (space group Pnma) are a class of cathode materials and viable alternatives to the conventional cathode LiCoO$_2$.[58] By means of the mixed-space approach, a X–C functional of GGA+U (refs 59,60; U is used to account for the strong on-site Coulomb interaction in transition metals), and a 112-atom supercell, a comparative phonon study has been performed for the anti-ferromagnetic LiMPO$_4$, and in turn, the associated thermodynamics and bonding strength (the strongest one being P–O bonding) have been reported.[58]

Li$_2$CO$_3$ with space group C2/c and band gap 5.0 eV has been identified as a main component of the solid electrolyte interphase —a passivating film that forms on Li-ion battery anode surfaces.[61] Phonons of Li$_2$CO$_3$ were predicted using the mixed-space approach (including the LO–TO splitting and using a 96-atom supercell).

TiO$_2$ with band gap ~ 3.0 eV has extensive applications such as solar cells, photocatalysts and storage capacitors. Phonon and associated thermodynamics of six TiO$_2$ polymorphs including rutile (space group P4$_2$/mnm), anatase (I4$_1$/amd), TiO$_2$-II phase (Pbcn), baddeleyite (P2$_1$/c), orthorhombic I (Pbca), and cotunnite (Pnma) were obtained using the mixed-space approach and the X–C of the local-density approximation (LDA),[62] indicating that all TiO$_2$ polymorphs are dynamically stable[62] and the pressure-induced phase transitions and the pressure–temperature phase diagrams of TiO$_2$ were predicted.[63]

Besides phonons in energy materials, a long-standing issue has been resolved regarding the occurrence of imaginary phonon frequencies in cubic perovskites. These are in fact spurious as they result from the methodology employed.[64] For example, in perovskites EuTiO$_3$ (ref. 64) and SrTiO$_3$,[65] a dynamic short-range ordering model using the mixed-space approach as well as the cubic force constants calculated from the low-temperature tetragonal phases was used for phonon calculations. It was seen that the spurious imaginary phonon frequencies in SrTiO$_3$ and EuTiO$_3$ disappear, resulting in a remarkably good agreement with experiments. A similar idea was used for Mott–Hubbard insulators MnO and NiO,[66] i.e., the dynamic matrices with the ideal cubic symmetry recovered from the distorted antiferromagnetic structures, the LO–TO splittings estimated using the mixed-space approach, and the strong electron correlations accounted for by the GGA+U method,[59] which produced accurate phonon dispersions for MnO and NiO. For room temperature multiferroic BiFeO$_3$,[67] the challenge in calculating its phonon properties is due to the fact that BiFeO$_3$ is a Mott–Hubbard insulator with band gap ~ 2.5 eV, involving a polar effect and strong correlation among the d electrons of Fe. The mixed-space approach together with the GGA+U method[60] accurately predicted the phonon dispersions of BiFeO$_3$ and suggested that no gapped magnon modes[68] exist and

contribute to the heat capacity of BiFeO$_3$ in the temperature range 5–30 K.

Accurate phonon properties have also been predicted for CaF$_2$ and CeO$_2$ with the fluorite structure.[69] CaF$_2$ is a typical superionic conductor, its phonons have been studied by the PBE exchange-correlation functional[70] using a 192-atom supercell. CeO$_2$ has been used in catalytic converters in automotive applications and as an electrolyte in fuel cells because of its relatively high oxygen ion conductivity. In particular for considering the f-electron system, phonon dispersions of CeO$_2$ have been studied by a HSE06 hybrid functional,[71,72] showing better accuracy[69] than the previous predictions from e.g., PWSCF[73] and ABINIT.[74]

In addition to the prototype α-Al$_2$O$_3$,[37] phonon and associated thermodynamics were successfully predicted using LDA+U and the mixed-space approach for another dense and continuous coating material of Cr$_2$O$_3$.[75] Furthermore, phonon-related properties have been investigated using the mixed-space approach for many other polar solids such as nanograined half-Heusler semiconductors,[76] calcium fluoride at high pressure,[32] the phase diagram of bismuth ferrite,[33] cubic SiC and hexagonal BN,[37] ZnO,[77] UN,[78,79] Bi$_2$S$_3$,[80] Si, Ge, InAs and GaAs,[38] GaN,[81] CrN,[82] WS$_2$,[83] ZnSe,[84] ZnS, ZnSe and ZnTe,[85] SnO$_2$,[86] TaO$_3$,[87] UO$_2$,[52,53,88] CaCO$_3$,[89] Bi$_2$SiO$_5$ (ref. 90) and BaZrO$_3$.[91] layered antimony telluride,[92] Phonon transport in SrTiO$_3$,[93] self-consistent phonon calculations for cubic SrTiO$_3$ (ref. 94) and Si(Se$_x$S$_{1-x}$)$_2$.[95]

Thermal conductivities of polar solids

The main heat carriers in nonmagnetic crystals are phonons and electrons, with phonons dominating in semiconductors and insulators. The phonon contribution to the total thermal conductivity is the lattice thermal conductivity. One important approach to studying phonon transport in solids is the Boltzmann transport equation (BTE).[96] However, many solutions of BTE rely on the relaxation time approximation along with the Debye approximation, neglecting the true phonon dispersions, and several parameters are introduced to treat different scattering mechanisms. Li et al.[8] and Chernatynskiy et al.[7] implemented parameter-free iterative solutions in their software packages (ShengBTE[8] and PhonTS[7]) to solve the BTE based on the inputs from firs principles. The programs compute converged sets of phonon scattering rates and use them to obtain the lattice thermal conductivity and many related quantities. The two main inputs needed by their software packages are sets of second-order (harmonic) and third-order (anharmonic) interatomic force constants (IFCs) for a given crystal structure. Our mixed-space approach has been implemented in both software packages to derive the second-order IFCs, so as to account for the long-range electrostatic interactions in polar compounds. To date, ab initio calculations of lattice thermal conductivity have been applied by Li et al. for many bulk systems such as Mg$_2$Si, Mg$_2$Sn and Mg$_2$Si$_x$Sn$_{1-x}$,[97] two-dimensional systems such as MoS$_2$ (ref. 98) and nanowires made of Si, diamond,[99] InAs, AIN and BeO[100] under the diffusive boundary conditions. All these applications show excellent agreement with experimental measurements, and an accurate description of polar–polar interactions was found crucial for theoretical predictions to be in line with experiments. As claimed by Li et al.[8,100] in their work for InAs of a well-known direct-band gap III–V semiconductor, 'Therefore it is ideally well suited for validating our approach when isotope scattering and polar bonds are introduced into the picture'. More examples of lattice thermal conductivity calculations have been reported for SnSe,[101] MgO, GaAs, SiC, BN, BP, BSb, BAs, BeTe, and BeSe,[102] InN,[103] phosphorene[104,105] and Si, Ge, InAs and GaAs alloys.[106]

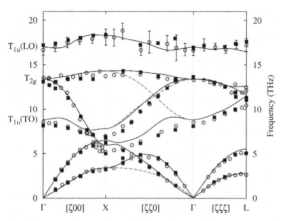

Figure 2. Phonon dispersions of UO_2. Open circles: measured data by Pang *et al.*;[52] solid squares: measured data by Dolling *et al.*[110] The lines represent the present calculations. The (red) dot-dashed lines along the [ζζ0] direction emphasize the three phonon dispersions that were not reported by the calculation of Pang *et al.*[52]

Phonon dispersions for systems with symmetry broken by magnetism

There are systems such as MnO, NiO, and UO_2 (refs 53,66) for which the high-temperature phase is paramagnetic, and the low temperature phase is antiferromagnetic. For these systems, it is too expensive to calculate the phonons of the high-temperature phase accurately. An alternative approximation to calculate the phonons of the high-temperature phase is to use the force constants calculated from the corresponding low-temperature phase. In doing so, one primary problem to solve is the symmetry broken by the magnetic degree of freedom. Using UO_2 as an example, the primitive cell of the antiferromagnetic structure contains 6 atoms resulting in 18 phonon dispersions, whereas the primitive cell of the paramagnetic structure contains 3 atoms resulting in 9 phonon dispersions. A solution to the problem is to restore the symmetry by a transformation as

$$\Phi_{st}^{\alpha\beta}(\text{high symmetry}) = \frac{1}{S}\sum_{r=1}^{N}\mathbf{O}_r^{-1}\Phi_{st}^{\alpha\beta}(\text{low symmetry})\mathbf{O}_r, \quad (22)$$

where $\Phi_{st}^{\alpha\beta}$ (high symmetry) represents the force constant matrix of the high-symmetry structure, $\Phi_{st}^{\alpha\beta}$ (low symmetry) the force constant matrix calculated from the low symmetry structure, $\mathbf{O}_r$ the space group operation of the high-symmetry structure, and S the number of $\mathbf{O}_r$'s. Figure 2 shows the phonon dispersions of UO_2. It is observed that our calculated phonon dispersions show great improvements over the previous calculations[52,107–109] by comparing them with the experimental data.[52,110]

OTHER PHONON SOFTWARE PACKAGES

PHON is an open-source code developed by Alfè[21] to calculate phonon frequencies following the direct approach by Parliński *et al.*[17] ATAT is a generic name that refers to a collection of open-source alloy theory tools developed by van de Walle *et al.*[4,22] For phonon calculations, it appears that neither codes can yet handle the vibration-induced polarization effects.

PHONON is a commercial code for phonon and thermal properties developed by Parliński *et al.*[17,25–27] As an example, Figure 3 illustrates the phonon dispersions of α-Al_2O_3 calculated using the PHONON package by Lodziana and Parliński[111] in comparison with those calculated using YPHON[37] and measured from the inelastic neutron scattering by Schober *et al.*[112] It can be seen that the application of PHONON code for phonon dispersion calculations of polar solids appears inaccurate due to an artificial implementation of the Gaussian smear extrapolation[25–27]

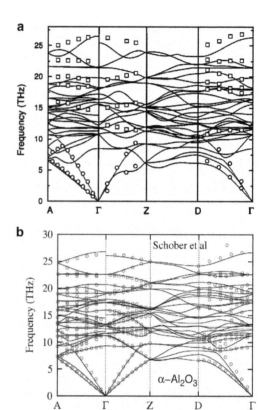

Figure 3. Phonon dispersions for Al_2O_3. (**a**) Calculated using PHONON by Lodziana and Parliński.[111] (**b**) Calculated using YPHON by Wang *et al.*[37] The symbols represent the inelastic neutron scattering data by Schober *et al.*[112]

in accounting for the vibration-induced polarization effects. Practically, PHONON code was applied to many polar solids including, for example, ZrO_2,[17] alkaline-earth metals and their hydrides,[113] $LiBeH_3$,[114] CoO,[115] Fe_2SiO_4 spinel,[116] Li_2O and Li_2CO_3,[117] LiF,[118] CsCl and $BaCl_2$,[119] hafnia and zirconia,[120] PuO_2,[121] CeO_2, ThO_2 and $(Ce,Th)O_2$ alloys,[122,123] $LiFePO_4$,[124] MnO,[125] ZnO,[126] BeO[127] and $BiFeO_3$.[128]

SOFTWARE PACKAGES FOR BOTH ELECTRONIC AND PHONON CALCULATIONS

The codes widely in use include the open-source packages PWSCF/QUANTUM ESPRESSO,[14] and ABINIT,[15] and the commercial packages CASTEP,[13] CRYSTAL[28] and VASP.[11,12] For phonon calculations of polar materials, PWSCF, QUANTUM ESPRESSO, ABINIT and CASTEP employ the linear-response approach. If one wants to calculate phonons of a polar solid using CRYSTAL (starting from CRYSTAL14[28]), then the mixed-space approach is the only choice.

VASP only calculates the phonon frequencies at the exact wave-vector points. Although VASP can calculate the dielectric constant and Born effective charge tensors, the current version of VASP (version 5.4.1.05 Feb16) does not report the LO–TO splitting.[34,40]

PWSCF is one of the core packages of open-source QUANTUM ESPRESSO.[14] We employed PWSCF to perform the linear-response calculations of phonon properties of Ni, Al, NiAl and Ni_3Al,[5] and MgO.[129] Linear-response approach is also employed for to phonon calculations by ABINIT,[15] which is an open-source package using pseudopotentials and a planewave basis. The linear-response approach relies on the availability of pseudopotentials in specific formats. For example, the CASTEP[13] code requires the use of the Norm-conserving pseudopotentials.[130] To account for the polar effects on phonon calculations, these linear-response codes commonly use the following computational procedure[14,19,130] for the evaluation of the dynamical matrix at an arbitrary **q** points:

(i) Calculate the dynamical matrices at a predetermined reference coarse grid

(ii) Remove the long-ranged coulombic contribution from the dynamical matrices that are calculated at the coarse wave-vector grid

(iii) Make backward Fourier-transform of the dynamical matrices to obtain the interatomic interaction constants and

(iv) For an arbitrary **q** point outside the coarse wave-vector grid, make forward Fourier-transform of the interatomic interaction constants followed by re-adding the long-ranged coulombic contribution to get the dynamical matrix at the arbitrary **q** points.

These steps were based on the belief from previous conclusions[14,19] that for solids having TO–LO splitting (e.g.: polar semiconductors), as the force constants in real space involves long-ranged interatomic interactions due to the nonanalytic term, Fourier interpolation is no longer possible. In contrast to this belief, we note that by using the mixed-space approach, the Fourier interpolation is possible and computationally efficient. One example was given by Zhao et al.[80] in their phonon calculations of Bi_2S_3 nanostructures using QUANTUM ESPRESSO. Instead of using the linear-response approach as implemented in QUANTUM ESPRESSO, Zhao et al.[80] employed the mixed-space approach and they found the mixed-space approach is more efficient without sacrificing the accuracy.

SUMMARY

This paper reviews a mixed-space formulation to treat the contribution of vibration-induced polarization to phonon dispersions in the direct or supercell approach. It decomposes the interatomic force constants into two contributions: one for all internal interactions within the supercell under zero macroscopic electric field and the other one accounts for the effects of nonzero macroscopic electric field arising from supercells that are incommensurate with the wavelengths of the lattice vibrations. The theory naturally gives rise to the analytic and the nonanalytic contributions to the total force constants at any wave-vector points, without the assumption of the long-wavelength limit made by Cochran and Cowley.[34] It provides a useful methodology separating the dipole–dipole interaction from those short-range interactions for Fourier interpolation.[16,20,49] It has been successfully applied in calculating the phonon and thermal properties of a wide range of polar materials and implemented in several broadly used software packages for calculating phonon properties.

ACKNOWLEDGEMENTS

This work was supported by the U.S. Department of Energy, Office of Basic Energy Sciences, Division of Materials Sciences and Engineering under Award DE-FG02-07ER46417 (Wang and Chen) and by National Science Foundation (NSF) through Grant Nos. DMR-1310289 and CHE-1230924 (Wang, Shang, Fang, and Liu). First-principles calculations were carried out partially on the LION clusters at the Pennsylvania State University, partially on the resources of NERSC supported by the Office of Science of the U.S. Department of Energy under contract No. DE-AC02-05CH11231, and partially on the resources of XSEDE supported by NSF with Grant No. ACI-1053575.

COMPETING INTERESTS

The authors declare no conflict of interest.

REFERENCES

1. Tamm, I. Über die quantentheorie der molekularen lichtzerstreuung in festen körpern. *Z. Phys.* **60**, 345–363 (1930).
2. Born, M. & Huang, K. *Dynamical Theory of Crystal Lattices* (Clarendn, Oxford, 1954).
3. Mittal, R., Chaplot, S. L. & Choudhury, N. Modeling of anomalous thermodynamic properties using lattice dynamics and inelastic neutron scattering. *Prog. Mater. Sci.* **51**, 211–286 (2006).
4. van de Walle, A., Asta, M. & Ceder, G. The alloy theoretic automated toolkit: a user guide. *Calphad* **26**, 539–553 (2002).
5. Wang, Y., Liu, Z. K. & Chen, L. Q. Thermodynamic properties of Al, Ni, NiAl, and Ni3Al from first-principles calculations. *Acta Mater.* **52**, 2665–2671 (2004).
6. Wang, Y. et al. A first-principles approach to finite temperature elastic constants. *J. Phys. Cond. Mat.* **22**, 225404 (2010).
7. Chernatynskiy, A. & Phillpot, S. R. Phonon transport simulator (PhonTS). *Comput. Phys. Commun.* **192**, 196–204 (2015).
8. Li, W., Carrete, J., Katcho, N. A. & Mingo, N. ShengBTE: a solver of the Boltzmann transport equation for phonons. *Comput. Phys. Commun.* **185**, 1747–1758 (2014).
9. Togo, A., Chaput, L. & Tanaka, I. Distributions of phonon lifetimes in Brillouin zones. *Phys. Rev. B* **91**, 094306 (2015).
10. Wallace, D. C. *Thermodynamics of Crystals* (Joha Wiley & Sons, Inc., 1972).
11. Kresse, G. & Furthmuller, J. Efficiency of ab-initio total energy calculations for metals and semiconductors using a plane-wave basis set. *Comput. Mater. Sci.* **6**, 15–50 (1996).
12. Kresse, G. & Joubert, D. From ultrasoft pseudopotentials to the projector augmented-wave method. *Phys. Rev. B* **59**, 1758–1775 (1999).
13. Clark, S. J. et al. First principles methods using CASTEP. *Z. Kristallogr.* **220**, 567–570 (2005).
14. Giannozzi, P. et al. QUANTUM ESPRESSO: a modular and open-source software project for quantum simulations of materials. *J. Phys. Condens. Matter* **21**, 395502 (2009).
15. Gonze, X. et al. ABINIT: first-principles approach to material and nanosystem properties. *Comput. Phys. Commun.* **180**, 2582–2615 (2009).
16. Baroni, S., de Gironcoli, S., Dal Corso, A. & Giannozzi, P. Phonons and related crystal properties from density-functional perturbation theory. *Rev. Mod. Phys.* **73**, 515–562 (2001).
17. Parliński, K., Li, Z. Q. & Kawazoe, Y. First-principles determination of the soft mode in cubic ZrO2. *Phys. Rev. Lett.* **78**, 4063–4066 (1997).
18. Togo, A., Oba, F. & Tanaka, I. First-principles calculations of the ferroelastic transition between rutile-type and CaCl_{2}-type SiO_{2} at high pressures. *Phys. Rev. B* **78**, 134106 (2008).
19. Giannozzi, P., Baroni, S. in *Handbook of Materials Modeling* (ed. Yip S.) 195–214 (Springer Science & Business Media, 2007).
20. Gonze, X. & Lee, C. Dynamical matrices, born effective charges, dielectric permittivity tensors, and interatomic force constants from density-functional perturbation theory. *Phys. Rev. B* **55**, 10355–10368 (1997).
21. Alfè, D. PHON: a program to calculate phonons using the small displacement method. *Comput. Phys. Commun.* **180**, 2622–2633 (2009).
22. van de Walle, A. & Ceder, G. The effect of lattice vibrations on substitutional alloy thermodynamics. *Rev. Mod. Phys.* **74**, 11–45 (2002).
23. Wang, Y., Chen, L.-Q. & Liu, Z.-K. YPHON: a package for calculating phonons of polar materials. *Comput. Phys. Commun.* **185**, 2950–2968 (2014).
24. Tadano, T. ALAMODE Documentation Release 0.9.7. Available at https://media.readthedocs.org/pdf/alamode/latest/alamode.pdf (2016).
25. Adeagbo, W. A. & Entel, P. Influence of dipole interactions on the lattice dynamics of crystalline ice. *Phase Transitions* **78**, 799–810 (2005).
26. Parliński, K., Lazewski, J. & Kawazoe, Y. Ab initio studies of phonons in MgO by the direct method including LO mode. *J. Phys. Chem. Solids* **61**, 87–90 (2000).
27. Parliński, K., Li, Z. Q. & Kawazoe, Y. Long-range Coulomb interaction in ZrO2 (reply to Parlinski, K., Z. Q. Li, and Y. Kawazoe). *Phys. Rev. Lett.* **81**, 3298–3298 (1998).
28. Dovesi, R. et al. CRYSTAL14: A program for the *ab initio* investigation of crystalline solids. *Int. J. Quantum Chem.* **114**, 1287–1317 (2014).
29. Ye, Z.-Y. et al. The origin of electronic band structure anomaly in topological crystalline insulator group-IV tellurides. *npj Comput. Mater.* **1**, 15001 (2015).
30. Scott, J. F. Searching for new ferroelectrics and multiferroics: a user's point of view. *npj Comput. Mater* **1**, 15006 (2015).
31. Alharbi, F. H. et al. An efficient descriptor model for designing materials for solar cells. *npj Comput. Mater* **1**, 15003 (2015).
32. Cazorla, C. & Errandonea, D. Superionicity and Polymorphism in calcium fluoride at high pressure. *Phys. Rev. Lett.* **113**, 235902 (2014).
33. Cazorla, C. & Iniguez, J. Insights into the phase diagram of bismuth ferrite from quasiharmonic free-energy calculations. *Phys. Rev. B* **88**, 214430 (2013).

34. Cochran, W. & Cowley, R. A. Dielectric constants and lattice vibrations. *J. Phys. Chem. Solids* **23**, 447–450 (1962).

35. Tulip, P. R. Dielectric and lattice dynamical properties of molecular crystals via density functional perturbation theory: implementation within a first principles code (PhD Thesis, Durham University, 2004).

36. Wang, Y., Shang, S. L., Liu, Z. K. & Chen, L. Q. Mixed-space approach for calculation of vibration-induced dipole-dipole interactions. *Phys. Rev. B* **85**, 224303 (2012).

37. Wang, Y. *et al.* A mixed-space approach to first-principles calculations of phonon frequencies for polar materials. *J. Phys. Condes. Matter* **22**, 202201 (2010).

38. Vermeersch, B., Carrete, J., Mingo, N. & Shakouri, A. Superdiffusive heat conduction in semiconductor alloys. I. Theoretical foundations. *Phys. Rev. B* **91**, 085202 (2015).

39. Herman, F. Lattice vibrational spectrum of germanium. *J. Phys. Chem. Solids* **8**, 405–418 (1959).

40. Kunc, K. & Martin, R. M. Ab initio force constants of GaAs: a new approach to calculation of phonons and dielectric properties. *Phys. Rev. Lett.* **48**, 406–409 (1982).

41. Churchill, R. V., Brown, J. W. & Verhey, R. F. *Complex Variables and Applications* (McGraw-Hill, 1974).

42. Gajdoš, M., Hummer, K., Kresse, G., Furthmuller, J. & Bechstedt, F. Linear optical properties in the projector-augmented wave methodology. *Phys. Rev. B* **73**, 045112 (2006).

43. Durman, R., Favre, P., Jayasooriya, U. A. & Kettle, S. F. A. Longitudinal optical transverse optical (LO-TO) splitting on internal-modes in the RAMAN-spectra of noncentric crystals. *J. Crystallogr. Spectrosc. Res.* **17**, 431–484 (1987).

44. Resta, R. Macroscopic polarization in crystalline dielectrics: the geometric phase approach. *Rev. Mod. Phys.* **66**, 899–915 (1994).

45. Kittel, C. *Introduction to solid state physics* 8th edn (Wiley, Hoboken, NJ, 2005).

46. Xie, J. J., de Gironcoli, S., Baroni, S. & Scheffler, M. First-principles calculation of the thermal properties of silver. *Phys. Rev. B* **59**, 965–969 (1999).

47. Kern, G., Kresse, G. & Hafner, J. Ab initio calculation of the lattice dynamics and phase diagram of boron nitride. *Phys. Rev. B* **59**, 8551–8559 (1999).

48. Parliński, K. Phonon Software, MedeA 1.8. Materials Design (2007).

49. Giannozzi, P., de Gironcoli, S., Pavone, P. & Baroni, S. Ab initio calculation of phonon dispersions in semiconductors. *Phys. Rev. B* **43**, 7231–7242 (1991).

50. Kunc, K. & Martin, R. M. Ab initio force-constants of GaAS—a new approach to calculation of phonons and dielectric-properties. *Phys. Rev. Lett.* **48**, 406–409 (1982).

51. Pick, R. M., Cohen, M. H. & Martin, R. M. Microscopic theory of force constants in adiabatic approximation. *Phys. Rev. B* **1**, 910–920 (1970).

52. Pang, J. W. L. *et al.* Phonon lifetime investigation of anharmonicity and thermal conductivity of UO2 by neutron scattering and theory. *Phys. Rev. Lett.* **110**, 157401 (2013).

53. Pang, J. W. L. *et al.* Phonon density of states and anharmonicity of UO2. *Phys. Rev. B* **89**, 115132 (2014).

54. Shang, S. *et al.* Insight into structural, elastic, phonon, and thermodynamic properties of alpha-sulfur and energy-related sulfides: a comprehensive first-principles study. *J. Mater. Chem. A* **3**, 8002–8014 (2015).

55. Shang, S., Wang, Y. & Liu, Z.-K. First-principles thermodynamics at finite temperatures: Perspective on ordered and disordered phases. *Mater. China* **34**, 297–304 (2015).

56. Shang, S. *et al.* Cation disorder regulation by microstate configurational entropy in photovoltaic absorber materials Cu2ZnSn(S,Se)(4). *J. Phys. Chem. C* **118**, 24884–24889 (2014).

57. Paier, J. *et al.* Screened hybrid density functionals applied to solids. *J. Chem. Phys.* **124**, 154709 (2006).

58. Shang, S. L., Wang, Y., Mei, Z. G., Hui, X. D. & Liu, Z. K. Lattice dynamics, thermodynamics, and bonding strength of lithium-ion battery materials LiMPO4 (M = Mn, Fe, Co, and Ni): a comparative first-principles study. *J. Mater. Chem.* **22**, 1142–1149 (2012).

59. Dudarev, S. L., Botton, G. A., Savrasov, S. Y., Humphreys, C. J. & Sutton, A. P. Electron-energy-loss spectra and the structural stability of nickel oxide: An LSDA +U study. *Phys. Rev. B* **57**, 1505–1509 (1998).

60. Liechtenstein, A. I., Anisimov, V. I. & Zaanen, J. Density-functional theory and strong-interactions—orbital ordering in Mott-Hubbard insulators. *Phys. Rev. B* **52**, R5467–R5470 (1995).

61. Shang, S. L. *et al.* Lattice dynamics, thermodynamics and elastic properties of monoclinic Li2CO3 from density functional theory. *Acta Mater.* **60**, 5204–5216 (2012).

62. Mei, Z. G., Wang, Y., Shang, S. L. & Liu, Z. K. First-principles study of the lattice dynamics and thermodynamics of TiO2 polymorphs. *Inorg. Chem.* **50**, 6996–7003 (2011).

63. Mei, Z.-G., Wang, Y., Shang, S. & Liu, Z.-K. First-principles study of the mechanical properties and phase stability of TiO2. *Comput. Mater. Sci.* **83**, 114–119 (2014).

64. Wang, Y., Shang, S., Chen, L.-Q. & Liu, Z.-K. Density functional theory-based database development and CALPHAD automation. *JOM* **65**, 1533–1539 (2013).

65. Wang, Y. *et al.* A first-principles scheme to phonons of high temperature phase: no imaginary modes for cubic SrTiO3. *Appl. Phys. Lett.* **97**, 162907 (2010).

66. Wang, Y. *et al.* Broken symmetry, strong correlation, and splitting between longitudinal and transverse optical phonons of MnO and NiO from first principles. *Phys. Rev. B* **82**, 081104 (2010).

67. Wang, Y. *et al.* First-principles lattice dynamics and heat capacity of BiFeO3. *Acta Mater.* **59**, 4229–4234 (2011).

68. Lu, J. *et al.* On the room temperature multiferroic BiFeO3: magnetic, dielectric and thermal properties. *Eur. Phys. J. B* **75**, 451–460 (2010).

69. Wang, Y., Zhang, L. A., Shang, S., Liu, Z.-K. & Chen, L.-Q. Accurate calculations of phonon dispersion in CaF2 and CeO2. *Phys. Rev. B* **88**, 024304 (2013).

70. Perdew, J. P., Burke, K. & Ernzerhof, M. Generalized gradient approximation made simple. *Phys. Rev. Lett.* **77**, 3865–3868 (1996).

71. Heyd, J., Scuseria, G. E. & Ernzerhof, M. Hybrid functionals based on a screened Coulomb potential. *J. Chem. Phys.* **118**, 8207–8215 (2003).

72. Heyd, J., Scuseria, G. E. & Ernzerhof, M. Hybrid functionals based on a screened Coulomb potential. **118**, 8207 (2003) *J. Chem. Phys* (Erratum in Hybrid functionals based on a screened Coulomb potential, *J. Chem. Phys.* **124**, 219906 (2006).

73. Verstraete, M. & Gonze, X. First-principles calculation of the electronic, dielectric, and dynamical properties of CaF2. *Phys. Rev. B* **68**, 195123 (2003).

74. Gürel, T. & Eryiğit, R. Ab initio pressure-dependent vibrational and dielectric properties of CeO2. *Phys. Rev. B* **74**, 014302 (2006).

75. Wang, Y. *et al.* First-principles lattice dynamics, thermodynamics, and elasticity of Cr2O3. *Surf. Sci.* **606**, 1422–1425 (2012).

76. Carrete, J., Mingo, N., Wang, S. & Curtarolo, S. Nanograined half-Heusler semiconductors as advanced thermoelectrics: an ab initio high-throughput statistical study. *Adv. Eng. Mater.* **24**, 7427–7432 (2014).

77. Calzolari, A. & Nardelli, M. B. Dielectric properties and Raman spectra of ZnO from a first principles finite-differences/finite-fields approach. *Sci. Rep.* **3**, 2999 (2013).

78. Mei, Z.-G. & Stan, M. Pressure-induced phase transitions in UN: a density functional theory study. *J. Alloys Compd* **588**, 648–653 (2014).

79. Mei, Z.-G., Stan, M. & Pichler, B. First-principles study of structural, elastic, electronic, vibrational and thermodynamic properties of UN. *J. Nucl. Mater.* **440**, 63–69 (2013).

80. Zhao, Y. *et al.* Phonons in Bi2S3 nanostructures: Raman scattering and first-principles studies. *Phys. Rev. B* **84**, 205330 (2011).

81. Zhou, Y., Wang, S., Wang, R. & Jiang, N. Ab initio calculation of the thermodynamic properties and phase diagram of gallium nitride. *Physica B* **431**, 115–119 (2013).

82. Zhou, L. *et al.* Structural stability and thermodynamics of CrN magnetic phases from ab initio calculations and experiment. *Phys. Rev. B* **90**, 184102 (2014).

83. Gandi, A. N. & Schwingenschloegl, U. WS2 as an excellent high-temperature thermoelectric material. *Chem. Mater.* **26**, 6628–6637 (2014).

84. Zhang, X. *et al.* Zincblende-wurtzite phase transformation of ZnSe films by pulsed laser deposition with nitrogen doping. *Appl. Phys. Lett.* **103**, 082111 (2013).

85. Katre, A., Togo, A., Tanaka, I. & Madsen, G. K. H. First principles study of thermal conductivity cross-over in nanostructured zinc-chalcogenides. *J. Appl. Phys.* **117**, 045102 (2015).

86. Dou, M. & Persson, C. Comparative study of rutile and anatase SnO2 and TiO2: Band-edge structures, dielectric functions, and polaron effects. *J. Appl. Phys.* **113**, 083703 (2013).

87. Ravi, C., Kaur, G. & Bharathi, A. First-principles study of lattice stability of ReO3-type hypothetical TaO3. *Comput. Mater. Sci.* **90**, 177–181 (2014).

88. Mei, Z.-G., Stan, M. & Yang, J. First-principles study of thermophysical properties of uranium dioxide. *J. Alloys Compd* **603**, 282–286 (2014).

89. Zhang, Z. & Liu, Z. High pressure equation of state for molten CaCO3 from first principles simulations. *Chinese J. Geochem.* **34**, 13–20 (2015).

90. Taniguchi, H. *et al.* Ferroelectricity driven by twisting of silicate tetrahedral chains. *Angew. Chem., Int. Ed.* **52**, 8088–8092 (2013).

91. Bjørheim, T. S., Kotomin, E. A. & Maier, J. Hydration entropy of BaZrO 3 from first principles phonon calculations. *J. Mater. Chem. A* **3**, 7639–7648 (2015).

92. Stoffel, R. P., Deringer, V. L., Simon, R. E., Hermann, R. P. & Dronskowski, R. A density-functional study on the electronic and vibrational properties of layered antimony telluride. *J. Phys.: Condens. Matter* **27**, 085402 (2015).

93. Feng, L., Shiga, T. & Shiomi, J. Phonon transport in perovskite SrTiO3 from first principles. *Appl. Phys. Express* **8**, 071501 (2015).

94. Tadano, T. & Tsuneyuki, S. Self-consistent phonon calculations of lattice dynamical properties in cubic SrTiO3 with first-principles anharmonic force constants. *Phys. Rev. B* **92**, 054301 (2015).

95. Chen, C. *et al.* Synthesis, characterization and chemical stability of silicon dichalcogenides, Si (Se x S 1-x) 2. *J. Cryst. Growth.* doi:10.1016/j.jcrysgro.2015.12.005 (2016).

96. Ziman, J. M. *Electrons and Phonons: The Theory of Transport Phenomena in Solids* (Clarendon Press, 1960).

97. Li, W., Lindsay, L., Broido, D. A., Stewart, D. A. & Mingo, N. Thermal conductivity of bulk and nanowire Mg2SixSn1-x alloys from first principles. *Phys. Rev. B* **86**, 174307 (2012).

98. Li, W., Carrete, J. & Mingo, N. Thermal conductivity and phonon linewidths of monolayer MoS2 from first principles. *Appl. Phys. Lett.* **103**, 253103 (2013).

99. Li, W. *et al.* Thermal conductivity of diamond nanowires from first principles. *Phys. Rev. B* **85**, 195436 (2012).

100. Li, W. & Mingo, N. Thermal conductivity of bulk and nanowire InAs, AlN, and BeO polymorphs from first principles. *J. Appl. Phys.* **114**, 183505 (2013).

101. Carrete, J., Mingo, N. & Curtarolo, S. Low thermal conductivity and triaxial phononic anisotropy of SnSe. *Appl. Phys. Lett.* **105**, 101907 (2014).

102. Lindsay, L., Broido, D. A., Carrete, J., Mingo, N. & Reinecke, T. L. Anomalous pressure dependence of thermal conductivities of large mass ratio compounds. *Phys. Rev. B* **91**, 121202 (2015).

103. Ma, J., Li, W. & Luo, X. Intrinsic thermal conductivity and its anisotropy of wurtzite InN. *Appl. Phys. Lett.* **105**, 082103 (2014).

104. Qin, G. *et al.* Anisotropic intrinsic lattice thermal conductivity of phosphorene from first principles. *Phys. Chem. Chem. Phys.* **17**, 4854–4858 (2015).

105. Zhu, L., Zhang, G. & Li, B. Coexistence of size-dependent and size-independent thermal conductivities in phosphorene. *Phys. Rev. B* **90**, 214302 (2014).

106. Zhang, J. *et al.* Phosphorene nanoribbon as a promising candidate for thermoelectric applications. *Sci. Rep.* **4**, 6452 (2014).

107. Sanati, M., Albers, R. C., Lookman, T. & Saxena, A. Elastic constants, phonon density of states, and thermal properties of UO2. *Phys. Rev. B* **84**, 014116 (2011).

108. Yin, Q. & Savrasov, S. Y. Origin of low thermal conductivity in nuclear fuels. *Phys. Rev. Lett.* **100**, 225504 (2008).

109. Yun, Y., Legut, D. & Oppeneer, P. M. Phonon spectrum, thermal expansion and heat capacity of UO2 from first-principles. *J. Nucl. Mater.* **426**, 109–114 (2012).

110. Dolling, G., Cowley, R. A. & Woods, A. D. B. Crystal dynamics of uranium dioxide. *Can. J. Phys.* **43**, 1397–1413 (1965).

111. Lodziana, Z. & Parliński, K. Dynamical stability of the alpha and theta phases of alumina. *Phys. Rev. B* **67**, 174106 (2003).

112. Schober, H., Strauch, D. & Dorner, B. Lattice dynamics of aapphire (Al2O3). *Z. Phys. B Condens. Mat.* **92**, 273–283 (1993).

113. Hector, L. G., Herbst, J. F. & Kresse, G. Ab Initio thermodynamic and elastic properties of alkaline-earth metals and their hydrides. *Phys. Rev. B* **76**, 014121 (2007).

114. Hu, C. H. *et al.* Crystal structure prediction of LiBeH3 using ab initio total-energy calculations and evolutionary simulations. *J. Chem. Phys.* **129**, 234105 (2008).

115. Wdowik, U. D. & Parliński, K. Lattice dynamics of cobalt-deficient CoO from first principles. *Phys. Rev. B* **78**, 224114 (2008).

116. Derzsi, M. *et al.* Effects of Coulomb interaction on the electronic structure and lattice dynamics of the Mott insulator Fe2SiO4 spinel. *Phys. Rev. B* **79**, 205105 (2009).

117. Duan, Y. H. & Sorescu, D. C. Density functional theory studies of the structural, electronic, and phonon properties of Li2O and Li2CO3: application to CO2 capture reaction. *Phys. Rev. B* **79**, 014301 (2009).

118. Evarestov, R. A. & Losev, M. V. All-electron LCAO calculations of the LiF crystal phonon spectrum: influence of the basis set, the exchange-correlation functional, and the supercell size. *J. Comput. Chem.* **30**, 2645–2655 (2009).

119. Jiang, C., Stanek, C. R., Marks, N. A., Sickafus, K. E. & Uberuaga, B. P. Predicting from first principles the chemical evolution of crystalline compounds due to radioactive decay: the case of the transformation of CsCl to BaCl. *Phys. Rev. B* **79**, 132110 (2009).

120. Luo, X. H., Zhou, W., Ushakov, S. V., Navrotsky, A. & Demkov, A. A. Monoclinic to tetragonal transformations in hafnia and zirconia: A combined calorimetric and density functional study. *Phys. Rev. B* **80**, 134119 (2009).

121. Minamoto, S., Kato, M., Konashi, K. & Kawazoe, Y. Calculations of thermodynamic properties of PuO2 by the first-principles and lattice vibration. *J. Nucl. Mater.* **385**, 18–20 (2009).

122. Sevik, C. & Cagin, T. Mechanical and electronic properties of CeO2, ThO2, and (Ce,Th)O-2 alloys. *Phys. Rev. B* **80**, 014108 (2009).

123. Shi, S. Q. *et al.* First-principles investigation of the bonding, optical and lattice dynamical properties of CeO2. *J. Power Sources* **194**, 830–834 (2009).

124. Shi, S. Q. *et al.* First-principles study of lattice dynamics of LiFePO4. *Phys. Lett. A* **373**, 4096–4100 (2009).

125. Wdowik, U. D. & Legut, D. Ab initio lattice dynamics of MnO. *J. Phys. Condes. Matter* **21**, 275402 (2009).

126. Wrobel, J., Kurzydlowski, K. J., Hummer, K., Kresse, G. & Piechota, J. Calculations of ZnO properties using the Heyd-Scuseria-Ernzerhof screened hybrid density functional. *Phys. Rev. B* **80**, 155124 (2009).

127. Wdowik, U. D. Structural stability and thermal properties of BeO from the quasiharmonic approximation. *J. Phys. Condes. Matter* **22**, 045404 (2010).

128. Wei, L. *et al.* Lattice dynamics of bismuth-deficient BiFeO 3 from first principles. *Comput. Mater. Sci.* **111**, 374–379 (2016).

129. Wang, Y., Liu, Z. K., Chen, L. Q., Burakovsky, L. & Ahuja, R. First-principles calculations on MgO: Phonon theory versus mean-field potential approach. *J. Appl. Phys.* **100**, 023533 (2006).

130. Kleinman, L. & Bylander, D. M. Efficacious form for model pseudopotentials. *Phys. Rev. Lett.* **48**, 1425–1428 (1982).

131. Gonze, X., Charlier, J. C., Allan, D. C. & Teter, M. P. Interatomic force-constants from first principles—the case of alpha-quartz. *Phys. Rev. B* **50**, 13035–13038 (1994).

8

Super square carbon nanotube network: a new promising water desalination membrane

Ligang Sun[1], Xiaoqiao He[1] and Jian Lu[2,3]

Super square (SS) carbon nanotube (CNT) networks, acting as a new kind of nanoporous membrane, manifest excellent water desalination performance. Nanopores in SS CNT network can efficiently filter NaCl from water. The water desalination ability of such nanoporous membranes critically depends on the pore diameter, permitting water molecule permeatration while salt ion obstruction. On the basis of the systematical analysis on the interaction among water permeability, salt concentration limit and pressure on the membranes, an empirical formula is developed to describe the relationship between pressure and concentration limit. In the meantime, the nonlinear relationship between pressure and water permeability is examined. Hence, by controlling pressure, optimal plan can be easily made to efficiently filter the saltwater. Moreover, steered molecular dynamics (MD) method uncovers bending and local buckling of SS CNT network that leads to salt ions passing through membranes. These important mechanical behaviours are neglected in most MD simulations, which may overestimate the filtration ability. Overall, water permeability of such material is several orders of magnitude higher than the conventional reverse osmosis membranes and several times higher than nanoporous graphene membranes. SS CNT networks may act as a new kind of membrane developed for water desalination with excellent filtration ability.

INTRODUCTION

Water scarcity remains a highlighted issue worldwide and becomes more serious due to the various factors such as the growth of population, the increasing global warming and consumptions in industry and agriculture. Although 71% of the earth is covered with water, only ~ 3% of the water all over the world is fresh and suitable for human consumption.[1] In spite of the wide availability of seawater, environmentally friendly and economic purification techniques are needed to produce fresh water from seawater for sustainability. Nowadays, the most widely adopted commercial desalination techniques are electrodialysis,[2] nanofiltration[3] and reverse osmosis (RO).[4] Although ROs, which use high pressure to force water through porous membranes, are believed to have the greatest practical potential, the water permeability is still not satisfactory, meanwhile energy and capital consumption is intensive. In contrast to classical RO membranes, nanoporous membranes can realise fast water permeation. Nanoporous graphene is one kind of nanoporous membranes. Small molecules can pass through them, whereas larger ones cannot.[5–7] In addition, nanoporous graphene membranes also can be used as selective molecular sieves for gas purification.[8,9] Therefore, nanoporous graphene is a newly developed membrane for filtration technology. On the other hand, molecular dynamics (MD) studies have predicted[10] and experiments have demonstrated fast transport of gases[11,12] and liquids[13–16] through carbon nanotube (CNT) membranes. The well-aligned CNTs can serve as robust pores in membranes for water desalination and decontamination applications.[17–19] Nanoporous membranes open a bright future for water desalination technology.

Super CNT networks can be obtained by assembling single-walled CNTs (SWCNTs) into ordered micronetworks, which is first designed for application in composites.[20,21] Theoretical and experimental studies show that CNTs can be merged covalently using controlled electron irradiation at high temperatures[22] or atomic welders during heat treatment.[23] Some kinds of CNT networks have already successfully been made and exhibit special mechanical properties.[24–26] Along with the production of CNT networks, theoretical models consisting of covalently bonded two- and three-dimensional networks have also been developed.[27] MD simulations, atomic finite element method, continuum models and molecular structure mechanics method are developed to study super CNTs.[28–31] Although existing studies have already shown the potential application of super CNT networks in composites, the potential role of such material for water desalination remains unexplored.

In this work, super square (SS) CNT network is selected as a representative for the study on the water desalination ability. To be specific, the SS CNT networks are constructed from SWCNT arms (6,6), denoted as SS@(6,6) in this work. The detailed introduction can be found in section 'Methods'. The computational results indicate that SS@(6,6) can effectively separate salt from water in the desalination system. On the basis of the results obtained from classical MD simulations, optimal plan can be made to filter saltwater with specific salt concentration by controlling the pressure on SS@(6,6). Furthermore, the deformation and failure mode is also discussed under different pressure on SS@(6,6), which cannot be taken into consideration in other MD simulations[5–10] because the membranes are completely fixed in their systems. Our calculations demonstrate that water molecules

[1]Department of Architecture and Civil Engineering, City University of Hong Kong, Kowloon, Hong Kong, China; [2]Department of Mechanical and Biomedical Engineering, City University of Hong Kong, Kowloon, Hong Kong, China and [3]Centre for Advanced Structural Materials, City University of Hong Kong, Shenzhen Research Institute, Shenzhen Hi-Tech Industrial Park, Shenzhen, China.
Correspondence: J Lu (jianlu@cityu.edu.hk)

can flow across SS@(6,6) with very high permeability (from 171 to 421 l/cm²/day/MPa) while still rejecting salt ions perfectly. This work will be helpful in developing a new kind of excellent nanoporous membrane based on super CNT networks.

RESULTS

Different filtration stages with the variation of pressure

We first investigate the pressure on SS@(6,6) (P_{ss}) and the distance between the two layers of SS@(6,6) (d_{ss}) during water desalination. The variation of d_{ss} and P_{ss} with the increase of time under constant pulling velocity are shown in Figure 1a. There are mainly three stages during the simulation. In stage I, the filtration velocity gradually reduces accompanied by the increase of pressure (green segment). Note that only very small amount of water molecules are filtered across SS@(6,6) in stage I, which is called pre-filtration stage. In stage II, the P_{ss} keeps in a stable range and the filtration velocity is almost a constant and equal to $v = 0.023$ Å/ps (orange segment, filtration velocity is equal to $\partial d_{ss}/\partial t$), which is called stable pressure filtration (SPF) stage, the corresponding stable pressure is denoted as P_c. Note that the salt rejection is 100% in stage II. When the salt concentration reaches a critical value (defined as SPF concentration limit), filtration enters into stage III. The filtration velocity gradually reduces once again, accompanied by the increase of pressure (yellow segment). This stage is named as the ascending pressure filtration (APF) stage. However, the salt ions may pass through SS@(6,6) with the increase of pressure in stage III. Therefore, some rules can be summarised: (1) filtration velocity can be kept as a constant under P_c. (2) There is a SPF concentration limit under constant filtration velocity v and corresponding stable pressure P_c. (3) High-concentration salt solution beyond the SPF concentration limit can be filtered by gradually increasing pressure provided that no salt ion passes through SS@(6,6). As increasing pressure could induce mechanical problems such as bending, buckling and failure of the SS CNT networks, stage II is an important regime to be paid attention to establish an optimisation criterion. For salt solution with different concentration, we can choose suitable P_c to filter it under relative high filtration velocity.

Pressure, concentration limit and water permeability

To investigate the relationship between the stable pressure P_c and the SPF concentration limit, a series of constant pulling velocities

are selected for saltwater desalination. Figure 1b shows the relationship between d_{ss} and P_{ss} under different constant pulling velocities. All of the cases have a relatively stable pressure (P_c) region corresponding to a series of constant filtration velocities from 0.023 to 0.082 Å/ps. Although the velocity is very fast. The filtration velocity can be considered as is equivalent to the water velocity, as the filtration velocity is almost uniform during the stable filtration stage. The water velocity ~2.3 to 8.2 m/s is achievable in real life. In addition, we choose the high filtration velocities to control the pressure on SS@(6,6) in a desired range (115–170 MPa) for comparison with the applied pressure (100–200 MPa) on nanoporous graphene.[6] As shown in Figure 1b, it is obvious that there is a negative correlation between P_c and SPF concentration limit. However, there are two kinds of circumstances needed to be distinguished. (1) Low SPF concentration limit induced by low P_c can be further filtered in APF stage. (2) Low SPF concentration limit induced by high P_c cannot be improved in APF stage. Such difference occurs as the salt ions can pass through SS@(6,6) easier under large constant filtration velocities, which can be verified in combination with the data in Table 1 and Figure 1b. For the curve of constant filtration velocity $v = 0.032$ Å/ps, the salt ions pass through SS@(6,6) when $d_{ss} = 1.6$ nm, the corresponding P_{ss} is 249.2 MPa, which is much higher than P_c. Furthermore, if the constant filtration velocity is even smaller ($v = 0.023$ Å/ps), yet no salt ions flow across SS@(6,6) even when $d_{ss} = 1.44$ nm. However, when the constant filtration velocity is large, salt ions can pass through SS@(6,6) at the very beginning of APF stage. The critical pressure for ion passage is ~175 MPa, much smaller than the slow filtration cases that are over 200 MPa (Table 1). So APF is not very suitable for large filtration velocity. Besides, it is discovered that Na⁺ ions always flow across SS@(6,6) earlier than Cl⁻ ions. As the size of Na⁺ ions is smaller than that of Cl⁻ ions, we believe ion size should be an important reason for earlier escape of Na⁺ ions.

Figure 2a further exhibits the relationship between stable pressures P_c and SPF concentration limits. Note that the salt rejection is 100% for all the MD simulation cases under different P_c in Figure 2a. Regions **R1** and **R2** are consistent to the low SPF concentration limits (as mentioned above) induced by low P_c and high P_c. If P_c is in region **R1**, the filtration limit can be increased by APF, whereas APF hardly works for further filtration if P_c is in region **R2**. In addition, it can be observed that there exists a critical P_c^{crt} corresponding to a critical SPF concentration limit.

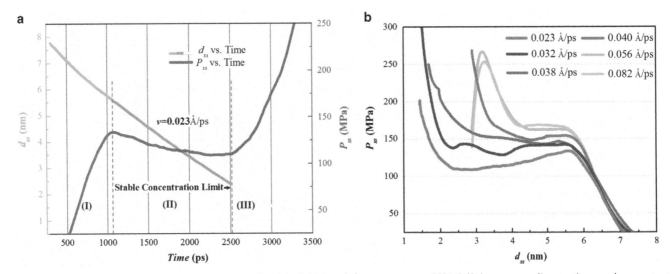

Figure 1. (a) The distance between the two layers of SS@(6,6) (d_{ss}) and the pressure on SS(6,6) (P_{ss}) corresponding to time under constant pulling velocity. (I) The pre-filtration stage. (II) The SPF stage. (III) The APF stage. (b) The relationship between the distance between the two layers of SS@(6,6) (d_{ss}) and the pressure on SS(6,6) (P_{ss}) under different constant pulling velocities. The velocities from 0.023 to 0.082 Å/ps are the velocities corresponding to different stable pressures.

Using the curve fitting to the solution of the SPF concentration limit (Figure 2a), an empirical formula is established to describe the relationship between the stable pressure and the SPF concentration limit as follow:

$$C(P_c) = \frac{k\mu}{A^P P_c} \cdot \exp\left[h\left(\frac{ln\left(\frac{P_c}{c_0}\right)}{A^P}\right)^2\right] \quad (1)$$

Where A^P is the area of each pore on SS@(6,6), μ is the total pore area fraction of one layer of SS@(6,6), and $k = 2.69 \times 10^6$, $h = -2.78 \times 10^4$ and $c_0 = 130.0$ are the fitting parameters.

The optimum pressure for water desalination can be achieved using the formula, depending on two key factors: (1) pore size on the SS CNT network; (2) total pore area fraction on the SS CNT network. It can be seen that: (1) the curve based on equation (1) well coincides with the MD simulation results (Figure 2a). (2) μ is positive correlated with the SPF concentration limit (C), whereas A^P is negative correlated with the SPF concentration limit (C). It is easy to imagine that if there are more pores on unit area of the SS CNT network, the filtration should be easier with no sacrifice of salt rejection ability, as there is no change of pore size. On the other hand, if the area of each pore (A^P) is enlarged, the salt rejection ability will reduce and result in the reduction of SPF concentration limit (C). Besides, if the pore size A_p is extremely small, SPF concentration limit (C) approaches infinity. This situation can be imagined as no salt ions can pass through the membrane without pores, nothing to do with the salt concentration. (3) As a new kind of RO membrane, applied pressure is needed to be exerted on the SS CNT networks to overcome osmotic pressure. Thus, if P_c is close to zero, SPF concentration limit (C) converges to zero in equation (1). Furthermore, if P_c approaches infinity, the SPF concentration limit (C) also converges to zero in equation (1). Therefore,

equation (1) is a reasonable empirical formula to predict the relationship between stable pressures P_c and SPF concentration limits. On the basis of the equation (1), the critical stable pressure P_c^{crt} is equal to 124.5 MPa and the critical SPF concentration limit is equal to 99.5 g/l for SS@(6,6). Moreover, all the SPF concentration limits in Figure 2a exceed the salinity of seawater (35 g/l). Hence, such kind of the SS CNT networks is with the prospect of seawater purification.

To estimate the water filtration efficiency of SS@(6,6), we calculate the effective water permeability (W_p) achieved under different P_c (Figure 2b). The water permeability, expressed in liters of output per square centimetre of SS@(6,6) per day and per unit of P_c, ranges from 171 to 421 l/cm²/day/MPa. The filtration efficiency is remarkable. Besides, the permeability scales non-linearly with P_c. There exists the relation that the filtration velocity v rises accompanied by the increase of P_c (Figure 1b). In addition, we can easily find that the increase of P_c results in the improvement of W_p (Figure 2b). Thus, the water permeability should be gradually improved along with the increase of filtration velocity. On the other hand, it is observed that the area of each pore during filtration (A_1) becomes larger than the size after relaxation (A_0). In addition, the variation of pore area ΔA ($\Delta A = A_1 - A_0$) is larger and more intensively under higher P_c. Therefore, the relationship between W_p and P_c is close related to two key factors: (1) filtration velocity v under different P_c; (2) ΔA under different P_c during filtration. The combined effect of them results in the nonlinear relationship between P_c and W_p.

Hence, the relatively optimal filtration pressure can be selected with consideration of the relationship among pressure, salt concentration and water permeability discovered in this work as mentioned above.

Table 1. The critical filtration distance (d_{ss}^{crt}) and pressure (P_{ss}) for 100% salt rejection corresponding to different constant filtration velocity

v (Å/ps)	0.023	0.032	0.038	0.040	0.056	0.082
d_{ss}^{crt} (nm)	Null	1.6	2.45	3.41	3.82	4.24
P_{ss} (MPa)	>210	249.2	173.9	174.6	184.9	167.3

Table 2. Four representative samples with different filtration velocities and the corresponding pressures (P_c)

Sample	I	II	III	IV
v (Å/ps)	0.023	0.038	0.056	0.082
P_c (MPa)	116.2	137.2	149.5	169

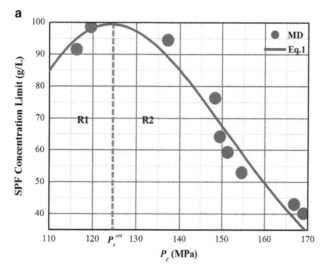
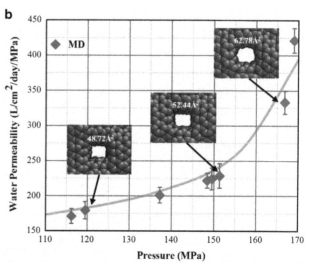

Figure 2. (a) The relationship between stable pressures (P_c) and filtration concentration limits under stable pressure for SS CNT SS@(6,6). Filtration limit can be increased in region R1, whereas cannot be increased in region R2. There exists a critical stable pressure P_c^{crt} corresponding to a critical stable pressure filtration concentration limit. (b) The nonlinear relationship between stable pressures (P_c) and water permeability for SS CNT SS@(6,6). Error bars represent the s.d. of pressures in the stable pressure filtration stage. The illustrations for nanopore size in CNT SS@(6,6) under different pressure. The pore sizes vary more intensively under higher pressure.

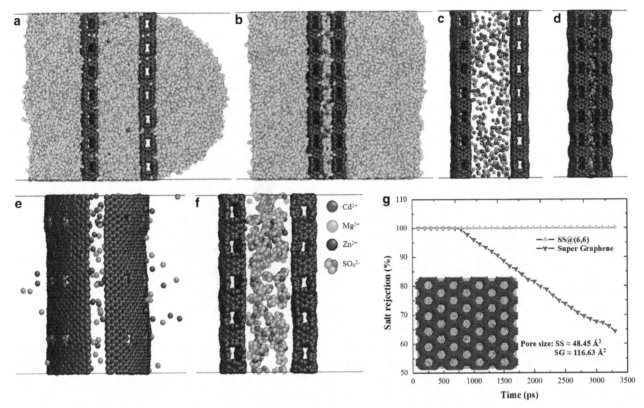

Figure 3. (**a, b**) The snapshots of SS@(6,6) during water desalination at stable and ascending pressure filtration stage, respectively, under filtration velocity $v = 0.023$ Å/ps. (**c, d**) The corresponding configuration excluding water molecules in **a, b**. (**e**) Salt ions pass through deformed nanoporous graphene during water desalination under filtration velocity $v = 0.023$ Å/ps. (**f**) Heavy ions filtration by SS@(6,6). (**g**) The comparison of salt rejection ability between SS@(6,6) and super graphene.

Structure evolution of SS@(6,6) during water desalination

Four representative samples with different filtration velocities and corresponding pressures (P_c) are selected to analyse the nanostructure evolution during water desalination (Table 2). Figure 3 shows the nanostructure evolution of sample I during water desalination compared with nanoporous graphene. Figure 3a is one of the pictures obtained during water desalination at SPF stage and Figure 3c is the corresponding figure in which water molecules are invisible. Figure 3b is one of the pictures obtained during water desalination at APF stage and Figure 3d is the corresponding figure only showing salt ions. It can be easily observed that the salt ions can be strictly filtered by SS@(6,6). Although the nanoporous graphene has the similar pore size and the same loading procedure as SS@(6,6), the salt ions cannot be impeded by it due to easier mechanical deformation of graphene under compression (Figure 3e). Moreover, the perfect heavy ions rejection ability (Cd^{2+}, Mg^{2+} and Zn^{2+}) of SS@(6,6) is also verified by similar filtration process (Figure 3f). Thus, SS CNT network has more excellent mechanical properties acting as a kind of nanoporous membrane compared with nanoporous graphene. On the other hand, the low SPF concentration limit can be improved in APF stage that is consistent with the R1 region in Figure 2a, as the filtration velocity in sample I is relatively slow. To improve the permeability of SS@(6,6), the filtration velocity is increased by raising P_c. However, increasing pressure could induce mechanical problems such as bending, local buckling and failure of the SS CNT networks (Figure 4). All these deformations occur in the APF stage through observation of nanostructure evolution during filtration. In addition, the faster the filtration velocity, the sooner the deformations occur in APF stage. The pore size is enlarged after bending or local buckling. Thus, the salt ions can flow across SS@(6,6) in the APF stage and SS@(6,6) lose the

filtration effectiveness. Note that the salt ions can pass through SS@(6,6) only when it deforms. Hence, even better filtration capability can be achieved by improving the strength of SS CNT networks through structure design. As the deformation and failure mode of the SS CNT networks is affected by the chiralities and sizes of the SWCNTs,[32] they can be taken into consideration to design the SS CNT networks with higher strength in the future study.

Outstanding water permeability and salt rejection ability

Overall, our results indicate that the SS CNT networks could act as a high-permeability desalination membrane with guarantee of 100% salt rejection and water passage. High-concentration salt solution can be purified efficiently by the SS CNT networks. Moreover, to illustrate the water permeability and salt rejection ability of the SS CNT networks, the water desalination performance of the SS@(6,6) from $P_c = 116.2$ to 169 MPa is plotted along with the performance of RO membranes[33] and nanoporous graphenes[6] in Figure 5. Note that although the MD simulation method adopted in our study is different from most of other studies,[5–10] we only compare the data in SPF stage (Figure 2b) with other membranes. As the filtration velocity in SPF stage is almost uniform under constant stable pressure, we think the pressure on membranes can approximately be regarded as the applied hydrostatic pressure. Exactly as the performance of nanoporous graphenes, the salt rejection of SS@(6,6) maintains at 100%, both are superior than conventional RO membranes. On the other hand, the water permeability of SS@(6,6) is several orders of magnitude higher than RO membranes and several times higher than nanoporous graphene. It should be emphasised that the water permeability of SS@(6,6) can be improved by increasing pressure with no sacrifice of salt rejection (Figure 2b).

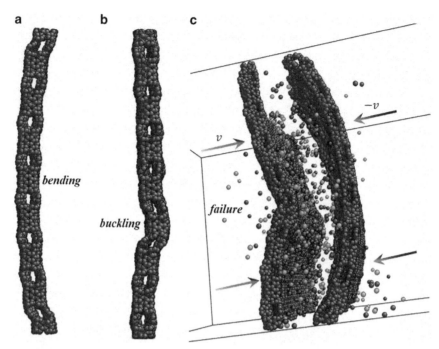

Figure 4. (**a**) Bending of SS@(6,6) in sample II with filtration velocity $v = 0.038$ Å/ps in ascending pressure filtration (APF) stage. (**b**) Local buckling of SS@(6,6) in sample III with $v = 0.056$ Å/ps in APF stage. (**c**) Bending and buckling combined failure of SS@(6,6) in sample IV with $v = 0.082$ Å/ps in APF stage. A lot of salt ions pass through deformed SS@(6,6).

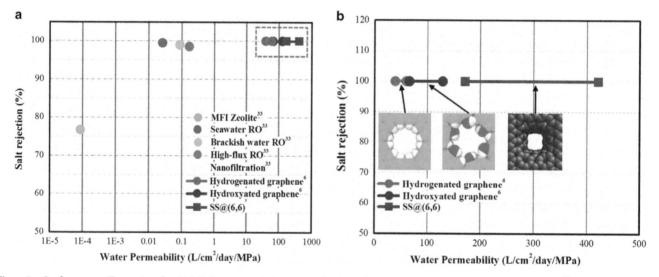

Figure 5. Performance illustration for SS@(6,6) versus existing technologies. The data is adapted from Pendergast et al.[33] and Cohen-Tanugi et al.[6] (**a**) The water permeability of passivated nanoporous graphene[6] and SS@(6,6) is expressed by the interval based on the simulation results. The SS@(6,6) examined in this work could fully reject salt ions with a water permeability several orders of magnitude higher than commercial RO and also have better performance than functionalised nanoporous graphene membranes. (**b**) The amplified red region in **a** clearly exhibits the outstanding water permeability of SS@(6,6) compared with nanoporous graphene membranes.

Moreover, the functionalisation by hydrophilic chemical group (hydroxyl, OH–) instead of hydrophobic ones (hydrogen atoms, H–) on nanoporous graphene can roughly double the water flux.[6] We think sidewall functionalisation of SS CNT networks to improve hydrophilicity could be an alternative to further improvement the performance of water desalination since both CNTs and CNT networks are chemically hydrophobic.[34] Therefore, the SS CNT networks could be a new kind of nanoporous membrane possessing outstanding water permeability.

Regarding the question of whether the SS CNT networks could be practical for commercial-scale desalination, we think that the fabrication of well-aligned SS CNT networks with the extremely narrow pore size is a main challenge. The well-aligned SS CNT networks with the strict controlled pore size and shape are not achievable due to the limitation of the current fabrication technology. Nowadays, the experimentally available CNT networks[22–26] are composed of a random distribution of pores, in both size and geometric shape. The randomness of pore size will inevitably result in the existence of unfavourable large pores. As shown in Figure 3g, the super graphene CNT network is developed and tested by MD simulation. The worse salt rejection is observed due to the enlargement of the pore size. Thus, it is essential to fabricate well-aligned networks with the uniform pore size for practical application in the future. Although challenging,

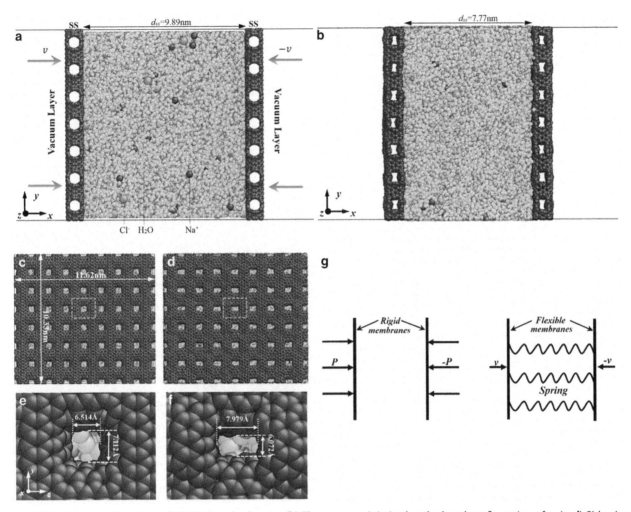

Figure 6. (**a**) The initial configuration of SS@(6,6) and saltwater. (**b**) The energy minimised and relaxed configuration of **a**. (**c, d**) Side views of **a**, **b**. (**e, f**) Pore sizes of **c, d** on SS@(6,6). (**g**) Schematic diagrams of two different simulation methods: the rigid membranes with pressure exerted on them (left) and the flexible membranes under traction by springs (right).

we believe it can be overcome and the CNT networks could offer important advantages over existing RO technology.

DISCUSSION

Acting as a new kind of nanoporous membrane, the SS CNT networks are demonstrated possessing excellent water permeability and salt rejection ability in our simulations. The water permeability is several orders of magnitude higher than the conventional RO membranes and also several times higher than nanoporous graphene membranes. We find that the water permeability with guarantee of 100% salt rejection is both affected by the salt concentration of water and the pressure on the SS@(6,6). To systematically describe the interaction among pressure, salt concentration and water permeability, the following two efforts are made: (1) an empirical formula is developed to describe the relationship between pressure and concentration limit; (2) the nonlinear relationship between pressure and water permeability is uncovered. Thus, optimal plan can easily be made to efficiently filter saltwater with specific salt concentration by controlling the pressure on the SS@(6,6). On the other hand, steered MD simulation method is first adopted to analyse the mechanical behaviour of membranes during water desalination, which cannot be realised due to the limitation of methods adopted by other researchers.[5–10] On the basis of this method, we discover that the bending and buckling of SS@(6,6) under high

pressure, which results in enlargement of the pore size, is the main reason for salt ions passing through SS@(6,6). Thus, better filtration capability can be achieved by improving the strength of SS CNT networks through structure design and optimising the pore size in the SS CNT networks. Therefore, a new kind of nanoporous membrane is studied in this work for water desalination that yields significant improvement over existing technological methods. We expect that our study will promote the understanding of the excellent performance of well-designed SS CNT networks for water desalination technology.

MATERIALS AND METHODS

All the MD simulations are performed using the LAMMPS package.[35] Two layers of the SS CNT network are constructed from SWCNT arms (6,6), denoted as SS@(6,6) (Figure 6). The X junction is built by joining four SWCNTs at an angle of 90° using pentagons and heptagons. Thus, uniformly distributed nanopores are fabricated in SS@(6,6) (Figure 6c) with 25,000 water molecules, 250 Na$^+$ ions and 250 Cl$^-$ ions between the two SS@(6,6) layers, corresponding to the salt concentration of 37.125 g/l (similar to the salinity of seawater 35 g/l). The SS@(6,6) has dimensions of 11.62×10 nm (Figure 6c). The periodic boundary condition is imposed along X, Y and Z directions with the two vacuum layers along X direction. CHARMM force field[36] is adopted to model the interactions between the atoms in the system. Water is modelled by TIP3P potential[37] and interactions between the atoms are modelled using Lennard–Jones and Coulombic terms.

Steered MD method[38] is adopted to simulate desalination process under constant pulling velocity. Note that the constant pulling velocity represents the moving speed of equilibrium position of the spring attached on the SS@(6,6), thus the SS@(6,6) are pulled under a traction force $F = k\Delta x$, where k is the spring constant and is set to 50 kcal/mol/Å, Δx is equal to $L - L_0$, L and L_0 are the actual length and equilibrium length of the spring respectively. As the SS@(6,6) endure resistance from salt solution, the filtration velocity should be slower than the constant pulling velocity. Moreover, unlike other methods in which the rigid condition is applied on the membranes,[5–10] the SS@(6,6) atoms are not held fixed during the steered MD simulation, so the nanostructure evolution of SS@(6,6) under compression can be detected (Figure 6g). Besides, the salt concentration between SS@(6,6) changes continuously during the desalination process. We can easily know the performance of SS@(6,6) for water desalination with specific salt concentration under different pressure according to Figure 1. Thus, the desalination process affected by salt concentration can also be analysed.

Initially, the created system is subjected to energy minimisation using the conjugate gradient method, and then relaxed in the Nose/Hoover isobaric–isothermal ensemble (NPT) under both the pressure 0 bar and the desired temperature (300 K) with an initial Gaussian velocity distribution consistent with the temperature for 300 ps. As shown in Figure 6, the distance between the two layers of SS@(6,6) changes from $d_{ss} = 9.89$ to 7.77 nm after relaxation. The pore size changes from 46.33 to 48.45 Å^2. Note that there exists trivial size difference among the pores on the relaxed SS@(6,6). Hereafter, desalination simulation is executed along X direction under a canonical ensemble (NVT). The total simulation time is 1,500–3,500 ps with a time step of 1 fs. A series of constant pulling velocity v are selected to simulate the desalination process. The average stress of all the carbon atoms is calculated during the simulation to detect the pressure on SS@(6,6) (P_{ss}). The distance d_{ss} is also recorded to calculate the filtration velocity and salt concentration variation.

ACKNOWLEDGEMENTS

The financial support from the Research Grants Council of the Hong Kong Special Administrative Region of China under grants (CityU-114111) is gratefully acknowledged. J.L. acknowledges the financial supports provided by the Grant 2012CB932203 of the National Key Basic Research Program of the Chinese Ministry of Science and Technology and from the Croucher Foundation CityU9500006.

CONTRIBUTIONS

J.L. and X.H. conceived the project and provided guidance. L.S. performed the theoretical computations, analysis and wrote the manuscript. J.L. and X.H. also reviewed the manuscript and offered helpful suggestions during the work.

COMPETING INTERESTS

The authors declare no conflict of interest.

REFERENCES

1. Gleick P. H. (ed.). *Water in Crisis: A Guide to the World's Fresh Water Resource* (Oxford Univ. Press, 1993).
2. Davis T. A. & Porter M. C. (eds). *Electrodialysis* (Noyes Publications, 1990).
3. Pontie, M., Derauw, J. S., Plantier, S., Edouard, L. & Bailly, L. Seawater desalination: nanofiltration-a substitute for reverse osmosis? *Desalin. Water Treat.* **51**, 485–494 (2013).
4. Fritzmann, C., Löwenberg, J., Wintgens, T. & Melin, T. State-of-the-art of reverse osmosis desalination. *Desalination* **216**, 1–76 (2007).
5. Wang, E. N. & Karnik, R. Water desalination: Graphene cleans up water. *Nat. Nanotechnol.* **7**, 552–554 (2012).
6. Cohen-Tanugi, D. & Grossman, J. C. Water desalination across nanoporous graphene. *Nano Lett.* **12**, 3602–3608 (2012).
7. Konatham, D., Yu, J., Ho, T. A. & Striolo, A. Simulation insights for graphene-based water desalination membranes. *Langmuir* **29**, 11884–11897 (2013).
8. Hauser, A. W. & Schwerdtfeger, P. Methane-selective nanoporous graphene membranes for gas purification. *Phys. Chem. Chem. Phys.* **14**, 13292–13298 (2012).
9. Koenig, S. P., Wang, L. D., Pellegrino, J. & Bunch, J. S. Selective molecular sieving through porous graphene. *Nat. Nanotechnol.* **7**, 728–732 (2012).
10. Hummer, G., Rasaiah, J. C. & Noworyta, J. P. Water conduction through the hydrophobic channel of a carbon nanotube. *Nature* **414**, 188–190 (2001).
11. Skoulidas, A. I., Ackerman, D. M., Johnson, J. K. & Sholl, D. S. Rapid transport of gases in carbon nanotubes. *Phys. Rev. Lett.* **89**, 185901 (2002).
12. Holt, J. K. *et al.* Fast mass transport through sub-2-nanometer carbon nanotubes. *Science* **312**, 1034–1037 (2006).
13. Majumder, M., Chopra, N., Andrews, R. & Hinds, B. J. Nanoscale hydrodynamics: enhanced flow in carbon nanotubes. *Nature* **438**, 44–44 (2005).
14. Wang, Z. *et al.* Polarity-dependent electrochemically controlled transport of water through carbon nanotube membranes. *Nano Lett.* **7**, 697–702 (2007).
15. Zhou, J. J., Noca, F. & Gharib, M. Flow conveying and diagnosis with carbon nanotube arrays. *Nanotechnology* **17**, 4845–4853 (2006).
16. Whitby, M. & Quirke, N. Fluid flow in carbon nanotubes and nanopipes. *Nat. Nanotechol.* **2**, 87–94 (2007).
17. Elimelech, M. & Phillip, W. A. The future of seawater desalination: energy, technology, and the environment. *Science* **333**, 712–717 (2011).
18. Hughes, Z. E., Shearer, C. J., Shapter, J. & Gale, J. D. Simulation of water transport through functionalized single-walled carbon nanotubes (SWCNTs). *J. Phys. Chem. C* **116**, 24943–24953 (2012).
19. Das, R., Ali, M. E., Abd Hamid, S. B., Ramakrishna, S. & Chowdhury, Z. Z. Carbon nanotube membranes for water purification: A bright future in water desalination. *Desalination* **336**, 97–109 (2014).
20. Song, Y. S. & Youn, J. R. Influence of dispersion states of carbon nanotubes on physical properties of epoxy nanocomposites. *Carbon* **43**, 1378–1385 (2005).
21. Coleman, J. N., Khan, U., Blau, W. J. & Gun'ko, Y. K. Small but strong: a review of the mechanical properties of carbon nanotube-polymer composites. *Carbon* **44**, 1624–1652 (2006).
22. Terrones, M. *et al.* Molecular junctions by joining single-walled carbon nanotubes. *Phys. Rev. Lett.* **89**, 075505 (2002).
23. Endo, M. *et al.* Atomic nanotube welders: boron interstitials triggering connections in double-walled carbon nanotubes. *Nano Lett.* **5**, 1099–1105 (2005).
24. Snow, E. S., Novak, J. P., Campbell, P. M. & Park, D. Random networks of carbon nanotubes as an electronic material. *Appl. Phys. Lett.* **82**, 2145–2147 (2003).
25. Hall, L. J. *et al.* Sign change of Poisson's ratio for carbon nanotube sheets. *Science* **320**, 504–507 (2008).
26. Ma, W. J. *et al.* Directly synthesized strong, highly conducting, transparent single-walled carbon nanotube films. *Nano Lett.* **7**, 2307–2311 (2007).
27. Romo-Herrera, J. M., Terrones, M., Terrones, H., Dag, S. & Meunier, V. Covalent 2D and 3D networks from 1D nanostructures: designing new materials. *Nano Lett.* **7**, 570–576 (2007).
28. Qin, Z., Feng, X. Q., Zou, J., Yin, Y. J. & Yu, S. W. Superior flexibility of super carbon nanotubes: molecular dynamics simulations. *Appl. Phys. Lett.* **91**, 043108 (2007).
29. Liu, B., Huang, Y., Jiang, H., Qu, S. & Hwang, K. C. The atomic-scale finite element method. *Comput. Methods Appl. Mech. Eng.* **193**, 1849–1864 (2004).
30. Wang, M., Qiu, X. M., Zhang, X. & Yin, Y. J. Equivalent parameter study of the mechanical properties of super carbon nanotubes. *Nanotechnology* **18**, 295708 (2007).
31. Li, Y., Qiu, X. M., Yin, Y. J., Yang, F. & Fan, Q. S. The elastic buckling of super-graphene and super-square carbon nanotube networks. *Phys. Lett. A* **374**, 1773–1778 (2010).
32. Liu, X., Yang, Q. S., He, X. Q. & Mai, Y. W. Molecular mechanics modeling of deformation and failure of super carbon nanotube networks. *Nanotechnology* **22**, 475701 (2011).
33. Pendergast, M. M. & Hoek, E. M. V. A review of water treatment membrane nanotechnologies. *Energy Environ. Sci.* **4**, 1946–1971 (2011).
34. Scarselli, M. *et al.* Applications of three-dimensional carbon nanotube networks. *Beilstein J. Nanotechnol.* **6**, 792–798 (2015).
35. Plimpton, S. Fast parallel algorithms for short-range molecular dynamics. *J. Comput. Phys.* **117**, 1–19 (1995).
36. MacKerell, A. D. *et al.* All-atom empirical potential for molecular modeling and dynamics studies of proteins. *J. Phys. Chem. B* **102**, 3586–3616 (1998).
37. Jorgensen, W. L., Chandrasekhar, J., Madura, J. D., Impey, R. W. & Klein, M. L. Comparison of simple potential functions for simulating liquid water. *J. Chem. Phys.* **79**, 926–935 (1983).
38. Izrailev S. *et al.* (eds). *Computational Molecular Dynamics: Challenges, Methods, Ideas*, 39–65 (Springer-Verlag, 1998).

9

Sequential slip transfer of mixed-character dislocations across Σ3 coherent twin boundary in FCC metals: a concurrent atomistic-continuum study

Shuozhi Xu[1], Liming Xiong[2], Youping Chen[3] and David L McDowell[1,4]

Sequential slip transfer across grain boundaries (GB) has an important role in size-dependent propagation of plastic deformation in polycrystalline metals. For example, the Hall–Petch effect, which states that a smaller average grain size results in a higher yield stress, can be rationalised in terms of dislocation pile-ups against GBs. In spite of extensive studies in modelling individual phases and grains using atomistic simulations, well-accepted criteria of slip transfer across GBs are still lacking, as well as models of predicting irreversible GB structure evolution. Slip transfer is inherently multiscale since both the atomic structure of the boundary and the long-range fields of the dislocation pile-up come into play. In this work, concurrent atomistic-continuum simulations are performed to study sequential slip transfer of a series of curved dislocations from a given pile-up on Σ3 coherent twin boundary (CTB) in Cu and Al, with dominant leading screw character at the site of interaction. A Frank-Read source is employed to nucleate dislocations continuously. It is found that subject to a shear stress of 1.2 GPa, screw dislocations transfer into the twinned grain in Cu, but glide on the twin boundary plane in Al. Moreover, four dislocation/CTB interaction modes are identified in Al, which are affected by (1) applied shear stress, (2) dislocation line length, and (3) dislocation line curvature. Our results elucidate the discrepancies between atomistic simulations and experimental observations of dislocation-GB reactions and highlight the importance of directly modeling sequential dislocation slip transfer reactions using fully 3D models.

INTRODUCTION

The strength of polycrystalline face-centered cubic (FCC) metals varies characteristically with the average grain size.[1] For grain size above 10 nm, the Hall–Petch effect that a smaller average grain size results in a higher yield stress, is confirmed by experiments,[2] constitutive modelling,[3] and atomistic simulations.[4] For a FCC polycrystal with sufficiently large grains and few short-range dislocation interactions (e.g., during stage I work hardening), the Hall–Petch effect can be rationalised in terms of the dislocation pile-up model.[1] Since the stress on the leading dislocation is proportional to the applied stress and to the number of dislocations in the pile-up, the tip of the pile-up in a smaller grain that accommodates fewer dislocations experiences a lower stress.[5] Thus, a higher applied stress is required to reach the critical stress level on the incoming side of the grain boundary (GB) to activate dislocation motion on the other side. Generally referred to as slip transfer processes, four possible lattice dislocation/GB reactions have been identified: direct transmission of the incoming dislocation, absorption of the incoming dislocation into an extrinsic GB dislocation, desorption of the GB dislocation into a neighbouring grain, and reflection back into the original grain.[6] At low levels of plastic strain, plastic deformation within individual grains is mainly carried by multiplication/generation of lattice dislocations, the resistance to which is manifested as the strength of polycrystal.[7] For polycrystalline metals with an average grain size above 100 nm or so, strength is mainly controlled by the generation of lattice dislocations. Thus, dislocation pile-ups and associated slip transfer at GBs have an important role in size dependent initiation and propagation of plastic deformation in polycrystalline metals.[7] At large plastic strain, however, significant dislocation network density dominates the work hardening; dislocation multiplication is restricted, pile-up is relaxed, and the dependence of strength on grain size then diminishes relative to the dependence on the dislocation substructure scale(s).[1]

On the basis of experimental observations, Lee et al.[6] formulated the Lee–Robertson–Birnbaum (LRB) slip transfer criteria, which take into account geometry, resolved shear stress and residual GB dislocations. In recent years, a series of in situ transmission electron microscopy (TEM) studies have been conducted in sequential slip transfer through GBs/interphase boundaries in FCC metals, hexagonal close-packed (HCP) metals and body-centered cubic (BCC) metals containing a variety of dislocations and boundaries with the influence of impurities or irradiation at different strain rate/temperature.[8] While the local stress is considered to be important for slip transfer, other factors including dislocation/GB types, lattice orientation, loading direction, dislocation impingement sites and the nature of neighboring grains can also be influential.[8] For a symmetric Σ3 coherent twin boundary (CTB) on a {111} plane, screw dislocation pile-ups can either cross slip onto a plane close and parallel to the twin plane or be absorbed by the CTB before being emitted into the twinned

[1]Woodruff School of Mechanical Engineering, Georgia Institute of Technology, Atlanta, GA, USA; [2]Department of Aerospace Engineering, Iowa State University, Ames, IA, USA; [3]Department of Mechanical and Aerospace Engineering, University of Florida, Gainesville, FL, USA and [4]School of Materials Science and Engineering, Georgia Institute of Technology, Atlanta, GA, USA.
Correspondence: DL McDowell (david.mcdowell@me.gatech.edu)

grain.[9] In this work, we focus on a Σ3 CTB because that it is a dominant feature in twinned FCC metals and manifests excellent mechanical properties;[10] moreover, it is among the most prevalent GBs in FCC polycrystals.[1]

Atomistic simulations have been performed to quantify dislocation/GB reactions.[9-14] Such simulations have found that in FCC pure metals, a screw dislocation can either directly transmit through the CTB by the Fleischer (FL) mechanism,[10,11] be absorbed and then desorbed into the twinned grain by the Friedel–Escaig (FE) mechanism,[10-12] or glide on the twin boundary plane.[12] The process that controls the reaction mechanism is subject to debate.[9] Ezaz et al.[14] proposed that the energy barrier in slip-CTB interaction is proportional to the magnitude of the Burgers vector of the residual dislocation, with the screw dislocation directly transmitting through the CTB and non-screw dislocations leaving a residual dislocation on the CTB, which then elevates the local stress and energy barrier for further dislocation transmission. Using the climbing image nudged elastic band (CINEB) method, Zhu et al.[10] found for Cu that at low applied stress, the activation energy for absorption is lower than that for direct transmission, the latter of which temporarily leaves a stair-rod dislocation on the CTB; further, the desorption energy barrier is much higher because two TB Shockley partials which were widely separated during absorption need to be constricted. Chassagne et al.[9] showed that there exists a critical reaction stress below which the screw dislocation glides on the CTB and above which the dislocation is absorbed and then desorbed into the nanotwin. In their study and that of Jin et al.,[12] no FL type direct transmission of screw dislocation through a CTB was observed.

However, typical atomistic simulations employ only an isolated, short, straight dislocation segment associated with a periodic image in a quasi 2D specimen,[9,12,13] or situate the source very near the interface with a very limited volume of a periodic unit cell.[14] In such cases of confined volumes and highly constrained simulation cell, image forces originating from the interaction between periodic or non-periodic boundaries and dislocations are typically non-negligible.[15-17] In some atomistic simulations,[18] a short crack in adjacent to GB is introduced to nucleate a 3D dislocation network which, like an indenter or a void, is not suitable for our work where a series of dislocations in a single pile-up is desired over extended distances. Moreover, experiments show that the slip transfer with an increasing number of incoming dislocations can activate additional slip systems or alter the dislocation emission.[8] It is therefore difficult to use results obtained for a single dislocation/GB interaction to extrapolate to the practical case of sequential slip transfer of multiple dislocations in a pile-up.

A commonly stated goal to enhance dislocation-based continuum modelling of polycrystals is to include effects of sequential slip transfer. Such models include the crystal plasticity finite element method (CPFEM),[19] the discrete dislocation dynamics (DDD),[20] the field dislocation mechanics[21] and the phase field method.[22] However, these continuum models do not naturally incorporate the necessary degrees of freedom associated with the GBs and other evolving internal state variables that relate to slip-transfer criteria.[23] In CPFEM, for example, a core-mantle type of approach can be employed where, compared with dislocation motion in grain interior, the slip transfer overcomes either a higher energy barrier, a larger slip resistance, a higher work hardening rate or a high dislocation density region.[19] Moreover, details of the dislocation structure, including stacking fault and core structures that affect GB slip transfer,[9,12] are not explicitly addressed in CPFEM or other continuum approaches such as DDD. While DDD updates the positions and velocities of all dislocation segments at each instant and tracks the long-range elastic interactions of dislocations, the short-range dislocation interaction follows pre-scribed rules.[20] Further, in DDD, high-angle GBs are typically avoided, considered as impenetrable obstacles, or the stress field

at the leading dislocation in the pile-up activates dislocation sources that are embedded either at GBs or in the neighbouring grain. Low-angle GBs are described by arrays of dislocations.[24] DDD largely ignores the role of GBs as dislocation sources even in the absence of pile-ups, as well as GB sliding/migration and elastic anisotropy.[25] For these reasons, atomistic simulations are preferred to understand GB structure-specific slip-transfer responses.

The sequential slip transfer process in a dislocation pile-up bypass of GBs is inherently multiscale; the GB structure evolution requires explicit atomistic treatment, whereas the dislocation pile-up itself has long-range character.[25] In this spirit, multiscale modelling has been pursued using the quasicontinuum (QC) method[26] and the coupled atomistic and discrete dislocation (CADD) method[27] to investigate sequential slip transfer. Although simulations following both methods employed a sufficiently large continuum domain to incorporate long-range fields of dislocation pile-ups and an atomistic domain for representation of GBs, to our knowledge this work has considered either quasi 2D approximations for segments or periodic boundary conditions (PBCs) along the dislocation line direction. In these models, (1) curved dislocations of mixed character have been excluded because of the 2D setup and (2) both ends of the outgoing dislocation line are forced to be 'reconnected' across the periodic boundary, effectively changing the length and forcing unrealistic constriction events. Moreover, the continuum domain in the QC approach associated with local nodes does not admit dislocations, so the finite elements must be adaptively remeshed to full atomistic resolution along the slip propagation path, even well away from interfaces; glide dislocations nucleated from either nanoindentation or a crack tip must pass through a fully resolved atomistic domain, which is relatively small due to its high computational cost and the corresponding dislocation density is much higher than that in experiments. The 2D CADD method adopts a 'detection band' to transfer dislocations between atomistic and continuum domains, yet a full 3D coupling of atomistic and discrete dislocation methods allowing curved dislocations has not yet been achieved or is in early stages of consideration.[28]

Therefore, we turn our attention to approaches which (1) describe interface reactions using fully resolved atomistics, (2) preserve the net Burgers vector and associated long-range stress fields of curved, mixed-character dislocations in a sufficiently large continuum domain in a fully 3D model, and preferably, (3) employ the same governing equations and interatomic potentials in both domains to avoid the usage of phenomenological parameters, essential remeshing operations or criteria and ad hoc procedures for passing dislocation segments between atomistic and coarse-grained atomistic domains. One such approach is the concurrent atomistic-continuum (CAC) method, a coarse-grained finite element integral formulation for the balance equations that admits propagation of displacement discontinuities (dislocations) through a lattice while employing only the underlying interatomic potential as a constitutive relation. Building on the foundation of a unified atomistic-continuum formulation,[29] CAC simulations admit descriptions of dislocations and stacking faults without adaptive refinement in the coarse-grained domain; the displacement fields of line/planar defects (e.g., Burgers vector) are smeared at interelement discontinuities. Both quasistatic[30] and dynamic CAC[31] have demonstrated capabilities to reproduce complex dislocation phenomena in FCC metals such as curved dislocation loop nucleation/migration and dislocation-void interactions.[32] Using nonlocality and an accurate representation of generalised stacking fault energy in both atomistic and coarse-grained domains, dislocations can pass through the domain interface smoothly without ghost forces or the need of overlapping pad regions.[30] The success of these calculations suggests the viability of using

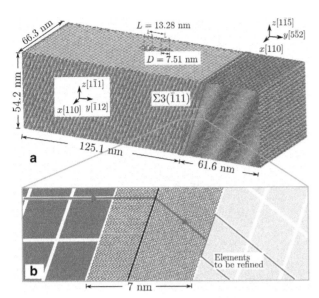

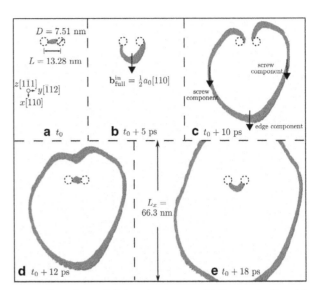

Figure 1. (**a**) Bicrystal simulation cell to study sequential slip across a Σ3 CTB in FCC metals. A pair of cylindrical holes is introduced as an FR source for dislocation multiplication. An atomistic domain is meshed in the vicinity of CTB, FR source, and at the otherwise zigzag boundaries; away from the CTB, holes, and boundaries are those of the coarse-grained finite elements, each of which contains 2,197 atoms. All boundaries are assumed stress free to allow full 3D description. (**b**) A zoom-in of the CTB region shows that the elements marked by the blue lines need to be refined for outgoing dislocations to propagate as they exit the atomistically resolved domain of the CTB. Note that all distances labelled here are for Cu; the size of the model for Al differs by the ratio of their lattice parameters.

Figure 2. Snapshots of dislocation loop multiplication in Cu subject to the applied shear stress σ_{xz} between a pair of cylindrical holes, which serve as an FR source. Atoms are colored by adaptive common neighbor analysis:[49] red are of HCP local structure, blue are not coordinated as either FCC or HCP, and all FCC atoms are deleted. In **a**, a straight edge dislocation is introduced between two cylindrical holes. In **b**, the dislocation reaches the critical semi-elliptical configuration; then it continues growing in **c** until a dislocation loop is formed in **d**. In **e**, the segments of dislocation loop with edge component are swept out at the stress-free boundaries, leaving a curved dislocation moving along the positive y direction towards the CTB. Similar phenomena are observed for Al.

CAC simulations in the context of sequential slip transfer across GBs in FCC metals.

In this paper, we perform CAC simulations for Cu and Al to clarify the mechanisms of the interactions between a number of curved incoming dislocations in a single pile-up and a Σ3 CTB, to our knowledge the first attempt to do so in the literature. Dewald and Curtin[27] conducted simulations of straight pure screw dislocation segments with a Σ3 CTB for Al using the CADD method, and had elucidated several additional criteria for slip transfer beyond those of the classical LRB criteria;[6] however, we are interested in exploring whether the findings of that kind of quasi-2D study hold up for cases of larger scale full 3D simulations of sequential dislocation slip transfer reactions, which prevail in *in situ* TEM experiments.[33] The differences between experimental and computational studies are analysed. These two FCC systems are selected based on the works of Chassagne *et al.*[9] and Jin *et al.*[12,13] as they have significantly different stable/unstable stacking/twin fault energies that affect the slip transfer reactions; moreover, they have planar dislocation cores,[34] which can be well accommodated along interelement boundaries in CAC. Dislocation reactions upon transmission and the corresponding CTB structure evolution will be investigated for a sequence of reactions of successive dislocations in a pile-up. We explore three scientific questions:

(1) How do the interactions of curved, mixed character dislocations with the CTB in a full 3D simulation differ from those of straight dislocation/CTB interactions in a quasi 2D simulation cell?

(2) How does the dislocation/CTB reaction change subject to different applied shear stress, dislocation line length, and dislocation line curvature?

(3) How do initial slip transfer events influence subsequent interactions of pile-up dislocations with the CTB?

While large-scale atomistic simulations are desirable in studying slip transfer across GBs, our purpose in this paper is to demonstrate the efficacy of coarse-graining in facilitating parametric studies of dislocation/GB reactions concerning a wide range of dislocations and GBs for the same computational resources. It is anticipated that this kind of coarse-grained modelling may assist in formulating constitutive laws and rules in describing slip transfer that may be useful upstream in DDD and CPFEM simulations.

RESULTS

The bicrystal simulation cell containing a Σ3 CTB is shown in Figure 1, with a pair of cylindrical holes introduced in the incoming grain throughout the specimen along the z direction. Full atomistic resolution is applied within 7 nm in the vicinity of CTB, around the holes, and at the otherwise zigzag boundaries;[30] away from the CTB, holes, and boundaries are the interfaces of discontinuous elements. The cylindrical holes effectively serve as a Frank–Read (FR) source to generate a series of curved dislocations in a single-ended pile-up as they are shown to strongly pin dislocations in FCC metals in the presence of stress-free boundaries.[35] Such a simulation cell is introduced because we are interested in promoting a full 3D effect and modeling specimen thicknesses that are comparable to those of TEM foils to facilitate more direct comparison with *in situ* TEM experiments. Effects of free surfaces on dislocation curvature are of interest; to the best of our knowledge, these effects have not been pursued in prior atomistic simulations. Details of the simulations are supplied in Materials and Methods.

In both Cu and Al, the initially straight edge dislocation segment that was anchored at two cylindrical holes in the absence of external stress responds to application of the shear stress σ_{xz} by

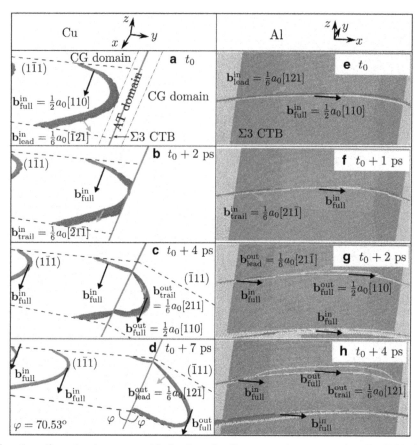

Figure 3. Snapshots of dislocation pile-up impingement with dominant leading screw character against $\Sigma3$ CTB in both Cu (**a–d**) and Al (**e–h**). Atoms are colored in the same way as in Figure 2. A fully atomistic domain is meshed in the vicinity of the CTB, as shown in **a**. In both materials, the incoming dislocation is constricted at the CTB, where two 30° Shockley partial dislocations are recombined into a full dislocation. In Cu, the dislocation cuts into the outgoing twinned grain and is redissociated into two partials; in Al, the redissociated dislocation is absorbed by the CTB, with two partials gliding on the twin plane in the same direction. Different views of the CTB are taken for Cu and Al, which are illustrated in the first row of each column.

bowing out, as shown in Figure 2. In Figure 2b, the critical configuration of a semi-ellipse is reached; then the segment continues growing until two parts contact and annihilate each other, as shown in Figure 2c. In Figure 2d, a dislocation loop is formed, leaving a straight edge dislocation segment behind as the new FR source. The segment of the dislocation loop with edge component has a wider core than that with screw component, consistent with both elasticity theory[5] and atomistic simulations.[36] As the dislocation loop continues to expand, the segment with edge component exits at the stress-free surface, leaving a screw dislocation dipole in the incoming grain, as shown in Figure 2e. Subject to a shear stress σ_{xz}, one curved dislocation segment moves towards the CTB, whereas the other eventually exits at the leftmost stress-free boundary.

In both materials, the core of the dislocation is split into two 30° Shockley partial dislocations, i.e.,

$$\frac{1}{2}a_0[110]^{in} \rightarrow \frac{1}{6}a_0[121]^{in}_{lead} + \frac{1}{6}a_0[21\overline{1}]^{in}_{trail} \tag{1}$$

Here and throughout the remainder of this paper, the superscripts 'in' and 'out' are used to distinguish the two grains in which the dislocation is located, whereas the subscripts 'lead' and 'trail' refer to leading and trailing partial dislocations, respectively. The reaction is energetically favourable according to Frank's rule.[5] In the coarse-grained domain, the two partials are separated by 29 and 19 Å in Cu (Figure 3a) and Al (Figure 3e), respectively, somewhat wider than separations computed using full atomistic simulations. Once the dislocation migrates into the atomistic

domain in the vicinity of the CTB, it obtains the same stacking fault width and core structure as those from full atomistics[30] and feels a repulsive force from the CTB in Cu but an attractive force in Al because of the relatively low CTB shear strength in Al.[37] At a shear stress of 1.2 GPa, two dislocation-CTB interaction modes are found:

In Cu, the leading partial in the incoming grain is stopped at the CTB, with the stacking fault width constricting up to the point where the trailing partial also reaches the CTB, as shown in Figure 3b. Then the full dislocation is redissociated into two 30° Shockley partials and transmitted into the outgoing grain, i.e.,

$$R \cdot \frac{1}{2}a_0[110]^{in} \rightarrow \frac{1}{2}a_0[110]^{out} \rightarrow \frac{1}{6}a_0\left[12\overline{1}\right]^{out}_{lead} + \frac{1}{6}a_0[211]^{out}_{trail} \tag{2}$$

Here the slip plane in the outgoing grain is $(\overline{1}11)$ and the rotation matrix between two grains is

$$R = \frac{1}{3}\begin{pmatrix} 1 & 2 & 2 \\ 2 & 1 & -2 \\ -2 & 2 & -1 \end{pmatrix} \tag{3}$$

In Al, the leading and trailing partial dislocations are spontaneously absorbed by and constricted at the CTB because of the attractive force. However, instead of transmitting into the outgoing grain, the full dislocation is re-dissociated to two 30° Shockley partials in the twin plane, i.e.,

$$R \cdot \frac{1}{2}a_0[110]^{in} \rightarrow \frac{1}{2}a_0[110]^{out} \rightarrow \frac{1}{6}a_0\left[21\overline{1}\right]^{out}_{lead} + \frac{1}{6}a_0[121]^{out}_{trail} \tag{4}$$

where the slip plane (i.e., twin plane) for the outgoing grain is $(1\overline{1}1)$.

In both materials, up to three dislocation loops are nucleated from the FR source and sequentially interact with the CTB. Each time a dislocation loop is generated, the cylindrical holes acquire steps at the free surface that potentially affects the source behavior; however, this is not of significant concern here as our intent is only to introduce successive dislocations on the same slip plane. Driven by the shear stress, the first dislocation migrates toward the CTB but stops at a distance when the applied shear stress, dislocation self force, repulsive force from the CTB, and the force required to create surface steps are balanced during energy minimisation. In the subsequent quenched dynamics, the second dislocation loop is formed, which drives the first dislocation towards the CTB. The CTB structure is found by energy minimisation. The last dislocation is nucleated from the FR source after the first but before the second dislocation/CTB interaction occurs. As more dislocation loops are formed, the FR source becomes progressively exhausted for a given applied stress as dislocations interact with upstream dislocations in the pile-up.

To better understand the geometric conditions involved in our simulations and compare our results with quasi 2D simulations and in situ TEM experiments in the literature, we also studied cases of different applied shear stresses, dislocation line length, and dislocation line curvature.

DISCUSSION

The critical stress for an edge dislocation bow-out between two cylindrical holes can be estimated by Scattergood–Bacon equation,[38] which considers the image force on dislocations due to the hole surface and is confirmed by atomistic simulation[39] to accurately predict the critical stress at the nanometer scale, i.e.,

$$\tau_c = \frac{\mu b}{2\pi(L-D)}\left[\ln\left(\frac{r_0}{D} + \frac{r_0}{L-D}\right)^{-1} + 1.52\right] \tag{5}$$

where μ is the shear modulus for $[110]\langle 1\bar{1}1\rangle$ slip system, L the distance between the centers of two cylindrical holes, D the diameter of each hole, and r_0 the dislocation core radius. If we set r_0 to b, equation (5) gives $\tau_c = 1.18$ GPa for Cu and 803 MPa for Al. In our simulations, after generating the first dislocation, the applied stress is ramped to and maintained at 1.2 GPa for both materials to bow out subsequent dislocations from the FR source and drive curved dislocations towards the CTB. The image forces on a dislocation with dominant screw character caused by stress-free boundaries, which are naturally captured in CAC via nonlocal, nonlinear interatomic force instead of by superposition via linear elastic solutions as in DDD,[17] are much smaller than those on edge and mixed type dislocations.[36] Particularly for the top and bottom surfaces, the image forces are negligible because the Burgers vector is parallel to these boundaries.[36]

Our CAC simulations suggest that for both Cu and Al, the dissociated dislocations in the incoming grain are constricted into a full dislocation at the CTB. This is because the arrangement of the leading and trailing dislocations must be switched before either entering the outgoing grain or gliding on the twin plane. Using a dislocation extraction algorithm (DXA),[40] we find that the reaction of dislocation at the CTB in terms of the transformed Burgers vector is

$$[1.28, 0.74, 0]_{\text{lead}}^{\text{in}} + [1.28, -0.74, 0]_{\text{trail}}^{\text{in}} \rightarrow [1.28, -0.32, 0.54]_{\text{lead}}^{\text{out}}$$
$$+ [1.28, 0.32, -0.54]_{\text{trail}}^{\text{out}} \tag{6}$$

for Cu, and

$$[1.43, 0.83, 0]_{\text{lead}}^{\text{in}} + [1.43, -0.83, 0]_{\text{trail}}^{\text{in}} \rightarrow [1.43, -0.26, 0.65]_{\text{lead}}^{\text{out}}$$
$$+ [1.43, -0.26, -0.65]_{\text{trail}}^{\text{out}} \tag{7}$$

for Al. Equations (6) and (7) show that for at least one of the incoming partials, the y component of the transformed Burgers vector, i.e., the pure edge component, changes sign, a process not

possible without recombination and redissociation since no stair-rod dislocation (as in the FL mechanism) is observed at the CTB. This requirement of two partial dislocations to exchange their order at the CTB is attributed to the twin symmetry.[12,13] Particularly for a 3D curved dislocation, part of the segment with pure screw component is constricted first, leaving the dissociated segment with mixed type component behind in the incoming grain, as shown in Figures 3c,g. This phenomenon, similar to the twinning dislocation multiplication at a CTB,[16] cannot be described using quasi 2D simulations but is reported in experiments.[8]

For Cu, the leading dislocation does not penetrate into the outgoing grain when the trailing partial is still in the incoming grain; the direct transmission of dislocations by the FL mechanism is not observed. In our simulation, the dislocation reaction is of FE type, i.e., the dissociated dislocation is always recombined at the CTB before any further motion can proceed, a phenomenon supported by atomistic simulations.[9,12] Interestingly, we did not find any CTB dislocations in the process of dislocation constriction, while NEB[10] predicts that the dislocations are absorbed to form CTB dislocations followed by desorption. We note that molecular statics is used by practitioners of NEB to determine initial and final replicates and molecular dynamics (MD) is sometimes employed to explore candidate transition states between these two replicates, before the 0 K energy minimised NEB method is used to identify the correct saddle point on the minimum-energy pathway. In the case of extended defects with complex reactions, this can be problematic when using overdriven dynamics of MD since the pathway taken in such simulations may be away from the near equilibrium trajectory associated with thermally activated, low-stress regime. After the first dislocation passes the CTB, the elements with edges along the blue lines in Figure 1 are refined, because the outgoing dislocation path is not aligned with the interelement boundaries and it would therefore be impeded, requiring cross-slip along the atomistic/coarse-grained interface and posing an aphysical back stress acting on subsequent dislocation reactions. The outgoing dislocation then continues migrating on the $(\bar{1}11)$ plane until it exits the rightmost stress-free boundary.

For Al, the incoming dislocation is always spontaneously absorbed and constricted by the CTB. After recombination, the full dislocation is split onto the CTB instead of passing it because of its relatively higher stable stacking fault energies.[9,12] In other words, the screw dislocation cross-slips on the CTB following the FE mechanism. The 30° CTB partials, with Burgers vectors parallel to the CTB, belong to the displacement shift complete lattice for the CTB and therefore migrate freely in the same direction. Our result differs from those obtained using quasi 2D atomistic[9,12] and multiscale simulations via the CADD method,[27] where two CTB partials move in opposite directions, adding one layer of atoms to the outgoing grain at the expense of the incoming grain, a process termed detwinning. In our simulation, only a local detwinning process that grows the incoming grain is temporarily observed; the CTB remains perfect after both partials glide to and are eventually swept out at the top stress-free boundary.

The different dislocation reactions at the CTB for Cu and Al under the same applied stress may relate to the fact that the normalised Hall–Petch coefficient for Cu is about twice that for Al.[1] For both materials, since each dislocation-CTB slip transfer event does not leave residual Burgers vector in the CTB interface, the interaction mechanism for subsequent dislocations is found to be precisely the same as for the first dislocation for each same material. We emphasise that this finding is particular to the Σ3 CTB and to the nature of the incoming dislocations considered here.

To further explore the differences between our simulations and those in the literature, we vary the applied stress from 100 MPa up to 2.4 GPa for Al, with an increment of 100 MPa. Note that a shear stress lower than 803 MPa is reached using the Parrinello–Rahman

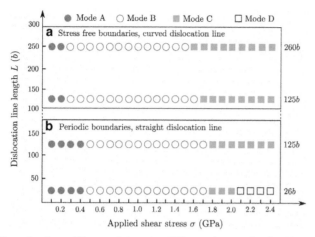

Figure 4. Four different dislocation-CTB reaction modes for Al, as a function of applied shear stress, dislocation line length, and dislocation line curvature. Prior atomistic simulations and multiscale methods in the literature only reported mode A reaction while other modes are observed in *in situ* TEM experiments. The surface steps at the stress-free boundaries in **a** retain curved dislocations while the PBCs applied on the dislocation line direction in **b** result in straight dislocation segments, regardless of the dislocation line length.

(PR) approach after each dislocation loop is emitted from the source at a higher stress, past the saddle point of the transition. With a Peierls stress of about 16 MPa[36] for a screw dislocation in Al, the stress levels employed in our simulations are high enough to retain a curved dislocation induced by the surface steps at stress-free boundaries. To study the influence of dislocation line length which is 125b shown in Figure 1, we investigate models with smaller dislocation line lengths L_x of 125b and 26b, respectively, where b is the magnitude of Burgers vector of a full dislocation. Since 26b is too small to accommodate the FR source, we introduce three screw dislocations spaced 50 nm from each other on the mid plane normal to z axis via a Volterra displacement followed by energy minimisation. For $L_x = 125b$, we also study the effect of dislocation line curvature, which is adjusted by varying the boundary conditions along the dislocation line direction: applying PBCs and stress-free boundary conditions result in a straight and a curved dislocation line, respectively.

It is found that four dislocation-CTB interaction modes exist for Al, which relate to applied shear stress, dislocation line length, and dislocation line curvature, as shown in Figure 4. The detailed stress-dependent dislocation-CTB reactions in the case of $L_x = 26b$ are shown in Figure 5, which agree with the equivalent fully atomistic simulations with a combined quenched dynamics and quasistatic periodic energy minimisation scheme. In mode A, a screw dislocation is first absorbed by the CTB subject to 100 MPa applied stress, then the two CTB partials move in opposite directions, migrating the CTB and growing the outgoing grain, as shown in Figure 5b. This mode is reported in both atomistic[9,12] and multiscale simulations,[26] but is not observed using *in situ* TEM to the best of our knowledge. At a higher stress of 500 MPa, the screw dislocation cross-slips onto the CTB via the FE mechanism, forming two CTB partials that move in the same direction, as shown in Figure 5c. This interaction result, referred to as mode B, has been reported in a high resolution TEM experiment.[41] When the applied stress increases to 1.8 and 2.1 GPa, dislocations are desorbed by the CTB into either the outgoing or both grains, referred to as modes C and D, respectively, as shown in Figures 5d,e. The penetration of a curved dislocation across a CTB in Al is observed in *in situ* TEM experiments in the presence of a large dislocation pile-up.[33] Note that for all four modes, the incoming dislocations are recombined before being redissociated at the CTB. The stress dependence of the dislocation/GB reaction agrees

with atomistic[9,12] and QC simulations,[26] as well as *in situ* TEM experiments.[8]

For the same dislocation line length of 125b at a shear stress of 300 or 400 MPa, the reaction of a quasi 2D straight dislocation segment at the CTB follows mode A, whereas the 3D curved dislocation line interacts with the CTB via mode B. Although our simulations consider all contributions to energy change, including surface steps at the stress-free boundaries, it is not the surface steps but the dislocation line curvature that leads to different reaction modes, because the surface steps are far from the center of the dislocation line where the dislocation/CTB interaction is locally initiated. In quasi 2D, the dislocation only has pure screw component along its straight segment; in a fully 3D model, an initially curved dislocation first encounters the CTB with dominant leading screw character, whereas the remainder of the dislocation line is of mixed type. This dislocation line curvature dependence of the dislocation/CTB reaction has not been previously reported in the literature to the best of our knowledge.

With the same boundary conditions, dislocations with different line lengths interact with the CTB differently. In quasi 2D, the mode D reaction is only observed for $L_x = 26b$ but not for $L_x = 125b$. In a 3D model at a shear stress of 1.6 GPa, dislocations with a line length of 125b and 260b lead to mode B and C reactions with the CTB, respectively. Previous NEB calculations in Al show that a screw dislocation segment shorter than 22b cross-slips via the FL mechanism, whereas a segment longer than 22b cross-slips via the FE mechanism.[42] The dislocation line length dependence of the dislocation/CTB reaction can be attributed to the energetics of dislocation constriction and dissociation, which are length dependent. Similarly, constraints on the twin boundary length also affect the type of dissociation/absorption events that are possible. This is an important result, as it indicates limitations on veracity of computed results for slip transfer using quasi 2D simulations in which the incoming straight dislocation line is of pure screw character in confined volumes. We remark that in reality the probability of such an encounter is exceedingly low.

There are four possible reasons why atomistic[9,12] and multiscale simulations based on CADD,[27] either involving a single dislocation or dislocation pile-up, predict only mode A interaction, but not other modes that are observed in *in situ* TEM experiments. The first reason is the accuracy of interatomic potential. Refs 12 and 27 use Mishin's embedded atom method (EAM) potential[43] and Ercolessi-Adams EAM potential,[44] respectively, whereas ref 9 shows that both EAM potentials predict the same interaction mode. The second reason is the stress level; in ref 12 a 100 MPa shear stress is applied, whereas in refs 9 and 27 no external shear stress is applied, and the screw dislocation moves towards the CTB because the attractive stress between them overcomes the Peierls stress. On the other hand, while it is difficult to measure local stress in TEM experiments,[8] it is expected that the stress level at the tip of the pile-up is much higher than 100 MPa because a large number of dislocations exist in a pile-up. The third reason is the atomic scale CTB structure: both TEM experiments[9] and MD simulations[45] show that a defected CTB responds differently from a perfect one. The last possible reason is the boundary condition. In most atomistic and multiscale simulations, PBCs are imposed along the dislocation line direction to enforce a short, straight dislocation segment, whereas dislocations in experiments are usually much longer and curved.

In this paper, 0 K quenched dynamic CAC simulations with periodic energy minimisation are employed to study 3D sequential slip transfer across a Σ3 CTB in Cu and Al to render interface reactions that may be considered close to minimum energy pathways for thermally activated processes. A series of curved dislocations are nucleated from an FR source, which then move towards the CTB subject to a constant applied shear stress. Although the leading screw segment cuts into the twinned grain in Cu, it is absorbed and glides on the CTB in Al. In particular for Al,

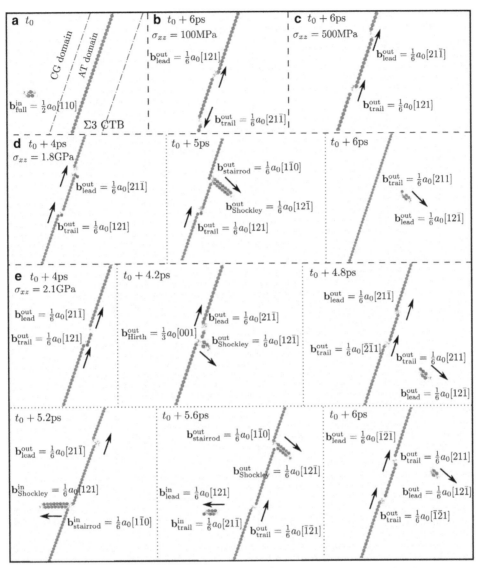

Figure 5. Snapshots of a straight screw dislocation interacting with a Σ3 CTB in a quasi 2D model for Al. Atoms are colored in the same manner as in Figure 2. The length of the dislocation line is reduced from 260b in Figure 1 to 26b. (**a**) Screw dislocation is in the coarse-grained domain; (**b**) at an applied stress of 100 MPa, the incoming dislocation splits into two Shockley partials, which then move in opposite directions; (**c**) at 500 MPa applied stress, the two CTB partial dislocations move in the same directions, leaving no CTB migration behind; (**d**) at 1.8 GPa applied stress, the incoming dislocation is first absorbed by the CTB then desorbed into the twinned grain, leaving behind a perfect boundary, similar to the NEB prediction for Cu;[10] (**e**) at the largest stress of 2.1 GPa, dislocations are desorbed into both incoming and outgoing grains.

four dislocation/CTB interaction modes are identified, which are affected by (1) applied shear stress, (2) dislocation line length, and (3) dislocation line curvature. Although we do not explicitly investigate the effect of the number of incoming dislocations, the effect of a dislocation pile-up can be manifested by varying the applied stress according to recent MD simulations.[17] In all cases studied in this paper, the dislocation/CTB reactions always follow the recombination-redissociation process, without forming any CTB dislocations in process of recombination. We emphasise that the results obtained for a single dislocation/GB interaction cannot be directly extrapolated to understand the practical case of slip transfer of dislocations in a pile-up in which the leading dislocation experiences a higher stress. The discrepancies between prior computational studies and experiments highlight the significance of the current work: it is important to directly model dislocation pile-ups, to let dislocations evolve freely in 3D, and to probe the mechanisms of slip transfer in polycrystalline and

twinned metals using sufficiently large models. The slip transfer of more general dislocation types with different curvatures interacting with the Σ3 CTB, other GBs, and the influence of FR source exhaustion will be addressed in future work.

We note that these quasi-2D limitations are not confined to slip transfer processes. For example, there exists a characteristic length of a screw dislocation segment in BCC Fe above which more than one kink-pair can form which then strongly influences the dislocation mobility.[34] These observations suggest that 3D modeling using faithful interatomic potentials, though much less prevalent in the literature than quasi 2D simulations, is potentially of high utility in exploring more realistic dislocation behaviour. In this regard, the fully 3D CAC method opens a promising avenue to explore sequential slip transfer reactions, allowing the possibility for direct comparison of numerical simulations with quasi 4D (spatial correlation with time) characterisation of dislocations reconstructed from *in situ* TEM experiments.[8]

MATERIALS AND METHODS

For FCC Cu and Al, the EAM potentials of Mishin et al.[43,46] (with appropriate approximation in coarse-grained elements[30]) are employed because the evaluated stable and unstable stacking fault energies are close to experimentally measured values.[9] In the coarse-grained domain, 3D rhombohedral elements are employed with surfaces corresponding to {111} slip planes. Within each element, piecewise continuous first order shape and interpolation functions are employed. Between elements, neither displacement continuity nor interelement compatibility is required. We choose a uniform element size of 2197 atoms solved by first order Gaussian quadrature because it strikes a balance between high accuracy and high efficiency; approximations employed in CAC simulations have been explained and evaluated.[30] The lattice orientations are $x[110], y[\bar{1}12]$, and $z[1\bar{1}\bar{1}]$ in the incoming grain and $x[110], y[\bar{5}\bar{5}2]$, and $z[11\bar{5}]$ in the outgoing grain, respectively. The domain around the FR source has full atomistic resolution because (1) the elements are of rhombohedral shape and (2) dislocation nucleation from a free surface can't be accurately described by the coarse-grained domain alone.[30] Similar to ref. 35 we find that the Lomer–Cottrell type sessile lock is either unstable or does not act as dislocation source in our model. The simulation cell size of the two materials differs only by their lattice parameter a_0, which is 3.615 Å for Cu and 4.05 Å for Al. The length along each direction in each grain is chosen by taking several factors into account: (1) curved dislocations should evolve freely in 3D—the box length along the x direction L_x for both grains needs to be large enough, e.g., $L_x = 66.3$ nm $= 260b$ for Cu is sufficient to distinguish between quasi 2D and 3D[34] and allow dislocation segments to have a non-negligible curvature between stress-free surfaces, (2) the possible CTB dislocations need to migrate far from the initial site of dislocation/CTB interaction to minimise the back stress on subsequent slip transfer, so the box length along z direction for Cu is assigned as $L_z = 54.2$ nm $= 213b$, which is larger than $196b$ used in a QC simulation;[26] (3) the outgoing dislocations should also be allowed to travel far from the interaction site on slip plane, so the outgoing grain length along y direction is selected for Cu as $L_y^{out} = 61.6$ nm $= 242b$; (4) to minimise the image forces from the leftmost boundary, the incoming grain length along the y direction L_y^{in} is chosen to be 125.1 nm for Cu. As a result, the simulation cell contains 21,812 elements and about 9 million atoms, with 9,076,792 degrees of freedom in total compared with otherwise 56,823,260 in an equivalent full atomistic model. Our previous work shows that the coarse-graining efficiency of quasistatic CAC is about 20% higher than that estimated based on the number of degrees of freedom because the outer iteration in each energy minimisation step converges faster in the coarse-grained domain.[30] The dislocation density in the incoming grain containing one dislocation loop is about $2.95 \times 10^{14}/m^2$.

To account for both dislocation migration over relatively long distances and CTB structure evolution, we take a combined approach of quenched dynamic CAC accompanied by periodic quasistatic energy minimisation in the simulations. The quasistatic CAC[30] employs the conjugate gradient method to minimise energy at increments of applied loading; the convergence criterion corresponds to the ratio of absolute energy variation between successive iterations to the energy magnitude to lie below 10^{-6}. In quenched dynamic CAC, the atomic/nodal velocities are adjusted by atomic/equivalent nodal forces[30] following the 'quick-min' MD approach,[47] which, like dynamic CAC,[31] performs damped dynamics but without the damping term. Specifically, at each step in quenched dynamics, the velocity is projected in the direction of the force, with its component that is normal to the force discarded; then if the new velocity is antiparallel to the force, it is zeroed. The idea of quenched dynamics is to gradually drain energy by occasionally zeroing certain velocity components such that the system energy is forced towards a minimum at 0 K.[47] Employed in CAC for the first time, this is important in the current simulations since multiple dislocations are generated from FR sources at applied stress that progressively exceeds the critical stress, and then driven towards the CTB interface. As a result, the dislocated ensemble evolves away from equilibrium and periodic energy minimisation must therefore be regarded as a means of constrained optimisation for a sequence of nonequilibrium configurations. The quenched dynamics and quasistatic methods are shown to adequately avoid typical issues related to overdriven kinetics in dynamic simulations;[47] in particular, the 'quick-min' method is employed to find the minimum energy path in NEB calculations.[10] In this paper, we employ zero temperature CAC simulations to alleviate the finite temperature effect and focus on obtaining trajectories that are close to minimum energy pathways for evolution of structure during slip transfer because even MD cannot account for thermally activated processes that are important in explaining TEM results.[9] The Velocity Verlet algorithm is used to integrate the equation of motion, with a constant time step of 2 fs for both domains. Note that the time advancement in this paper is for quenched dynamics only. For all practical purposes, we regard the simulations to pertain to 0 K (or very nearly so) conditions.

In preparation for the simulations, we first compute the 0 K equilibrium structure of the CTB, which due to its simplicity can be achieved using energy minimisation without trying different in-plane translations.[1] A straight edge dislocation is introduced on the mid plane normal to z axis by moving atoms/nodes between two cylindrical holes by Burgers vector $b = \frac{1}{2}a_0[110]$, followed by energy minimisation. Only one dislocation segment is created to avoid the interaction between a dislocation dipole on parallel slip planes.[39] Once the equilibrium atomic configuration of the FR source is achieved, quenched dynamic CAC simulations are carried out with an increment of applied shear stress $\Delta\sigma_{xz} = 4$ MPa for each step until the shear stress reaches 1.2 GPa, which is then maintained using the PR method.[47] To prevent the simulation cell from rotating around the y axis, all nodes in the coarse-grained domain are constrained within the x–y plane, while the atoms are free to move to relax the stress free boundaries and obtain an energy minimised boundary structure. Note that while a large shear stress is necessary to bow out dislocations from the FR source and drive the dislocation pile-up towards the CTB, the high stress and accompanying inertia effect favour transmission over absorption.[9] Note that the quenched dynamic CAC still has some, albeit small, inertia effects because it is not strictly static minimisation. To alleviate inertia effects and accurately reproduce the CTB structure during dislocation impingement, the quasistatic CAC simulation (0 K energy minimisation) is performed every 500 steps before and every 100 steps after the first dislocation/CTB reaction. In the coarse-grained domain, the post-processing is performed after the atomic positions are interpolated from the nodal positions. Simulation results are visualised using Paraview[48] and OVITO;[49] the DXA[40] is also employed to identify dislocations and associated Burgers vector. Some runs are completed using Blacklight and Comet on Extreme Science and Engineering Discovery Environment (XSEDE).[50]

ACKNOWLEDGEMENTS

These results are based upon work supported by the National Science Foundation as a collaborative effort between Georgia Tech (CMMI-1232878) and University of Florida (CMMI-1233113). Any opinions, findings, and conclusions or recommendations expressed in this material are those of the authors and do not necessarily reflect the views of the National Science Foundation. The work of L.X. was supported in part by the Department of Energy, Office of Basic Energy Sciences under Award Number DE-SC0006539. The authors would like to thank Dr Zhaohui Jin, Dr Chaitanya Deo, Dr Josh Kacher, Dr Shreevant Tiwari, Dr Garritt Tucker, Dr Shengfeng Yang, Mr Matthew Priddy, Mr Zhi Zeng, and Mr Rui Che for helpful discussions, and Dr Alexander Stukowski for providing the dislocation extraction algorithm code. This work used the Extreme Science and Engineering Discovery Environment (XSEDE), which is supported by National Science Foundation grant number ACI-1053575.

CONTRIBUTIONS

D.L.M. and Y.C. conceived the research and provided guidance. S.X. performed computer simulations and wrote the paper. S.X. and L.X. developed the model and analysed the data. All authors discussed the results and revised the paper.

COMPETING INTERESTS

The authors declare no conflict of interest.

REFERENCES

1. Ramesh, K. T. Nanomaterials: Mechanics and Mechanisms (Springer, 2009).
2. Ratanaphan, S. et al. Grain boundary energies in body-centered cubic metals. Acta Mater. 88, 346–354 (2015).
3. Counts, W. A., Braginsky, M. V., Battaile, C. C. & Holm, E. A. Predicting the Hall-Petch effect in fcc metals using non-local crystal plasticity. Int. J. Plast. 24, 1243–1263 (2008).
4. Spearot, D. E. & Sangid, M. D. Insights on slip transmission at grain boundaries from atomistic simulations. Curr. Opin. Solid State Mater. Sci. 18, 188–195 (2014).
5. Hirth, J. P. & Lothe, J. Theory of Dislocations (John Wiley & Sons, 1982).
6. Lee, T. C., Robertson, I. M. & Birnbaum, H. K. TEM in situ deformation study of the interaction of lattice dislocations with grain boundaries in metals. Philos. Mag. A 62, 131–153 (1990).

7. Cottrell, A. H. in *Dislocations in Solids* Vol. 11 (eds Nabarro F. R. N. & Duesday M. S.) vii–xvii (Elsevier, 2002).
8. Kacher, J., Eftink, B. P., Cui, B. & Robertson, I. M. Dislocation interactions with grain boundaries. *Curr. Opin. Solid State Mater. Sci.* **18**, 227–243 (2014).
9. Chassagne, M., Legros, M. & Rodney, D. Atomic-scale simulation of screw dislocation/coherent twin boundary interaction in Al, Au, Cu and Ni. *Acta Mater.* **59**, 1456–1463 (2011).
10. Zhu, T., Li, J., Samanta, A., Kim, H. G. & Suresh, S. Interfacial plasticity governs strain rate sensitivity and ductility in nanostructured metals. *Proc. Natl Acad. Sci. USA* **104**, 3031–3036 (2007).
11. Zheng, Y. G., Lu, J., Zhang, H. W. & Chen, Z. Strengthening and toughening by interface-mediated slip transfer reaction in nanotwinned copper. *Scripta Mater.* **60**, 508–511 (2009).
12. Jin, Z. H. et al. The interaction mechanism of screw dislocations with coherent twin boundaries in different face-centred cubic metals. *Scripta Mater.* **54**, 1163–1168 (2006).
13. Jin, Z. H. et al. Interactions between non-screw lattice dislocations and coherent twin boundaries in face-centered cubic metals. *Acta Mater.* **56**, 1126–1135 (2008).
14. Ezaz, T., Sangid, M. D. & Sehitoglu, H. Energy barriers associated with slip-twin interactions. *Philos. Mag.* **91**, 1464–1488 (2011).
15. Szajewski, B. A. & Curtin, W. A. Analysis of spurious image forces in atomistic simulations of dislocations. *Model. Simul. Mater. Sci. Eng.* **23**, 025008 (2015).
16. Li, N. et al. Twinning dislocation multiplication at a coherent twin boundary. *Acta Mater.* **59**, 5989–5996 (2011).
17. Wang, J. Atomistic simulations of dislocation pileup: grain boundaries interaction. *JOM* **67**, 1515–1525 (2015).
18. de Koning, M., Miller, R., Bulatov, V. V. & Abraham, F. F. Modelling grain-boundary resistance in intergranular dislocation slip transmission. *Philos. Mag. A* **82**, 2511–2527 (2002).
19. Mayeur, J. R., Beyerlein, I. J., Bronkhorst, C. A. & Mourad, H. M. Incorporating interface affected zones into crystal plasticity. *Int. J. Plast.* **65**, 206–225 (2015).
20. Quek, S. S., Wu, Z., Zhang, Y. W. & Srolovitz, D. J. Polycrystal deformation in a discrete dislocation dynamics framework. *Acta Mater.* **75**, 92–105 (2014).
21. Puri, S., Acharya, A. & Rollett, A. Controlling plastic flow across grain boundaries in a continuum model. *Metall. Mater. Trans. A* **42**, 669–675 (2011).
22. Levitas, V. I. & Javanbakht, M. Phase transformations in nanograin materials under high pressure and plastic shear: nanoscale mechanisms. *Nanoscale* **6**, 162–166 (2014).
23. McDowell, D. L. A perspective on trends in multiscale plasticity. *Int. J. Plast* **26**, 1280–1309 (2010).
24. Liu, B. et al. Dislocation interactions and low-angle grain boundary strengthening. *Acta Mater.* **59**, 7125–7134 (2011).
25. McDowell, D. L. Viscoplasticity of heterogeneous metallic materials. *Mater. Sci. Eng. R Rep.* **62**, 67–123 (2008).
26. Shimokawa, T., Kinari, T. & Shintaku, S. Interaction mechanism between edge dislocations and asymmetrical tilt grain boundaries investigated via quasicontinuum simulations. *Phys. Rev. B* **75**, 144108 (2007).
27. Dewald, M. P. & Curtin, W. A. Multiscale modelling of dislocation/grain boundary interactions. II. Screw dislocations impinging on tilt boundaries in Al. *Philos. Mag.* **87**, 4615–4641 (2007).
28. Pavia, F. & Curtin, W. A. Parallel algorithm for multiscale atomistic/continuum simulations using LAMMPS. *Modelling Simul. Mater. Sci. Engl.* **23**, 055002 (2015).
29. Chen, Y. Reformulation of microscopic balance equations for multiscale materials modeling. *J. Chem. Phys.* **130**, 134706 (2009).
30. Xu, S., Che, R., Xiong, L., Chen, Y. & McDowell, D. L. A quasistatic implementation of the concurrent atomistic-continuum method for FCC crystals. *Int. J. Plast.* **72**, 91–126 (2015).
31. Xiong, L., Tucker, G., McDowell, D. L. & Chen, Y. Coarse-grained atomistic simulation of dislocations. *J. Mech. Phys. Solids* **59**, 160–177 (2011).
32. Xiong, L., Xu, S., McDowell, D. L. & Chen, Y. Concurrent atomistic-continuum simulations of dislocation-void interactions in fcc crystals. *Int. J. Plast.* **65**, 33–42 (2015).
33. Kashihara, K. & Inoko, F. Effect of piled-up dislocations on strain induced boundary migration (SIBM) in deformed aluminum bicrystals with originally Σ3 twin boundary. *Acta Mater.* **49**, 3051–3061 (2001).
34. Cai, W., Bulatov, V. V., Chang, J. P., Li, J. & Yip, S. in *Dislocations in Solids* Vol. 12 (eds Nabarro F. R. N. & Hirth J. P.) **64**, 1–80 (Elsevier, 2004).
35. Weinberger, C. R. & Tucker, G. J. Atomistic simulations of dislocation pinning points in pure face-centered-cubic nanopillars. *Model. Simul. Mater. Sci. Engl.* **20**, 075001 (2012).
36. Olmsted, D. L., Hardikar, K. Y. & Phillips, R. Lattice resistance and Peierls stress in finite size atomistic dislocation simulations. *Modelling Simul. Mater. Sci. Eng.* **9**, 215–247 (2001).
37. Chen, Z., Jin, Z. & Gao, H. Repulsive force between screw dislocation and coherent twin boundary in aluminum and copper. *Phys. Rev. B* **75**, 212104 (2007).
38. Scattergood, R. O. & Bacon, D. J. The strengthening effect of voids. *Acta Metall.* **30**, 1665–1677 (1982).
39. Shimokawa, T. & Kitada, S. Dislocation multiplication from the Frank-Read source in atomic models. *Mater. Trans.* **55**, 58–63 (2014).
40. Stukowski, A., Bulatov, V. V. & Arsenlis, A. Automated identification and indexing of dislocations in crystal interfaces. *Modelling Simul. Mater. Sci. Eng.* **20**, 085007 (2012).
41. Yang, Z. Q., Chisholm, M. F., He, L. L., Pennycook, S. J. & Ye, H. Q. Atomic-scale processes revealing dynamic twin boundary strengthening mechanisms in face-centered cubic materials. *Scripta Mater.* **67**, 911–914 (2012).
42. Jin, C., Xiang, Y. & Lu, G. Dislocation cross-slip mechanisms in aluminum. *Philos. Mag.* **91**, 4109–4125 (2011).
43. Mishin, Y., Farkas, D., Mehl, M. J. & Papaconstantopoulos, D. A. Interatomic potentials for monoatomic metals from experimental data and *ab initio* calculations. *Phys. Rev. B* **59**, 3393–3407 (1999).
44. Ercolessi, F. & Adams, J. B. Interatomic potentials from first-principles calculations: the force-matching method. *Europhys. Lett.* **26**, 583–588 (1994).
45. Wang, Y. M. et al. Defective twin boundaries in nanotwinned metals. *Nat. Mater.* **12**, 697–702 (2013).
46. Mishin, Y., Mehl, M. J., Papaconstantopoulos, D. A., Voter, A. F. & Kress, J. D. Structural stability and lattice defects in copper: Ab initio, tight-binding, and embedded-atom calculations. *Phys. Rev. B* **63**, 224106 (2001).
47. Tadmor, E. B. & Miller, R. E. *Modeling materials: continuum, atomistic and multiscale techniques* (Cambridge Univ. Press, 2011).
48. Schroeder, W., Martin, K. & Lorensen, B. *The Visualization Toolkit: An Object Oriented Approach to 3D Graphics* (Kitware, 2003).
49. Stukowski, A. Visualization and analysis of atomistic simulation data with OVITO --- the Open Visualization Tool. *Modelling Simul. Mater. Sci. Eng.* **18**, 015012 (2010).
50. Towns, J. et al. XSEDE: Accelerating scientific discovery. *Comput. Sci. Eng.* **16**, 62–74 (2014).

10

Mechanochemical spinodal decomposition: a phenomenological theory of phase transformations in multi-component, crystalline solids

Shiva Rudraraju[1], Anton Van der Ven[2] and Krishna Garikipati[3,4]

We present a phenomenological treatment of diffusion-driven martensitic phase transformations in multi-component crystalline solids that arise from non-convex free energies in mechanical and chemical variables. The treatment describes diffusional phase transformations that are accompanied by symmetry-breaking structural changes of the crystal unit cell and reveals the importance of a mechanochemical spinodal, defined as the region in strain–composition space, where the free-energy density function is non-convex. The approach is relevant to phase transformations wherein the structural order parameters can be expressed as linear combinations of strains relative to a high-symmetry reference crystal. The governing equations describing mechanochemical spinodal decomposition are variationally derived from a free-energy density function that accounts for interfacial energy via gradients of the rapidly varying strain and composition fields. A robust computational framework for treating the coupled, higher-order diffusion and nonlinear strain gradient elasticity problems is presented. Because the local strains in an inhomogeneous, transforming microstructure can be finite, the elasticity problem must account for geometric nonlinearity. An evaluation of available experimental phase diagrams and first-principles free energies suggests that mechanochemical spinodal decomposition should occur in metal hydrides such as ZrH_{2-2c}. The rich physics that ensues is explored in several numerical examples in two and three dimensions, and the relevance of the mechanism is discussed in the context of important electrode materials for Li-ion batteries and high-temperature ceramics.

INTRODUCTION

Spinodal decomposition is a continuous phase transformation mechanism occurring throughout a solid that is far enough from the equilibrium for its free-energy density to lose convexity with respect to an internal degree of freedom. The latter could include the local composition as in classical spinodal decomposition described by Cahn and Hilliard,[1] or a suitable non-conserved order parameter as in the theory by Allen and Cahn[2] for spinodal ordering. A key requirement for continuous transformations is that order parameters can be formulated to uniquely describe continuous paths connecting the various phases of the transformation. These phases then correspond to local minima on a single, continuous free-energy density surface in that order parameter space. For classical spinodal decomposition inside a miscibility gap, all phases have the same crystal structure and symmetry, and the order parameter is simply the local composition. The existence of a single, continuous free-energy density surface for all phases participating in a transformation implies, by geometric necessity, the presence of domains in order parameter space, where the free-energy density is non-convex. Reaching those domains through supersaturation (by externally varying temperature or composition) makes the solid susceptible to a generalised spinodal decomposition.

Many important multi-component solids undergo phase transformations that couple diffusional redistribution of their components with a structural change of the crystallographic unit cell. One prominent example is the decomposition that occurs when cubic yttria-stabilised zirconia $Zr_{1-c}Y_cO_{2-c/2}$ is quenched into a two-phase equilibrium region between tetragonal $Zr_{1-c}Y_cO_{2-c/2}$ having a low-Y composition and cubic $Zr_{1-c}Y_cO_{2-c/2}$ having a high-Y composition. Another occurs in Li-battery electrodes made of spinel $Li_cMn_2O_4$. Discharging to low voltages causes the compound to transform from cubic $LiMn_2O_4$ to tetragonal $Li_2Mn_2O_4$ through a two-phase diffusional phase transformation mechanism. As with simple diffusional phase transformations, these coupled diffusional/martensitic phase transformations can occur either through a nucleation and growth mechanism or, if certain symmetry requirements are met, through a continuous mechanism due to an onset of an instability with respect to the composition and/or a structural order parameter.

Here we present a treatment of coupled diffusional/martensitic phase transformations triggered by instabilities with respect to both strain and composition. These phase transformations are characterised by a mechanochemical spinodal that is defined as a non-convex region of the free-energy density function in the strain–composition space. The possibility of a mechanochemical spinodal decomposition is motivated by the recent first-principles studies of martensitic transformations in transition metal hydrides, where a high-temperature cubic phase is predicted to display negative curvatures with respect to strain, thus making strain a natural order parameter to distinguish the cubic parent phase

[1]Department of Mechanical Engineering, University of Michigan, Ann Arbor, MI, USA; [2]Materials Department, University of California, Santa Barbara, CA, USA; [3]Department of Mechanical Engineering, University of Michigan, Ann Arbor, MI, USA and [4]Department of Mathematics, University of Michigan, Ann Arbor, MI, USA.
Correspondence: K Garikipati (krishna@umich.edu)

from its low-temperature tetragonal daughter phases.[3] In addition, the coupling with composition degrees of freedom (e.g., through the introduction of hydrogen vacancies in the metal hydrides) allows for the possibility that the free energy also exhibits a negative curvature with respect to the composition. Mechanochemical spinodal decomposition, therefore, is a phenomenon that is likely present in many multi-component materials but has to date been overlooked as a mechanism by which a high-symmetry phase can decompose martensitically and through diffusional redistribution upon quenching into a two-phase regime. Structural phase transformations in solids driven by instability with respect to an internal shuffle of the atoms within the unit cell have been treated rigorously in the literature with coupled Cahn–Hilliard and Allen–Cahn approaches,[4–7] but mechanochemical spinodal decomposition is fundamentally different and necessitates a coupled treatment of both the strain and composition instabilities.

Our treatment is based on a generalised, Landau-type free-energy density function that couples strain and composition instabilities. The governing equations of mechanochemical spinodal decomposition, obtained by variational principles, generalise the classical equations of the Cahn–Hilliard formulation,[1] and of nonlinear gradient elasticity[8,9] by coupling these systems. The ability to solve this complex, nonlinear, strain- and composition-gradient-driven, mechanochemical system for sufficiently general initial and boundary value problems in two and three dimensions also has been lacking heretofore. We introduce the computational framework to obtain such solutions in general, three-dimensional solids. This new-found capability allows us to then reinforce our discussion of the phenomenology with dynamics predicted by the numerical solutions.

The mechanochemical spinodal in two dimensions

For accessibility of the arguments, we first consider the two-dimensional analogue of the cubic-to-tetragonal transformation: the square-to-rectangle transformation. The high-symmetry square lattice will serve as the reference crystal relative to which strains are measured. Lower-symmetry lattices that can be derived from the square lattice by homogeneous strain include the rectangle, the diamond and lattices, where there are no constraints on the cell lengths and their angles.

We use the composition c, varying between 0 and 1, as our order parameter for the chemistry of our binary two-dimensional solid. Symmetry-breaking structural changes are naturally described by the strain relative to the high-symmetry square lattice. The elastic free-energy density is also a function of strain. In both contexts, strain will in general be of finite magnitude. Following Barsch and Krumhansl,[8] we use the Green–Lagrange strain tensor, $\boldsymbol{E}$, relative to the square lattice. Rotations are exactly neutralised in this strain measure; for any rigid body motion, $\boldsymbol{E} = 0$ (Supplementary Information). In two dimensions, the relevant strain components are E_{11}, E_{22} and $E_{12} = E_{21}$. However, it is more convenient to use linear combinations of these components that transform according to the irreducible representations of the point group of the high-symmetry square lattice. In two dimensions, these include $e_1 = (E_{11} + E_{22})\sqrt{2}$, $e_2 = (E_{11} - E_{22})\sqrt{2}$ and $e_6 = \sqrt{2}E_{12}$. Here e_1 and e_6 reduce to the dilatation and shear strain, respectively, in the infinitesimal strain limit. The strain measure e_2 uniquely maps the square lattice into the two rectangular variants (Figure 1a): positive and negative e_2 generate the rectangles elongated along the global X_1 and X_2 directions, respectively. The equivalence of the rectangular variants under the point group symmetry of the square lattice ($e_2 = 0$) restricts the free-energy density to even functions of e_2.

If the crystalline solid has multiple chemical species, its free-energy density dependence on e_1, e_2 and e_6 can change with composition, c. Figure 2a illustrates a free-energy density

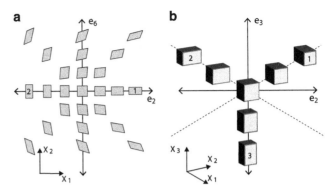

Figure 1. (**a**) Square-to-rectangle transition (2D) in reparametrised strain space. (**b**) Cubic-to-tetragonal transition (3D) in reparametrised strain space (equation 1). Lattice vectors are labelled by the corresponding coordinate directions x_1, x_2 and x_3, and the corresponding lower-symmetry phases are labelled 1, 2 and 3. The deformations shown in the sub-figures are area/volume preserving, respectively.

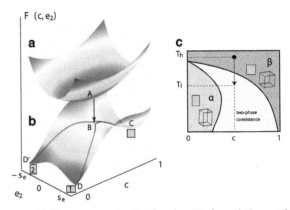

Figure 2. (**a**) Free-energy density for the 2D formulation at high temperature and (**b**) at low temperature, showing the mechanochemical spinodal along with two energy-minimising paths (brown lines) and their corresponding minimum energy-strained structures. (**c**) Temperature–composition phase diagram.

surface, $\mathcal{F}(c, e_2)$, for a binary solid that, at higher temperature, forms a solid solution having square symmetry. In this case, $\mathcal{F}$ is convex, a condition made precise by specifying positive eigenvalues of its Hessian matrix over the $\{c, e_2\}$ space. In addition, $\mathcal{F}(c, 0)$ is a minimum with respect to e_2 for fixed c, making the square phase stable for all c at this temperature. However, at a lower temperature, $\mathcal{F}$ may lose convexity, inducing the notion of a mechanochemical spinodal region. We define it as the domain in $\{c, e_2\}$ space over which the Hessian matrix admits negative eigenvalues, as illustrated in Figure 2b. Here we focus on the conditions $\partial^2 \mathcal{F}/\partial c^2 < 0$ and $\partial^2 \mathcal{F}/\partial e_2^2 < 0$. The square phase ($e_2 = 0$) remains stable at high composition ($c \sim 1$) with positive eigenvalues of the Hessian (Figure 2b). But it is mechanically unstable at low composition, with $\partial^2 \mathcal{F}/\partial e_2^2 < 0$ for $(c, e_2) \sim (0, 0)$, and has two symmetrically equivalent, stable, rectangular phases, with $\partial^2 \mathcal{F}/\partial e_2^2 > 0$ at $(c, e_2) = (0, \pm s_e)$. Although not shown in Figure 2b, we assume that $\mathcal{F}$ is convex with respect to e_1 and e_6. Figure 2c illustrates a schematic temperature–composition phase diagram consistent with the free-energy densities of Figure 2a,b. The square/cubic phase, β, forms at high composition or at high temperature; the rectangle/tetragonal phase, α, forms at low composition and low temperature. A large two-phase region separates them.

Zuzek et al.[10] have obtained phase diagrams experimentally that are topologically equivalent to Figure 2c for the ZrH_{2-2c} system. Recent first-principles calculations[3] have demonstrated the existence of a mechanical instability that exists in this system at low c via non-convexity with respect to strain. Furthermore, the two-phase coexistence at low temperature also implies non-convexity with respect to the composition, as we have demonstrated in Supplementary Information. Figure 2a,b therefore represents a two-dimensional analogue of the free energy for such systems.

Consider our model binary solid, annealed at high temperature T_h, to form a solid solution in the square phase, β. Its state is at point A in Figure 2a, with $e_2 = 0$. It is then quenched into the two-phase region (Figure 2c) with free-energy density at point B in Figure 2b. For a quench at sufficiently high rate, the dimensions of the square lattice, controlled by strains e_1, e_2 and e_6, and the composition remain momentarily unchanged. However, as the state at point B satisfies $\partial^2 \mathcal{F}/\partial e_2^2 < 0$ and $\partial^2 \mathcal{F}/\partial c^2 < 0$, there exist thermodynamic driving forces for segregation by strain and composition within the mechanochemical spinodal.

Diffusion being substantially slower than elastic relaxation, the solid immediately deforms to either positive or negative e_2, driven towards a local minimum at constant, c. These deformations due to the mechanical instability will happen like many martensitic transformations, where a mix of symmetrically equivalent rectangular variants coexist to minimise macroscopic strain energy. For finite but moderate strain, the transformation could proceed coherently,[11] even if the two symmetrically equivalent rectangular variants coexist. We neglect non-essential complexities of this process and assume that, instantly upon quenching, finitely sized neighbourhoods of the solid deform homogeneously into one of these rectangular variants at the original composition.

The solid also becomes susceptible to uphill diffusion because $\partial^2 \mathcal{F}/\partial c^2 < 0$ implies a negative diffusion coefficient. However, it does not occur at constant e_2, as the valleys traversing the local minima, $\partial \mathcal{F}/\partial e_2 = 0$, between the square lattice at $c = 1$ and the rectangular lattices at $c = 0$ span intervals of negative and positive e_2. Mechanochemical spinodal decomposition sets in. Composition modulations are amplified: high-c regions strive to be more square (point C), whereas low-c regions strive to be more rectangular (points D or D'). However, as coherency is maintained, some neighbourhoods in the solid will be frustrated from attaining strains that ensure minima, $\partial \mathcal{F}/\partial e_2 = 0$, for local values of c. Coherency strain-induced free-energy penalties arise to alter the driving forces for purely chemical spinodal decomposition. Supplementary Movie S5 in the Supplementary Information shows the evolution of the state (c, e_2) of the material points on the free-energy manifold $\mathcal{F}$.

Mathematical formulation: three dimensions

Armed with the insight conveyed by the two-dimensional study, we next lay out the three-dimensional treatment, in which setting the considered phase transformations proceed via lattice deformation and diffusion. The arguments are made more concrete by considering the general mathematical form of the free-energy density.

Strain order parameters. The strain measures e_1–e_6 are first redefined using the full Green–Lagrange strain tensor components in three dimensions:

$$e_1 = \frac{1}{\sqrt{3}}(E_{11} + E_{22} + E_{33}), e_2 = \frac{1}{\sqrt{2}}(E_{11} - E_{22}),$$
$$e_3 = \frac{1}{\sqrt{6}}(E_{11} + E_{22} - 2E_{33}), e_4 = \sqrt{2}E_{23} = \sqrt{2}E_{32}, \quad (1)$$
$$e_5 = \sqrt{2}E_{13} = \sqrt{2}E_{31}, e_6 = \sqrt{2}E_{12} = \sqrt{2}E_{21}$$

In the limit of infinitesimal strains, e_1 describes the dilatation, whereas e_4, e_5 and e_6 reduce to shears. The point group operations of the cubic crystal leave e_1 invariant, as it is the trace of $\mathbf{E}$, while collecting each of the subsets $\{e_2, e_3\}$ and $\{e_4, e_5, e_6\}$ into a symmetry-invariant subspace whose elements transform into each

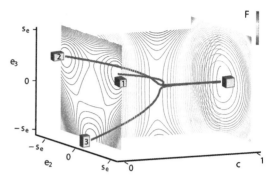

Figure 3. Mechanochemical spinodal for the 3D formulation depicted by contours of the free-energy manifold along the e_2–e_3 and e_3–c planes. The three energy-minimising paths (brown lines) and their corresponding energy-minimising-strained structures are also shown.

other. The measures e_2 and e_3 are especially suited as order parameters to describe cubic-to-tetragonal distortions. All three tetragonal variants that emerge from the cubic reference crystal can be uniquely represented by these two measures. See Figure 1b, where tetragonal distortions of the cubic crystal along the X_1, X_2 and X_3 axes have been labelled as 1, 2 and 3, respectively. Deviations from the dashed lines in the e_2–e_3 space of Figure 1b correspond to orthorhombic distortions of the cubic reference crystal. This role of e_2 and e_3 as structural order parameters to denote the degree of tetragonality, and to distinguish between the three tetragonal variants, complements their fundamental purpose as arguments of the elastic free-energy density.

The mechanochemical spinodal in three dimensions. For brevity, we write the strains as $\mathbf{e} = \{e_1, ..., e_6\}$. We introduce further phenomenology by specifying that $\mathcal{F}(c, \mathbf{e})$ is only one part of the total free-energy density. It is a homogeneous contribution whose composition and strain dependence cannot be additively separated. Figure 3, for example, illustrates contour plots of $\mathcal{F}(c, \mathbf{e})$ on the e_2–e_3 and e_3–c planes for a binary solid having a temperature–composition phase diagram similar to that of Figure 2c. To generate the tetragonal α phase as illustrated in that diagram, the homogeneous free-energy density as a function of strain must qualitatively follow the contours of the e_2–e_3 plane at $c = 0$ in Figure 3. Here the tetragonal variants are local minimisers of $\mathcal{F}$ in e_2–e_3 space, with equal minima. In turn, to obtain the cubic β phase, $\mathcal{F}(c, \mathbf{e})$ must follow the contours of the e_2–e_3 plane at $c = 1$. Thus, $\mathcal{F}$ changes smoothly from convex with respect to e_2 and e_3 on the $c = 1$ plane to non-convex at $c = 0$ to form the three variants of the tetragonal α phase. Importantly, the planes at $c = 0$ and $c = 1$ themselves must be minimum energy surfaces to represent the tetragonal α and cubic β phases, respectively. Supplementary Movie S8 in the Supplementary Information shows the evolution of the state (c, e_2, e_3) of the material points on the free-energy manifold $\mathcal{F}$.

Gradient regularisation of the free-energy density: Following van der Waals,[12] and Cahn and Hilliard[1] in their treatment of non-uniform composition fields, we can extend the total free-energy density beyond $\mathcal{F}$ by writing it as a Taylor series, retaining terms that depend on the composition gradient, ∇c. We extend this gradient dependence to the strain measures $\nabla \mathbf{e}$, as did Barsch and Krumhansl[8] following Toupin.[9] (Also see Karatha *et al.*,[13,14] for treatments using infinitesimal and finite strain, respectively.) These gradient dependences appear in a non-uniform free-energy density $\mathcal{G}(c, \mathbf{e}, \nabla c, \nabla \mathbf{e})$. Frame invariance of $\mathcal{F}$ and $\mathcal{G}$ is guaranteed, as the members of $\mathbf{e}$ are linear combinations of the tensor components of $\mathbf{E}$. They also must be invariant under point group operations of the cubic reference crystal.

Free-energy functional. The crystal occupies a reference (undeformed) configuration Ω with boundary Γ. The total free-energy, Π, is the integral of the free-energy density $\mathscr{F} + \mathscr{G}$ over the solid with boundary contributions included. Thus, Π is a functional of the composition c and the displacement vector field $\boldsymbol{u}$, from which the strains are derived (Supplementary Information):

$$\prod [c, \boldsymbol{u}] = \int_\Omega (\mathscr{F} + \mathscr{G}) dV - \sum_{i=1}^3 \int_{\Gamma_{T_i}} u_i T_i dS. \quad (2)$$

where traction vector component T_i is specified on the boundary subset $\Gamma_{T_i} \subset \Gamma$. Following the above authors, we include only quadratic terms in the gradients, but in a generalisation we also allow cross terms between ∇c and ∇e in the non-uniform contribution $\mathscr{G}$, which can therefore be written as

$$\mathscr{G}(c, \boldsymbol{e}, \nabla c, \nabla \boldsymbol{e}) = \frac{1}{2} \nabla c \cdot \boldsymbol{\kappa}(c, \boldsymbol{e}) \nabla c + \sum_{\alpha, \beta} \frac{1}{2} \nabla e_\alpha \cdot \boldsymbol{\gamma}^{\alpha\beta}(c, \boldsymbol{e}) \nabla e_\beta$$

$$+ \sum_\alpha \nabla c \cdot \boldsymbol{\theta}^\alpha(c, \boldsymbol{e}) \nabla e_\alpha. \quad (3)$$

Here $\boldsymbol{\kappa}$ is a symmetric tensor of composition-gradient energy coefficients, each $\boldsymbol{\gamma}^{\alpha\beta}$ ($\alpha, \beta = 1, ..., 6$) is a tensor of strain gradient energy coefficients, and each $\boldsymbol{\theta}^\alpha$ is a tensor of the mixed, composition–strain gradient energy coefficients. Note that, in general, these coefficients will be functions of the local composition and strain. The point group symmetry of the cubic reference crystal imposes constraints on the tensor components of $\boldsymbol{\kappa}$, $\boldsymbol{\gamma}^{\alpha\beta}$ and $\boldsymbol{\theta}^\alpha$ as well as on the form of $\mathscr{F}$.

Although the gradient energies bestow greater accuracy upon the free-energy description of solids with non-uniform composition and strain fields, they are essential at a more fundamental level if the homogeneous free-energy density is non-convex. At compositions that render $\mathscr{F}$ non-convex, the absence of a gradient energy term will allow spinodal decomposition characterised by composition fluctuations of arbitrary fineness, thus leading to non-unique microstructures— a fundamentally unphysical result.[15] With $\boldsymbol{\kappa} \neq 0$, the composition-gradient energy penalises the interfaces wherein composition varies rapidly between high and low limits. This ensures physically realistic results, manifesting in unique microstructures with a mathematically well-posed formulation.

An essentially analogous situation exists with respect to the negative curvatures of $\mathscr{F}$ in the e_2–e_3 plane at low c, which drive the cubic lattice to distort into the tetragonal variants corresponding to the three free-energy wells. Consider a solid with a homogeneous free-energy density as in Figure 3 and a strain state lying between the valleys in the e_2–e_3 plane at $c = 0$. Absent the strain gradient energy, mechanochemical spinodal decomposition would allow tetragonal variants of arbitrary fineness—an unphysical result, reflecting further mathematical ill-posedness. Retention of the strain gradient energy, ($\boldsymbol{\gamma}^{\alpha\beta} \neq 0$) penalises interfaces of sharply varying strain between tetragonal variants to ensure physically realistic results and unique microstructures from a mathematically well-posed formulation. This is well understood in the literature that studies the formation of martensitic microstructures from non-convex free-energy density functions.[16–18]

Governing equations of non-equilibrium chemistry. The free energy for non-homogeneous composition and strain fields (equation 2), must be a minimum at equilibrium. The state of a solid out of equilibrium will evolve to reduce the free energy $\Pi [c, \boldsymbol{u}]$. In formulating a kinetic equation for the redistribution of atomic species through diffusion, we are guided by variational extremisation of the free energy to identify the chemical potential, μ. Details of this calculation appear as Supplementary Information.

The result follows:

$$\mu = \frac{\partial \mathscr{F}}{\partial c} - \nabla \cdot (\boldsymbol{\kappa} \nabla c) + \nabla c \cdot \frac{\partial \boldsymbol{\kappa}}{\partial c} \nabla c + \sum_{\alpha, \beta} \frac{1}{2} \nabla e_\alpha \cdot \frac{\partial \boldsymbol{\gamma}^{\alpha\beta}}{\partial c} \nabla e_\beta$$

$$+ \sum_\alpha \left(\nabla c \cdot \frac{\partial \boldsymbol{\theta}^\alpha}{\partial c} \nabla e_\alpha - \nabla \cdot (\boldsymbol{\theta}^\alpha \nabla e_\alpha) \right). \quad (4)$$

For solids where c tracks the composition of an interstitial element within a chemically inert host, such as Li in $Li_c Mn_2 O_4$, μ in equation (4) corresponds to the chemical potential of the interstitial element. If c tracks the composition of a substitutional species, such as in alloys or on sublattices of complex compounds (e.g., the cation sublattice of yttria-stabilised zirconia), μ is equal to the chemical potential difference between the substitutional species.

The common phenomenological relation for the flux is $\boldsymbol{J} = -\boldsymbol{L}(c, \boldsymbol{e}) \nabla \mu$, where $\boldsymbol{L}$ is the Onsager transport tensor (see de Groot and Mazur[19]). For an interstitial species, $\boldsymbol{L}$ is related to a mobility,[20] whereas it is a kinetic, intermixing coefficient for a binary substitutional solid.[21] Inserting the flux in a mass conservation equation yields the strong form of the governing partial differential equation for time-dependent mass transport. It is of fourth order in space due to the composition-gradient dependence of μ in equation (4). See Supplementary Information for strong and weak forms of this partial differential equation.

Governing equations of mechanical equilibrium: strain gradient elasticity. Mechanical equilibrium is assumed as elastic wave propagation typically is a much faster process than diffusional relaxation in crystalline solids. Equilibrium is imposed by extremising the free-energy functional with respect to the displacement field. Standard variational techniques lead to the weak and strong forms of strain gradient elasticity. The treatment is technical, for which reason we restrict ourselves to the constitutive relations that are counterparts to the chemical potential equation (4) for chemistry. Coordinate notation is used for transparency of the tensor algebra, and summation is implied over repeated spatial index, I. Details appear in Supplementary Information. The final form of the equations is complementary to that mentioned in Toupin,[9] as our derivation is relative to the reference crystal, Ω.

With the deformation gradient $\boldsymbol{F}$ being related to the Green–Lagrange strain as $E_{KL} = \frac{1}{2}(F_{iK}F_{iL} - \delta_{KL})$, the first Piola–Kirchhoff stress tensor and the higher-order stress tensor, respectively, are given by

$$P_{iJ} = \sum_\alpha \frac{\partial(\mathscr{F} + \mathscr{G})}{\partial e_\alpha} \frac{\partial e_\alpha}{\partial F_{iJ}} + \sum_\alpha \frac{\partial \mathscr{G}}{\partial e_{\alpha,l}} \frac{\partial e_{\alpha,l}}{\partial F_{iJ}} \quad (5)$$

$$B_{iJK} = \sum_\alpha \frac{\partial \mathscr{G}}{\partial e_{\alpha,l}} \frac{\partial e_{\alpha,l}}{\partial F_{iJ,K}} \quad (6)$$

The higher-order stress $\boldsymbol{B}$, which is absent in classical, non-gradient elasticity[22] (and in earlier treatments of mechanochemistry[23,24]), makes the strong form of gradient elasticity a fourth order, nonlinear partial differential equation in space (Supplementary Information). The first three-dimensional solutions to general boundary value problems of Toupin's strain gradient elasticity theory at finite strains were recently presented by the authors.[25]

RESULTS

Two-dimensional examples

We first consider a two-dimensional solid to better visualise the microstructures that can emerge from mechanochemical spinodal decomposition. Plane strain elasticity is assumed, for which E_{13}, E_{23} and $E_{33} = 0$, giving e_4 and $e_5 = 0$, and $e_3 = e_1/\sqrt{2}$, reducing

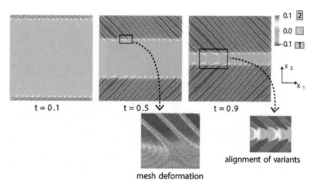

Figure 4. Evolution of 2D microstructure during outflux from the top and bottom surfaces of a solid under plane strain. Contours show strain e_2. Note the legend and corresponding square/rectangular variant crystal structures. The deformation and accompanying twinned microstructure are seen clearly in the distorted mesh.

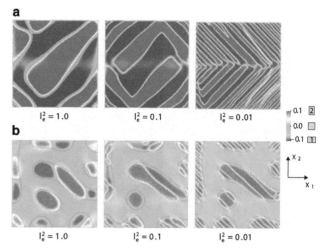

Figure 5. Microstructure controlled by elastic gradient length scale parameter (l_e). Shown are the final contours of e_2 for the simulation of (**a**) outflux from top and bottom surfaces, and (**b**) quenching.

equations (1, 2, 3, 4, 5 and 6) to two dimensions. The discussion on two-dimensional mechanochemical spinodal decomposition (Figure 2) holds: as a particular parameterisation of $\mathscr{F}(c, e_1, e_2, e_6)$, we consider a regular solution model as a function of c at zero strain. At $c = 0$, $\mathscr{F}(c, e_1, e_2, e_6)$ is double-welled in e_2, corresponding to the two rectangular variants. The gradient energy is $\mathscr{G}(\nabla c, \nabla e_2)$ with a constant isotropic $\kappa \neq 0$ and constant $\gamma^{22} \neq 0$, all other gradient coefficients being zero. Although the fullest possible complexity of the coupling is not revealed by these simplifications, the aim here is to present the essential physics that is universal to mechanochemical spinodal decomposition, postponing details of the more complex couplings and tensorial forms to future communications, where they will be derived for specific materials systems. See Supplementary Information for specific forms of $\mathscr{F}(c, e_1, e_2, e_6)$ and $\mathscr{G}(\nabla c, \nabla e_2)$.

Figure 4 shows the evolution of microstructure over a 0.01×0.01 domain whose reference (initial) state has the square crystal structure and $c = 1$. The displacement component $u_1 = 10^{-5}$ on the right boundary ($X_1 = 0.01$), with the remaining boundaries fixed ($\boldsymbol{u} = 0$). An outward flux is imposed on the top and bottom boundaries causing a decrease in composition, c, starting from the boundaries. As the composition falls, the homogeneous free-energy density $\mathscr{F}$ loses convexity and the state of the material (c, e_2) enters the mechanochemical spinodal along $e_2 = 0$ (Figure 2b). The continuing outward flux first drives material near the top and bottom boundaries fully into the regime, where the rectangular crystal structure is stable. As explained in Figure 2b, the negative curvature $\partial^2 \mathscr{F} / \partial e_2^2 < 0$ creates thermodynamic driving forces that distort the square structure, shown in green, into rectangular variants, which form as red/blue laminae (all e_2 values). The coexistence of the parent, square structure with the daughter, rectangular variants also can give rise to cross-hatched microstructures that parallel the tweed microstructures described in the work of Karatha et al.[13,14] Supplementary Movie S9 in the Supplementary Information shows the formation of such a microstructure.

A laminar, micro-twinned microstructure develops as the two rectangular variants form, distinguished by the sign of e_2 (see legend). Note that e_2 remains continuous because of the penalisation of strain gradients ∇e_2, although discontinuities can develop in ∇e_2 itself. Because our numerical framework uses basis functions that are continuous up to their first derivatives (see Methods and Supplementary Information), ∇e_2 is indeed discontinuous in the computations. The lamination accommodates the strain difference between the two rectangular variants to minimise the free energy. If strains were infinitesimal ($|e_1|, |e_2|, |e_6| \gtrsim 0$) e_2 would correspond to shear along the directions that are rotated $\pi/4$ radians from the crystal axes. But the strains in these microstructures can be finite (reflected in the range $-0.1 \leqslant e_2 \leqslant 0.1$

in Figure 4) and involve large rotations to accommodate the rectangular variants; this necessitates a finite-strain formulation, as shown in the recent studies by Alphonse et al.[26] comparing infinitesimal-strain and finite-strain formulations for polytwinned microstructure evolution in martensitic alloys, and by Hildebrand and Miehe.[7]

The micro-twinning and coherency strains are seen in the distorted mesh of discretisation (inset of top centre panel). The undeformed mesh cells are squares, and hence the numerical discretisation strikingly delineates the kinematics of the cubic-to-rectangular transformation, and highlights the rectangular twins as well as the concentrated distortions along the twin boundaries. Supplementary Movie S6 in the Supplementary Information shows such mesh distortion and formation of the rectangular twins with a different form of the function $\mathscr{F}$.

In some cases, the long-range nature of elasticity forces like rectangular variants to align even when separated by an untransformed square phase. This is first seen in the finger-like extensions of strain contours from the rectangular variants into the square phase in Figure 4, followed by their alignment and eventual incorporation into laminae of the same variant (top right panel and its evolution shown as an inset). In this instance, although the micro-twins end at an invariant habit plane that is common with the untransformed square phase, the lattice parameters of the rectangular variants differ from those of the square phase, inducing elastic strain energy in the latter. The alignment lowers this strain energy, and also is seen in Figure 5a. Note, however, that this is not a universal feature. Supplementary Movie S2 shows instances where unlike variants come close to alignment.

The fineness of laminae depends on the strain gradient elasticity length scale $l_e \sim \sqrt{\gamma^{22} / \sqrt{\sum_{\alpha, \beta = 1,2,6} (\partial^2 \mathscr{F} / \partial e_\alpha \partial e_\beta)^2}}$, as explored in Figure 5. In Figure 5a, the initial and boundary conditions are the same as in the example of Figure 4. Decreasing l_e weakens the penalty on the strain gradient ∇e_2 across neighbouring, unlike rectangular variants and allows more twin boundaries. Notably, self-similarity is not maintained between microstructures for different l_e, even for the same initial and boundary value problem. We understand this to be the influence of elastic strain accommodation: to minimise the total free energy when the strain gradient penalty changes, the physics optimises twin boundaries via laminations of different sizes as well as different patterns. Importantly, however, crystal symmetry admits non-vanishing strain gradient energy coefficients beyond $\gamma^{22} \neq 0$

used in these simulations (Supplementary Information). Furthermore, the composition dependence of $\mathscr{F}$ could be more complex than the simple regular solution model used here. Given the already strong effect of l_e alone, we conjecture that varying these forms will have a significant influence on the resulting coherent twin microstructure. The proper form, while guided by crystal symmetry arguments must ultimately be determined experimentally or by first-principles statistical mechanical methods.

The example in Figure 5b further explores the influence of l_e. In this suite of computations, the unstrained material with an initially square microstructure, convex free-energy density and composition having random fluctuations about $c = 0.45$ was quenched to a low temperature and into the mechanochemical spinodal. The local composition and strain evolve under the thermodynamic driving forces detailed in the context of Figure 2. We draw attention, once again, to the changing identity of rectangular variants, their shapes and sizes depending on l_e. Also, note the progressively finer lamination of rectangular domains with decreasing l_e. Such studies suggest how dynamic mechanochemical spinodal decomposition can lead to an atlas of microstructures, which in turn will determine material properties. Supplementary Movies S1–S4 in the Supplementary Information show the evolution of some of these microstructures.

A three-dimensional study

This final example displays the full three-dimensional complexity of microstructures resulting from the mechanochemical spinodal. We persist with the above simplifications aimed at presenting the essential physics of mechanochemical spinodal decomposition, and postpone a full exploration of the coupling and anisotropies in coefficients to future work on specific materials systems. The free-energy density function used for the three-dimensional study appears as Supplementary Information.

Figure 6 shows the equilibrium microstructure that results in a solid that is initially in the cubic phase, $c = 1$, subject to an outflux on all surfaces. The domain is a unit cube with displacement components $u_2, u_3 = 0.01$ on the boundary $X_1 = 1$, with zero displacement, $\boldsymbol{u} = 0$, on the boundary $X_1 = 0$. The cubic-to-tetragonal transformation takes place as $c \to 0$. Under these conditions, all three tetragonal variants form, with an intricately interleaved microstructure for strain accommodation. Note the finer microstructures and changing pattern for smaller l_e. The inset shows the surface straining around a corner, delineated by the distorted mesh lines. Observe the three, oriented, tetragonal variants formed by twinning deformation from the initial cubic structure. See Supplementary Movie S7 in the Supplementary Information for a detailed view of the three-dimensional structure

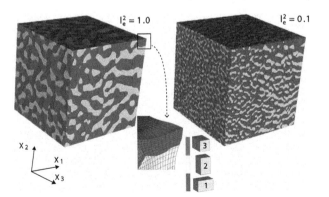

Figure 6. Microstructure observed in 3D for different values of the elastic gradient length scale parameter (l_e). The three tetragonal variants appear in blue (variant 1), yellow (variant 2) and red (variant 3) for $c = 0$. The transformation strains are easily discerned in the distorted mesh.

of these individual tetragonal variants. Supplementary Movie S10 in Supplementary Information shows other three-dimensional microstructures whose cross-sectional planes bear closer resemblance to the plate-like structures predicted by two-dimensional computations. To the best of our knowledge, such computations of a cubic-to-tetragonal transformation with twinned variants whose microstructure is controlled by nonlinear, strain gradient interface energies have not been previously presented.

DISCUSSION

Several kinetic mechanisms have been proposed to describe the decomposition of a solid upon entering a two-phase region.[27,28] A commonly observed mechanism occurs through localised nucleation followed by diffusional growth that is mediated by the migration of interfaces separating the daughter phases from the parent phase. Nucleation and growth mechanisms are common when the phases participating in the transformation have little or no crystallographic relation to each other. The transformation then proceeds reconstructively, where the interphase interfaces are disordered or at best only semi-coherent. Numerical treatments of this mechanism rely on sharp-interface methods such as a level set approach.[29]

Other kinetic mechanisms of decomposition involving some degree of coherency are also possible when the participating phases share sufficient crystallographic commonality that order parameters can be defined describing continuous pathways connecting the parent and daughter phases. The structural changes of the crystal that accompany these transformations can fall in one of two categories. One subclass of structural transformations is driven by an internal shuffle, where the atomic arrangement within the unit cell of the crystal undergoes a symmetry-breaking change. The unit cell vectors of the crystal may also undergo a change and lower the symmetry of the lattice, but this effect is secondary and in response to the atomic shuffle within the unit cell. The order parameters for the structural change therefore describe the atomic shuffles within the unit cell and are non-conserved. The second subclass of structural transformations is driven by a symmetry-reducing change of the lattice vectors of the crystal, with any atomic rearrangement of the basis of the crystallographic unit cell occurring in response to the symmetry-breaking changes of the unit cell vectors. The natural order parameters describing these transitions are strains. The free energy in the first subclass will exhibit negative curvatures with respect to the non-conserved shuffle order parameters, whereas the free energy in the second subclass will have negative curvatures with respect to the relevant strain order parameters, as was recently predicted for the cubic-to-tetragonal transition of ZrH_2.[3]

Decomposition reactions that require both diffusional redistribution and a structural change due to an *internal shuffle* have been treated successfully and rigorously with phase field approaches that combine Cahn–Hilliard with Allen–Cahn. The Cahn–Hilliard description accounts for the composition degrees of freedom, whereas the Allen–Cahn component describes the evolution of the non-conserved shuffle order parameters required to distinguish the various phases of the transformation.[4–7] These treatments have included strain energy as a secondary effect serving only as a positive contribution to the overall free energy due to coherency strains. The approach is rigorous if the free-energy density remains convex with respect to strain, with instabilities in the free energy appearing as a function of concentration and the non-conserved shuffle order parameters.

Here we have presented a mathematical formulation and an accompanying computational framework for decomposition reactions that combine diffusional redistribution with a structural change driven by a symmetry-breaking strain of the crystallographic unit cell as opposed to an internal shuffle within the unit cell. Strains have a primary role, explicitly serving as order parameters to distinguish variants of a daughter phase that has a symmetry subgroup–group relationship with its parent phase due to a structural change of the crystallographic unit cell. Crucial to the description is that the driving force for the formation of the daughter from a supersaturated parent phase emerges from an instability with respect to composition. The treatment therefore combines Cahn–Hilliard for composition with a description of martensitic transformations at fixed concentration introduced by Barsch and Krumhansl.[8] The existence of simultaneous instabilities with respect to strain and composition has not been considered, as a mechanism to describe decomposition reactions upon quenching into a two-phase region.

The possibility of such a mechanochemical spinodal mechanism is motivated by recent first-principles studies of the cubic-to-tetragonal phase transformations of ZrH_2, where the free energy of the high-temperature cubic form was predicted to become unstable with respect to strain upon cooling below the cubic-to-tetragonal second-order transition temperature.[3] The ZrH_{2-2c} hydride can accommodate large concentrations of hydrogen vacancies, c, and has a phase diagram that is topologically identical to that depicted in Figure 2c, with a two-phase region separating a hydrogen-rich tetragonal form of $ZrH_{2-2c\alpha}$ from a cubic form of $ZrH_{2-2c\beta}$ (with $c_\alpha < c_\beta$). See Zuzek et al.[10] (in their phase diagrams, Zuzek et al. have an inverted composition axis relative to our notation, the tetragonal phase is labelled ε and the cubic phase is δ). To be consistent with the predicted free energies for stoichiometric ZrH_2 (i.e., $c=0$) and the experimental T versus c phase diagram with the form of Figure 2c, the free energy of this hydride as a function of composition and strain (i.e., e_2 and e_3) should be similar to those depicted in Figures 2a,b and 3. Decomposition upon quenching cubic ZrH_{2-2c} into the two-phase region can therefore proceed through a mechanochemical spinodal decomposition mechanism.

Not only does the phenomenological description introduced in this work describe a new mechanism of decomposition that is qualitatively distinct from previous combined Cahn–Hilliard and Allen–Cahn approaches, its numerical solution also proves to be substantially more complex due to contributions from strain gradient terms. Indeed, even numerical solutions to general, three-dimensional boundary value problems of gradient elasticity at finite strain were not available until presented recently by the authors.[25] Furthermore, as other authors have pointed out before, the use of strain metrics as order parameters makes a reliance on infinitesimal strains untenable due to the rigid rotations that accompany the finite strains characterising most structural transformations.[7,26] The use of finite-strain metrics introduces geometric nonlinearity into the problem, which although having been treated in past Cahn–Hilliard and Allen–Cahn approaches,[7] presents additional numerical challenges when also considering strain gradient contributions. These challenges have been overcome in this work as demonstrated by our three-dimensional numerical examples.

Mechanochemical spinodal decomposition as described here can be expected in solids forming high-temperature phases that exhibit dynamical instabilities at low temperature. An accumulating body of first-principles calculations of the Born–Oppenheimer surfaces have shown that many high-temperature phases are dynamically unstable at low temperature,[30,31] becoming stable at high temperature through large anharmonic vibrational excitations,[3,32–35] usually through a second-order phase transition. Although such instabilities are frequently dominated by phonon modes describing an internal shuffle, a subset of chemistries become dynamically unstable at low temperature with respect to phonon modes that break the symmetry of the crystal unit cell,[3,34] an instability that can be described phenomenologically with strain order parameters. If, upon alloying such compounds, the

high-symmetry phase becomes stable by passing through a two-phase region, its free-energy surface will resemble that of Figures 2b and 3, and thereby make the solid susceptible to the mechanochemical spinodal decomposition mechanism described here.

Although first-principles and experimental evidence suggest that mechanochemical spinodal decomposition should occur in ZrH_{2-2c}, we expect it to occur in a wide range of other chemistries as well. One possible example, as described in the Introduction, is the decomposition of cubic yttria stabilised zirconia[36,37] upon quenching from the high-temperature cubic phase into a two-phase region separating a low-Y tetragonal phase from a Y-rich cubic phase. Past treatments of this transformation[4–6] relied on non-conserved order parameters to distinguish the different tetragonal variants from each other and from the cubic parent phase. The elastic part of the free-energy density function was parameterised using linearised elasticity and infinitesimal strains, as other authors have also done.[38,39] The chemical part of the free-energy density was assumed to have a negative curvature as a function of the non-conserved order parameter at Zr-rich compositions, but made up of convex (and quadratic) potentials with respect to strain. We reiterate that such a treatment rests on the implicit assumption that internal shuffles drive the cubic-to-tetragonal transformation, whereas the free energy as a function of strain remains convex for all relevant values of the non-conserved order parameters.

The possibility also exists that a coarse-grained free-energy density of cubic ZrO_2 exhibits negative curvatures with respect to strain below the cubic-to-tetragonal transition temperature making the formalism developed here an accurate description of decomposition reactions of yttria-stabilised zirconia. Although the precise mechanism and nature of the cubic-to-tetragonal transition of pure ZrO_2 remain to be resolved,[33] first-principles calculations predict that the cubic form of ZrO_2 is dynamically unstable with respect to the transformation to the tetragonal variant.[40] A rigorous statistical mechanics treatment[3,32] is required to determine whether the cubic-to-tetragonal transition of pure ZrO_2 is accompanied by a change in the sign of the curvature of the free-energy density with respect to e_2 and e_3. If this proves to be the case, yttria-stabilised zirconia should also be susceptible to the mechanochemical spinodal decomposition upon quenching, consistent with the coherent spinodal microstructures between tetragonal and cubic phases observed in single-crystal regions of quenched cubic $Zr_{1-c}Y_xO_{2-c/2}$.[37]

A mechanochemical spinodal may also have a role in a variety of important electrode materials for Li-ion batteries, and intercalation compounds considered for two-dimensional nanoelectronics. These include cubic $LiMn_2O_4$ transforming to tetragonal $Li_2Mn_2O_4$.[41] Although most mechanochemical spinodal transitions will occur in three dimensions, many of the qualitative features of these transitions are more conveniently illustrated in two dimensions. Our two-dimensional studies should also prove relevant to understanding mechanochemical phase transformations in two-dimensional layered materials for nanoelectronics. Materials such as TaS_2, are susceptible to Peierls instabilities upon variation of the composition of adsorbed or intercalated guest species that donate to or extract electrons from the sheet-like host.[42] Our phenomenological treatment by introduction of the concept of a mechanochemical spinodal, coupled with gradient stabilisation of the ensuing non-convex free energy in strain–composition space, thus offers a framework with potential for extension to a wide range of phase transformation phenomena.

With regard to the fundamental thermodynamic underpinnings of analogous processes, the phenomenological model introduced here can also be used to describe temperature-driven martensitic transformations. The composition, c, would then be analogous to the internal energy density, $\mathscr{U}$, whereas the chemical potential, μ, would be analogous to the temperature, T. Due to the presence of temperature gradients throughout the solid, though, the starting point must be a formulation of the entropy density, $\mathscr{S}$, (as opposed to the Helmholtz free energy) as a functional of spatially varying $\mathscr{U}$ and displacements, $\boldsymbol{u}$. In analogy with equation (2), the total entropy can be written as a volume integral over a homogeneous entropy density that depends on the local internal energy density and strain $\overline{\mathscr{S}}(\mathscr{U}, e)$, as well as a non-uniform entropy density contribution expressed in terms of gradients in internal energy density and strain ($\nabla\mathscr{U}$, ∇e). Variational maximisation of the entropy will yield mechanical equilibrium equations as well as an expression for the temperature T (strictly speaking, for its reciprocal, $1/T$), similar to equation (4) for the chemical potential. Due to the presence of gradient terms, $\nabla\mathscr{U}$ and ∇e, the temperature will not only be a function of the local internal energy density and strain, but also gradients in these field variables. As with the diffusion problem treated here, heat flow can be described with an Onsager expression relating the heat flux to a gradient in temperature. The possibility exists that the homogeneous entropy $\overline{\mathscr{S}}$ exhibits instabilities with respect to internal energy density $\mathscr{U}$, allowing for spinodal decomposition with respect to the redistribution of $\mathscr{U}$ in a manner that is similar to the well understood problem of spinodal decomposition with respect to composition. The treatment introduced here can therefore describe (upon replacing c with $\mathscr{U}$ and μ with T) temperature-driven martensitic transformations for solids that exhibit instabilities with respect to both internal energy density and strain. Past treatments of this phenomenon also relied on a Barsch and Krumhansl approach, solving the mechanical problem together with Fourier's law of heat conduction.[38,39] However, these studies were restricted to two dimensions,[26,38,39] with some neglecting geometric nonlinearity by relying on infinitesimal strains.[38,39] At a more fundamental level, they did not consider the possibility of spinodal instabilities with respect to the redistribution of internal energy density.

MATERIALS AND METHODS

The governing fourth-order partial differential equations of mass transport and gradient elasticity are solved numerically in weak form, wherein second-order spatial gradients appear on the trial solutions and their variations. Numerical solutions require basis functions that are continuous up to their first spatial gradients, at least. Our numerical framework (see Rudraraju et al.[25]) uses isogeometric analysis[43] and the PetIGA code framework[44] with spline basis functions, which can be constructed for arbitrary degree of continuity (Supplementary Information). This framework has proven pivotal to the current work. The code used for the numerical examples in this paper can be downloaded at the University of Michigan Computational Physics Group open-source codes webpage: http://www.umich.edu/~compphys/codes.html

ACKNOWLEDGEMENTS

The mathematical formulation for this work was carried out under an NSF CDI Type I grant: CHE1027729 'Meta-Codes for Computational Kinetics' and NSF DMR 1105672. The numerical formulation and computations have been carried out as part of research supported by the US Department of Energy, Office of Basic Energy Sciences, Division of Materials Sciences and Engineering under award no. DE-SC0008637 that funds the PRedictive Integrated Structural Materials Science (PRISMS) Center at University of Michigan.

COMPETING INTERESTS

The authors declare no conflict of interest.

REFERENCES

1. Cahn, J. W. & Hilliard, J. E. Free energy of a nonuniform system. i interfacial energy. *J. Chem. Phys.* **28**, 258–267 (1958).
2. Allen, S. M. & Cahn, J. W. A microscopic theory for antiphase boundary motion and its application to antiphase boundary coarsening. *Acta Metallurgica* **27**, 1085–1091 (1979).
3. Thomas, J. C. & Van der Ven, A. Finite-temperature properties of strongly anharmonic and mechanically unstable crystal phases from first principles. *Phys. Rev. B* **88**, 214111 (2013).
4. Wang, Y., Wang, H., Chen, L. Q. & Khatchaturyan, A. G. Shape evolution of a coherent tetragonal precipitate in partially stabilized cubic zirconia: a computer simulation. *J. Am. Ceram. Soc.* **76**, 3029–3033 (1993).
5. Fan, D. & Chen, L. Q. Computer simulation of twin formation during the displacive $c \rightarrow t'$ phase transformation in the Zirconia-Yttria system. *J. Am. Ceram. Soc.* **78**, 769–773 (1995).
6. Fan, D. & Chen, L. Q. Possibility of spinodal decomposition in $ZrO_2 - Y_2O_3$ alloys: a theoretical investigation. *J. Am. Ceram. Soc.* **78**, 1680–1686 (1995).
7. Hildebrand, F. E. & Miehe, C. A phase field model for the formation and evolution of martensitic laminate microstructure at finite strains. *Philos. Mag.* **92**, 4250–4290 (2012).
8. Barsch, G. R. & Krumhansl, J. A. Twin boundaries in ferroelastic media without interface dislocations. *Phys. Rev. Lett.* **53**, 1069–1072 (1984).
9. Toupin, R. A. Elastic materials with couple-stresses. *Arch. Ration. Mech. Anal.* **11**, 385–414 (1962).
10. Zuzek, E., Abriata, J. P., San-Martin, A. & Manchester, F. D. The H-Zr (hydrogen-zirconium) system. *Bull. Alloy Phase Diagr.* **11**, 385 (1990).
11. Bhattacharya, K., Conti, S., Zanzotto, G. & Zimmer, J. Crystal symmetry and the reversibility of martensitic transformations. *Nature* **428**, 55–59 (2004).
12. van der Waals, J. D. Thermodynamische theorie der capillariteit in onderstelling van continue dichtheidsverandering. *Verhandelingen der Koninklijke Nederlandse Akademie* **1**, 3–56 (1893).
13. Kartha, S., Castan, T., Krumhansl, J. A. & Sethna, J. P. Spin-glass nature of tweed precursors in martensitic transformations. *Phys. Rev. Lett.* **67**, 3630–3633 (1991).
14. Kartha, S., Krumhansl, J. A., Sethna, J. P. & Wickham, L. K. Disorder-driven pretransitional tweed pattern in martensitic transformations. *Phys. Rev. B* **52**, 803–827 (1995).
15. Hilliard, J. E. in *Phase Transformations* (ed Aronson, H. I.) 470–560 (ASM, 1970).
16. Ball, J. M. & James, R. D. Fine phase mixtures as minimizers of energy. *Arch. Ration. Mech. Anal.* **100**, 13–52 (1987).
17. Müller, S. *Calculus of Variations and Geometric Evolution Problems (Cetraro, 1996).* Lecture Notes in Mathematics, vol. 1713 (Springer, 1999).
18. Ball, J. M. & Crooks, E. C. M. Local minimizers and planar interfaces in a phase-transition model with interfacial energy. *Calc. Var. Partial. Differ. Equ.* **40**, 501–538 (2011).
19. de Groot, S. R. & Mazur, P. *Non-Equilibrium Thermodynamics* (Dover, 1984).
20. Van der Ven, A., Bhattacharya, J & Belak, A. A. Understanding li diffusion in li-intercalation compounds. *Acc. Chem. Res.* **46**, 1216–1225 (2013).
21. Van der Ven, A., Yu, H. C., Ceder, G. & Thornton, K. Vacancy mediated substitutional diffusion in binary crystalline solids. *Prog. Mater. Sci.* **55**, 61–105 (2010).
22. Truesdell, C. & Noll, W. *The Nonlinear Field Theories of Mechanics* (Springer Verlag, 1965).
23. Voorhees, P. W. & Johnson, W. C. The thermodynamics of elastically stressed crystals. *Solid State Phys. Adv. Res. Appl.* **59**, 1–201 (2004).
24. Larche, F. & Cahn, J. W. Non-linear theory of thermochemical equilibrium of solids under stress. *Acta Metallurgica* **26**, 53–60 (1978).
25. Rudraraju, S., Van der Ven, A. & Garikipati, K. Three-dimensional isogeometric solutions to general boundary value problems of Toupin's gradient elasticity theory at finite strains. *Comput. Methods Appl. Mech. Eng.* **278**, 705–728 (2014).
26. Alphonse, F., Le Bouar, Y., Gaubert, A. & Salman, U. Phase field methods: microstructures, mechanical properties and complexity. *Comptes Rendus Physique* **11**, 245–256 (2010).
27. Balluffi, R. W., Allen, A. M. & Carter, W. C. *Kinetics of Materials* (John Wiley & Sons, 2005).
28. Porter, D. A. & Easterling, K. E. *Phase transformations in metals and alloys* (Nelson Thorne, Ltd., 2001).
29. Sethian, J. A. *Level Set Methods and Fast Marching Methods: Evolving Interfaces in Computational Geometry, Fluid Mechanics, Computer Vision and Materials Science.* (Cambridge Univ. Press, 1999).
30. Craievich, P. J., Weinert, M., Sanchez, J. M. & Watson, R. E. Local stability of nonequilibrium phases. *Phys. Rev. Lett.* **72**, 3076–3079 (1994).
31. Grimvall, G., Magyair-Koepe, B., Ozolins, V. & Persson, K. A. Lattice instabilities in metallic elements. *Rev. Mod. Phys.* **84**, 945–986 (2012).
32. Zhong, W., Vanderbilt, D. & Rabe, K. A. Phase transitions in $BaTiO_3$ from first-principles. *Phys. Rev. Lett.* **73**, 1861–1864 (1994).
33. Fabris, S., Paxton, A. T. & Finnis, M. W. Free energy and molecular dynamics calculations for the cubic-tetragonal phase transition in zirconia. *Phys. Rev. B* **63**, 094101 (2001).
34. Bhattacharya, J. & Van der Ven, A. Mechanical instabilities and structural phase transitions: The cubic to tetragonal transformation. *Acta Materiala* **56**, 4226–4232 (2008).

35. Souvatzis, P., Eriksson, O., Katsnelson, M. I. & Rudin, S. P. Entropy driven stabilization of energetically unstable crystal structures explained from first principles theory. *Phys. Rev. Lett.* **100**, 095901 (2008).

36. Chien, F. R., Ubic, F. J., Prakash, V. & Heuer, A. H. Stress-induced martensitic transformation and ferroelastic deformation adjacent microhardness indents in tetragonal zirconia single crystals. *Acta Materialia* **46**, 2151–2171 (1998).

37. Krogstad, J. A. *et al.* Phase stability of t′-Zirconia-based thermal barrier coatings: mechanistic insights. *J. Am. Ceram. Soc.* **94**, S168–S177 (2011).

38. Bouville, M. & Ahluwalia, R. Effect of lattice-mismatch-induced strains on coupled diffusive and displacive phase transformations. *Phys. Rev. B* **75**, 054110 (2007).

39. Maraldi, M., Wells, G. & Molari, L. Phase field model for coupled displacive and diffusive microstructural processes under thermal loading. *J. Mech. Phys. Solids* **59**, 1596–1612 (2011).

40. Jomard., G. *et al.* First-principles calculations to describe zirconia pseudopoly-morphs. *Phys. Rev. B* **59**, 4044–4052 (1999).

41. Thackeray, M. M. Manganese oxides for lithium batteries. *Prog. Solid State Chem.* **25**, 1–71 (1997).

42. Rossnagel, K. Suppression and emergence of charge-density waves at the surfaces of layered $1T-TiSe_2$ and $1T-TaS_2$ by *in situ* Rb deposition. *New J. Phys.* **12**, 125018 (2010).

43. Cottrell, J., Hughes, T. J .R. & Bazilevs, Y. *Isogeometric Analysis: Toward Integration of CAD and FEA* (Wiley, 2009).

44. Collier, N., Dalcín, L. & Calo, L. M. Petiga: high-performance isogeometric analysis. Preprint at http://arXiv:1305.4452 (2013).

11

Prediction of pressure-promoted thermal rejuvenation in metallic glasses

Narumasa Miyazaki[1], Masato Wakeda[1], Yun-Jiang Wang[2] and Shigenobu Ogata[1,3]

Rejuvenation is the structural excitation of glassy materials, and is a promising approach for improving the macroscopic deformability of metallic glasses. This atomistic study proposes the application of compressive hydrostatic pressure during the glass-forming quenching process and demonstrates highly rejuvenated glass states that have not been attainable without the application of pressure. Surprisingly, the pressure-promoted rejuvenation process increases the characteristic short- and medium-range order, even though it leads to a higher-energy glassy state. This 'local order'–'energy' relation is completely opposite to conventional thinking regarding the relation, suggesting the presence of a well-ordered high-pressure glass/high-energy glass phase. We also demonstrate that the rejuvenated glass made by the pressure-promoted rejuvenation exhibits greater plastic performance than as-quenched glass, and greater strength and stiffness than glass made without the application of pressure. It is thus possible to tune the mechanical properties of glass using the pressure-promoted rejuvenation technique.

INTRODUCTION

Rejuvenation is the opposite phenomena to ageing, and it has been investigated for decades in relation to molecular and polymeric glasses,[1] and more recently in relation to metallic glasses (MGs).[2–8] In contrast to rejuvenation, ageing is a structural relaxation of the glassy material that changes both the internal structure and mechanical properties of the glass. Well-aged glass usually exhibits brittle fracture triggered by shear band formation[9] at an ambient temperature and usual strain rate. As brittle fracture is a fatal flaw in terms of the application of MGs as structural and mechanical materials, realising a highly rejuvenated state without losing strength and stiffness, and preventing both ageing and brittle fracture (i.e., maintaining the highly rejuvenated state for a long time) are key factors in expanding the applicability of MGs. Some mechanical approaches, such as the application of high-pressure torsion[3] and shot peening[5] to mechanically realise a rejuvenated state in MGs, have been proposed, but significant shape change or heterogeneous property distribution is unavoidable. Recently, thermal-based approaches such as thermal rejuvenation consisting of recovery annealing[10] conducted above the glass transition temperature T_g followed by fast cooling have been proposed.[2,6] The thermal approach can control not only the surface properties but also the interior of any glass component shape. It can be also utilised to maintain plastic performance in thermoplastic nanoimprinting of MG,[11] which has been developed to realise low-cost fabrication of micro- and nano-sized devices. However, thermal rejuvenation requires rapid cooling after annealing, which should be faster than that of the initial glass-forming quenching process, and moreover it may lose intrinsic strength and elastic stiffness.[2,6] These are serious drawbacks in terms of practical application of this approach. To overcome the practical drawbacks, we propose a pressure-promoted thermal rejuvenation technique here, in which pressure is applied during the glass-forming quenching process.

In this study, we performed molecular dynamics (MD) simulations that demonstrate the feasibility of the proposed pressure-promoted thermal rejuvenation process, and we discuss the underlying physics based on diffusivity, α-relaxation, T_g, and internal structure analyses. We found that pressure-promoted rejuvenation is achieved only when a comprehensive influence of three pressure effects on glass-forming process, i.e., effects of temperature-dependent liquid inherent structure energy, pressure-dependent diffusively slowdown (relaxation slowdown) and T_g increase, gains ascendancy over an increase in the Arrhenius–non-Arrhenius diffusivity transition temperature. Then, using MD uniaxial tensile test, we demonstrate that the pressure-promoted thermally rejuvenated glass actually has a weak strain localisation tendency, resulting in better plastic performance. We also examine the feasibility of the pressure-promoted thermal rejuvenation on various alloy systems. Note that some of the examined alloy systems do not exhibit pressure-promoted thermal rejuvenation because the comprehensive influence of three pressure effects fails to override the diffusivity transition temperature rise effect in these alloy systems.

RESULTS

Prediction of pressure-promoted thermal rejuvenation map

We constructed MG models via a melt-quenching process (A → D in Figure 1a) as follows. First, the molten alloy was equilibrated at 3,000 K for 100 ps (A → B) with zero external pressure; then, the model was quenched to 0 K at the same constant cooling rate of $K = 1.0$ K/ps (B → C). Next, the potential energy was minimised using the steepest decent (SD) method at 0 K under zero pressure $P = 0$ GPa (C → D). The constructed MG model (state D) is here

[1]Department of Mechanical Science and Bioengineering, Graduate School of Engineering Science, Osaka University, Osaka, Japan; [2]State Key Laboratory of Nonlinear Mechanics, Institute of Mechanics, Chinese Academy of Sciences, Beijing, China and [3]Center for Elements Strategy Initiative for Structural Materials (ESISM), Kyoto University, Kyoto, Japan.
Correspondence: M Wakeda (wakeda@me.es.osaka-u.ac.jp) or S Ogata (ogata@me.es.osaka-u.ac.jp)

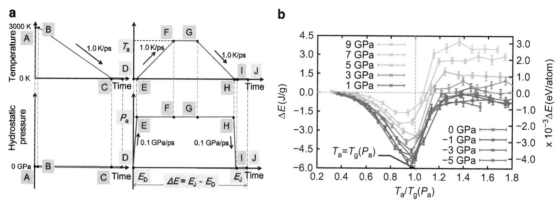

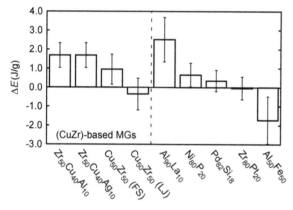

Figure 1. (a) Schematic illustration of initial melt-quenching (A → D) and subsequent thermal-pressure loading (D → J) processes. (b) The change in potential energy ΔE of the $Zr_{50}Cu_{40}Al_{10}$ model induced by the thermal-pressure loading process (D → J). In **b**, the error bars represent the s.e. of 10 different simulations. The positive and negative pressures indicate compressive and tensile pressures, respectively. The horizontal axis represents the annealing temperature, which is normalised by $T_g(P_a)$ for each pressure condition. In general, T_g depends on the applied pressure condition.[31] Therefore, we conducted melt-quenching simulations under different pressure conditions, and computed $T_g(P_a)$ for each P_a from the kink in the quenching process volume–temperature curve

Figure 2. The change in potential energy ΔE of $Zr_{50}Cu_{40}Al_{10}$, $Zr_{50}Cu_{40}Ag_{10}$, $Cu_{50}Zr_{50}$ (FS), $Cu_{50}Zr_{50}$ (LJ), $Al_{90}La_{10}$, $Ni_{80}P_{20}$, $Pd_{82}Si_{18}$, $Zr_{80}Pt_{20}$ and $Al_{50}Fe_{50}$. For each alloy system, we conducted 10 independent simulations with different initial atomic configurations and velocities. The averaged ΔE values over 10 simulations with s.e. bars are shown. Similar to Figure 1b, detailed analyses of effect of annealing temperature and pressure were conducted for $Cu_{50}Zr_{50}$ (LJ) model

referred to as the 'as-quenched model.' Using the as-quenched model, we then conducted thermal-pressure loading simulations (D → J) comprising the following steps: (1) loading of the hydrostatic pressure up to the applied pressure P_a at 0 K (D → E); (2) heating from 0 K to the annealing temperature T_a (E → F); (3) isothermal annealing at T_a and P_a for 1 ns (F → G); (4) cooling from T_a to 0 K (G → H) at a constant cooling rate of the $K = 1.0$ K/ps as B → C, followed by minimisation of the potential energy using the SD method; (5) unloading of all normal stresses (σ_{xx}, σ_{yy} and σ_{zz}) independently from P_a to 0 GPa under zero velocity (H → I); and (6) minimising the potential energy under zero pressure using the SD method (I → J). The final obtained model (state J) will be hereafter referred to as the 'annealed model.' Cooling (B → C and G → H) is conducted at a constant cooling rate of $K = 1.0$ K/ps and heating rate (E → F) is also 1.0 K/ps. Pressure loading (D → E) and unloading (H → I) are conducted at a constant rate of 0.1 GPa/ps. It is worth noting that the technique of thermal loading under high-pressure condition, such as 10 GPa and more, has been already realised in experiments.[12]

In this study, we primarily use the $Zr_{50}Cu_{40}Al_{10}$ MG model, where interatomic interactions are computed using the embedded atom method (EAM) potentials.[13] Figure 1b shows

the change in the potential energy induced by the thermal-pressure loading process (D → J) $\Delta E = E_J - E_D$, where E_D and E_J are the potential energies of the as-quenched and annealed models, respectively. ΔE represents the change in the glass state during the thermal-pressure loading process; ΔE becomes positive if rejuvenation occurs and falls below zero in the case of ageing.

Under zero external pressure ($P_a = 0$ GPa), ΔE is solely induced by thermal loading. In Figure 1b, ΔE under the zero-pressure condition is negative below $1.3T_g$ and becomes almost zero above $1.3T_g$. This result demonstrates that rejuvenation cannot be realised under zero pressure if the quenching process (G → H) cooling rates are equal to those of the initial melt-quenching process (B → C). Our previous work[2] demonstrated the feasibility conditions for rejuvenation induced by thermal loading, and the thermal rejuvenation was found to be realised only when T_a was above $1.1T_g$ and when the cooling rate after isothermal annealing was higher than that of the initial melt-quenching process. This was because the fast cooling suppresses ageing during the glass-forming quenching process. Thus, the ΔE profile under zero external pressure shown in Figure 1b supports the conclusions of our previous work,[2] even though the glass systems and employed interatomic potentials differ. On the other hand, ΔE profiles under 'non-zero' pressure conditions shift from those under zero pressure. The ΔE values for higher annealing temperatures $T_a > 1.1$–$1.2T_g$ clearly demonstrate that tensile pressure enhances ageing, whereas compressive pressure suppresses it. Thus, in this alloy system, the thermal rejuvenation is realised by the application of compressive external pressure instead of rapid quenching, but only when the annealing temperature is above the critical annealing temperature, 1.1–$1.2T_g$. The degree of rejuvenation increases with increased compressive pressure and/or annealing temperature up to ~$1.3T_g$, implying that the employed compressive pressure has the same role in increasing the cooling rate in the heating–annealing–quenching process (D → H) from a viewpoint of the degree of rejuvenation ΔE. To clarify effect of the cooling rate K of the processes B → C and G → H on ΔE, we also conducted the same initial melt-quenching (A → D) and subsequent thermal-pressure loading simulations (D → H) with the other several different cooling rates, such as $K = 0.01$, 0.1 and 10 K/ps for $P_a = 5$ GPa, $T_a = 1.3T_g(P_a)$, and confirmed that thermal rejuvenation was actually realised in all the cooling rates tested here, and moreover no significant changes in ΔE can be found in the cases of cooling rate $K \leqslant 1.0$ K/ps (Supplementary Figure S3). It is worth noting that ΔE has the minimum value, in other words, ageing is the most striking, at around $T_a \sim T_g(P_a)$ in Figure 1b. This is the result of a

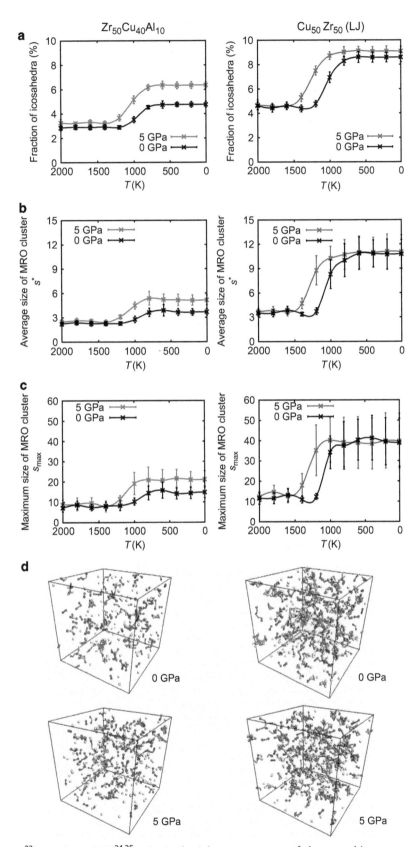

Figure 3. (a) Icosahedral SRO[23] and (**b, c**) MRO[24,25] exist in the inherent structure of the quenching process. The left figure shows the $Zr_{50}Cu_{40}Al_{10}$ alloy results, whereas the right one shows the $Cu_{50}Zr_{50}$ (LJ) alloy results. (**a**) A fraction of icosahedral SRO, and (**b, c**) the average and maximum size of the MRO cluster composed of interpenetrating s^*, s_{max}, respectively. The definitions of average and maximum size are given in ref. 25. (**d**) Spatial distribution of icosahedral SRO and MRO[24] in the 0 K inherent structure of the $Zr_{50}Cu_{40}Al_{10}$ (left) and $Cu_{50}Zr_{50}$ (LJ; right) models constructed via the quenching process with ($P = 5$ GPa) and without pressure ($P = 0$ GPa).

temperature dependency of two factors; (i) an allowable amount of relaxation (energy reduction) bounded by the energy state of equilibrium liquid, and (ii) an amount of relaxation during certain annealing time. The factor (i) decreases with increasing T_a, and rapidly decreases especially at above T_g, whereas the factor (ii) increases with increasing T_a, and rapidly increases especially at above T_g. As a result at $T_a \sim T_g(P_a)$, the amount of ageing maximises because the relaxation rate is fast and moreover the large amount of relaxation is allowed (see a schematic illustration in Supplementary Figure S4).

To verify the feasibility of pressure-promoted thermal rejuvenation in other glass systems, we constructed eight different MG models via melt-quenching process (A → D) and applied the same thermal-pressure loading process (D → J) to the MG models, where the interatomic interactions were calculated from different interatomic potentials, such as $Zr_{50}Cu_{40}Ag_{10}$ (EAM[14]), $Cu_{50}Zr_{50}$ (Finnis–Sinclair: FS[15]), $Cu_{50}Zr_{50}$ (Lennard–Jones: LJ[16]), $Al_{90}La_{10}$ (EAM[17]), $Ni_{80}P_{20}$ (EAM[18]), $Pd_{82}Si_{18}$ (EAM[19]), $Zr_{80}Pt_{20}$ (EAM[20]) and $Al_{50}Fe_{50}$ (FS[21]). Here we set $P_a = 5$ GPa and $T_a = 1.3 T_g(P_a)$. The resultant changes in the potential energy ΔE are summarised in Figure 2. Pressure-promoted thermal rejuvenation can be seen in many of the examined alloy systems, whereas pressure-promoted 'ageing' can be observed in $Cu_{50}Zr_{50}$ (LJ), $Zr_{80}Pt_{20}$ and $Al_{50}Fe_{50}$. This fact reveals that the effect of pressure on the atomic diffusion, which dominates the structural relaxation in the quenching process, is strongly dependent on alloy composition; in other words, it depends on atomic sizes and interatomic interactions. Similar alloy dependence of pressure effects on structural relaxation has been reported for an experimental tensile test conducted on a notched MG sample.[22]

Changes in internal structure by compressive pressure

To reveal the internal structure change induced by compressive pressure, we analysed icosahedral short-range order (SRO)[23] and medium-range order (MRO) composed of interpenetrating icosahedra[2,24,25] existing in the inherent structures of the quenching process from far above T_g to 0 K, under a constant cooling rate K and a constant pressure P. The inherent structure is defined as the structure of local minimum on the potential energy landscape at $P = 0$ GPa, and it is calculated using the following process: first, the potential energy from an instantaneous atomic

configuration in the cooling process is minimised using the SD method. Then, the pressure is unloaded from P to 0 GPa, and finally the potential energy is again minimised at 0 GPa. The results of the changes in internal structure are summarised in Figure 3, in which icosahedral SRO and MRO are surprisingly increased by compressive pressure in the case of $Zr_{50}Cu_{40}Al_{10}$, and thus these ordered local structures are supersaturated and more tightly filled in by the compressive pressure application in cooling process, even though the compressive pressure leads to an energetically unstable (i.e., rejuvenated) glass state. These results reverse the conventional understanding of the relationship between characteristic topological order and energy state, in which much icosahedral SRO and MRO lead to an energetically more stable glass state and vice versa.[23,24,26,27] Moreover, in the case of $Cu_{50}Zr_{50}$ (LJ), we cannot see a clear increase in the icosahedral SRO and MRO, even though the pressure leads to an energetically more stable (i.e., aged) glassy state (for other MG systems, see Supplementary Figure S6). These results suggest the presence of a well-ordered high-energy glassy state. It is worth noting that volume of the pressure-promoted thermally rejuvenated $Zr_{50}Cu_{40}Al_{10}$ model with the compressive pressure application decreases with increasing the degree of rejuvenation ΔE (i.e., increasing the compressive pressure), and thus the volume is always smaller than that of as-quenched model (Supplementary Figure S7). On the other hand, the volume of thermally rejuvenated model with a rapid cooling rate of the G → H faster than that of the B → C, without compressive pressure application (pure thermal rejuvenation model[2]), increases with increasing the ΔE (i.e., increasing the cooling rate of the G → H process), and thus the volume is always larger than the as-quenched model (Supplementary Figure S7). This means that although if these rejuvenated glasses have the same degree of rejuvenation ΔE, these volumes are totally different; the former one has smaller and the latter one has larger volume than as-quenched model. More details of the volume and SRO analyses are plotted in Supplementary Figure S7. We show the schematic illustration of characteristics of glasses after the pure thermal rejuvenation and the pressure-promoted thermal rejuvenation in Figure 4, in which energy, density and SRO of as-quenched, pure thermally aged and rejuvenated, pressure-promoted thermally rejuvenated glasses are summarised.

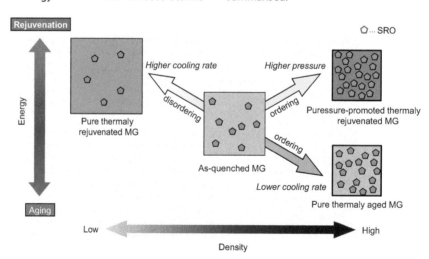

Figure 4. Schematic illustration of rejuvenation and ageing with and without pressure. The horizontal axis represents density, which increases form left to right, whereas the vertical axis represents energy, which increases from bottom up. The left, centre, right lower and right upper cartoons of glass structure represent pure thermally rejuvenated glass, as-quenched glass, pure thermally aged glass and pressure-promoted thermally rejuvenated glass, respectively. The pure thermally rejuvenated glass has lesser SRO and lower density, whereas the pure thermally aged glass has more SRO and higher density than the as-quenched glass. On the other hand, the pressure-promoted thermally rejuvenated glass has more SRO but higher density than the as-quenched glass. The colour of the glass cartoon represents energy state; blue shows lower energy, whereas red shows higher energy.

a

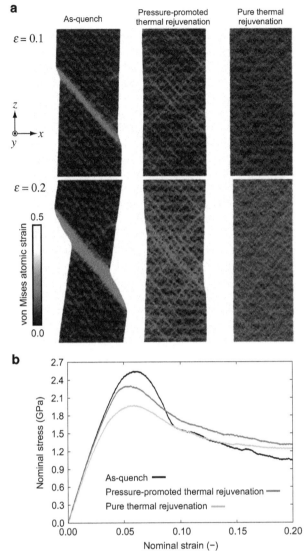

b

Figure 5. (**a**) von Mises atomic strain[37] distributions in the uniaxal tensile tests at nominal strain ε of 0.1 (upper) and 0.2 (lower) for the as-quenched model (left), the pressure-promoted thermal rejuvenation model (centre) and the pure thermal rejuvenation model (right). (**b**) Nominal stress–strain relations along the loading direction (z direction) during uniaxial loading tests. Black, red and green curves represent the nominal stress–nominal strain relation of the as-quenched model, the pressure-promoted thermal rejuvenation model and the pure thermal rejuvenation model, respectively.

Changes in deformation behaviour by pressure-promoted thermal rejuvenation

To demonstrate the pressure-promoted rejuvenation effect on the plastic performance, we performed uniaxial tensile simulations for three different $Zr_{50}Cu_{40}Al_{10}$ models: the as-quenched model and the thermally rejuvenated models with and without pressure application, i.e., pressure-promoted thermal rejuvenation model and pure thermal rejuvenation model. Figure 5a shows the von Mises atomic strain evolution during uniaxial tensile tests with a constant strain rate $\dot{\varepsilon}$ of 10^8 1/s at 300 K. We can see marked strain localisation in the 'as-quenched model', whereas more homogeneous deformations with less strain localisation are observed in the two rejuvenated models (Supplementary Movies S5a, S5b and S5c). Thus, the rejuvenation induced by the thermal-pressure loading process can realise more homogeneous deformation than the as-quenched model. Figure 5b represents nominal stress and

nominal strain relations along the loading direction (z direction) during the uniaxial loading tests. The maximum stress and elastic stiffness, slope of the stress–strain relation of the thermally rejuvenated model with pressure application are higher than those of the thermally rejuvenated model without pressure application. Moreover, the flow stress of the pressure-promoted thermal rejuvenation model is higher than those of the other two models. Thus, the pressure-promoted thermal rejuvenation process can realise higher strength and stiffness glass without significant loss of plastic deformability than a pure thermal rejuvenation process. We also conducted simple shear and nanoindentation simulations using the same $Zr_{50}Cu_{40}Al_{10}$ models, which predicts that the excellent elastic and plastic performances of the thermally rejuvenated glass with pressure application is maintained even under different loading and boundary conditions (for the simple shear and nanoindentation simulations see Supplementary Figures S8 and S9, and also Supplementary Movies S8a, S8b, S8c, S9a and S9b).

The MGs constructed under high-pressure condition have rich SRO and MRO, but exhibit homogeneous deformation as shown in Figures 3 and 5. This trend is interesting, but not intuitive. As we revealed in our previous work,[2] the pure thermal rejuvenation model shows more homogeneous deformation than as-quenched model. This can be understood by thinking of activation energy of local plastic deformation process, which is often called as shear transformation.[28] A lower activation energy of the local plastic deformation process leads to more homogeneous deformation under a certain strain rate and temperature because the process is more equally activated everywhere in MG due to the less local stress state change by the activation of each process and then the MG has less opportunity to have a localised deformation, which will grow to shear band. The pure thermal rejuvenation model may have a lower activation energy of the plastic deformation process because the model is already in the high-energy rejuvenated state. The same way of thinking may be applicable also to the pressure-promoted thermal rejuvenation model. The pressure-promoted thermal rejuvenation model is also in a high-energy rejuvenated state, although the structural details are different from the pure thermal rejuvenation model, and thus it may have lower activation energy of the plastic deformation process. Therefore, the pressure-promoted thermal rejuvenation exhibits the homogeneous deformation. To confirm this scenario, we also performed nanoindentation simulations for several different pressure-promoted thermal rejuvenation (or ageing) models; $Cu_{50}Zr_{50}$(LJ) (ageing), $Ni_{80}P_{20}$ (rejuvenation) and $Al_{90}La_{10}$ (rejuvenation) in addition to $Cu_{50}Zr_{40}Al_{10}$ (rejuvenation). All of the rejuvenated models actually exhibit more homogeneous deformation than that of as-quenched model, whereas the aged LJ model exhibits less homogeneous deformation (Supplementary Figures S9b, S10b, S11b, S12b; Supplementary Movies S9a, S9b, S10a, S10b, S11a, S11b, S12a and S12b). We should note that, regarding elastic stiffness and maximum load, these are maintained in $Zr_{50}Cu_{40}Al_{10}$ (rejuvenation) or even increased in $Cu_{50}Zr_{50}$(LJ) (ageing), whereas decreased in $Ni_{80}P_{20}$ (rejuvenation) and $Al_{90}La_{10}$ (rejuvenation). Thus, among these glasses, only $Zr_{50}Cu_{40}Al_{10}$ (rejuvenation) departs from the intuitive trend, that is, rejuvenation usually leads to lower stiffness and strength and ageing leads to higher. The excellent elastic performance of $Zr_{50}Cu_{40}Al_{10}$ (rejuvenation) may arise from the significant increase of the icosahedral SRO by the pressure-promoted thermal rejuvenation (Figure 3a), because the icosahedral SRO structure has higher stiffness.[24,27] The other rejuvenated $Ni_{80}P_{20}$ and $Al_{90}La_{10}$ glasses do not exhibit such significant SRO change. See again Supplementary Figure S6.

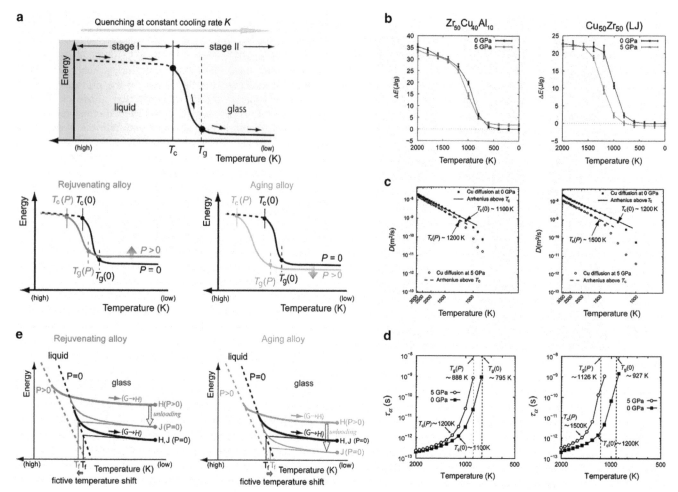

Figure 6. (a) Schematic of inherent structure energy change during quenching process. The horizontal axis represents temperature, which increases from right to left. Upper figure defines cooling stages I and II divided at the critical temperature T_c, at which transition from Arrhenius behaviour (stage I; $T \geqslant T_c$) to non-Arrhenius (stage II; $T < T_c$) occurs. The lower figures show a typical inherent structure energy change in the quenching process for alloy exhibits rejuvenation (left) or ageing (right). (b) Inherent structure energy change during the quenching process both with and without pressure for the $Zr_{50}Cu_{40}Al_{10}$ alloy (left) and for $Cu_{50}Zr_{50}$ (LJ) alloy (right). A crossover between the two curves can be seen in $Zr_{50}Cu_{40}Al_{10}$, whereas no crossover in $Cu_{50}Zr_{50}$ (LJ). Details are mentioned in Supplementary Figure S14. (c) The left and right figures show the pressure effect on the temperature-dependent Cu diffusivity of $Zr_{50}Cu_{40}Al_{10}$ and $Cu_{50}Zr_{50}$ (LJ), respectively. (d) The left and right figures show the pressure effect on the temperature-dependent α-relaxation time of $Zr_{50}Cu_{40}Al_{10}$ and $Cu_{50}Zr_{50}$ (LJ), respectively. In the figures, the critical temperature T_c detected by the diffusively analysis (c) is indicated. The glass transition temperature T_g is also denoted by the vertical dotted lines; $T_g = 795$ K (0 GPa), 888 K (5 GPa) for $Zr_{50}Cu_{40}Al_{10}$, $T_g = 927$ K (0 GPa), 1,126 K (5 GPa) for $Cu_{50}Zr_{50}$ (LJ). (e) Schematic of change in the fictive temperature T_f due to pressure in the quenching process for alloys exhibits rejuvenation (left) and ageing (right). The vertical and horizontal axes are energy and temperature, respectively. Thick black curves represent the energy profile in G → H at $P = 0$ GPa, whereas thick red (left) and blue (right) curves represent that at $P > 0$ GPa. Broken lines represent the energy profile of the equilibrium liquid, whereas solid curves always represent non-equilibrium liquid or glass state. Energy of state H is reduced to state J by unloading process (H → J). The thin red (or blue) curves are fictitious cooling curves at $P = 0$ GPa that reach the glass state J after unloading in actual pressure-promoted thermal rejuvenation (or ageing) process. We defined the fictive temperature as the intersection point between the equilibrium liquid line of $P = 0$ GPa and the tangent line of the fictitious cooling curve, which ends state J obtained under $P = 0$ or $P > 0$ conditions and has the same tangent with the cooling glass curve of $P = 0$ condition at state J. The fictive temperature T_f of pressure-promoted thermal rejuvenation (left) and ageing (right) are represented by red and blue characters, respectively.

DISCUSSION

To understand the interesting dependency of the effects of pressure on the alloy composition seen in Figure 2, we discuss structural relaxation in the quenching process (G → H) both with and without the effects of pressure below. Figure 6a is a schematic of the change in the inherent structure energy[29] of an alloy system during the cooling process from far above T_g to 0 K, under a constant cooling rate K. The inherent structure energy is the energy of inherent structure defined above. Figure 6b shows the actual change in the inherent structure energy of the $Zr_{50}Cu_{40}Al_{10}$ and $Cu_{50}Zr_{50}$ (LJ) systems. In Figure 6b, we may reasonably define a transition temperature T_c as a kink temperature on the curve of

the inherent energy change. Above the transition temperature T_c (quenching stage I; $T \geqslant T_c$), the system can be always in an equilibrium liquid state because of very short relaxation time of the high-temperature liquid. Below T_c (quenching stage II; $T < T_c$), owing to a lesser free volume, a finite activation energy for atomic structure relaxation is expected, leading to a finite relaxation time competes against the time of cooling process specified by the cooling rate. To examine the physical meaning of T_c, we computed the temperature-dependent diffusivity D and α-relaxation time τ_α of the $Zr_{50}Cu_{40}Al_{10}$ and $Cu_{50}Zr_{50}$ (LJ) glasses, as shown in Figure 6c,d. Above $T \sim 1,100$ ($P = 0$ GPa) and $T \sim 1,200$ K ($P = 5$ GPa) in the case of $Zr_{50}Cu_{40}Al_{10}$ and $T \sim 1,200$ K ($P = 0$ GPa) and $T \sim 1,500$ K ($P = 5$ GPa) in the case of $Cu_{50}Zr_{50}$ (LJ), the diffusion kinetics have a constant activation enthalpy for diffusion ΔG_0, and obeys Arrhenius-type behaviour $D(T) \propto \exp\left(\frac{\Delta G_0}{k_B T}\right)$, where k_B is the Boltzmann constant. In contrast, below these temperatures, the diffusion kinetics deviate from Arrhenius-type linear behaviour, indicating a change in the activation enthalpy for diffusion. Because these critical temperatures agree well with T_c's obtained from the kink in Figure 6b, T_c could be understood as a transition temperature between Arrhenius and non-Arrhenius behaviours in diffusivity.[30] Because a higher compressive pressure mechanically reduces the free volume, the pressure actually slows down the diffusion in both the alloys over the whole temperature range, as shown in Figure 6c, and these T_c tends to be higher, i.e., $T_c(P) > T_c(0)$ for $P > 0$, as shown in Figure 6b–d. It is worth noting that the same trend can be found in the pressure dependency of T_g[31] (Supplementary Figure S2). In Figure 6d, α-relaxation time τ_α also drastically increases with decreasing temperature at the temperatures ranging from T_g to T_c. This is consistent with the diffusivity behaviour in Figure 6c. It should be noted that if the annealing time is too short compared with a typical relaxation time, such as α-relaxation time τ_α, the rejuvenation/ageing during the thermal rejuvenation are not sufficiently proceed regardless of with and without pressure application.[2] The 1 ns annealing time used in this study is equivalent to or even longer than α-relaxation time τ_α at temperatures $T > T_g$ (necessary temperature condition for the rejuvenation; see Figure 1b) as shown in Figure 6d. Thus, the 1 ns annealing time is long enough and does not affect on the rejuvenation.

The final inherent structure energy per atom of the state after the quenching process is completed (i.e., state H) at a given P can be expressed as

$$E_H(P) = E_{\text{stage I}}(P, T_c(P)) + K^{-1} \int_{T_c(P)}^{0} \dot{E}(P, T) dT, \quad (1)$$

where $K = dT/dt$ with t being time. Here $\dot{E}$ is the rate of change of the inherent structure energy and $E_{\text{stage I}}(P, T_c(P))$ is the inherent structure energy at the end of stage I (i.e., at the beginning of stage II). Hereafter, the inherent structure energy will be simply referred to as the 'energy'. The degree of rejuvenation is estimated from the final energy difference $\Delta E(P) = E_H(P) - E_H(0)$, which can also be expressed as the sum of (I) the energy difference at the end of stage I (i.e., at the beginning of state II) $\Delta E_{\text{stage I}}(P)$ between the $P > 0$ and $P = 0$ GPa conditions and (II) the difference in the total energy change in stage II $\Delta E_{\text{stage II}}(P)$ between the $P > 0$ and $P = 0$ GPa states, with

$$\Delta E(P) = \Delta E_{\text{stage I}}(P) + \Delta E_{\text{stage II}}(P), \quad (2)$$

where

$$\Delta E_{\text{stage I}}(P) = E_{\text{stage I}}(P, T_c(P)) - E_{\text{stage I}}(0, T_c(0)), \quad (3)$$

$$\Delta E_{\text{stage II}}(P) = K^{-1} \int_{T_c(P, T)}^{0} \dot{E}(P, T) dT - K^{-1} \int_{T_c(0, T)}^{0} \dot{E}(0, T) dT. \quad (4)$$

From the equation and also schematic illustration shown in Figure 6a, we can find the following five factors can dominate the final energy difference $\Delta E(P)$. First, there are three factors in the stage I ($T \geq T_c$). (i) Temperature dependency of the energy. Usually there exists temperature dependency of liquid energy, and higher-temperature liquid tends to have higher energy,[32] i.e., $\partial E(P)/\partial T > 0$ ($T \geq T_c(P)$), although alloy dependency can be found as shown in Figure 6b. Larger temperature dependency leads to larger $\Delta E_{\text{stage I}}$, and then contributes to enhance the rejuvenation. (ii) Pressure dependency of energy at $T \geq T_c(P)$. The change of energy by pressure can also contributes to the degree of rejuvenation $\Delta E(P)$ via changing $\Delta E_{\text{stage I}}(P)$, but usually the pressure dependency should be small, actually error bars of the data with and without

pressure are overlapping as seen in Figure 6b and Supplementary Figure S13a, because the liquid states at $T \geq T_c(P)$ should be always well equilibrated regardless of with and without pressure application. (iii) Pressure dependency of $T_c(P)$. The T_c has not only a strong pressure dependency but also a strong alloy dependency as seen in Figure 6c,d and Supplementary Figure S13b, because it should be closely related to the potential energy surface that determines the starting temperature of the α-relaxation.[32] The higher $T_c(P)$ causes α-relaxation to be initiated at higher temperature, then a rapid energy reduction occurs at earlier stage of the cooling process but below T_c. As a result the higher $T_c(P)$ enhances more ageing. Second, there are two factors in the stage II ($T < T_c$). (iv) Pressure dependency of the relaxation time constant (energy reduction rate), which is the slope of the energy curve in Figure 6b, i.e., $\partial E(P)/\partial T (T < T_c(P))$. Usually the relaxation time becomes longer and thus the energy reduction rate becomes slower with increasing pressure as seen in Figure 6b–d and Supplementary Figure S13a,b, because compressive pressure induces densification (see volume change in Supplementary Figure S13c, induces a less space for moving atom, may effectively suppress the local atomic shuffling and diffusion. (v) Pressure dependency of $T_g(P)$. $T_g(P)$ is related to the temperature at which the energy reduction rate significantly slows down as shown in Figure 6b,d and Supplementary Figure 13Sa,b. The pressure dependency of $T_g(P)$ also has a strong alloy dependency as shown in Figure 6d. The reason should be the same as that of pressure dependency of $T_c(P)$. Hence, the factors (i), (iv) and (v) enhance the rejuvenation, whereas the factor (iii) enhances the ageing. Because of the factor (iii), the energy curve under compressive pressure application is always below that of zero pressure at temperatures just below $T_c(P)$. Therefore, the two energy curves (i.e., with and without pressure) should have a crossover to realise the rejuvenation eventually. As shown in Figure 6b, we actually observed the crossover between the energy–temperature curves both with and without pressure for the $Zr_{50}Cu_{40}Al_{10}$, but not for the $Cu_{50}Zr_{50}$ (LJ) (details are given in Supplementary Figure S14). As the crossover temperature increases with increasing the $\partial E(P)/\partial T > 0 (T \geq T_c(P))$ (factor (i)), the energy reduction rate (factor (iv)) and/or $T_g(P)$ (factor (v)), and decreases with increasing $T_c(P)$ (factor (iii)), the pressure induces two competing effects in the rejuvenation; (A) the factors (i), (iv) and (v), and (B) the factor (iii). In the case of $Cu_{50}Zr_{50}$ (LJ), the latter effect (B) overcomes the former effect (A), results in the pressure induces ageing. In contrast, in the case of $Zr_{50}Cu_{40}Al_{10}$, as T_c increase is much smaller than that of $Cu_{50}Zr_{50}$ (LJ) (Figure 6c), the former effect overcomes the later effect, results in the pressure induces rejuvenation. Therefore, rejuvenation and ageing by thermal-pressure loading process can be predicted from these factors. It should be noted that all of the data determine that these factors are obtainable and predictable by MD simulation based on certain potential description of target alloy system as we performed above.

To clarify underlying physics of the pressure-promoted thermal rejuvenation process and characteristics of the pressure-promoted thermally rejuvenated glass, we additionally conducted energy decomposition analyses, which may provide origin of the pressure-promoted higher energy state (rejuvenation) of $Zr_{50}Cu_{40}Al_{10}$ and lower-energy state (ageing) of $Cu_{50}Zr_{50}$ (LJ) from energetics and geometrical ordering viewpoints (Supplementary Figures S15 and S16). We found that in the pressure-promoted thermally rejuvenated and aged glass models, the icosahedral SRO has a significant lower-energy state than the other local structures as similar to the conventional MGs constructed under zero pressure (Supplementary Figure S16). Thus, in the energetics and geometrical ordering viewpoints, the high-energy state of the rejuvenated $Zr_{50}Cu_{40}Al_{10}$ glass is attributed to the pressure-promoted significant energy increment of the local structures with lower ordering (i.e., non-icosahedral local structures) as schematically shown in Figure 4. The energy increment of the local structure with lower ordering is a result of the less diffusive and mechanical relaxation during the cooling process under the compressive pressure as discussed above Supplementary Figure S17). On the other hand, in the pressure-promoted thermally aged $Cu_{50}Zr_{50}$ (LJ) glass, the energy increment of the non-icosahedral local structures is not so significant (Supplementary Figure S15), because a mechanical relaxation (atomic rearrangement) unfortunately may proceed in the LJ central force field under the compressive pressure, whereas the diffusive relaxation is well suppressed. As a results, the total energy decreases in the $Cu_{50}Zr_{50}$ (LJ) glass.

It is worth noting that the less relaxation induced by pressure application may have great benefit also to the glass-forming ability. It effectively prevents a crystallisation of alloy during the quenching process, and thus may improve the apparent glass-forming ability of alloy. We should also note that although such a high cooling rate ($K = 1.0$ K/ps) was used in the MD simulation because of the time-scale limitations of MD, the above discussion for rejuvenation and ageing should be maintained even at practical cooling rates, because by experiments we have confirmed that the ratio of cooling rate of the final quenching process (G → I) to the initial melt-quenching process (B → C) is the dominant factor for the degree of rejuvenation and ageing[2] rather than the absolute cooling rates. In addition, the relationship between energy state and deformation behaviour observed in Figure 5 is also consistent with the critical fictive temperature concept, which identifies the brittle–ductile transition state of MG.[33] At the fictive temperature T_f, the extrapolated glass line intersects the equilibrium liquid line (Figure 6e), thus the T_f represents the glass state and then more rejuvenated glass has higher T_f. The critical fictive temperature T_{fc} represents a transition glass state at between glass states showing brittle and ductile behaviours, i.e., a MG having $T_{fc} > T_f$ is brittle, whereas having $T_{fc} < T_f$ is ductile. Therefore, the critical fictive temperature concept naturally predicts that the pressure-promoted thermal rejuvenation, which leads to higher T_f, can improve the ductility of MGs. As we have shown in Figure 5, the rejuvenated glass having higher T_f exhibits more homogeneous deformation, supporting the validity of the fictive temperature-based discussion of ductility.

We have demonstrated in this study that the feasibility of pressure-promoted thermal rejuvenation for various alloy systems. It was found that rejuvenation is actually enhanced by applying external pressure in many alloy systems, whereas some alloy systems do not exhibit this response. Further, we clarified the underlying physics of pressure-promoted rejuvenation based on diffusivity, α-relaxation, T_g and internal structure analyses. The pressure-promoted rejuvenation is achieved only when the comprehensive influence of three pressure effects on glass-forming process, i.e., the effects of temperature-dependent liquid energy, pressure-dependent diffusively slowdown (relaxation slowdown) and pressure-dependent T_g increase, overcomes the effect of pressure-dependent increase in the Arrhenius–non-Arrhenius diffusivity transition temperature. Pressure-promoted thermal rejuvenation increases the icosahedral SRO and MRO, even though it leads to an energetically unstable glassy state. This result could change the conventional understanding of the relationship between energy and local structure in MGs, and it strongly indicates the necessity of new energetics of SRO and MRO under significant pressure. We found that the rejuvenated $Zr_{50}Cu_{40}Al_{10}$ MG made by the pressure-promoted thermal rejuvenation process exhibits greater plastic performance than as-quenched glass, and greater strength and stiffness than glass made by the pure thermal rejuvenation process, and then we believe that this glass rejuvenation engineering technique could introduce a new avenue of research towards the development of glass-state control, and elastic and plastic performance control techniques by thermal processing. Moreover, the pressure effects allow us to increase the upper limit of reachable rejuvenation,

which has been restricted by the practical upper limit of a cooling rate in previous rejuvenation-level control techniques with no pressure application.[2] It also may allow us to extrinsically improve the apparent glass-forming ability of alloys by delaying crystallisation.

MATERIALS AND METHODS

Models for thermal-pressure loading process shown in Figures 1,2,3 and 6 consist of 10,000 atoms. It is worth noting that a bigger model composed of 100,000 atoms showed quantitatively the same results as the 10,000-atom model (Supplementary Figure S18). Ten different as-quenched models were prepared and then 10 independent thermal-pressure loading simulations were conducted to reduce the statistical errors of the data, which are due to the finite model size. Temperature and pressure during quench, anneal and heat process are controlled by the isothermal–isobaric ensemble.[34–36]

Models for uniaxial tensile simulations shown in Figure 5 consist of 1,280,000 atoms by organising 10,000-atom unit models ('As-quench', 'Pressure-promoted thermal rejuvenation' and 'Pure thermal rejuvenation') in an $8 \times 1 \times 16$ array (x, y, z directions, respectively). The as-quenched unit model was constructed via melt-quenching from 3,000 to 0 K at a relatively slow cooling rate of 0.01 K/ps. The pressure-promoted thermal rejuvenation and pure thermal rejuvenation unit models were constructed via the thermal-pressure loading process of $P_a = 21$ GPa and $P_a = 0$ GPa, respectively, with $T_a = 1.3 T_g(P_a)$. In case of $P_a = 0$ GPa, thermal rejuvenation was realised by the relatively fast cooling rate of the final quenching process (G → H) 0.20 K/ps, whereas in case of $P_a = 21$ GPa, the cooling rate of the final quenching process (G → H) is set to 0.01 K/ps, which is the same as that of the as-quenched unit model, and thus in this cooling rate condition the pressure application is necessary for rejuvenation. The assembled 1,280,000-atom simulation models were heated from 0 to 300 K at a rate of 1.0 K/ps and then relaxed at 300 K for 1.0 ns under full (i.e., x, y and z) periodic boundary and additionally relaxed at 300 K for 100 ps with partial periodic boundary condition only in y and z (free surface boundary in x). Using this model, we conducted uniaxial tensile simulations. Note that after the relaxation processes both the 1,280,000-atom simulation models, i.e., pressure-promoted thermal rejuvenation model and pure thermal rejuvenation model, had higher-energy states than the as-quenched model by 9.6 and 10.6 J/g, respectively. Energy reductions in the relaxation process for totally 1.1 ns were < 1.0 J/g, implying that rejuvenated states are maintained during the deformation tests for 2 ns after the relaxation process.

Diffusivity D is calculated from the mean-square displacement of the atoms. α-Relaxation time τ_α is defined by the time needed for the self-intermediate scattering function at first peak of structural factor decays to $1/e$.[30]

ACKNOWLEDGEMENTS

This work is supported by the following funding awards: Grants-in-Aid for Scientific Research in Innovative Area (no. 22102003), Scientific Research (A) (no. 23246025), Challenging Exploratory Research (no. 25630013) and the Elements Strategy Initiative for Structural Materials (ESISM).

CONTRIBUTIONS

N.M. conducted almost all calculation, Y.-J.W. performed the diffusivity computation, and M.W. performed α-relaxation time analyses and MRO analyses. M.W. and S.O. assisted with data interpretation and designed the paper. S.O. designed and organised the study. All authors discussed the results and participated in writing the paper.

COMPETING INTERESTS

The authors declare no conflict of interest.

REFERENCES

1. Struik, L. C. E. *Physical Aging in Amorphous Polymers and Other Materials*. (Elsevier, Houston, TX, USA, 1978).
2. Wakeda, M., Saida, J., Li, J. & Ogata, S. Controlled rejuvenation of amorphous metals with thermal processing. *Sci. Rep.* **5**, 10545 (2015).

3. Meng, F., Tsuchiya, K., Ii, S. & Yokoyama, Y. Reversible transition of deformation mode by structural rejuvenation and relaxation in bulk metallic glass. *Appl. Phys. Lett.* **101**, 121914 (2012).

4. Ketov, S. V. *et al.* Rejuvenation of metallic glasses by non-affine thermal strain. *Nature* **524**, 200–203 (2015).

5. Concustell, A., Méar, F. O., Surinach, S., Baró, M. D. & Greer, A. L. Structural relaxation and rejuvenation in a metallic glass induced by shot-peening. *Philos. Mag. Lett.* **89**, 831–840 (2009).

6. Saida, J., Yamada, R. & Wakeda, M. Recovery of less relaxed state in Zr-Al-Ni-Cu bulk metallic glass annealed above glass transition temperature. *Appl. Phys. Lett.* **103**, 221910 (2013).

7. Adachi, N., Todaka, Y., Yokoyama, Y. & Umemoto, M. Improving the mechanical properties of Zr-based bulk metallic glass by controlling the activation energy for β-relaxation through plastic deformation. *Appl. Phys. Lett.* **105**, 131910 (2014).

8. Kumar, G., Prades-Rodel, S., Blatter, A. & Schroers, J. Unusual brittle behavior of Pd-based bulk metallic glass. *Scr. Mater.* **65**, 585–587 (2011).

9. Xi, X. K. *et al.* Fracture of brittle metallic glasses: brittleness or plasticity. *Phys. Rev. Lett.* **94**, 125510 (2005).

10. Kumar, G., Rector, D., Conner, R. D. & Schroers, J. Embrittlement of Zr-based bulk metallic glasses. *Acta Mater.* **57**, 3572–3583 (2009).

11. Kumar, G., Hong, X. T. & Schroers, J. Nanomoulding with amorphous metals. *Nature* **457**, 868–872 (2009).

12. Wang, W. H. *et al.* Effect of pressure on nucleation and growth in the $Zr_{46.75}Ti_{8.25}Cu_{7.5}Ni_{10}Be_{27.5}$ bulk glass-forming alloy investigated using *in situ* x-ray diffraction. *Phys. Rev. B* **68**, 184105 (2003).

13. Cheng, Y. Q., Ma, E. & Sheng, H. W. Atomic level structure in multicomponent bulk metallic glass. *Phys. Rev. Lett.* **102**, 245501 (2009).

14. Fujita, T. *et al.* Coupling between chemical and dynamic heterogeneities in a multicomponent bulk metallic glass. *Phys. Rev. B* **81**, 140204 (2010).

15. Mendelev, M. I. *et al.* Development of suitable interatomic potentials for simulation of liquid and amorphous Cu-Zr alloys. *Philos. Mag.* **89**, 967–987 (2009).

16. Kobayashi, S., Maeda, K. & Takeuchi, S. Computer simulation of deformation of amorphous $Cu_{57}Zr_{43}$. *Acta Metall.* **28**, 1641–1652 (1980).

17. Sheng, H. W., Cheng, Y. Q., Lee, P. L., Shastri, S. D. & Ma, E. Atomic packing in multicomponent aluminum-based metallic glasses. *Acta Mater.* **56**, 6264–6272 (2008).

18. Sheng, H. W., Ma, E. & Kramar, M. J. Relating dynamic properties to atomic structure in metallic glasses. *JOM* **64**, 856–881 (2012).

19. Ding, J., Cheng, Y. Q., Sheng, H. W. & Ma, E. Short-range structural signature of excess specific heat and fragility of metallic-glass-forming supercooled liquids. *Phys. Rev. B* **85**, 060201 (2012).

20. Hirata, A. *et al.* Geometric frustration of icosahedron in metallic glasses. *Science* **341**, 376–379 (2013).

21. Mendelev, M. I., Srolovitz, D. J., Ackland, G. J. & Han, S. Effect of Fe segregation on the migration of a non-symmetric Σ5 tilt grain boundary in Al. *J. Mater. Res.* **20**, 208–218 (2005).

22. Wang, Z. T., Pan, J., Li, Y. & Schuh, C. A. Densification and strain hardening of a metallic glass under tension at room temperature. *Phys. Rev. Lett.* **111**, 135504 (2013).

23. Frank, F. C. Supercooling of liquids. *Proc. R. Soc. Lond. A Math. Phys. Sci.* **215**, 43–46 (1952).

24. Wakeda, M. & Shibutani, Y. Icosahedral clustering with medium-range order and local elastic properties of amorphous metals. *Acta Mater.* **58**, 3963–3969 (2010).

25. Wakeda, M., Shibutani, Y. & Ogata, S. Atomistic study on medium-range order structures in amorphous metals. *J. Soc. Mater. Sci. Japan* **64**, 156–162 (2015).

26. Ma, E. Tuning order in disorder. *Nat. Mater.* **14**, 547 (2015).

27. Wakeda, M., Shibutani, Y., Ogata, S. & Park, J. Relationship between local geometrical factors and mechanical properties for Cu-Zr amorphous alloys. *Intermetallics* **15**, 139–144 (2007).

28. Argon, A. S. Plastic deformation in metallic glasses. *Acta Metall.* **27**, 47–58 (1979).

29. Stillinger, F. H. & Weber, T. A. Packing structures and transitions in liquids and solids. *Science* **225**, 983–989 (1984).

30. Jaiswal, A., Egami, T. & Zhang, Y. Atomic-scale dynamics of a model glass-forming metallic liquid: Dynamical crossover, dynamical decoupling, and dynamical clustering. *Phys. Rev. B* **91**, 134204 (2015).

31. Debenedetti, P. G. *Metastable Liquid: Concepts And Principles.* (Princeton Univ. Press, Princeton, NJ, USA, 1996).

32. Debenedetti, P. G. & Stillinger, F. H. Supercooled liquids and the glass transition. *Nature* **410**, 259–267 (2001).

33. Kumar, G., Neibecker, P., Liu, Y. H. & Schroers, J. Critical fictive temperature for plasticity in metallic glasses. *Nat. Commun.* **4**, 1536 (2013).

34. Nosé, S. A molecular dynamics method for simulations in the canonical ensemble. *Mol. Phys.* **52**, 255–268 (1984).

35. Hoover, W. G. Canonical dynamics: equilibrium phase-space distributions. *Phys. Rev. A* **31**, 1695–1697 (1985).

36. Parrinello, M. & Rahman, A. Polymorphic transitions in single crystals: a new molecular dynamics method. *J. Appl. Phys.* **52**, 7182–7190 (1981).

37. Shimizu, F., Ogata, S. & Li, J. Theory of shear banding in metallic glasses and molecular dynamics calculations. *Mater. Trans.* **48**, 2923–2927 (2007).

12

On the tuning of electrical and thermal transport in thermoelectrics: an integrated theory–experiment perspective

Jiong Yang[1], Lili Xi[2], Wujie Qiu[2,3], Lihua Wu[1], Xun Shi[2], Lidong Chen[2], Jihui Yang[4], Wenqing Zhang[1,2], Ctirad Uher[5] and David J Singh[6]

During the last two decades, we have witnessed great progress in research on thermoelectrics. There are two primary focuses. One is the fundamental understanding of electrical and thermal transport, enabled by the interplay of theory and experiment; the other is the substantial enhancement of the performance of various thermoelectric materials, through synergistic optimisation of those intercorrelated transport parameters. Here we review some of the successful strategies for tuning electrical and thermal transport. For electrical transport, we start from the classical but still very active strategy of tuning band degeneracy (or band convergence), then discuss the engineering of carrier scattering, and finally address the concept of conduction channels and conductive networks that emerge in complex thermoelectric materials. For thermal transport, we summarise the approaches for studying thermal transport based on phonon–phonon interactions valid for conventional solids, as well as some quantitative efforts for nanostructures. We also discuss the thermal transport in complex materials with chemical-bond hierarchy, in which a portion of the atoms (or subunits) are weakly bonded to the rest of the structure, leading to an intrinsic manifestation of part-crystalline part-liquid state at elevated temperatures. In this review, we provide a summary of achievements made in recent studies of thermoelectric transport properties, and demonstrate how they have led to improvements in thermoelectric performance by the integration of modern theory and experiment, and point out some challenges and possible directions.

INTRODUCTION

Thermoelectric (TE) materials are materials that can generate useful electric potentials when subjected to a temperature gradient (known as the Seebeck effect). Conversely, they also transfer heat against the temperature gradient when a current is driven against this potential (known as the Peltier effect). They are promising energy materials with many applications, such as waste heat harvesting, radioisotope TE power generation, and solid state Peltier refrigeration, all of which are driving growing research interest. A key challenge is to improve the TE properties in order to obtain more efficient energy conversion and in turn enable new practical applications. Good TE materials must have excellent electrical transport properties, measured by the TE power factor ($=S^2\sigma$, where S is the Seebeck coefficient and σ is the electrical conductivity), and also a very low thermal conductivity κ (composed of the electronic contribution κ_e, the lattice contribution κ_L, and the bipolar contribution κ_{bi}). Combining the two aspects gives us the dimensionless figure of merit ZT,

$$ZT = \frac{S^2\sigma T}{\kappa},\qquad(1)$$

where T is the absolute temperature. ZT is the key parameter that characterises the energy conversion efficiency that is possible using a given TE material.

Optimisation of the TE performance of materials has proved to be difficult since the three transport parameters, i.e., S, σ and κ, are intercorrelated in a way that works against the occurrence of high ZT. For instance, increasing the Seebeck coefficient S usually means lowering the electrical conductivity σ, and vice versa; the electronic thermal conductivity κ_e also relates with the electrical conductivity σ via the Wiedemann–Franz law $\kappa_e = L\sigma T$ (L is the Lorenz number). The correlations among the parameters were studied in the early 1960s, and the best TE materials were identified to be narrow band gap semiconductors with some amount of doping (the optimal carrier concentration usually ranging from 10^{19} to 10^{21} cm^{-3}), leading to the first step forward in the TE research.[1,2] The carrier mobility μ and the band effective mass m^* present another important consideration. High m^* is beneficial to S; however, under single parabolic band (SPB) model and acoustic phonon scattering mechanism, high m^* reduces μ. Novel band engineering is necessary to break up this relationship. Several excellent reviews have discussed interrelations in the electrical transport and outlined possible solutions.[3,4] The lattice thermal conductivity κ_L is more or less independent of the electronic parameters. However, the traditional ways of achieving low κ_Ls, such as forming solid solutions or making composite structures, typically have a detrimental effect on the carrier mobility μ. A judicious choice of the grain size between the critical lengths for the electrical and the heat transport, which could preferentially scatter phonons, and/or introducing appropriate structural units that tend to optimise the two types of transport separately, were regarded as best solutions for the optimisation of μ/κ_L. Before 1990s, the exact recipes how to break the above-mentioned dilemmas of optimisation were either unclear, or hard

[1]Materials Genome Institute, Shanghai University, Shanghai, China; [2]State Key Laboratory of High Performance Ceramics and Superfine Microstructure, Shanghai Institute of Ceramics, Chinese Academy of Sciences, Shanghai, China; [3]Department of Physics, East China Normal University, Shanghai, China; [4]Material Science and Engineering Department, University of Washington, Seattle, WA, USA; [5]Department of Physics, University of Michigan, Ann Arbor, MI, USA and [6]Department of Physics and Astronomy, University of Missouri, Columbia, MO, USA.
Correspondence: L Chen (cld@mail.sic.ac.cn) or J Yang (jihuiy@uw.edu) or W Zhang (wqzhang@mail.sic.ac.cn)

to achieve experimentally and the best values of ZT stagnated near unity for about 30 years.

Some 20 years ago, two big ideas shed light on new directions in the field of thermoelectricity. Hicks and Dresselhaus proposed to utilise quantum well structures, aiming to enhance the power factor by altering the electronic density of states (DOS) in lower dimensions.[5,6] Although the quantum well structures are hard to fabricate even today, their main features can be imported into the realm of bulk materials by the state-of-the-art band engineering strategies. Moreover, lower dimensional structures with numerous interfaces, very effectively scatter the heat-carrying phonons and thus significantly reduce the lattice thermal conductivity.[3,7,8] The other important idea was the 'Phonon Glass Electron Crystal' (PGEC) paradigm, proposed by Slack.[9] It describes an ideal architecture of new TE materials with potentially independent electrical and thermal transport networks. The concept gave rise to intense studies on new complex compounds (e.g., filled skutterudites and clathrates) with atomic-level heterogeneity, different from the traditional binary TE compounds. Inspired by these two ideas, the field of thermoelectricity received a second major impulse in the late 1990s, which continues unabated till today.

The first major advance some half a century ago was theoretically underpinned by the classical SPB band model for electrical transport optimisation and by heavy elements, solid solutions and so on, for thermal transport reduction. The second and the more recent impetus mentioned above is based on new concepts, such as the role of a nanostructure, band (or scattering) engineering and complex structures. It is obvious that understanding the microscopic phenomena is critical, and this presents new challenges for the experimentalists. Take the band engineering as an example that relies on accurate information regarding the band structure and its momentum-dependence. The regular experimental characterisation of the transport parameters, such as the carrier concentrations and the effective masses, is usually extracted based on analytic band models, particularly the well-accepted SPB model. The main characters of the band of a material are reflected by the effective mass in an averaged way. Being simple and analytically clear, the SPB approach is considered to be very useful in understanding the general trend of transport properties and their dependence on carrier concentration, such as those reflected in the Pisarenko curves. Efforts beyond the SPB model by considering multiple bands and even non-parabola makes those expressions very complex and are only taken to explain the experimental data for very specific systems. The direct techniques of determining the band structure in experiments, e.g., angle-resolved photoemission spectroscopy, are seldom used in TEs due to some technical limitations, e.g., the requirements of single-crystal sample, extremely low temperatures and sensitivity to the surface contamination and so on.[10] Effective way of band structure characterisation for complex TE materials is still in a developing stage. Even if such measurements were available, it would still be difficult to touch upon more fundamental information concerning the nature of chemical bonds in order to reliably determine any band variation.

Theoretical work, including modelling and *ab initio*-based calculations, naturally span length scales from sub-angstrom to tens of nanometer and provide the bottom–up information. Band structure calculations can help to rationalise the band engineering, in a more direct way.[11] Additional information the theoretical work can offer includes the total energy, electronic and phonon dispersions, as well as atomic movement in a larger scale simulation. Although theoretical calculations cannot fully represent the experimental reality owing to some approximations and technical difficulties (e.g., the neglect of thermal motion in standard density functional calculations, the difficulties on the treatment of defects, as well as carrier scatterings), they still become a complementary tool to experimental techniques in modern research on thermoelectrics. Over the years, integrated theory–experiment studies have greatly improved understanding of TE transport and at the same time have greatly speeded up and advanced the optimisation of materials leading to enhancements in ZT values. The increasing numbers of integrated research efforts bode very well for the future success of TE research.

It has been exactly 20 years since Slack proposed the PGEC concept and since that time one has witnessed many important theoretical contributions providing new ideas that have advanced the field of thermoelectricity in a fundamental way. It is necessary then to provide a bird's eye view of the theoretical understanding and ideas in TEs and its integration with experiment. This is the main aim of this paper. We basically focus on the electrical and thermal transport processes. In the electrical transport ('Electrical Transport in Thermoelectrics'), we start from engineering both the band and scattering phenomena, as well as describing the concept of conductive networks emerging in complex TE materials. For the thermal transport ('Thermal Transport in Thermoelectrics'), we summarise the traditional approaches for κ_L reduction in conventional solids, including strong phonon-phonon interactions (PPIs) and the influence of nanostructures. Then, we review the work on thermal transport in complex materials possessing chemical-bond hierarchy. A few TE materials with synergistically optimised electrical and thermal transports will be briefly discussed in 'Synergistic Optimisation for High ZTs'.

ELECTRICAL TRANSPORT IN THERMOELECTRICS

The electrical transport properties are determined by the electronic structures of the materials and various carrier-scattering processes. The commonly used formulas based on Boltzmann transport theory for the electrical conductivity and the Seebeck coefficient are (tensor notions are omitted),[12]

$$\sigma(T) = \frac{1}{\Omega} \int \overline{\sigma}(\varepsilon) \left[-\frac{\partial f_0(T,\varepsilon)}{\partial \varepsilon} \right] d\varepsilon, \tag{2}$$

$$S(T) = \frac{1}{eT\Omega\sigma} \int \overline{\sigma}(\varepsilon)(\varepsilon - \varepsilon_F) \left[-\frac{\partial f_0(T,\varepsilon)}{\partial \varepsilon} \right] d\varepsilon. \tag{3}$$

Here ε_F, Ω, f_0, e are the Fermi level, the volume of the cell calculated, the Fermi–Dirac distribution, and the electron charge, respectively. The essential part in equations (2) and (3) is the transport distribution function (TDF),

$$\overline{\sigma}(\varepsilon) = \frac{e^2}{N} \sum_{i,k} \tau_{i,k} \mathbf{v}_{i,k}^2 \frac{\delta(\varepsilon - \varepsilon_{i,k})}{d\varepsilon}, \tag{4}$$

where N and i are the number of **k** points sampling and band index, respectively. **v** is the group velocity and τ is the relaxation time. The latter can be approximately expressed in the energy dependent form,

$$\tau = \tau_0 \varepsilon^r, \tag{5}$$

where τ_0 and r are constants for a given scattering mechanism.

The 'effective' electronic structures responsible for the electrical transport are within a narrow energy span determined by $-\partial f_0/\partial \varepsilon$. The term $(\varepsilon - \varepsilon_F)$, appearing in the numerator of the Seebeck coefficient (equation (3)), is of particular importance. It means the TDF on different sides of the Fermi level has an opposite contribution to the Seebeck coefficient. Rapidly changing TDF around the Fermi level is then favourable for a large Seebeck coefficient, which could be achieved by the manipulation of either DOS $N(\varepsilon)$ or relaxation time τ around ε_F, as shown in Figure 1. These manipulations can also be rationalised by the Mott relation,

$$S = \frac{\pi^2}{3} \frac{k_B^2}{e} T \left\{ \frac{1}{n} \frac{dn(\varepsilon)}{d\varepsilon} + \frac{1}{\mu} \frac{d\mu(\varepsilon)}{d\mu} \right\}_{\varepsilon = \varepsilon_F}, \tag{6}$$

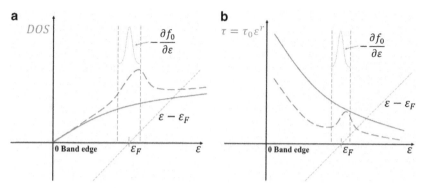

Figure 1. Schematic plots for (**a**) band engineering and (**b**) scattering engineering, with the blue dashed curves representing the manipulated cases.

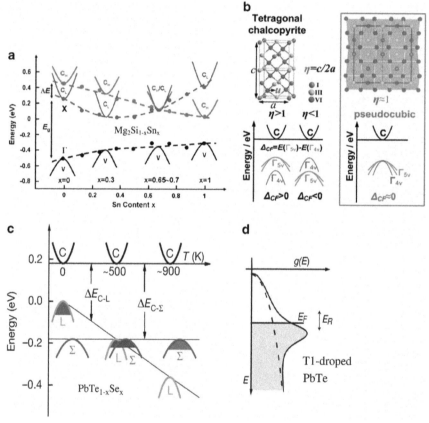

Figure 2. Several reported strategies for band engineering. (**a**) Proper solid solution composition in $Mg_2Si_{1-x}Sn_x$;[14] (**b**) The pseudocubic structure in tetragonal chalcopyrite compounds;[17] (**c**) temperature induced band order evolution in $PbTe_{1-x}Se_x$;[22] (**d**) Enhanced DOS in PbTe by the resonant states induced by the doping of Tl.[38]

where k_B is the Boltzmann constant. We can see that the enhanced energy dependence of the carrier concentration n (can be from a local increase in $N(\varepsilon)$) and the mobility μ is beneficial for the Seebeck coefficient. These are the bases for band and scattering engineering, which will be detailed in 'Band engineering for enhanced power factors' and 'Manipulation of carrier scattering', respectively.

In 'Carrier conductive network', we will review the concept of 'conductive network' in complex compounds. This concept applies to those multinary compounds where only part of the components is responsible for the electrical transport, which is distinct from the traditional binary TE materials. This will be elucidated in detail by several theory–experiment efforts in certain model systems, and some well-accepted qualitative features have been concluded for conductive networks.

Band engineering for enhanced power factors
The beneficial effects of a large-band degeneracy have long been recognised, as discussed by Goldsmid.[1] The energy-degenerate band edges (or band convergence in another words[13]) instead of energy separated ones can enhance the energy dependence of DOS at the Fermi level (Figure 1a), favourable for achieving large Seebeck coefficients for a fixed carrier density. On the other hand, the band shape for each energy pocket is unchanged, and thus the group velocities will not be largely altered. The twofold influence is the basis for enhancing the power factor of many TE materials by a large-band degeneracy. Several strategies for band engineering will be reviewed in this subsection.

Solid solutions between compounds with different band orders can serve as an effective strategy to achieve band convergence at proper compositions, as shown in Mg_2X (X = Si and Sn) solid solution. The conduction band of Mg_2Si and Mg_2Sn shows a typical dual band structure (one light and one heavy band) with an inverted band order in the two compounds. *Ab initio* calculations done by Liu *et al.*[14] revealed that the position of the band edges varies with the composition as shown in Figure 2a, and the band edges of the light and heavy conduction bands coincide at the composition $Mg_2Si_{0.35}Sn_{0.65}$. The enhanced Seebeck coefficient and the unaffected electrical conductivity resulted in a significantly higher power factor over the parent compounds. The result was reproduced in the analytical work by Bahk *et al.*[15] As the usage of solid solutions is very common in TEs for the purpose of reducing κ_L, it is useful then to examine the band structure of the parent compounds and their solid solutions for the possibility of realising band convergence.

Tetragonal chalcopyrite semiconductors demonstrate another way how to achieve large band degeneracy by tuning the structural parameters. The valence band structure of binary cubic zinc-blende compounds consists of degenerate Γ_{5V} and Γ_{4V} bands.[16] After the cation cross-substitution into ternary or quaternary compositions with the tetragonal structure, the crystal-field splitting $\Delta_{CF} = \Gamma_{5V} - \Gamma_{4V}$ and a non-unity structural parameter $\eta = c/2a$ are observed (Figure 2b). By performing systematic calculations of transport properties, Zhang *et al.*[17] revealed a direct link between Δ_{CF} and η in ternary tetragonal chalcogenides, with $\eta \approx 1$ (pseudocubic structure) resulting in converged bands and enhanced power factors. The simple unity-η rule is useful for the screening of new candidates of tetragonal chalcopyrite semiconductors. It also suggests a direction of optimisation in these compounds by altering the η towards unity, which can be achieved through doping, composition tuning and solid solutions between compounds with $\eta > 1$ and $\eta < 1$, as shown in the experimental efforts in ternary and quaternary chalcogenides.[17,18]

In the compounds with multiple bands, the different band evolution with temperature may cause a crossing of band extrema at some point, together with enhanced transport properties. The temperature-induced band convergence has only been studied in PbTe and the related rock-salt IV–VI compounds (Figure 2c). These compounds possess non-parabolic edge states with the second or even the third energy pocket close to band edges, as revealed by Singh[19] and Chen *et al.*[20] The pockets, denoted as L, Σ and so on, have different temperature dependence, leading to the variations of both band gap[21] and the energy difference between pockets $\Delta E (= E_L - E_\Sigma)$ at finite temperatures.[22] The theoretical studies based on quasiharmonic approximation[23,24] show that the band evolution cannot be fully captured by the temperature-induced volume change. Using *ab initio* molecular dynamics (AIMD) and taking snapshots of structures for band calculations, Kim *et al.*[25] found that the L and Σ pockets converged at 450 K. The results of Gibbs *et al.*[26] further supported to the twofold contribution (lattice expansion and atomic displacement) to the L–Σ convergence, with the converged temperature was 700 K for PbTe. Generally speaking, the mechanism of temperature-induced band convergence is at the stage of rationalisation and more efforts are needed to get a better understanding of the phenomenon.

Defects, including vacancies, antisites, interstitials and elemental doping, can have a strong influence on both the electronic structure and charge-carrier scattering. From the band engineering point of view, defects offer another strategy to achieve band convergence by altering the bands. Explicitly, there are at least two types of band alterations by the defects: (i) altering relative band positions of the matrix (type-I) and (ii) introducing the so-called resonant levels via defect–host interactions (type-II). Numerous experiments[27–32] with IV–VI compounds have suggested that the tuning of the relative positions of different pockets by doping on the IV-site (type-I alterations). Some of the beneficial dopants were verified by *ab initio* band structure calculations,

without looking into the mechanisms.[27,29,31–34] Tan *et al.* revealed that the smaller ΔE in Mn-doped PbTe is due to the anti-bonding of Te-p and Mn-d orbitals which push the second Σ pocket upwards.[34] The mechanism for other dopants with no d states near the Fermi level, such as Mg, Cd, is still unknown. The type-II band alteration—the resonant level—can also increase the band edge DOS as shown in Figure 2d. The aim is to have a dopant that forms the defect state in the vicinity of the Fermi level.[35,36] Such dopants usually have their electronic configuration very close to that of the host, such as when doping by elements from the neighbouring columns of the periodic table. Examples are IIIA elements doping in the rock-salt IV–VI structure and functioning as p-type dopants,[36–39] and IVA elements in V_2VI_3 compounds.[40] Pb doping on the Bi-site in BiCuSeO,[41] Sb-doping on the Te-site in $CuGaTe_2$,[42] and even antisite defects in ZrNiSn[43] are all known to form resonant levels at the respective band edges, illustrating a variety of resonant levels one can achieve in TE materials.

Beside the above-discussed mechanisms, the lower dimensional DOS caused by the Rashba spin-splitting effect and thus the larger S in comparison with predictions based on the SPB model were reported recently,[44] indicating another interesting approach to modify the DOS distribution around the Fermi level. Having more compounds with converged bands enhances chances of discovering TE materials with higher power factors. As the experimental determinations of the DOS are subject to uncertainties and technical challenges,[14,45] accurate *ab initio* band structure calculations are a convenient approach to inquire about the prospect of band convergence in a particular TE material and, no doubt, such calculations could have an even more prominent role in future studies.

Manipulation of carrier scattering

Carrier scattering is an indispensable part of electrical transport. The carrier relaxation time is determined by several scattering mechanisms, including electron–phonon scattering (both acoustic and optic phonon modes may participate), impurity scattering (both neutral and ionised), energy barrier scattering, piezoelectric scattering and electron–electron scattering. As shown in equation (5) and Figure 1b, each scattering mechanism can be expressed in terms of an energy-independent prefactor τ_0 and an exponential term ε^r where r is the scattering parameter. Different scattering mechanisms are expressed through different forms of the scattering parameter. For instance, for charge carriers scattered by acoustic phonons the scattering parameter $r = -1/2$, whereas for their scattering by ionised impurities $r = 3/2$. The carrier scattering can be altered by either the intensity of scattering events (prefactor τ_0) or by changing its energy dependence (scattering parameter r), both manipulations referred to as scattering engineering.

Given the expected operational range of TE materials used for power generation, acoustic phonon scattering is usually the dominant mechanism. Its scattering rate under the SPB model is,

$$\frac{1}{\tau} = \frac{1}{\tau_0}\varepsilon_F^{\frac{1}{2}} = \frac{\sqrt{2}E_{def}^2(m^*)^{\frac{3}{2}}k_BT}{\pi\hbar^4 v_s^2\rho}\varepsilon_F^{\frac{1}{2}} = \frac{\pi E_{def}^2 k_BT}{\hbar v_s^2\rho}N(\varepsilon_F),\tag{7}$$

where E_{def}, $\hbar$, v_s, and ρ are the effective deformation potential for the electronic states, the reduced Planck constant, the sound velocity and the density of the material, respectively. In order to increase τ_0, the focus is usually on the reduction of the band effective mass m^* or DOS at the Fermi levels, the parameters that are more easily manipulated than the deformation potential or the sound velocity. Yang *et al.*[46] studied possible solutions for the relatively low mobility in Fe-based p-type skutterudites, and theoretically proposed the usage of $4d$ or $5d$ elements with which to form p-type skutterudites. The $4d$ or $5d$ electrons are spatially more delocalised, resulting in a lighter valence band and higher

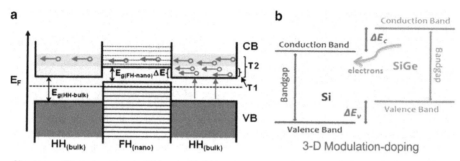

Figure 3. (**a**) Energy filtering effect in the half-Heusler/full-Heusler composites consisting of the half-Heusler matrix with the full-Heusler nanophase.[60] (**b**) Modulation doping in a two-phase nanocomposite consisting of the $Si_{95}Ge_5$ matrix and a large fraction of phosphorus-doped high Ge content $Si_{70}Ge_{30}P_3$ nanophase.[62]

mobilities. Fu *et al.* found experimentally that the *p*-type half-Heusler compound NbFeSb has a larger power factor enhancement than its isoelectronic cousin VFeSb due to the reduced effective mass and thus higher mobility.[47] The reduced effective mass in NbFeSb can be rationalised owing to stronger Nb 4*d*–Fe 3*d* interaction than 3*d*–3*d* one in VFeSb, similar to a situation in light valence band skutterudites. Other experiments emphasising the importance of low DOS in order to achieve high-carrier mobility are I-doped PbTe[48] and K-doped BiCuSeO.[49] Improved power factors can be obtained by a successful tradeoff between the mobility enhancement and a reduction in the Seebeck coefficient.

The enhancement of the Seebeck coefficient by enhancing the scattering exponent *r* is another aspect of the scattering engineering. The SPB expression for the Seebeck coefficient in the degenerate doping limit is,

$$S = \frac{k_B^2 \pi^2 T}{3 e \varepsilon_F} \cdot \left(\frac{3}{2} + r \right). \tag{8}$$

This effect is usually achieved by introducing additional scattering centers, e.g., ionised impurities that force the scattering parameter to change from − 1/2 to +3/2. Several experimental works have supported this idea, among them the research with doped skutterudites,[50–53] Ni cross-substituted type-I clathrates,[54] and Bi_2Te_3 with native defects.[55] This change of *r* brings about additional scattering, which inevitably reduces the mobility. Finding a proper ratio between the acoustic phonon scattering and the impurity scattering is essential to maximise the benefits. For instance, Dyck *et al.* adopted a two-scattering-mechanism model, considering both the acoustic phonon and ionised impurity scattering, to deduce the optimal power factors of Ni-doped Ba-filled $CoSb_3$.[50] Similar procedure has been adopted in Cr-doped Ce-filled $CoSb_3$.[53] Both the enhanced ionised impurity scatterings and the charge compensation effects (increasing the filling fraction limits of Ce) of the dopant Cr are beneficial to the averaged ZT values.

Energy barrier scattering also involves the use of strongly energy-dependent scattering to enhance the Seebeck coefficient. It can be done by introducing interfaces in a composite structure that have energy barriers, which block carriers at low energy but not those at high energy.[56] Energy-barrier scattering is conceptually similar to the scattering by ionised impurities; both processes scatter low-energy carriers more effectively than the high energy ones, but the implementation is different. The beneficial effect of energy barriers can be found in the systems having a semiconducting host matrix and metallic composites or nano-inclusions, such as the pristine PbTe matrix containing Ag-doped PbTe nanoparticles,[57] semimetallic ErAs particles in InGaAs/InGaAlAs superlattices,[58,59] and half-Heusler compounds with full-Heusler metallic inclusions (Figure 3a).[60,61] Another possible beneficial scheme relating to the interface between matrix and nanostructure is the modulation doping and δ-doping

techniques. In principle, it is different from the barrier scattering mentioned above since the dopants are only incorporated into certain areas, e.g., spatially separated nanograins, leading to reduced electron scattering and thus higher mobility. Improved power factors and figure of merit ZT are experimentally achieved in the $Si_{95}Ge_5$ matrix and the phosphorus-doped $Si_{70}Ge_{30}P_3$ nanoparticle phase, where the electrons can spill over from the doped nanoparticle phase into the essentially undoped matrix phase, as shown in Figure 3b.[62]

Another strategy on how to apply scattering engineering might be found in topological insulators having a large difference in the relaxation times corresponding to bulk and surface states. As the surface states are topologically protected, at the energy interface between surface and bulk states there is an abrupt change in the relaxation time. Xu *et al.*[63] proposed a dual-scattering time model for TE transport tuning in topological insulators. Their model calculations show that the ratio between the surface and bulk states has an important role in the figure of merit ZT, and ZT values of over 8 might be achieved when the ratio is $>10^3$ in two-dimensional topological insulators. Shi *et al.*[64] also connected the topological insulator property to TE performance based on calculations for Bi_2Te_3 and Bi_2Te_2Se, more from the band complexity point of view. Up to now, it is still an open question whether the topologically insulating behaviour is beneficial to thermoelectrics, as well as the origin of the benefits (from scattering or band variations) if there is any.

In general, scattering engineering offers exciting possibilities for enhancing TE performance. The power factor can be enhanced by a more rapid energy dependence of the relaxation time, by introducing defects or interfaces with energy barriers that filter out low energy carriers, and by using innovative modulation doping schemes that promote high carrier mobilities. Although the intent is clear, the realisation of the above scattering engineering approaches is quite challenging. More research is needed to understand the rationale regarding to the microscopic mechanisms leading to the desired changes in the scattering parameter.

Carrier conductive network

In 1995, Slack proposed an architecture for high-ZT ternary compounds in which the vast majority of atoms formed a conductive framework for charge carriers, while the chief role of the remaining atoms was to scatter phonons.[9] This PGEC paradigm was very influential in the discovery of several families of TE materials. It also presented a novel architecture, which was distinctly different from the traditional TE compounds such as PbTe, Bi_2Te_3 and so on, in which all the component elements were responsible for the electrical transport. It was not recognised initially that the PGEC paradigm also contained a concept of conductive networks. Strictly speaking, electronic structures of crystalline solids possess contributions from the electronic orbitals of all atoms. However, since the effective energy windows for

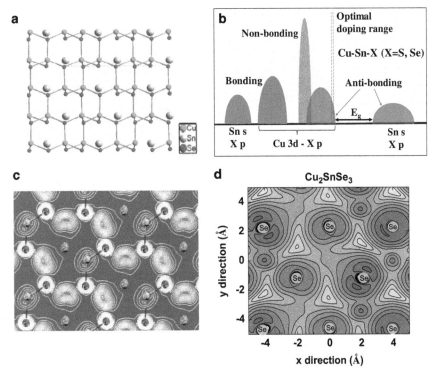

Figure 4. (a) The crystal structure of Cu_2SnSe_3. (b) The scheme plot of DOS in Cu_2SnX_3 (X = S, Se), with the bonding characters and the p-type optimal doping range labeled. (c) and (d) are the contour plots of the partial charge density within the p-type optimal doping range of Cu_2SnSe_3 (−0.2 to −0.3eV) on the Cu–Se–Sn plane,[72] and the close-packed Se–Se–Se plane, respectively.[71]

electrical transport are narrow (Figure 1), the concept of conductive networks is approximately valid when those atoms forming the network dominate the electronic states of interest. On the basis of the experience gained with various TE compounds in the recent years, several well-accepted qualitative features have emerged that speak strongly for the usefulness and validity of the concept of conductive networks based on both theory and experiment, as will be shown in the following.

N-type filled $CoSb_3$-based skutterudites are regarded as prototype compounds that embody the PGEC concept. The conduction band minimum (CBM) of pure $CoSb_3$ is composed of d–p hybridised orbitals of both framework elements.[65] The guest fillers, usually very metallic cations, form ionic bonds with the surrounding Sb atoms, and do not affect the CBM of $CoSb_3$ except for shifting the Fermi level position.[66,67] Experimentally, the rigidity of the CBM is reflected by the influence of fillers on the transport parameters. As shown in ref. 68, the experimental Seebeck coefficients of skutterudites with various fillers and filling fractions follow the same Pisarenko relation; and the electron mobilities have the same −3/2 temperature dependence. Thus there exists a universal optimal doping level for n-type filled skutterudites (~0.5 electrons per Co_4Sb_{12} cell at 850 K), simplifying the optimisation in experiments especially for multiple-filled systems.[67,69] The conductive networks in p-type Fe-based skutterudites were also studied by Yang et al.,[70] with the conductive network mainly composed of d orbitals from Fe.

The concept of conductive networks is not limited to the caged structures. Xi et al. studied electronic structures and chemical bonding in diamond-like compounds $Cu_2SnS(Se)_3$ (Figure 4a).[71] On the basis of the analysis of the bonding character, the bands constituting the valence-band maximum (VBM) are mainly composed of the anti-bonding between Cu d and Se p orbitals, and contain no obvious contribution from Sn atoms (Figure 4b). The energy range corresponding to the maximum p-type power factor is close to the VBM (around 0.1 hole per formula unit in the case of Cu_2SnSe_3), as labelled in Figure 4b. In this optimal doping

range, there exist a 3D [-Cu-Se(S)-Cu-] conductive network with Cu-Se(S) anti-bonding (Figure 4c),[72] as well as a closely packed chalcogenide framework (Figure 4d),[71] and both of them contribute to the hole transport. In either of the network, Sn is isolated. Doping on the reservoir element Sn in these compounds for maximising the electrical properties is desirable,[72] owing to the undisturbed conductive networks. The Cu 3d-chalcogen p hybridised bond forming the VBM is universal in Cu-based diamond-like compounds and similar conductive networks have been revealed by calculations for other diamond-like compounds.[73,74]

The above studies reveal some general qualitative features of conductive networks in TE materials. At the Fermi level, atoms on the conductive network form a clear bonding channel in real space, with out-of-network atoms being nearly isolated. In the energy space, the electronic states at the Fermi level are derived from the corresponding network atoms, with little contribution from other species. This makes it possible for the out-of-network atoms to serve as a carrier reservoir. Tuning and modification of the carrier reservoir will change neither the band structure nor the scattering mechanism. The optimal doping levels can be easily obtained based on the bulk bands, an approach which is very useful for experimental optimisation. From the chemical bond perspective, compounds with conductive networks usually possess the so-called chemical bond hierarchical structure. The out-of-network atoms are typically loosely bonded with the rest of the structure with large atomic displacement parameters (ADPs), and giving rise to low-frequency resonant phonon modes. The topic will be discussed in 'Resonant phonon scattering and low-frequency rattling modes' and 'Diverse lattice dynamics and part-crystalline part-liquid state in complex thermoelectrics'. Owing to these beneficial features, it is advisable to carry out the identification of conductive networks in more complex compounds. Besides the filled skutterudites and diamond-like compounds, many Zintl compounds are another type of structure

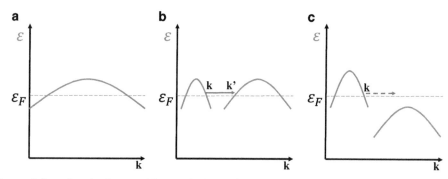

Figure 5. The comparison of three band schemes, (**a**) a single heavy band; (**b**) energy converged multiple bands; (**c**) energy separated multiple bands.

with a well-defined conductive network, in this case composed of the polyanion sublattice.[75–78]

Theoretical description of electrical transport and challenges

So far, we have reviewed three topics concerning the electrical transport: band engineering; scattering engineering; and conductive networks in complex compounds. The first two are primarily aimed at the power-factor enhancement by the manipulations on either the band structure or the carrier scattering. Conductive networks are an idealised architecture for complex TE materials, where only a part of the structure contributes to the electrical transport. Applications of these three topics in a variety of TE materials were discussed in 'Band engineering for enhanced power factors', 'Manipulation of carrier scattering' and 'Carrier conductive network'; here we will focus on the theoretical understanding and the challenges ahead.

Electrical transport could be described within the framework of the Boltzmann transport theory. One basically inquires how the distribution function evolves under external fields. Relevant formula are presented at the beginning of 'Electrical Transport in Thermoelectrics'. Comparing the definition of σ (equations (2) and (4)) with the macroscopic form of $\sigma = ne\mu$, the integral of the DOS is equivalent to n, and therefore the carrier mobility is directly related to the integral of the relaxation time $\tau_{i,k}$ and the group velocity,

$$\mathbf{v}_{i,k} = \frac{1}{\hbar}\nabla_k \varepsilon_{i,k}. \tag{9}$$

Dispersive bands and low scattering rates are favourable for high μ. The Seebeck coefficient S, due to the cancelling effect at both the denominator and numerator, is less sensitive to $\mathbf{v}_{i,k}$ or $\tau_{i,k}$. This is the foundation of the constant relaxation time approximation (CRTA) and thus τ-free S in many of the methodologies for electrical transport calculations nowadays. The term $(\varepsilon - \varepsilon_F)$ in equation (3), however, implies that S is sensitive to the changing rates of TDF $\overline{\sigma}(\varepsilon)$ on both sides of the Fermi level, as already mentioned in the context of band engineering and scattering engineering.

The influence of band engineering on power factors can be fully understood after taking both S and μ into account. For example, one single heavy band (Figure 5a) or multiple energy converged bands (Figure 5b) with equivalent DOS effective masses could have the similar Seebeck coefficients but rather different mobilities due to the different $\mathbf{v}_k$. For systems that already possess multiple bands, achieving the band convergence is beneficial from the DOS point of view, in which the system still keeps the same $\mathbf{v}_k$, as shown in Figure 5b,c. However, one thing that might be detrimental to the mobility is the carrier intervalley scattering (Figure 5b). The energy degenerate bands do increase the scattering channels for carriers, which could increase rates for all possible scattering mechanisms. Experimentally, the mobility values for large band degeneracy systems are lower than for

systems with lower band degeneracy at comparable carrier concentrations.[14,28]

A comprehensive band engineering approach to evaluate effects influencing the power factor should consider intervalley scattering processes caused by the high-band degeneracy. Direct calculations of the intrinsic electron–phonon scattering based on full electron and phonon dispersions have been realised only in the recent years.[79,80] The inverse relaxation time for an electron state at $\mathbf{k}$ being scattered by a phonon $\mathbf{q}$ can be written as,

$$\frac{1}{\tau_\mathbf{k}} = \frac{2\pi}{\hbar}\sum_{\mathbf{k}'}|\langle\mathbf{k}'|\partial_\mathbf{q}V|\mathbf{k}\rangle|^2[(f_{\mathbf{k}'} + n_\mathbf{q})\delta(\varepsilon_\mathbf{k} - \varepsilon_{\mathbf{k}'} + \hbar\omega_\mathbf{q})\delta_{\mathbf{k}+\mathbf{q},\mathbf{k}'+\mathbf{G}}$$
$$+ (1 - f_{\mathbf{k}'} + n_\mathbf{q})\delta(\varepsilon_\mathbf{k} - \varepsilon_{\mathbf{k}'} - \hbar\omega_\mathbf{q})\delta_{\mathbf{k}-\mathbf{q},\mathbf{k}'+\mathbf{G}}]\left(1 - \frac{\mathbf{v}_{\mathbf{k}'}\cdot\mathbf{v}_\mathbf{k}}{|\mathbf{v}_{\mathbf{k}'}||\mathbf{v}_\mathbf{k}|}\right). \tag{10}$$

Here $n_\mathbf{q}$ and $\omega_\mathbf{q}$ are the phonon number and the phonon frequency, respectively. $|\langle\mathbf{k}'|\partial_\mathbf{q}V|\mathbf{k}\rangle|$ is the scattering matrix, which is the key element in the calculation. Implementation of such method could be helpful for better quantification of the effects from band engineering. It is, however, a very demanding undertaking and rarely carried out. For other scattering mechanisms, to our best knowledge, there is no direct calculation reported so far.

The limited knowledge concerning various scattering processes also impacts attempts to do scattering engineering, which is currently beyond the theoretical capability. From experiments, the relative strength of multiple scattering mechanisms is commonly extracted by fitting the Pisarenko plot (S versus n). A more direct way to measure the scattering parameter r is to use the so-called 'four coefficients' method, i.e., from the measurements of four transport coefficients to determine another four microscopic parameters, including the scattering parameter r.[38,81,82] The 'four coefficients' method should be used with more systems to shed light on the underlying physical processes that cause variations in the scattering processes. Other strategies for scattering engineering, such as the carrier barrier scattering and nanoinclusions, currently lack any microscopic understanding.

The concept of conductive networks is an approximation in which, within the energy window of interest, only parts of the compound contribute to the electrical transport. Several qualitative descriptions have already been provided in 'Carrier conductive network'. The currently used theoretical approaches which describe conductive networks are the dominant contributions for in-network atoms from electronic structures. These, however, may not necessarily be the same as contributions to the electrical transport. A more accurate and definitive theoretical approach that can directly relate the electrical transport properties to the atomic species is needed. The task is interesting but challenging, especially when considering the fact that the loosely bonded out-of-network atoms may lead to structural fluctuations.

THERMAL TRANSPORT IN THERMOELECTRICS

As mentioned, the thermal conductivity κ has three components, the electronic part κ_e, the lattice part κ_L, and the bipolar part κ_{bi}. The κ_e links with σ via the Windeman–Franz law $\kappa_e = L\sigma T$. The Lorenz number L in semiconductors varies between 1.5 and 2.44×10^{-8} WΩ K^{-2}, spanning the range from non-degenerate semiconductors to strongly degenerate systems. The κ_{bi}, which arises as a consequence of electron–hole pair creation at the hot side of a sample and its recombination at the cold side, depends on the ratio of mobilities for majority and minority carriers in the non-degenerate case[83,84] and on the minority conduction only in degenerate case.[85] In order to suppress the bipolar contribution to the κ in degenerate semiconductors, it is desirable to reduce the minority carrier transport. This can be achieved by the band alteration, such as increasing the band gap, reducing the minority effective mass and so on, or by the preferential scattering of the minority carriers, as proposed by Wang et al.[85] In this section, however, we shall focus on the discussion of the lattice thermal conductivity κ_L, the dominant contribution to the heat transport in thermoelectrics for most of the cases.

In a conventional crystalline solid, the lattice thermal conductivity is determined by the specific heat capacity c, the sound velocity v_s, and the mean free path (MFP) l or phonon relaxation time τ_q according to $\kappa_L = \frac{1}{3}cv_sl = \frac{1}{3}cv_s^2\tau_q$. The respective parameters entering the above equation and, specifically, the phonon relaxation time τ_q are frequency dependent.[86,87] Figure 6 presents the phonon scattering times for various mechanisms as a function of phonon frequency. Among these, PPIs are an intrinsic mechanism. Of the extrinsic mechanisms, the grain boundary scattering is nearly independent of frequency and influence mainly the transport of long-wavelength phonons. Point defects are particularly effective in scattering high frequency phonons on account of their ω^4 frequency dependence (Rayleigh scattering). Nano-inclusions, behave approximately as point defects for low-frequency (long wavelength) phonons, and as grain boundaries for short wavelength phonons. Some of these traditional scattering mechanisms will be discussed in 'From the conventional phonon–phonon interactions to nanostructures'.

Large atomic displacements of guest fillers in compounds with open crystal structures make them interesting and unique. The physical origin of substantial reduction of the κ_L achieved in these materials with void fillers, i.e., the rattling induced resonant scattering at low frequencies (Figure 6) or merely alterations of the phonon spectrum, has been debated for a long time. These topics will be presented in 'Resonant phonon scattering and low-frequency rattling modes'. We will also review those compounds with some atoms having very large atomic displacements and even liquid-like behaviours in 'Diverse lattice dynamics and part-crystalline part-liquid state in complex thermoelectrics'. The dynamic behaviours of these atoms and the associated phonon

scattering mechanisms owing to the large atomic displacements fall beyond the conventional treatment relying on small parameter perturbative theory. How to properly describe the heat transport in these compounds is still an open question.

From the conventional phonon–phonon interactions to nanostructures

The traditional ways to achieve low κ_L in TE materials is to use heavy constituent elements or complex structures with low phonon velocities or reduced fractions of heat-carrying phonons.[1,88,89] Solid solutions are another time honored approach to enhance point defect scattering.[90–95] PPIs, as the intrinsic mechanism and basically the dominant one at high temperatures, have offered limited opportunities for regulation. Recently, thanks to advanced computational capabilities, the understanding on PPIs and their influences on lattice thermal conductivity has been greatly improved. Among the extrinsic scattering mechanisms, nanostructures have been widely used in TE materials due to the significant effect on scattering long-wavelength phonons which are difficult to scatter otherwise. In this subsection, the focus will be on PPIs and the role of nanostructures.

PPI is an intrinsic process taking place in all crystalline solids and is robust to doping, alloying, grain size and so on. Large lattice anharmonicity, characterised by the effective Grüneisen parameter γ, and low group velocity are common explanations for low κ_L under strong PPIs.[86,96] In addition, as revealed by the recent ab initio third-order interatomic force constant (IFC) calculations, a large scattering channel, analogue to the electronic DOS in the carrier scatterings (equation (7)), can serve as another approach to obtain a strong PPI.[97] Binary chalcogenide compounds have served as a target of many theoretical studies aimed at the understanding of their low κ_Ls.[24,98–102] An et al.[98] and Zhang et al.[100] revealed that the very soft transverse optic phonon at Γ point in PbTe, responsible for the low κ_L of the compound, is caused by the near ferroelectricity and the partially covalent character of Pb–Te bond. The covalent character was further highlighted by Lee et al.[102] who found strong resonant bonding in IV–VI rock-salt chalcogenides. These resonant bonds result in large anharmonicities, optic phonon softening and large scattering channels, all favourable for low lattice thermal conductivity.[101–103] The deformed rock-salt structure material SnSe was reported to show an exceptionally low in-plane κ_L, resulting in a record-high ZT = 2.62.[104] The ab initio calculations reveal that the strong PPI of SnSe originates from the large phonon scattering channel and the very anharmonic resonant bonds.[103,105]

Recent molecular dynamics calculations of PPI even account for the evolution of κ_L with temperature. The stiffening of the transverse optic phonon at Γ point in PbTe with increasing temperature is revealed by both inelastic neutron scattering and AIMD calculations.[106,107] The transverse optic phonon and longitudinal acoustic phonon are decoupled at 600 K, which reduces the scattering channel and causes the kink in high temperature κ_L. The opposite case is presented by Bi$_2$Te$_3$ where more optic phonons are involved in scattering when the temperature increases and thus the decrease of κ_L is faster at room temperature.[108]

The above studies emphasise the several factors causing strong PPI and low κ_L. Those zone center acoustic phonons, however, whose wavelengths are usually longer than the scales of unitcells, are hard to be scattered by PPI due to their low frequencies and small wave vectors. Nanostructures, including nanometer scale grains, nanoinclusions and nanocomposites, thus need to be considered in order to achieve interface scattering to long-wavelength phonons. A representative example is the concept of 'all-scale' phonon scattering by the integrated length scale of microstructures.[109] By using different amounts of SrTe, Biswas et al. introduced atomic scale solid solutions, nanoscale

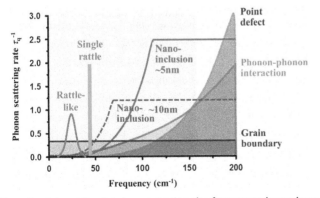

Figure 6. Schematic plot demonstrating the frequency dependence of phonon scattering rates for various mechanisms.

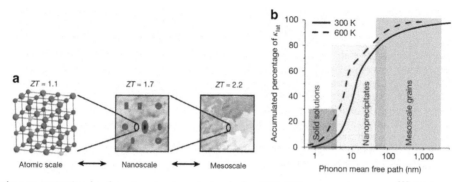

Figure 7.   (a) All scale phonon scattering by the microstructure design in the PbTe-SrTe composite system.[109] (b) Accumulated κ_L of PbTe as a function of phonon mean free path.[109]

precipitates and mesoscale grain boundaries into PbTe (Figure 7a), and increased the ZT values correspondingly. Over the years, the benefits of nanostructuring on κ_L reduction have been well-accepted. The vital issue with the use of nanostructures is the lack of accurate size and shape control of grains, as well as a quantitative and insightful theory. This makes the nanostructures only a qualitative explanation for the κ_L reduction.

The foundation element of the all-scale phonon scattering paradigm is the accumulated fraction of the thermal conductivity with respect to the phonon mean-free path (Figure 7b). On the other hand, charge carrier mobilities will also be interfered with by the much enhanced density of interfaces. Thus a proper control of the nanostructure scale is necessary to preferentially scatter phonons. It is essential then to know both the critical lengths (i.e., the wavelength or the mean free path) of electrons and phonons. Yang et al. calculated the room temperature values of the electron de Broglie wavelength λ,[110] and the values of λ for degenerate compounds fall within a narrow range ($\leqslant$5 nm). The phonon mean-free paths, if we define the mean values corresponding to 50 % κ_L reduction,[111] vary by 2–3 orders of magnitude according to ab initio IFC calculations.[94,102,105–108] Usually, low κ_L materials have low MFPs, sometimes even lower than the effective length of charge carriers. This makes the nanostructuring approach less effective for low κ_L materials, as verified by the theoretical work with nanoparticles embedded in (Bi, Sb)$_2$Te$_3$.[112] Only compounds with long MFPs of phonons, such as Si or SiGe alloys, half-Heusler alloys, and skutterudites, can sufficiently take the benefits from the nanostructures. Furthermore, owing to the decreasing of MFP of phonons as the temperature increases (Figure 7b),[109,111] the effect on the κ_L from the interface scattering will diminish at higher temperatures.

Another quantitative effort is the modelling of κ_L variations due to modulations of the nanostructure. Simulations on different length scales, using the methods from ab initio calculations to molecular dynamics with empirical potentials, have revealed a strong influence on κ_L by nanocomposites possessing complex geometry, usually fabricated within the Si-Ge system. The κ_L is typically determined by the surface area and the volumetric fraction of the composite, as shown by the study with different types of Si nanowires embedded in the Ge host.[113,114] Size effects in the phonon transport of Ge containing Si nanoparticles were emphasised by the Monte Carlo simulation.[115] Several studies documented the optimal sizes and distributions in the nanocomposites.[116–119] Mingo et al. studied 17 different silicide nanoparticles in the SiGe host,[116] and revealed the existence of the optimal nanoparticle size that minimises the nanocomposite's κ_L. Similarly, the optimal Si/Ge superlattice period of around 3.3 nm minimised the value of the κ_L.[117] Besides the focus on κ_L, calculations were also made to estimate ZT values for some composite systems such as SiGe composites, nanotubes and quantum dot superlattices.[120–122]

A great progress has been achieved in reducing the κ_L by scattering phonons over a wide range of frequencies. Advances in computational methodologies have been instrumental in promoting the understanding of the lattice dynamics in TE materials as exemplified by prolific studies of PbTe. The calculations also indicated the proper nanostructure size effective to preferentially scatter long-wavelength phonons. Modelling work has helped to rationalise the role of complex nanocomposites on κ_L. On the basis of the above studies, the understanding of the phonon transport in conventional materials has been clarified to a large extent even though the full quantitative prediction is still a challenge.

Resonant phonon scattering and low-frequency rattling modes

In 'From the conventional phonon–phonon interactions to nanostructures', we have seen the great influence on the κ_L exerted by extrinsic scattering mechanisms such as nanostructures, as well as the improved understanding of intrinsically strong PPIs. Then an interesting question arises concerning whether it is possible to find structures where some other 'intrinsic' process for low κ_L might exist, and do so with a high degree of adjustability like nanostructures? Moreover, if such desired structures existed, would it still be possible to use small-perturbative phonon approximation to describe the thermal transport properties in the structures?

As a kind of compounds transiting from the conventional to the unconventional materials are compounds with an open crystal structure, such as skutterudites and clathrates. These compounds possess large structural voids that can be filled with loosely bonded foreign species, which have a strong influence on the κ_L. Filled skutterudites were first synthesised by Jeitschko and Braun[123] and they became a target of intense scrutiny as TE materials in the late 1990s.[124–126] Compared with the conventional compounds with the chemical-bond homogeneity, the fillers possess much larger ADPs compared with the atoms forming the framework, leading to chemical-bond hierarchy and much reduced κ_L.[127] Nolas et al.[127] found that the point defect scattering is not sufficient to explain the very low κ_L in partially filled skutterudites. The resonant scattering mechanism, in which acoustic phonons are scattered by the resonant vibrations of the fillers, was thus proposed. A typical form of the scattering rate under this scattering mechanism is,[128]

$$\frac{1}{\tau_{Res}} = \frac{C\omega^2}{\left(\omega^2 - \omega_0^2\right)^2}. \tag{11}$$

Here C is a constant proportional to the concentration of the resonant centers, which oscillate with the frequency ω_0. Various filler species in Co$_4$Sb$_{12}$ with characteristic resonant frequencies were explored subsequently.[129,130] The reduction of κ_L by rattling fillers in the filled skutterudites could be achieved over a wide temperature range, as shown in Figure 8a. This is a very effective

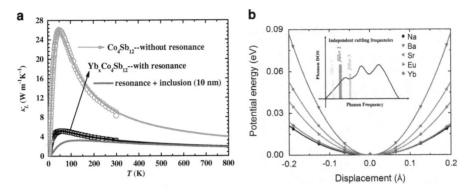

Figure 8. (**a**) The evolution of κ_L from pure Co_4Sb_{12} (without resonance), to $Yb_xCo_4Sb_{12}$ (with resonance) and to the calculated $Yb_xCo_4Sb_{12}$ containing 10 nm nano-inclusions. (**b**) Schematic plot for the resonant scattering and the potential well for different fillers in $CoSb_3$. Data are from ref. 131.

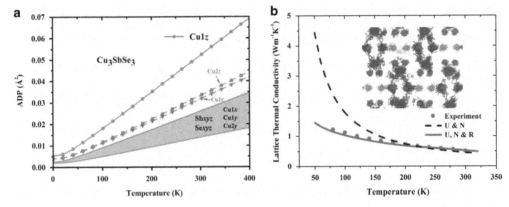

Figure 9. (**a**) Atomic displacement parameters of all constituent elements of Cu_3SbSe_3. (**b**) The lattice thermal conductivity with and without the rattle-like scattering term (R) in the same compound. Data are from ref. 147.

way of realising the thermal conductivity reduction; even augmenting nanoinclusions into a $CoSb_3$ matrix can hardly give rise to a comparable reduction in a broad temperature range (Figure 8a). In 2007, Yang et al.[131] theoretically calculated the potential well for various fillers in Co_4Sb_{12} (Figure 8b), and determined their resonant frequencies using a harmonic approximation. On the basis of the results, using multiple fillers with significantly different resonant frequencies was proposed as a method to obtain a further reduction in κ_L. Experimentally, the so-designed multiple-filled skutterudites do possess reduced κ_L and enhanced ZT values, as shown in refs. 68,69,132,133. Resonant frequencies of fillers in p-type skutterudites were also calculated by Liu et al.[134] and used to design high-performance p-type skutterudites.

Despite the great success in guiding the experiments, the resonant scattering mechanism itself has often been questioned. On the basis of the neutron spectroscopy experiments and ab initio lattice dynamic calculations for the fully filled La(Ce)Fe$_4$Sb$_{12}$, Koza et al.[135] argued that the guest fillers should couple with the host atoms in a coherent way, and show an ordinary phonon behaviour of filler species rather than the independent rattling motion. The κ_L reductions should be attributed to the zero group velocities of the low-frequency optic phonon branches. Christensen et al. demonstrated the acoustic-optic interaction and avoided-crossing behaviour of phonons in another cage structure, $Ba_8Ga_{16}Ge_{30}$ clathrate, using neutron triple-axis spectroscopy.[136] The reduction of phonon relaxation time caused by the additional optic phonons is inadequate to explain massive suppressions in κ_L, hence, the influence of the reduced group velocity was invoked. The importance of reduced group velocity of phonons was further discussed in refs. 88,137. Subsequent experimental and ab initio

work also confirmed the presence of acoustic–optic interactions in skutterudites.[138,139]

The above-mentioned controversy focuses mainly on the exact role of fillers in reducing the κ_L. With fillers introducing the nearly zero-velocity optic phonon modes at certain frequencies is phenomenologically equivalent to the resonant scattering mechanism proposed earlier. It is worth noting that suppressions of κ_L by the fillers cannot be fully accounted for by the introductions of additional optic phonons, based on the third-order IFC calculations.[111,140,141] The κ_L for the unfilled $CoSb_3$ can be perfectly reproduced by these calculations;[111] for fully filled skutterudites, however, the theoretically estimated κ_Ls turn out to be either too high such as that for $BaFe_4Sb_{12}$, or too low such as that for $YbFe_4Sb_{12}$.[141] In addition, the largest κ_L reduction due to the fillers are observed to appear at about 30 K, and this also cannot be rationalised solely by the variations of phonons, either. Heat transport in open structure materials with guest fillers seems to be complicated and still requires more research efforts.

The rattling motion of guest fillers in clathrates is usually viewed as similar to the case in filled skutterudites. Interestingly, relatively small fillers, such as Sr and Eu in Ge-based clathrates or Ba in Sn-based clathrates, have the off-center rattling character, with fillers tunneling among the off-center-sites.[142–144] The unique behaviour results in flat and broad range of optic phonons in the dispersion relations and glass-like κ_L at low temperatures.[144] Another good example of rattling modes was found in sodium cobaltate by Voneshen et al.[145] The Einstein-like rattling modes were caused by the sodium vacancies. The rattling modes suppress κ_L by a factor of six. All of these compounds belong to a more common chemical-bond hierarchy architecture. The identification of this hierarchy architecture in more compounds are desirable, both for practical and physical interests.

Diverse lattice dynamics and part-crystalline part-liquid state in complex thermoelectrics

The resonant term (equation (11)) describing the rattling behaviour indicates the phonon transport in caged open structure compounds has already gone beyond the conventional crystalline solids. Hierarchically bonded architectures are not limited to the caged compounds; several compounds with other crystalline symmetries are reported to have quite distinct strong–weak bonding hierarchy.

Qiu et al. investigated the lattice dynamics and thermal transport behaviour of Cu–Sb–Se compounds, Cu_3SbSe_4, $CuSbSe_2$ and Cu_3SbSe_3, containing identical constituent species in different crystal structures.[146] Their work pointed out differences between crystalline and part-crystalline materials. For Cu_3SbSe_4 and $CuSbSe_2$, all atoms are confined to vibrate around their equilibrium positions. In Cu_3SbSe_3, however, Cu atoms oscillate with very large ADPs along certain specific directions (Figure 9a). If adopting the Lindemann melting criterion, then specified by the dimensionless Lindemann parameter δ ($\delta = \sqrt{ADP}/R_{NN}$, R_{NN} is the nearest-neighbour distance) exceeding 0.07, the Cu-sublattice could be considered as melted above 400 K (Figure 9a). The entire compound is thus in a special part-crystalline part-liquid hybrid state at elevated temperature, representing the atomic-level heterogeneity.

The measured temperature dependence of the κ_L of Cu_3SbSe_3 (Figure 9b) strongly deviates from the classical T^{-1} behaviour.[147] An effective rattle-like thermal damping term (Figure 6) is phenomenologically added to the denominator of equation (11), which can describe the reduction and the unique temperature dependence of the κ_L perfectly (Figure 9b).[147] The local motion or vibration of the set of atoms leads to the collective broadband thermal damping term, different from the single rattling behaviour in filled skutterudites. Interestingly, Cu_3SbSe_3 is also reported to have an s^2 lone-electron pair in an earlier work, which seemingly offers another explanation of the low κ_L.[148,149] But the analysis of Qiu et al. revealed that the reduction in thermal conductivity is more from the structural factor instead of the lone-pair electrons.[147] In contrast, the κ_L of Cu_3SbSe_4 and $CuSbSe_2$ shows a T^{-1} relationship,[146] which can be successfully described by the traditional PPIs in perturbative theory.

An even larger Cu mobility can be found in the superionic Cu_2Se which has a liquid-like Cu sublattice.[150] The structure in Cu_2Se contains a static disordering as well as a fluctuating dynamic disordering due to its liquid-like structure feature. Using AIMD, Kim et al.[151] studied the temperature-dependent lattice dynamics of β-Cu_2Se. The Cu–Cu bond lengths are close to that in Cu metal, much shorter than those found in the ordered antifluorite structure. The calculated κ_L, based on the non-equilibrium AIMD, agrees well with the experiments. More results regarding Cu_2Se and related compounds will be reviewed in 'Synergistic optimisation for high ZTs'.

A part-crystalline part-liquid (amorphous) material, intrinsically exhibits the unusual lattice dynamics and attains an abnormally low κ_L due to its statically disordered substructure and the dynamically fluctuating sublattice. These new material types offer good TE properties and interesting physics, as well as challenges for theoretical and experimental descriptions of the transport issues. For instance, the perturbative theory for the κ_L treatment, and even the concept of a phonon, may become questionable for these materials. Proper theoretical methods accounting for the structural fluctuation should be developed. For experimentalists, the fluidity of the ions, such as Cu^+ in Cu_2Se, Cu_2S, presents an unpleasant realisation that the sample under study will likely degrade during measurements.[152] How to balance the extraordinary TE performance with the practicality is an open question.

Theoretical description of thermal transport in bulk thermoelectrics and challenges

The lattice thermal conductivity for conventional solids can be calculated based on the relaxation time approximation and the phonon gas model, considering contributions from each phonon q,

$$\kappa_L = \frac{1}{3}\sum_q \hbar\omega_q v_s^2 \tau_q \frac{\partial N_q^0}{\partial T}. \tag{12}$$

Here, ω_q, τ_q, and N_q^0 are the phonon frequency, the relaxation time of phonons, and the equilibrium Bose–Einstein distribution function, respectively. An analytical model, taking into account the perturbed phonon population caused by normal process, was proposed initially by Callaway[153] and widely adopted in the study the κ_L for solids over the years.[147,154,155] The resistive scattering processes include grain boundary scattering (τ_B), point defect scattering (τ_{PD}), phonon–phonon Umklapp processes (τ_U), electron–phonon interaction (τ_{e-p}), etc. The overall scattering rate is obtained by adding scattering rates of the individual scattering processes,

$$\frac{1}{\tau_q} = \frac{1}{\tau_B} + \frac{1}{\tau_{PD}} + \frac{1}{\tau_U} + \frac{1}{\tau_{e-p}} + \ldots. \tag{13}$$

Variations in κ_L come mostly from the relaxation time.

Point defect scattering can affect mid-to-high frequency phonons (Figure 6). The coefficient of the scattering rate can be altered by mass and strain field fluctuations with the former being expressed as,[87]

$$\frac{1}{\tau_{PD}} = \frac{\Omega}{4\pi v_g^3}\omega^4 \sum_i f_i \left(\frac{\overline{m}-m_i}{\overline{m}}\right)^2. \tag{14}$$

f_i is the fraction of atoms with mass m_i, and $\overline{m}$ is the average mass of all atoms. Point defect scattering is important for the κ_L variations in solid solution of TE material systems as already mentioned,[90–95] as well as in the isotope effect.[97,156] It is noteworthy that the solid solutions may also interfere electrical transport, which makes the net increase in ZT values rely on the competition between the mobility loss and the reduction on κ_L.[94] Grain boundaries can scatter long-wavelength phonons which otherwise are hard to be interfered with by other mechanisms, i.e., $\tau_B^{-1} = v_g/L$ (L is the grain size). For TE polycrystalline materials with micrometer grain sizes, grain boundary scattering dominates only at very low temperatures. However, the situation changes when the grain size is further reduced to nanoscale. We summarised some useful theoretical guidance for better quantification regarding nanostructures in 'From the conventional phonon–phonon interactions to nanostructures'.

PPIs, due to their rates increasing with the temperature, dominate the heat transport from room to high temperatures for the majority of TE materials. Slack has proposed a simple but useful model for high-temperature κ_L,[96]

$$\kappa_L = A \cdot \frac{\overline{m}\theta_D^3 \delta}{\gamma^2 N^{\frac{2}{3}} T}. \tag{15}$$

Here N is the number of atoms in the cell, δ^3 is the volume per atom, θ_D is the Debye temperature, and A is a γ-related parameter, respectively. Since the parameters can be approximately obtained from experiments, equation (15) is often used to give a qualitative description of the κ_L for simple TE compounds. The third-order IFC method in recent years has taken into account transitions among all possible combinations of three phonons and applies the Fermi's golden rule. Although these calculations are a textbook knowledge, the implementation of IFC calculations can provide a quantitative description of heat transport in most TE materials, as well as the choices of the nanostructure size ('From the conventional phonon–phonon interactions to nanostructures').

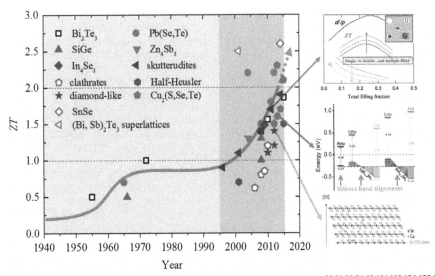

Figure 10. The chronological evolution of ZT values in various thermoelectric materials,[22,31,38,54,69,104,109,124,132,150,160–181] and the three detailed examples (PbSe matrix with band aligned nanocomposites,[31] n-type filled skutterudites,[69] and superionic compound Cu_2Se[150]).

In addition to the above discussed mechanisms, very recently, electron–phonon scattering was also found to limit κ_L. In Si, large amount of heat will be scattered if the carrier concentration is high due to a much enhanced electron–phonon interaction, as reported by Liao et al.[157] Considering that many TE materials have fairly high carrier concentrations (10^{20}–10^{21} cm^{-3}), the electron-phonon interaction is likely to be a player in reducing the κ_L. Experimentally, the noticeable influence of electron–phonon interaction has been pointed out in some heavily doped TE materials, such as SiGe[158] and Mo_3Se_7.[159]

When going beyond the conventional solids, the above mentioned small-perturbative methods have already become questionable or even invalid. The current implementation consists of introducing a phenomenological resonant scattering term as shown in equation (11) and treating it as another resistive scattering process. This is still done with well-defined phonons. A better description is necessary for providing the quantitative assessment in these systems. Possible solutions will depend on the extent of structural fluctuations. For compounds with modest fluctuations, such as the case with filled skutterudites, the temperature-dependent effective potential mentioned above might suffice. For compounds with structural fluctuations far beyond normal, statistical methods combined with classical and/or ab initio molecular dynamics simulations are possible ways with which to tackle the problem, including both the equilibrium molecular dynamics (based on the Green–Kubo formula), and the non-equilibrium molecular dynamics which traces the heat flux and temperature gradient.[151] Furthermore, the statistical methods can also be used to address and treat thermal transport processes on nano- or even mesoscales, which is useful for modelling of nanostructures.

SYNERGISTIC OPTIMISATION FOR HIGH ZTS

We have reviewed several different topics regarding both electrical and thermal transport processes so far. To achieve high ZT values in TE materials, a synergistic optimisation on both transport aspects is required. Armed with a deep understanding of transport phenomena, spectacular enhancements in the figure of merit ZT have been achieved in recent years across many families of TE materials,[22,31,38,54,69,104,109,124,132,150,160–181] as shown in Figure 10. Many traditional TE materials have doubled their ZT and new promising material systems have been identified and optimised. In this section, we will briefly summarise three different

compounds with optimised power factors and low lattice thermal conductivities, including PbSe matrix with band-aligned nanostructures, n-type filled $CoSb_3$-based skutterudites, and the superionic material Cu_2Se and its sister compounds. There are, of course, other compounds with equally good or perhaps even better TE performance than the examples above. We have chosen materials that have not only a tremendous potential as thermoelectrics but also because they illustrate interesting physical phenomena and an excellent synergy between the theory and experimental efforts.

As mentioned, nanostructures in bulk TE materials reduce the κ_L by the interface scattering. In order to improve the ZT by incorporating nanostructures, it is essential to preserve or perhaps even enhance the electrical transport properties. Making an assumption that the mobility is mainly affected by the band edge offset between the matrix and inclusions in IV–VI rock-salt compounds, Zhao et al. proposed a 'minimal valence band offset' strategy[30,31,182] to minimise scattering of carriers at interfaces. The case demonstrating this idea was the matrix of PbSe composited with alloys of $CdS_{1-x}Se_x/ZnS_{1-x}Se_x$.[31] Ab initio calculations indicated that cadmium and zinc selenides have positive band offsets with respect to PbSe and the corresponding sulfides have negative offset values. Thus, the alloys of selenide and sulfide were used to get almost zero offset inclusions, and maintained the transport of electrons. Furthermore, by substituting Cd for Pb, the energy separation between L and Σ pockets decreased, and the Seebeck coefficient was thus enhanced. The two beneficial effects in the electrical transport, together with the much reduced κ_L by the nanostructure, resulted in the figure of merit ZT ~ 1.6 at 923 K (Figure 10), the highest value for tellurium-free chalcogenides. In a composite system, the use of band engineering to optimise the electronic properties is a worthwhile effort. Whether the concept of 'minimal band offset' is limited to IV–VI compounds or whether it will serve as a general guidance, however, requires further research efforts.

N-type partially filled skutterudites are among the first generation of 'new' compounds that have emerged from the era of 'big ideas' in the mid-1990s. After the initial exploratory efforts, the experimentalists found that large structural voids in the skutterudite lattice can be partially filled with certain rare earth and alkaline earth metals and both the electron concentration and the lattice thermal conductivity could be tuned. In 2005, Shi et al. concluded that the filling fraction limit in $CoSb_3$-based skutterudites is determined by the competition between the filled phases and the secondary phases.[183] By calculating total energies of both phases, they not only reproduced the available experimental filling fraction limits at that time but also predicted new fillers among the alkaline metals and, in follow-up studies, pointed out the possibilities of multiple element filling.[66,184–186] As for the electrical transport, due to the large electronegativity difference between the guest and host species, the outermost s electrons of the filler atoms are donated to the conduction band of $CoSb_3$ without much band alteration.[67] Thus revealed rigidity of the band structure of filled $CoSb_3$-based skutterudites and the optimal doping levels make the carrier concentration tuning merely a simple electron counting exercise based on the charge state and the filling fraction of the fillers.[69] Power factors over $50 \, \mu W \, cm^{-1} \, K^{-2}$ can be easily obtained upon attaining the proper electron count (Figure 10). For the thermal transport, the resonant scattering mechanism has historically been used to describe variations in κ_L which is consistent with large displacements of the fillers. Theoretical determination of the resonant frequencies for a variety of fillers in 2007 was a milestone for the performance advance of filled $CoSb_3$ as it revealed the physical meaning of the previously merely fitting parameter.[131] Multiple filling was then proposed to scatter a broader range of phonons. Experimentally, by selecting rattling species such as Ba, La, Yb with frequencies of 93, 66, 42 cm^{-1}, respectively, the κ_L of such multiple filled $CoSb_3$-based skutterudites was gradually reduced from values corresponding to single, to double and to multiple filled structures.[69] Throughout the whole process one could witness a good integration of theoretical input and insights with experimental studies which enriched the understanding in this material system and resulted in enhanced values of ZT ~ 1.7 (Figure 10).[69]

In 2012, Liu et al. reported on the experimental work with Cu_2Se which had a ZT of 1.5 at 1,000 K, as shown in Figure 10.[150] This compound inherits features of the chemical bond hierarchy, namely a weakly bonded Cu ion sublattice and a rigid Se sublattice. However, the bonds between Cu and Se are so weak that they give rise to a liquid-like Cu movement in the high temperature antifluorite β phase. The ion fluidity is the unique aspect of Cu_2Se that immediately stimulated considerable research interest. The lattice thermal conductivities maintain their extremely low values (around 0.4–0.6 $W \, m^{-1} \, K^{-1}$ in the entire temperature range). The quantification of the lattice dynamics, however, is difficult using the standard IFC calculations due to fast movements of Cu ions, which breaks down the perturbation assumptions, and static methods have to be considered as shown.[151] As a promising TE material, Cu_2Se shows a good electrical conductivity ($10^4 \sim 10^5 \, S \, m^{-1}$) and a large Seebeck coefficient ($100 \sim 300 \, \mu V \, K^{-1}$). The deficiency of Cu seems to be a good source of holes. There crystal-like electrical transport properties in Cu_2Se suggest a conductive network, though variations in electrical transport properties have not yet been rationalised. The sister compounds, Cu_2S and Cu_2Te, also demonstrate interesting properties such as even lower κ_L of Cu_2S[179], and a very complicated phase diagram of Cu_2Te.[187] The solid solutions of Cu_2S and Cu_2Te form nanoscale mosaic crystals with ZT values of up to 2.1 at 1,000 K.[180] Overall, copper chalcogenide compounds with their large ion mobility offer high TE performance as well as new physics and rich chemistry, representing a new research direction to explore for both experimentalists and theorists.

CONCLUDING REMARKS AND OUTLOOK

In this work, we have reviewed several topics pertaining to the high TE performance of TE materials. The viewpoint was mostly theoretical. Indeed, the theoretical insight has helped with the rationalisation of experimental approaches and speeded up the TE materials research in many aspects. Explicitly, band engineering has been widely adopted for studies of the electrical transport,

including the band convergence, a reduction of the DOS to achieve weaker scattering of charge carriers, and the role of conductive networks in complex compounds. Interatomic force constant calculations, carried out in both harmonic and anharmonic approximations, can be used to advance understanding of the influence of nanostructures and of the PPI in conventional solids. Unconventional solids, such as those possessing phonon liquid features were also identified and studied both theoretically and experimentally. Because of advances in the theoretical understanding of transport phenomena, more powerful facilities combined with faster algorithms, and the integration of theory with experiment, TE research has evolved from a rather minor and sleepy field into an intense, worldwide and highly interdisciplinary activity, which gains from a unique perspective of physicists, chemists, engineers and computer modelers. The remaining challenges are the developments of methodologies for solving scattering problems and, specifically, problems concerning of carrier scatterings. Another challenge of the theoretical perspective is how to deal with transport in systems showing large atomic displacement parameters or even large structural fluctuations. Such structures encompass important families of TE materials and it is imperative to understand and predict their transport behaviour.

High throughput calculations are another area of considerable interest as it offers a timely and inexpensive identification of promising TE materials. The current standard is to screen potential compounds from a large compound library, usually a crystalline database, and then study them in detail with property descriptors. The screening for compounds with a lone-pair of electrons has been carried out in this way.[188] More high-throughput theoretical studies focus on half-Heusler alloys and related 1:1:1 compounds because of their simple unit cells and the large variety of elemental combinations. Yang et al. carried out electrical transport calculations under the CRTA to evaluate the power factor and the optimal doping level in 35 half-Heusler alloys.[189] Promising compounds are predicted and several of them are confirmed years later.[47,174] Anharmonic IFC calculations were used by Carrete et al. to evaluate the lattice thermal conductivity of 75 half-Heusler alloys out of 79,057 entries.[190] These results, combined with the electrical transport calculations, result to the ZT evaluations, with about 15 % of these compounds may have ZT>2 at high temperatures.[191] In general, current high throughput studies are still in their infancy and offer only limited accuracy. It remains indispensable to carry out trial-and-error studies for the best proposed compositions. In this regard, more accurate descriptors are needed to nail down the right composition.[192] On the experimental side, very little work is done currently on high-throughput screening.[193] The culprit is the critical conditions necessary to follow when synthesising TE bulk materials. New techniques for rapid synthesis, such as the self-propagating high-temperature synthesis,[194] could be a solution to high-throughput experimentation with TE materials.

Finally, we note that it was not so long ago when the highest ZT known in any material was less than unity, and there were even researchers who held that ZT of unity could never be exceeded. Now there are several families of TE materials with ZT values approaching and even exceeding 2, a fact that has enormous implications for applications. This change has not come about by chance, but in fact is a consequence of applications of modern theory and complementary experimental efforts that exploit new understanding and ideas in the search for high performance TE materials. It may be that to achieve another big jump in the development of TE materials, however, new 'big ideas' as influential as was the PGEC paradigm and the concept of nanostructures some 20 years ago will be needed. Many of the topics reviewed above are closely connected to these original concepts. With the help of theory, new and exciting phenomena have been discovered that have helped shape the direction of TE research. These include band convergence, scattering engineering, all-scale phonon scattering, minimum band offset for nanocomposites, partial crystalline-partial liquid compounds with structural fluctuations, to name a few. With the help of new 'big ideas' there is a good chance that novel structural forms of highly efficient TE materials would be discovered.

ACKNOWLEDGEMENTS

This work was supported by National Basic Research Program of China (973-program) under project number 2013CB632501, National Natural Science Foundation of China under contract number 11234012, the Key Research Program of Chinese Academy of Sciences (Grant No. KGZD-EW-T06), research grants (14DZ2261200 and 15JC1400301) from Science and Technology Commission of Shanghai Municipality, and International S&T Cooperation Program of China (2015DFA51050). Research of C. Uher was supported by the U.S. Department of Energy, Office of Basic Energy Sciences under award number DE-SC-0008574. Work at the University of Missouri (DJS) was supported by the Department of Energy through the S3TEC Energy Frontier Research Center award #DE-SC0001299/DE-FG02-09ER46577. Research of Jihui Yang was supported by the U.S. Department of Energy under corporate agreement DE-FC26-04NT42278, by GM, and by National Science Foundation under award number 1235535. WQ and LD also thank to the support from Shanghai Institute of Materials Genome.

COMPETING INTERESTS

The authors declare no conflict of interest.

REFERENCES

1. Goldsmid, H. J. *Introduction to Thermoelectricity* (Springer, 2010).
2. Ioffe, A. F. *Semiconductor thermoelements, and Thermoelectric cooling*, Reviewd and supplemented for the English edition. Translated from the Russian by A. Gelbtuch (Infosearch, 1957).
3. Snyder, G. J. & Toberer, E. S. Complex thermoelectric materials. *Nat. Mater.* **7**, 105–114 (2008).
4. Liu, W., Yan, X., Chen, G. & Ren, Z. Recent advances in thermoelectric nanocomposites. *Nano Energy* **1**, 42–56 (2012).
5. Hicks, L. & Dresselhaus, M. Effect of quantum-well structures on the thermoelectric figure of merit. *Phys. Rev. B* **47**, 12727 (1993).
6. Hicks, L. D., Harman, T. C., Sun, X. & Dresselhaus, M. S. Experimental study of the effect of quantum-well structures on the thermoelectric figure of merit. *Phys. Rev. B* **53**, 10493–10496 (1996).
7. Minnich, A. J., Dresselhaus, M. S., Ren, Z. F. & Chen, G. Bulk nanostructured thermoelectric materials: current research and future prospects. *Energy Environ. Sci.* **2**, 466–479 (2009).
8. Heremans, J. P., Dresselhaus, M. S., Bell, L. E. & Morelli, D. T. When thermoelectrics reached the nanoscale. *Nat. Nanotechnol.* **8**, 471–473 (2013).
9. Slack, G. A. in *CRC Handbook of Thermoelectrics* (ed Rowe D. M.) Ch. 9 (CRC Press, 1995).
10. Chen, Y. L. *et al.* Experimental realization of a three-dimensional topological insulator, Bi$_2$Te$_3$. *Science* **325**, 178–181 (2009).
11. May, A. F., Singh, D. J. & Snyder, G. J. Influence of band structure on the large thermoelectric performance of lanthanum telluride. *Phys. Rev. B* **79**, 153101 (2009).
12. Madsen, G. K. H. & Singh, D. J. BoltzTraP. A code for calculating band-structure dependent quantities. *Comput. Phys. Commun.* **175**, 67–71 (2006).
13. Pei, Y. Z., Wang, H. & Snyder, G. J. Band engineering of thermoelectric materials. *Adv. Mater.* **24**, 6125–6135 (2012).
14. Liu, W. *et al.* Convergence of conduction bands as a means of enhancing thermoelectric performance of n-Type Mg2Si1-xSnx solid solutions. *Phys. Rev. Lett.* **108**, 166601 (2012).
15. Bahk, J. H., Bian, Z. X. & Shakouri, A. Electron transport modeling and energy filtering for efficient thermoelectric Mg2Si1-xSnx solid solutions. *Phys. Rev. B* **89**, 075204 (2014).
16. Chen, S. Y., Gong, X. G., Walsh, A. & Wei, S. H. Electronic structure and stability of quaternary chalcogenide semiconductors derived from cation cross-substitution of II-VI and I-III-VI2 compounds. *Phys. Rev. B* **79**, 165211 (2009).
17. Zhang, J. W. *et al.* High-performance pseudocubic thermoelectric materials from non-cubic chalcopyrite compounds. *Adv. Mater.* **26**, 3848–3853 (2014).
18. Zeier, W. G. *et al.* Band convergence in the non-cubic chalcopyrite compounds Cu2MGeSe4. *J. Mater. Chem. C* **2**, 10189–10194 (2014).
19. Singh, D. J. Doping-dependent thermopower of PbTe from Boltzmann transport calculations. *Phys. Rev. B* **81**, 195217 (2010).
20. Chen, X., Parker, D. & Singh, D. J. Importance of non-parabolic band effects in the thermoelectric properties of semiconductors. *Sci. Rep.* **3**, 3168 (2013).
21. Tauber, R. N., Machonis, A. A. & Cadoff, I. B. Thermal and optical energy gaps in PbTe. *J. Appl. Phys.* **37**, 4855 (1966).
22. Pei, Y. Z. *et al.* Convergence of electronic bands for high performance bulk thermoelectrics. *Nature* **473**, 66–69 (2011).
23. Zhu, H., Sun, W. H., Armiento, R., Lazic, P. & Ceder, G. Band structure engineering through orbital interaction for enhanced thermoelectric power factor. *Appl. Phys. Lett.* **104**, 082107 (2014).
24. Skelton, J. M., Parker, S. C., Togo, A., Tanaka, I. & Walsh, A. Thermal physics of the lead chalcogenides PbS, PbSe, and PbTe from first principles. *Phys. Rev. B* **89**, 205203 (2014).
25. Kim, H. & Kaviany, M. Effect of thermal disorder on high figure of merit in PbTe. *Phys. Rev. B* **86**, 045213 (2012).
26. Gibbs, Z. M. *et al.* Temperature dependent band gap in PbX (X = S, Se, Te). *Appl. Phys. Lett.* **103**, 262109 (2013).
27. Wu, D. *et al.* Origin of the high performance in GeTe-based thermoelectric materials upon Bi2Te3 doping. *J. Am. Chem. Soc.* **136**, 11412–11419 (2014).
28. Pei, Y. Z., Wang, H., Gibbs, Z. M., LaLonde, A. D. & Snyder, G. J. Thermopower enhancement in Pb1-xMnxTe alloys and its effect on thermoelectric efficiency. *NPG Asia Mater.* **4**, e28 (2012).
29. Zhao, L. D. *et al.* All-scale hierarchical thermoelectrics: MgTe in PbTe facilitates valence band convergence and suppresses bipolar thermal transport for high performance. *Energy Environ. Sci.* **6**, 3346–3355 (2013).
30. Zhao, L. D., Dravid, V. P. & Kanatzidis, M. G. The panoscopic approach to high performance thermoelectrics. *Energy Environ. Sci.* **7**, 251–268 (2014).
31. Zhao, L. D. *et al.* High thermoelectric performance via hierarchical compositionally alloyed nanostructures. *J. Am. Chem. Soc.* **135**, 7364–7370 (2013).
32. Banik, A., Shenoy, U. S., Anand, S., Waghmare, U. V. & Biswas, K. Mg alloying in SnTe facilitates valence band convergence and optimizes thermoelectric properties. *Chem. Mater.* **27**, 581–587 (2015).
33. Lusakowski, A., Boguslawski, P. & Radzynski, T. Calculated electronic structure of Pb1-xMnxTe (0 < =x < 11%): The role of L and Sigma valence band maxima. *Phys. Rev. B* **83**, 115206 (2011).
34. Tan, X. J., Shao, H. Z., Hu, T. Q., Liu, G. Q. & Ren, S. F. Theoretical understanding on band engineering of Mn-doped lead chalcogenides PbX (X = Te, Se, S). *J. Phys. Condens. Matter* **27**, 095501 (2015).
35. Heremans, J. P., Wiendlocha, B. & Chamoire, A. M. Resonant levels in bulk thermoelectric semiconductors. *Energy Environ. Sci* **5**, 5510–5530 (2012).
36. Wiendlocha, B. Fermi surface and electron dispersion of PbTe doped with resonant Tl impurity from KKR-CPA calculations. *Phys. Rev. B* **88**, 205205 (2013).
37. Tan, G. J. *et al.* Codoping in SnTe: enhancement of thermoelectric performance through synergy of resonance levels and band convergence. *J. Am. Chem. Soc.* **137**, 5100–5112 (2015).
38. Heremans, J. P. *et al.* Enhancement of thermoelectric efficiency in PbTe by distortion of the electronic density of states. *Science* **321**, 554–557 (2008).
39. Xiong, K. *et al.* Behaviour of group IIIA impurities in PbTe: implications to improve thermoelectric efficiency. *J. Phys. D Appl. Phys.* **43**, 405403 (2010).
40. Jaworski, C. M., Kulbachinskii, V. & Heremans, J. P. Resonant level formed by tin in Bi2Te3 and the enhancement of room-temperature thermoelectric power. *Phys. Rev. B* **80**, 233201 (2009).
41. Lan, J. L. *et al.* Enhanced Thermoelectric Properties of Pb-doped BiCuSeO Ceramics. *Adv. Mater.* **25**, 5086–5090 (2013).
42. Cui, J. L., Li, Y. P., Du, Z. L., Meng, Q. S. & Zhou, H. Promising defect thermoelectric semiconductors Cu1-xGaSbxTe2 (x = 0-0.1) with the chalcopyrite structure. *J. Mater. Chem. A* **1**, 677–683 (2013).
43. Qiu, P. F., Yang, J., Huang, X. Y., Chen, X. H. & Chen, L. D. Effect of antisite defects on band structure and thermoelectric performance of ZrNiSn half-Heusler alloys. *Appl. Phys. Lett.* **96**, 152105 (2010).
44. Wu, L. *et al.* Two-dimensional thermoelectrics with Rashba spin-split bands in bulk BiTeI. *Phys. Rev. B* **90**, 202115 (2014).
45. Putley, E. H. *The Hall Effect And Related Phenomena* (Butterworths, 1960).
46. Yang, J. *et al.* Power factor enhancement in light valence band p-type skutterudites. *Appl. Phys. Lett.* **101**, 022101 (2012).
47. Fu, C. G., Zhu, T. J., Liu, Y. T., Xie, H. H. & Zhao, X. B. Band engineering of high performance p-type FeNbSb based half-Heusler thermoelectric materials for figure of merit zT > 1. *Energy Environ. Sci.* **8**, 216–220 (2015).
48. Pei, Y. Z., LaLonde, A. D., Wang, H. & Snyder, G. J. Low effective mass leading to high thermoelectric performance. *Energy Environ. Sci.* **5**, 7963–7969 (2012).
49. Lee, D. S. *et al.* Density of state effective mass and related charge transport properties in K-doped BiCuOSe. *Appl. Phys. Lett.* **103**, 232110 (2013).
50. Dyck, J. S. *et al.* Thermoelectric properties of the n-type filled skutterudite Ba0.3Co4Sb12 doped with Ni. *J. Appl. Phys.* **91**, 3698–3705 (2002).

51. Dyck, J. S., Chen, W., Yang, J. H., Meisner, G. P. & Uher, C. Effect of Ni on the transport and magnetic properties of Co1-xNixSb3. *Phys. Rev. B* **65**, 115204 (2002).

52. Li, X. Y. *et al.* Thermoelectric properties of Te-doped CoSb3 by spark plasma sintering. *J. Appl. Phys.* **98**, 083702 (2005).

53. Wang, S. *et al.* On intensifying carrier impurity scattering to enhance thermoelectric performance in Cr-doped CeyCo4Sb12. *Adv. Funct. Mater.* **25**, 6660–6670 (2015).

54. Shi, X. *et al.* On the design of high-efficiency thermoelectric clathrates through a systematic cross-substitution of framework elements. *Adv. Funct. Mater.* **20**, 755–763 (2010).

55. Suh, J. *et al.* Simultaneous enhancement of electrical conductivity and thermopower of Bi2Te3 by multifunctionality of native defects. *Adv. Mater.* **27**, 3681–3686 (2015).

56. Faleev, S. V. & Leonard, F. Theory of enhancement of thermoelectric properties of materials with nanoinclusions. *Phys. Rev. B* **77**, 214304 (2008).

57. Martin, J., Wang, L., Chen, L. & Nolas, G. S. Enhanced Seebeck coefficient through energy-barrier scattering in PbTe nanocomposites. *Phys. Rev. B* **79**, 115311 (2009).

58. Zeng, G. H. *et al.* Cross-plane Seebeck coefficient of ErAs: InGaAs/InGaAlAs superlattices. *J. Appl. Phys.* **101**, 034502 (2007).

59. Shakouri, A. Recent developments in semiconductor thermoelectric physics and materials. *Annu. Rev. Mater. Res.* **41**, 399–431 (2011).

60. Makongo, J. P. A. *et al.* Simultaneous large enhancements in thermopower and electrical conductivity of bulk nanostructured half-heusler alloys. *J. Am. Chem. Soc.* **133**, 18843–18852 (2011).

61. Liu, Y. *et al.* Large enhancements of thermopower and carrier mobility in quantum dot engineered bulk semiconductors. *J. Am. Chem. Soc.* **135**, 7486–7495 (2013).

62. Yu, B. *et al.* Enhancement of thermoelectric properties by modulation-doping in silicon germanium alloy nanocomposites. *Nano Lett.* **12**, 2077–2082 (2012).

63. Xu, Y., Gan, Z. X. & Zhang, S. C. Enhanced thermoelectric performance and anomalous Seebeck effects in topological insulators. *Phys. Rev. Lett.* **112**, 226801 (2014).

64. Shi, H., Parker, D., Du, M.-H. & Singh, D. J. Connecting thermoelectric performance and topological-insulator behavior: Bi$_2$Te$_3$ and Bi$_2$Te$_2$Se from first principles. *Phys. Rev. Appl.* **3**, 014004 (2015).

65. Sofo, J. O. & Mahan, G. D. Electronic structure of CoSb$_3$: a narrow-band-gap semiconductor. *Phys. Rev. B* **58**, 15620–15623 (1998).

66. Mei, Z. G., Yang, J., Pei, Y. Z., Zhang, W. & Chen, L. D. Alkali-metal-filled CoSb$_3$ skutterudites as thermoelectric materials: theoretical study. *Phys. Rev. B* **77**, 045202 (2008).

67. Yang, J., Xi, L., Zhang, W., Chen, L. D. & Yang, J. Electrical transport properties of filled CoSb3 skutterudites: a theoretical study. *J. Electron. Mater.* **38**, 1397–1401 (2009).

68. Bai, S. Q. *et al.* Enhanced thermoelectric performance of dual-element-filled skutterudites BaxCeyCo4Sb12. *Acta Mater.* **57**, 3135–3139 (2009).

69. Shi, X. *et al.* Multiple-filled skutterudites: high thermoelectric figure of merit through separately optimizing electrical and thermal transports. *J. Am. Chem. Soc.* **133**, 7837–7846 (2011).

70. Yang, J. *et al.* Trends in electrical transport of p-type skutterudites RFe4Sb12 (R = Na, K, Ca, Sr, Ba, La, Ce, Pr, Yb) from first-principles calculations and Boltzmann transport theory. *Phys. Rev. B* **84**, 235205 (2011).

71. Xi, L. *et al.* Chemical bonding, conductive network, and thermoelectric performance of the ternary semiconductors Cu2SnX3 (X = Se, S) from first principles. *Phys. Rev. B* **86**, 155201 (2012).

72. Shi, X., Xi, L., Fan, J., Zhang, W. & Chen, L. Cu-Se bond network and thermoelectric compounds with complex diamondlike structure. *Chem. Mater.* **22**, 6029–6031 (2010).

73. Li, W. *et al.* Cu2HgSnSe4 nanoparticles: synthesis and thermoelectric properties. *CrystEngComm* **15**, 8966–8971 (2013).

74. Wang, B. *et al.* Heterovalent Substitution to Enrich Electrical Conductivity in Cu2CdSn1-xGaxSe4 Series for High Thermoelectric Performances. *Sci. Rep.* **5**, 9365 (2015).

75. Aydemir, U. *et al.* Thermoelectric enhancement in BaGa2Sb2 by Zn doping. *Chem. Mater.* **27**, 1622–1630 (2015).

76. He, H., Stearrett, R., Nowak, E. R. & Bobev, S. BaGa2Pn2 (Pn = P, As): new semiconducting phosphides and arsenides with layered structures. *Inorgan. Chem.* **49**, 7935–7940 (2010).

77. Toberer, E. S., May, A. F., Melot, B. C., Flage-Larsen, E. & Snyder, G. J. Electronic structure and transport in thermoelectric compounds AZn(2)Sb(2) (A = Sr, Ca, Yb, Eu). *Dalton Trans.* **39**, 1046–1054 (2010).

78. Wang, X.-J. *et al.* Synthesis and high thermoelectric efficiency of Zintl phase YbCd2-xZnxSb2. *Appl. Phys. Lett.* **94**, 092106 (2009).

79. Tandon, N., Albrecht, J. D. & Ram-Mohan, L. R. Electron-phonon coupling and associated scattering rates in diamond. *Diamond Relat. Mater.* **56**, 1–5 (2015).

80. Liao, B., Zhou, J., Qiu, B., Dresselhaus, M. S. & Chen, G. Ab initio study of electron-phonon interaction in phosphorene. *Phys. Rev. B* **91**, 235419 (2015).

81. Heremans, J. P., Thrush, C. M. & Morelli, D. T. Thermopower enhancement in lead telluride nanostructures. *Phys. Rev. B* **70**, 115334 (2004).

82. Sun, P. J. *et al.* Large Seebeck effect by charge-mobility engineering. *Nat. Commun.* **6**, 7475 (2015).

83. Abeles, B. Thermal conductivity of germanium in the temperature range 300°-1080 °K. *J. Phys. Chem. Solids* **8**, 340–343 (1959).

84. Kettel, F. Die Wärmeleitfähigkeit von Germanium bei hohen temperaturen. *J. Phys. Chem. Solids* **10**, 52–58 (1959).

85. Wang, S. *et al.* Conductivity-limiting bipolar thermal conductivity in semiconductors. *Sci. Rep.* **5**, 10136 (2015).

86. Ziman, J. M. *Electrons and Phonons; The Theory of Transport Phenomena in Solids* (Clarendon Press, 1960).

87. Yang, J. in *Thermal Conductivity: Theory, Properties and Applications*, edTritt T. M. Ch. 1.1, 1–20 (Kluwer Academic/Plenum Publishers, 2004).

88. Toberer, E. S., Zevalkink, A. & Snyder, G. J. Phonon engineering through crystal chemistry. *J. Mater. Chem.* **21**, 15843–15852 (2011).

89. Nolas, G. S., Sharp, J. & Goldsmid, H. J. *Thermoelectrics: Basic Principles and New Materials Developments* (Springer, 2001).

90. Hu, L. P., Zhu, T. J., Liu, X. H. & Zhao, X. B. Point defect engineering of high-performance bismuth-telluride-based thermoelectric materials. *Adv. Funct. Mater.* **24**, 5211–5218 (2014).

91. Petersen, A., Bhattacharya, S., Tritt, T. M. & Poon, S. J. Critical analysis of lattice thermal conductivity of half-Heusler alloys using variations of Callaway model. *J. Appl. Phys.* **117**, 035706 (2015).

92. Yang, J., Meisner, G. P. & Chen, L. Strain field fluctuation effects on lattice thermal conductivity of ZrNiSn-based thermoelectric compounds. *Appl. Phys. Lett.* **85**, 1140–1142 (2004).

93. Yan, X. *et al.* Stronger phonon scattering by larger differences in atomic mass and size in p-type half-Heuslers Hf1-xTixCoSb0.8Sn0.2. *Energy Environ. Sci.* **5**, 7543–7548 (2012).

94. Wang, H., LaLonde, A. D., Pei, Y. Z. & Snyder, G. J. The criteria for beneficial disorder in thermoelectric solid solutions. *Adv. Funct. Mater.* **23**, 1586–1596 (2013).

95. Wang, H., Wang, J. L., Cao, X. L. & Snyder, G. J. Thermoelectric alloys between PbSe and PbS with effective thermal conductivity reduction and high figure of merit. *J. Mater. Chem. A* **2**, 3169–3174 (2014).

96. Morelli, D. T., Slack, G. A. in *High Thermal Conductivity Materials* (eds Shindé S. L. & Goela J. S.) Ch. 2, 37–68 (Springer, 2006).

97. Lindsay, L., Broido, D. A. & Reinecke, T. L. Ab initio thermal transport in compound semiconductors. *Phys. Rev. B* **87**, 165201 (2013).

98. An, J. M., Subedi, A. & Singh, D. J. Ab initio phonon dispersions for PbTe. *Solid State Commun.* **148**, 417–419 (2008).

99. Zhang, Y., Ke, X. Z., Chen, C. F., Yang, J. & Kent, P. R. C. Thermodynamic properties of PbTe, PbSe, and PbS: First-principles study. *Phys. Rev. B* **80**, 024304 (2009).

100. Zhang, Y., Ke, X., Kent, P. R. C., Yang, J. & Chen, C. Anomalous lattice dynamics near the ferroelectric instability in PbTe. *Phys. Rev. Lett.* **107**, 175503 (2011).

101. Tian, Z. *et al.* Phonon conduction in PbSe, PbTe, and PbTe1-xSex from first-principles calculations. *Phys. Rev. B* **85**, 184303 (2012).

102. Lee, S. *et al.* Resonant bonding leads to low lattice thermal conductivity. *Nat. Commun.* **5**, 4525 (2014).

103. Li, C. *et al.* Orbitally driven giant phonon anharmonicity in SnSe. *Nat. Phys.* **11**, 1063 (2015).

104. Zhao, L. D. *et al.* Ultralow thermal conductivity and high thermoelectric figure of merit in SnSe crystals. *Nature* **508**, 373–377 (2014).

105. Carrete, J., Mingo, N. & Curtarolo, S. Low thermal conductivity and triaxial phononic anisotropy of SnSe. *Appl. Phys. Lett.* **105**, 101907 (2014).

106. Romero, A. H., Gross, E. K. U., Verstraete, M. J. & Hellman, O. Thermal conductivity in PbTe from first principles. *Phys. Rev. B* **91**, 214310 (2015).

107. Li, C. W. *et al.* Phonon self-energy and origin of anomalous neutron scattering spectra in SnTe and PbTe thermoelectrics. *Phys. Rev. Lett.* **112**, 175501 (2014).

108. Hellman, O. & Broido, D. A. Phonon thermal transport in Bi2Te3 from first principles. *Phys. Rev. B* **90**, 134309 (2014).

109. Biswas, K. *et al.* High-performance bulk thermoelectrics with all-scale hierarchical architectures. *Nature* **489**, 414–418 (2012).

110. Yang, J., Yip, H.-L. & Jen, A. K. Y. Rational design of advanced thermoelectric materials. *Adv. Energy Mater.* **3**, 565 (2013).

111. Guo, R., Wang, X. & Huang, B. Thermal conductivity of skutterudite CoSb3 from first principles: substitution and nanoengineering effects. *Sci. Rep.* **5**, 7806 (2015).

112. Katcho, N. A., Mingo, N. & Broido, D. A. Lattice thermal conductivity of (Bi1-xSbx)(2)Te-3 alloys with embedded nanoparticles. *Phys. Rev. B* **85**, 115208 (2012).

113. Yang, R. G. & Chen, G. Thermal conductivity modeling of periodic two-dimensional nanocomposites. *Phys. Rev. B* **69**, 195316 (2004).

114. Yang, R. G., Chen, G. & Dresselhaus, M. S. Thermal conductivity of simple and tubular nanowire composites in the longitudinal direction. *Phys. Rev. B* **72**, 125418 (2005).

115. Jeng, M. S., Yang, R. G., Song, D. & Chen, G. Modeling the thermal conductivity and phonon transport in nanoparticle composites using Monte Carlo simulation. *J. Heat Transfer* **130**, 042410 (2008).

116. Mingo, N., Hauser, D., Kobayashi, N. P., Plissonnier, M. & Shakouri, A. 'Nanoparticle-in-Alloy' Approach to Efficient Thermoelectrics: Silicides in SiGe. *Nano Lett.* **9**, 711–715 (2009).

117. Garg, J. & Chen, G. Minimum thermal conductivity in superlattices: A first-principles formalism. *Phys. Rev. B* **87**, 140302(R) (2013).

118. Zhang, H. & Minnich, A. J. The best nanoparticle size distribution for minimum thermal conductivity. *Sci. Rep.* **5**, 8995 (2015).

119. Chan, M. K. Y. *et al.* Cluster expansion and optimization of thermal conductivity in SiGe nanowires. *Phys. Rev. B* **81**, 174303 (2010).

120. Minnich, A. J. *et al.* Modeling study of thermoelectric SiGe nanocomposites. *Phys. Rev. B* **80**, 155327 (2009).

121. Wei, J. *et al.* Theoretical study of the thermoelectric properties of SiGe nanotubes. *RSC Adv.* **4**, 53037–53043 (2014).

122. Fiedler, G. & Kratzer, P. Theoretical prediction of improved figure-of-merit in Si/Ge quantum dot superlattices. *New J. Phys.* **15**, 125010 (2013).

123. Jeitschko, W. & Braun, D. LaFe4P12 with filled CoAs3-type structure and isotypic lanthanoid-transition metal polyphosphides. *Acta Crystallographica Section B* **33**, 3401–3406 (1977).

124. Sales, B. C., Mandrus, D. & Williams, R. K. Filled skutterudite antimonides: A new class of thermoelectric materials. *Science* **272**, 1325–1328 (1996).

125. Sales, B. C., Mandrus, D., Chakoumakos, B. C., Keppens, V. & Thompson, J. R. Filled skutterudite antimonides: Electron crystals and phonon glasses. *Phys. Rev. B* **56**, 15081–15089 (1997).

126. Chen, B. X. *et al.* Low-temperature transport properties of the filled skutterudites CeFe4-xCoxSb12. *Phys. Rev. B* **55**, 1476–1480 (1997).

127. Nolas, G. S., Cohn, J. L. & Slack, G. A. Effect of partial void filling on the lattice thermal conductivity of skudderudites. *Phys. Rev. B* **58**, 164–170 (1998).

128. Pohl, R. O. Thermal Conductivity and Phonon Resonance Scattering. *Phys. Rev. Lett.* **8**, 481 (1962).

129. Puyet, M. *et al.* Low-temperature thermal properties of n-type partially filled calcium skutterudites. *J. Phys. Condens. Matter* **18**, 11301–11308 (2006).

130. Yang, J. *et al.* Effect of Sn substituting for Sb on the low-temperature transport properties of ytterbium-filled skutterudites. *Phys. Rev. B* **67**, 165207 (2003).

131. Yang, J., Zhang, W., Bai, S. Q., Mei, Z. & Chen, L. D. Dual-frequency resonant phonon scattering in BaxRyCo4Sb12 (R = La, Ce, and Sr). *Appl. Phys. Lett.* **90**, 192111 (2007).

132. Shi, X. *et al.* Low thermal conductivity and high thermoelectric figure of merit in n-type BaxYbyCo4Sb12 double-filled skutterudites. *Appl. Phys. Lett.* **92**, 182101 (2008).

133. Bai, S. Q., Shi, X. & Chen, L. D. Lattice thermal transport in BaxREyCo4Sb12 (RE = Ce, Yb, and Eu) double-filled skutterudites. *Appl. Phys. Lett.* **96**, 202102 (2010).

134. Liu, R. *et al.* p-Type skutterudites RxMyFe3CoSb12 (R, M = Ba, Ce, Nd, and Yb): Effectiveness of double-filling for the lattice thermal conductivity reduction. *Intermetallics* **19**, 1747–1751 (2011).

135. Koza, M. M. *et al.* Breakdown of phonon glass paradigm in La- and Ce-filled Fe(4)Sb(12) skutterudites. *Nat. Mater.* **7**, 805–810 (2008).

136. Christensen, M. *et al.* Avoided crossing of rattler modes in thermoelectric materials. *Nat. Mater.* **7**, 811–815 (2008).

137. Zebarjadi, M., Esfarjani, K., Yang, J. A., Ren, Z. F. & Chen, G. Effect of filler mass and binding on thermal conductivity of fully filled skutterudites. *Phys. Rev. B* **82**, 195207 (2010).

138. Koza, M. M. *et al.* Low-energy phonon dispersion in LaFe4Sb12. *Phys. Rev. B* **91**, 014305 (2015).

139. Thompson, D. R. *et al.* Rare-earth free p-type filled skutterudites: Mechanisms for low thermal conductivity and effects of Fe/Co ratio on the band structure and charge transport. *Acta Mater.* **92**, 152–162 (2015).

140. Li, W. & Mingo, N. Thermal conductivity of fully filled skutterudites: Role of the filler. *Phys. Rev. B* **89**, 184304 (2014).

141. Li, W. & Mingo, N. Ultralow lattice thermal conductivity of the fully filled skutterudite YbFe4Sb12 due to the flat avoided-crossing filler modes. *Phys. Rev. B* **91**, 144304 (2015).

142. Avila, M. A. *et al.* Ba(8)Ga(16)Sn(30) with type-I clathrate structure: Drastic suppression of heat conduction. *Appl. Phys. Lett.* **92**, 041901 (2008).

143. Takabatake, T., Suekuni, K. & Nakayama, T. Phonon-glass electron-crystal thermoelectric clathrates: Experiments and theory. *Reviews of Modern Physics* **86**, 669–716 (2014).

144. Nakayama, T. & Kaneshita, E. Significance of Off-Center Rattling for Emerging Low-Lying THz Modes in Type-I Clathrates. *Journal of the Physical Society of Japan* **80**, 104604 (2011).

145. Voneshen, D. J. *et al.* Suppression of thermal conductivity by rattling modes in thermoelectric sodium cobaltate. *Nat. Mater.* **12**, 1027–1031 (2013).

146. Qiu, W., Wu, L., Ke, X., Yang, J. & Zhang, W. Diverse lattice dynamics in ternary Cu-Sb-Se compounds. *Sci. Rep.* **5**, 13643 (2015).

147. Qiu, W. J. *et al.* Part-crystalline part-liquid state and rattling-like thermal damping in materials with chemical-bond hierarchy. *Proceedings of the National Academy of Sciences of the United States of America* **111**, 15031–15035 (2014).

148. Skoug, E. J. & Morelli, D. T. Role of Lone-Pair Electrons in Producing Minimum Thermal Conductivity in Nitrogen-Group Chalcogenide Compounds. *Phys. Rev. Lett.* **107**, 235901 (2011).

149. Zhang, Y. S. *et al.* First-principles description of anomalously low lattice thermal conductivity in thermoelectric Cu-Sb-Se ternary semiconductors. *Phys. Rev. B* **85**, 054306 (2012).

150. Liu, H. L. *et al.* Copper ion liquid-like thermoelectrics. *Nat. Mater.* **11**, 422–425 (2012).

151. Kim, H. *et al.* Ultralow thermal conductivity of b-Cu2Se by atomic fluidity and structure distortion. *Acta Mater.* **86**, 247–253 (2015).

152. Dennler, G. *et al.* Are binary copper sulfides/selenides really new and promising thermoelectric materials? *Adv. Energy Mater.* **4**, 1301581 (2014).

153. Callaway, J. Model for Lattice Thermal Conductivity at Low Temperatures. *Physical Review* **113**, 1046–1051 (1959).

154. Asen-Palmer, M. *et al.* Thermal conductivity of germanium crystals with different isotopic compositions. *Phys. Rev. B* **56**, 9431–9447 (1997).

155. Morelli, D. T., Heremans, J. P. & Slack, G. A. Estimation of the isotope effect on the lattice thermal conductivity of group IV and group III-V semiconductors. *Phys. Rev. B* **66**, 195304 (2002).

156. Ward, A., Broido, D. A., Stewart, D. A. & Deinzer, G. Ab initio theory of the lattice thermal conductivity in diamond. *Phys. Rev. B* **80**, 754 (2009).

157. Liao, B. L. *et al.* Significant reduction of lattice thermal conductivity by the electron-phonon interaction in silicon with high carrier concentrations: a first-principles study. *Phys. Rev. Lett.* **114**, 115901 (2015).

158. Vining, C. B. A model for the high-temperature transport-properties of heavily doped n-type silicon-germanium alloys. *J. Appl. Phys.* **69**, 331–341 (1991).

159. Shi, X., Pei, Y., Snyder, G. J. & Chen, L. Optimized thermoelectric properties of Mo3Sb7-xTex with significant phonon scattering by electrons. *Energy Environ. Sci.* **4**, 4086–4095 (2011).

160. Poudel, B. *et al.* High-thermoelectric performance of nanostructured bismuth antimony telluride bulk alloys. *Science* **320**, 634–638 (2008).

161. Xie, W. *et al.* Identifying the specific nanostructures responsible for the high thermoelectric performance of (Bi, Sb)2Te3 nanocomposites. *Nano Lett.* **10**, 3283–3289 (2010).

162. Kim, S. I. *et al.* Dense dislocation arrays embedded in grain boundaries for high-performance bulk thermoelectrics. *Science* **348**, 109–114 (2015).

163. Hsu, K. F. *et al.* Cubic AgPbmSbTe2+m: Bulk thermoelectric materials with high figure of merit. *Science* **303**, 818–821 (2004).

164. Joshi, G. *et al.* Enhanced thermoelectric figure-of-merit in nanostructured p-type silicon germanium bulk alloys. *Nano Lett.* **8**, 4670–4674 (2008).

165. Wang, X. *et al.* Enhanced thermoelectric figure of merit in nanostructured n-type silicon germanium bulk alloy. *Appl. Phys. Lett.* **93**, 193121 (2008).

166. Snyder, G. J., Christensen, M., Nishibori, E., Caillat, T. & Iversen, B. B. Disordered zinc in Zn4Sb3 with phonon-glass and electron-crystal thermoelectric properties. *Nat. Mater.* **3**, 458–463 (2004).

167. Rhyee, J.-S. *et al.* Peierls distortion as a route to high thermoelectric performance in In4Se3-delta crystals. *Nature* **459**, 965–968 (2009).

168. Chen, L. *et al.* Anomalous barium filling fraction and n-type thermoelectric performance of BayCo4Sb12. *J. Appl. Phys.* **90**, 1864 (2001).

169. Rogl, G. *et al.* n-Type skutterudites (R, Ba, Yb)yCo4Sb12 (R = Sr, La, Mm, DD, SrMm, SrDD) approaching ZT≈2.0. *Acta Mater.* **63**, 30–43 (2014).

170. Martin, J., Wang, H. & Nolas, G. S. Optimization of the thermoelectric properties of Ba8Ga16Ge30. *Appl. Phys. Lett.* **92**, 222110 (2008).

171. Wang, L., Chen, L.-D., Chen, X.-H. & Zhang, W.-B. Synthesis and thermoelectric properties of n-type Sr8Ga16-xGe30-y clathrates with different Ga/Ge ratios. *J. Phys. Appl. Phys.* **42**, 045113 (2009).

172. Fujita, I., Kishimoto, K., Sato, M., Anno, H. & Koyanagi, T. Thermoelectric properties of sintered clathrate compounds Sr8GaxGe46-x with various carrier concentrations. *J. Appl. Phys.* **99**, 093707 (2006).

173. Shen, Q. *et al.* Effects of partial substitution of Ni by Pd on the thermoelectric properties of ZrNiSn-based half-Heusler compounds. *Appl. Phys. Lett.* **79**, 4165–4167 (2001).

174. Fu, C. *et al.* Realizing high figure of merit in heavy-band p-type half-Heusler thermoelectric materials. *Nat. Commun.* **6**, 8144 (2015).

175. Shi, X. Y., Xi, L. L., Fan, J., Zhang, W. Q. & Chen, L. D. Cu-Se bond network and thermoelectric compounds with complex diamondlike structure. *Chem. Mater.* **22**, 6029–6031 (2010).

176. Liu, R. *et al.* Ternary compound CuInTe$_2$: a promising thermoelectric material with diamond-like structure. *Chemical Communications* **48**, 3818–3820 (2012).

177. Plirdpring, T. *et al.* Chalcopyrite CuGaTe2: A High-Efficiency Bulk Thermoelectric Material. *Adv. Mater.* **24**, 3622–3626 (2012).

178. Liu, H. L. *et al.* Ultrahigh thermoelectric performance by electron and phonon critical scattering in Cu2Se1-xlx. *Adv. Mater.* **25**, 6607–6612 (2013).

179. He, Y. *et al.* High thermoelectric performance in non-toxic earth-abundant copper sulfide. *Adv. Mater.* **26**, 3974–3978 (2014).

180. He, Y. *et al.* Ultrahigh thermoelectric performance in mosaic crystals. *Adv. Mater.* **27**, 3639–3644 (2015).

181. Venkatasubramanian, R., Siivola, E., Colpitts, T. & O'quinn, B. Thin-film thermoelectric devices with high room-temperature figures of merit. *Nature* **413**, 597–602 (2001).

182. Biswas, K. *et al.* Strained endotaxial nanostructures with high thermoelectric figure of merit. *Nat. Chem.* **3**, 160–166 (2011).

183. Shi, X., Zhang, W., Chen, L. D. & Yang, J. Filling fraction limit for intrinsic voids in crystals: Doping in skutterudites. *Phys. Rev. Lett.* **95**, 185503 (2005).

184. Xi, L., Yang, J., Zhang, W., Chen, L. & Yang, J. Anomalous dual-element filling in partially filled skutterudites. *J. Am. Chem. Soc.* **131**, 5560–5563 (2009).

185. Xi, L. *et al.* Systematic study of the multiple-element filling in caged skutterudite CoSb3. *Chem. Mater.* **22**, 2384–2394 (2010).

186. Zhang, W. *et al.* Predication of an ultrahigh filling fraction for K in CoSb3. *Appl. Phys. Lett.* **89**, 112105 (2006).

187. He, Y., Zhang, T., Shi, X., Wei, S.-H. & Chen, L. High thermoelectric performance in copper telluride. *NPG Asia Mater.* **7**, e210 (2015).

188. Nielsen, M. D., Ozolins, V. & Heremans, J. P. Lone pair electrons minimize lattice thermal conductivity. *Energy Environ. Sci.* **6**, 570–578 (2013).

189. Yang, J. *et al.* Evaluation of half-heusler compounds as thermoelectric materials based on the calculated electrical transport properties. *Adv. Funct. Mater.* **18**, 2880–2888 (2008).

190. Carrete, J., Li, W., Mingo, N., Wang, S. & Curtarolo, S. Finding unprecedentedly low-thermal-conductivity half-Heusler semiconductors via high-throughput materials modeling. *Phys. Rev. X* **4**, 011019 (2014).

191. Carrete, J., Mingo, N., Wang, S. & Curtarolo, S. Nanograined half-Heusler semiconductors as advanced thermoelectrics: an ab initio high-throughput statistical study. *Adv. Funct. Mater.* **24**, 7427–7432 (2014).

192. Yan, J. *et al.* Material descriptors for predicting thermoelectric performance. *Energy Environ. Sci.* **8**, 983–994 (2015).

193. Funahashi, R. *et al.* High-throughput screening of thermoelectric oxides and power generation modules consisting of oxide unicouples. *Meas Sci Technol* **16**, 70–80 (2005).

194. Su, X. L. *et al.* Self-propagating high-temperature synthesis for compound thermoelectrics and new criterion for combustion processing. *Nat. Commun.* **5**, 4908 (2014).

13

Multiple unpinned Dirac points in group-Va single-layers with phosphorene structure

Yunhao Lu[1,2], Di Zhou[1], Guoqing Chang[3,4], Shan Guan[5], Weiguang Chen[6], Yinzhu Jiang[1,2], Jianzhong Jiang[1,2], Xue-sen Wang[4], Shengyuan A Yang[5], Yuan Ping Feng[4], Yoshiyuki Kawazoe[7,8] and Hsin Lin[3,4]

Emergent Dirac fermion states underlie many intriguing properties of graphene, and the search for them constitutes one strong motivation to explore two-dimensional (2D) allotropes of other elements. Phosphorene, the ultrathin layers of black phosphorous, has been a subject of intense investigations recently, and it was found that other group-Va elements could also form 2D layers with similar puckered lattice structure. Here, by a close examination of their electronic band structure evolution, we discover two types of Dirac fermion states emerging in the low-energy spectrum. One pair of (type-I) Dirac points is sitting on high-symmetry lines, while two pairs of (type-II) Dirac points are located at generic k-points, with different anisotropic dispersions determined by the reduced symmetries at their locations. Such fully-unpinned (type-II) 2D Dirac points are discovered for the first time. In the absence of spin-orbit coupling (SOC), we find that each Dirac node is protected by the sublattice symmetry from gap opening, which is in turn ensured by any one of three point group symmetries. The SOC generally gaps the Dirac nodes, and for the type-I case, this drives the system into a quantum spin Hall insulator phase. We suggest possible ways to realise the unpinned Dirac points in strained phosphorene.

INTRODUCTION

Recent years have witnessed a surge of research interest in the study of Dirac fermions in condensed matter systems, ranging from graphene and topological insulator surfaces in two-dimensions (2D) to Dirac and Weyl semimetals in 3D,[1-4] which possess many intriguing physical properties owing to their relativistic dispersion and chiral nature. Especially, 2D Dirac fermion states have been extensively discussed in honeycomb lattices, commonly shared by group-IVa elements with graphene as the most prominent example,[5-9] for which Dirac points are pinned at the two inequivalent high-symmetry points K and K' of the hexagonal Brillouin zone (BZ), around which the dispersion is linear and isotropic. Later on, 2D Dirac points on high-symmetry lines were also predicted in some nanostructured materials,[10] including graphynes[11] and rectangular carbon and boron allotropes.[12,13] However, the possibility of 2D Dirac points at generic k-points has not been addressed, and such Dirac point has not been found so far.

Meanwhile, the exploration of 2D materials built of group-Va elements (P, As, Sb and Bi) has just started. Single- and few-layer black phosphorous, known as phosphorene, have been successfully fabricated, and was shown to be semiconducting with a thickness-dependent bandgap and a good mobility up to $\sim 10^3 cm^2/Vs$, generating intense interest.[14-21] While 2D allotropes with different lattice structures have been predicted and analysed for the other group-Va elements,[22-25] we note that the puckered lattice structure similar to phosphorene has been demonstrated experimentally for Sb (refs 26-28; W. Xu et al., unpublished) and Bi[29-32] (down to single-layer) grown on suitable substrates, and been predicted for As as well.[22] Motivated by these previous experimental and theoretical works, and in view of the ubiquitous presence of the Dirac fermions and the associated interesting physics, one may wonder: Is it possible to have Dirac fermion states hosted in such 2D puckered lattices? A simple consideration shows that here any possible Dirac point cannot occur at high-symmetry points. The reason is that each Dirac point at k must have a time reversal (TR) partner at $-k$ with opposite chirality, whereas the BZ of the puckered lattice has a rectangular shape, of which all the high-symmetry points are invariant under TR. Therefore, if Dirac states indeed exist in such systems, they must be of a type distinct from those in graphene.

In this work, we address the above question by investigating the electronic structures of group-Va 2D puckered lattices. We find that Dirac fermion states not only exist, but in fact occur with two different types: one type (referred to as type-I) of (two) Dirac points are located on high-symmetry lines; while the other type (referred to as type-II) of (four) Dirac points are located at generic k-points. Depending on their reduced symmetries, dispersions around these points exhibit different anisotropic behaviours. Points of each type can generate or annihilate in pairs of opposite chiralities, accompanying topological phase transitions from a band insulator to a 2D Dirac semimetal phase, and since they are not fixed at high-symmetry points, their locations can be moved around in the BZ. Particularly, to our best knowledge, the novel

[1]School of Materials Science and Engineering, Zhejiang University, Hangzhou, China; [2]State Key Laboratory of Silicon Materials, Zhejiang University, Hangzhou, China; [3]Centre for Advanced 2D Materials and Graphene Research Centre, National University of Singapore, Singapore, Singapore; [4]Department of Physics, National University of Singapore, Singapore, Singapore; [5]Research Laboratory for Quantum Materials, Singapore University of Technology and Design, Singapore, Singapore; [6]College of Physics and Electronic Engineering, Zhengzhou Normal University, Zhengzhou, China; [7]New Industry Creation Hatchery Center, Tohuku University, Sendai, Japan and [8]Institute of Thermophysics, Siberian Branch of Russian Academy of Sciences, Novosibirsk, Russia.
Correspondence: Y Lu (luyh@zju.edu.cn) or SA Yang (shengyuan_yang@sutd.edu.sg) or H Lin (nilnish@gmail.com)

fully-unpinned (type-II) 2D Dirac points are discovered here for the first time. In the absence of spin-orbit coupling (SOC), each Dirac node is protected from gap opening by a sublattice (chiral) symmetry, which can in turn be ensured by any one of three point group symmetries. The inclusion of SOC could gap the Dirac nodes, and in the case of type-I nodes it transforms the system into a quantum spin Hall (QSH) insulator phase. All these properties make the system distinct from graphene and other 2D materials. We further suggest that the novel unpinned Dirac points can be experimentally realized by the strain engineering of phosphorene. Our discovery therefore greatly advances our fundamental understanding of 2D Dirac points, and it also suggests a promising platform for exploring interesting effects with novel types of Dirac fermions.

RESULTS

A group-Va pnictogen atom typically forms three covalent bonds with its neighbours. As shown in Figure 1 for a single-layer phosphorene structure, the P atoms have strong sp^3-hybridisation character hence the three P–P bonds are more close to a tetrahedral configuration. This results in two atomic planes (marked with red and blue colours) having a vertical separation of h comparable to the bond length. In each atomic plane, the bonding between atoms forms zig-zag chains along y-direction. The unit cell has a four-atom basis, which we label as A_U, B_U, A_L and B_L (see Figure 1c), where U and L refer to the upper- and lower-plane, respectively. The structure has a non-symmorphic $D_{2h}(7)$ space group which includes the following elements that will be important in our discussion: an inversion centre i, a vertical mirror plane σ_v perpendicular to $\hat{y}$, and two twofold rotational axes c_{2y} and c_{2z}. Note that due to the puckering of the layer, the mirror planes perpendicular to $\hat{x}$ and $\hat{z}$ are broken. With the same valence electron configuration, As, Sb and Bi possess allotropes with similar puckered lattice structures.

To study the electronic properties, we performed first-principles calculations based on the density functional theory (DFT). The details are described in the materials and methods. The calculated geometric parameters of group-Va 2D puckered lattices with $D_{2h}(7)$ symmetry are summarised in the Supplementary Information. The obtained structures agree with the experiments and other theoretical calculations (refs 17,22,32; W. Xu et al., unpublished). The lattice constants $a > b$, reflecting that the inter-chain coupling is weaker than the coupling along the zig-zag chains. The angle θ_2 increases from ~70° for P to ~85° for

Bi, whereas θ_1 remains ~95°. The inter-plane separation h, as well as the bond lengths R_1 and R_3 increase by almost 1 Å; from P to Bi, while R_2, the distance between sites of neighbouring zig-zag chains, increases only slightly, implying that the inter-chain coupling becomes relatively more important with increasing atomic number.

We first examine their corresponding band structures without SOC, whose effect will be discussed later. The results are shown in Figure 2. The puckered lattice of P is a semiconductor with a bandgap around Γ-point. From P to Bi, the direct bandgap at Γ-point keeps decreasing, and a drastic change occurs from Sb to Bi where linear band crossings can be clearly spotted along the Γ-X_2 line. Examination of the band dispersion around the two points (labelled as D and D' in Figure 1d) shows that they are indeed Dirac points (see Figure 3a). Furthermore, along Γ-X_1 line, there gradually appear two sharp local band extremum points for both conduction and valence bands, where the local gap decreases from P to Sb with the two bands almost touching for Sb, yet the trend breaks for Bi. Remarkably, close examination reveals that for Sb and Bi, close to each extremum point there are actually two Dirac points on the two sides of the Γ-X_1 line (see Figures 1d and 3b). The energy dispersions around these Dirac points are shown in Figures 1e and f, clearly demonstrating the Dirac cone characters. Therefore, two types of Dirac points with distinct symmetry characters exist in this system: one pair of type-I Dirac points (D and D') sitting on high-symmetry lines and two pairs of type-II Dirac points (near F and F') at generic k-points.

The band evolution around Γ-point from Sb to Bi and the appearance of type-I Dirac points in Figure 2 are reminiscent of a band-inversion process. Indeed, by checking the parity eigenvalues at Γ, one confirms that the band order is reversed for Bi around Γ-point (see Supplementary Information). For a better understanding, we construct a tight-binding model trying to capture the physics around Γ-point. Since the low-energy bands are dominated with p_z-orbital character (Figure 2), we take one orbital per site, and include couplings along R_1 and R_2 in the same atomic plane (with amplitudes t_1 and t_2, respectively), as well as nearest-neighbour inter-plane hopping along R_3 (with amplitude $t_\perp$) (see Supplementary Information). Written in the basis of (A_U, A_L, B_U, B_L), the Hamiltonian takes the form:

$$\mathcal{H}(\boldsymbol{k}) = \begin{bmatrix} 0 & Q(\boldsymbol{k}) \\ Q^\dagger(\boldsymbol{k}) & 0 \end{bmatrix}, \qquad (1)$$

where $Q(\boldsymbol{k})$ is a 2×2 matrix of the Fourier transformed hopping terms (see Supplementary Information). The Hamiltonian (1) can

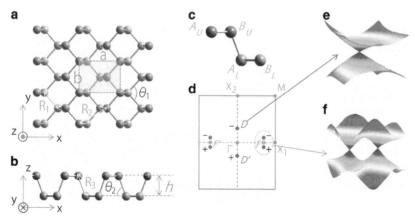

Figure 1. (**a** and **b**) Top- and side-view of the 2D puckered lattice structure. The green shaded region marks the unit cell. (**c**) Four-atom basis sites in a unit cell. (**d**) 2D Brillouin zone with high-symmetry points. The locations of Dirac points are schematically marked by the red dots: two type-I Dirac points at D and D' on Γ-X_2; and four type-II Dirac points around Γ-X_1 forming two mirror image pairs. F (F') on Γ-X_1 labels the mid-point of each pair. +(−) indicates the chirality of each point. **e** and **f** show the schematic energy dispersions around point D and point F, respectively, corresponding to the result in Figure 3.

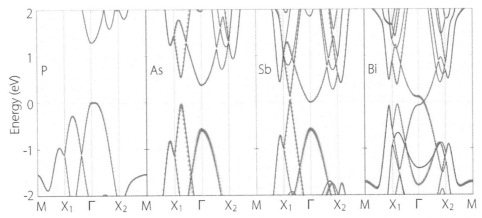

Figure 2. Band structures of group-Va elements with 2D phosphorene lattice structures in the absence of SOC. The size of red (green) dots denotes the weight of projection onto p_z (p_x) atomic orbitals.

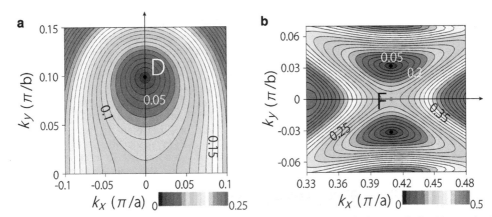

Figure 3. Dispersion around (a) type-I and (b) type-II Dirac points for 2D puckered Bi. Black dots mark the Dirac point locations. Energy is in unit of eV.

be diagonalized and possible band crossings can be probed by searching for the zero-energy modes, which exist when the condition $\lambda \equiv t_\perp/[2(t_1 + t_2)] < 1$ is satisfied, with two band touching points at $(0, \pm k_D)$ where $k_D = (2/b)\arccos(\lambda)$. The direct gap at Γ can be obtained as $\Delta = 2[t_\perp - 2(t_1 + t_2)]$. Hence this simple model indeed captures the emergence of two Dirac points D and D', along with a transition as parameter λ varies: when $\lambda > 1$, the system is a band insulator; when $\lambda < 1$, it is a 2D Dirac semimetal. The transition occurs at the critical value $\lambda_c = 1$ when the conduction and valence bands touch at Γ-point and the band order starts to be inverted. This corresponds to a quantum (and topological) phase transition,[33] during which there is no symmetry change of the system.

Equation (1) captures the trend observed in DFT results. The overlap between-p_z orbitals is larger along the R_3 bond, hence one expects that $t_\perp > t_1 > t_2$. By fitting the DFT bands around Γ-point, one finds that from P to Bi, $t_\perp$ decreases a lot, while t_2 increases and becomes relatively more important (Supplementary Information). The result shows that $(t_\perp, t_1, t_2)$ changes from (2.50, 0.77, 0.33) for Sb to (1.86, 0.63, 0.35) for Bi (units in eV). Hence λ crosses the critical value from Sb to Bi, indicating the band inversion at Γ and the appearance of two Dirac points.

The emergence of low-energy relativistic chiral modes is the most remarkable property of Dirac points.[33] To explicitly demonstrate this, we expand Hamiltonian (1) around each Dirac point, which leads to the low-energy Hamiltonian

$$H_\tau(\boldsymbol{q}) = v_x q_x \sigma_y + \tau v_y q_y \sigma_x, \qquad (2)$$

where $\boldsymbol{q}$ is the wave-vector measured from each Dirac point, $\tau = \pm 1$ for D and D', σ_is are Pauli matrices for the sub-space spanned by the two eigenstates at the Dirac point (apart from the Bloch phase factor): $|u_1\rangle = (0, 0, 1, -1)/\sqrt{2}$ and $|u_2\rangle = (1, -1, 0, 0)/\sqrt{2}$ and $v_x = a t_\perp(t_1 - t_2)/(t_1 + t_2)$ and $v_y = b\sqrt{4(t_1 + t_2)^2 - t_\perp^2}$ are the two Fermi velocities. The form of equation (2) may also be argued solely from symmetry. Compared with graphene, these type-I points are unpinned from the high-symmetry points. They can be shifted along Γ-X_2 (and even pair-annihilated) by varying system parameters such as λ, although they cannot go off the line as constrained by the symmetries. In addition, different from graphene,[6] the dispersion here is anisotropic, characterised by two different Fermi velocities.

Next, we turn to the fully-unpinned type-II Dirac points. The four type-II Dirac points start to appear for Sb in our DFT result, located close to the Γ-X_1 line. They can be more clearly seen for Bi (see Figure 3b). Again the band evolution implies a local band inversion near F and F'. Here F and F' (on Γ-X_1) are the mid-points of the lines connecting each pair of the type-II points. The low-energy bands are mainly of p_x-orbital character. To reproduce the fine features using a tight-binding model would require more hopping terms. Instead, we construct a low-energy effective Hamiltonian around point F (F') based on symmetry analysis. There the Hamiltonian is constrained by σ_v, which maps inside each pair (labelled by $\mu = \pm 1$ for F and F'), and by i, c_{2y}, c_{2z}, and TR that map between the two pairs. Expansion to leading order in each wave-vector component q_i gives (see Supplementary

Information)

$$H_\mu(\boldsymbol{q}) = wq_x\sigma_y + (-m_0 + \mu w'q_x + m_1 q_y^2)\sigma_x, \qquad (3)$$

where $\boldsymbol{q}$ is measured from F (or F'), w, w', m_0 and m_1 are expansion coefficients. Two Dirac points appear at $(0, \pm q_0)$ with $q_0 = \sqrt{m_0/m_1}$ when $\mathrm{sgn}(m_0/m_1) = 1$, corresponding to a local band inversion around $\boldsymbol{q} = 0$. Further expansion of the Hamiltonian around the Dirac point $(0, vq_0)$ $(v = \pm 1)$ leads to

$$\tilde{H}_\mu^\nu = wq_x\sigma_y + [2\nu q_0(q_y - \nu q_0) + \mu w'q_x]\sigma_x. \qquad (4)$$

This demonstrates that the two points at $v = \pm 1$ are of opposite chirality, as required by σ_v. The dispersion is highly anisotropic (at leading order, characterised by three parameters: w, q_0 and w') because the Dirac point is at a generic k-point with less symmetry constraint, as compared with type-I Dirac points.

Unlike in 3D systems, Dirac nodes in 2D have a co-dimension of 2 hence are generally not protected from gap opening.[33] In the absence of SOC, however, the Dirac nodes here are stable due to the protection by sublattice (chiral) symmetry between $\{A_i\}$ and $\{B_i\}$ $(i = U, L)$ sites, which allows the definition of a winding number[34,35] (that is, quantised Berry phase in units of π) along a closed loop ℓ encircling each Dirac point: $N_\ell = \oint_\ell \mathcal{A}_{\boldsymbol{k}} \cdot d\boldsymbol{k}/\pi = \pm 1$, where $\mathcal{A}_{\boldsymbol{k}}$ is the Berry connection of the occupied valence bands. And for a 2D Dirac point, the sign of N_ℓ (or the $\pm\pi$ Berry phase) is also referred to as the chirality.[6] Using DFT results, we numerically calculate the Berry phase for each Dirac point and indeed confirm that they are quantised as $\pm\pi$. The signs are indicated in Figure 1d.

More interestingly, in the puckered lattice with a four-atom basis in a non-coplanar geometry, the sublattice symmetry can be ensured by any one of three independent point group symmetries: i, c_{2y} and c_{2z}. The resulting protection of Dirac nodes can be explicitly demonstrated in low-energy models. For example, consider the type-I points described by equation (2). There the representations of i, c_{2y} and c_{2z} (denoted by $\mathcal{P}$, $\mathcal{R}_y$ and $\mathcal{R}_z$, respectively) are the same, which is, σ_x. Then the symmetry requirement $\mathcal{R}_y H_\tau(q_x, q_y)\mathcal{R}_y^{-1} = H_\tau(-q_x, q_y)$ by c_{2y} directly forbids the presence of a mass term $m\sigma_z$. Meanwhile, since i and c_{2z} map one valley to the other, they protect the Dirac nodes when combined with TR (or σ_v if it is unbroken), e.g.,

considering the combined symmetry of c_{2z} and TR (with representation $\mathcal{T} = K$ the complex conjugation operator): $(\mathcal{R}_z\mathcal{T})H_\tau(\boldsymbol{q})(\mathcal{R}_z\mathcal{T})^{-1} = H_\tau(\boldsymbol{q})$, which again forbids a mass generation. The underlying reason i, c_{2y} and c_{2z} each protects the Dirac node is that they each map between the two sublattices hence ensures the sublattice (chiral) symmetry. In comparison, the mirror plane σ_v maps inside each sublattice, hence it alone cannot provide such protection. This reasoning is general and applies to type-II points as well. (In equation (3), i, c_{2y} and c_{2z} have representations as σ_x by construction, and when combined with TR, again each forbids the generation of a mass term $\sim m\sigma_z$. See Supplementary Information.) We stress that the three symmetries i, c_{2y} and c_{2z} each protects the Dirac points independent of the other two. For example, we could disturb the system as in Figure 4 such that only one of the three symmetries survives. The corresponding DFT results confirm that the Dirac nodes still exist. Thus the crystalline symmetries actually offer multiple protections for the Dirac nodes in the current system.

SOC could break the sublattice symmetry. Hence when SOC is included, the Dirac nodes would generally be gapped.[36] For type-I points, treating SOC as a perturbation, its leading-order symmetry-allowed form is $H_{\mathrm{SOC}} = \tau\Delta\sigma_z s_z$, where s_z is Pauli matrix for real spin. This is similar to the intrinsic SOC term in graphene,[37] which opens a gap of $2|\Delta|$ at the Dirac points. For the type-II points, we obtain $H_{\mathrm{SOC}} = \eta q_y\sigma_z s_z$ in equation (3) hence a gap of $2q_0|\eta|$ is also opened at these Dirac points. Gap opening by SOC is closely related to the QSH insulator phase.[1,2,37] Here the band topology can be directly deduced from the parity analysis at the four TR invariant momenta.[38] This means that only the band inversion at Γ between the two type-I points contributes to a non-trivial $\mathbb{Z}_2$ invariant; whereas that associated with type-II points does not. It follows that Sb is topologically trivial since it has only type-II Dirac points, while Bi is non-trivial since it has additional type-I points. These results are in agreement with previous studies.[32]

Breaking all three symmetries i, c_{2y} and c_{2z} can also generate a trivial gap term $m\sigma_z$ at the Dirac points, which competes with the SOC gap. For example, this happens when each atomic plane forms additional buckling structure.[32] Nevertheless, as long as the

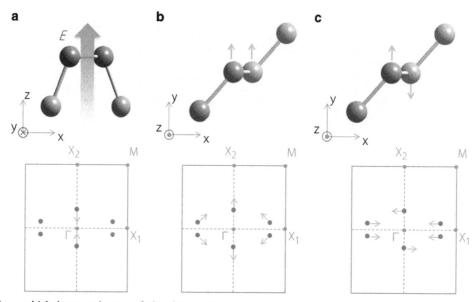

Figure 4. Perturbations which leave only one of the three symmetries preserved: (**a**) c_{2z} survives when applying E field in z-direction; (**b**) c_{2y} survives when shifting the two basis sites along $\hat{y}$ in the same direction; (**c**) i survives when shifting the two sites along $\hat{y}$ but in opposite directions. For each case, the Dirac nodes are still protected in the absence of SOC. The lower panel in each sub-figure schematically indicates the movement of the Dirac points under each perturbation. For the case in **a**, the type-II points do not show appreciable change in their locations under a weak electric field, while the two type-I points annihilate with each other when they meet at Γ-point.

trivial mass term does not close the SOC-induced gap, by adiabatic continuation the band topology will not change.

DISCUSSION

Due to their different locations and the associated symmetries, the two types of Dirac points here exhibit properties distinct from that of graphene. With preserved sublattice symmetry and in the absence of SOC, the Dirac nodes are topologically stable—they can only disappear by pair-annihilation between opposite chiralities. This is unlikely for graphene since the Dirac points are pinned at the high-symmetry points. In contrast, the two types of Dirac points here are less constrained. Pair-annihilation (pair-generation) indeed occurs during the quantum phase transition as observed from the band evolution.

It is noted that similar type-I points were also predicted in a few nanostructured materials.[11–13] Meanwhile the type-II points discovered here are completely new. They are fully-unpinned and have highly anisotropic dispersions. With this discovery, now we can have an almost complete picture: 2D Dirac points can occur at high-symmetry points, along high-symmetry lines and also at generic k-points.

It is possible to have Dirac points, originally sitting at high-symmetry points, to become unpinned when crystalline symmetry is reduced due to structural distortions. However, we stress that the type-II points here are distinct in that they are realized in a native crystalline structure with relatively high symmetry. Only in such a case, we can have a sharp contrast between generic k-points where the group of wave vectors is trivial and the high-symmetry k-points where the group is non-trivial, and accordingly the type-II point can move around (hence fully-unpinned) without any symmetry-breaking. More importantly, it is just because that type-II points occur in a state with high symmetry that the Dirac nodes can be protected (in the absence of SOC): as we discussed, the various crystalline symmetries ensure the protection of the Dirac nodes from gap opening.

It is remarkable that the two different types of Dirac points can coexist in the same 2D material. We emphasise that it is a result of the lattice structure and the valence character of the pnictogen elements. Our DFT result indeed shows that even starting from the P lattice, the two types of Dirac points can be separately tuned to appear or disappear by lattice deformations. For example, we find that the type-II Dirac points can be generated in phosphorene by applying uniaxial tensile strains along the y-direction. The DFT result in Figures 5 and 6 indeed shows the band inversion on Γ-X_1 and the formation of four type-II Dirac points around a strain of 16%. Since phosphorene has excellent mechanical properties and

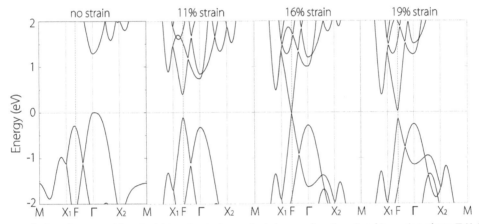

Figure 5. Band structure of phosphorene with uniaxial tensile strain applied along y-direction. Band inversion along Γ-X_1 line occurs around a value of 16%.

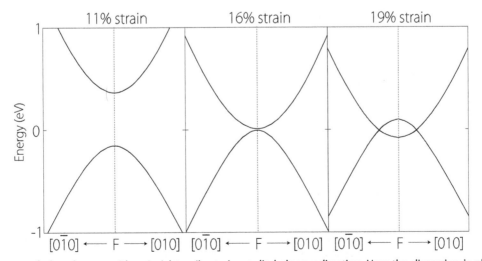

Figure 6. Band structure of phosphorene with uniaxial tensile strain applied along y-direction. Here the dispersion is plotted along a path crossing point F (as indicated in Figure 5) and perpendicular to the Γ-X_1 direction. Two type-II Dirac points near F can be clearly observed at strain value $>16\%$.

a critical strains >25% has been predicted,[39] it is promising that the novel strain-induced topological phase transitions and the appearance of type-II Dirac points can be directly observed in strained phosphorene. The lattice deformation that produces type-I points is discussed in the Supplementary Information. Similar scenarios occur for other group-Va elements as well.

So far, single-layer As, Sb and Bi in their free-standing form have not been realized yet. Nevertheless, in view of the rapid progress in experimental techniques, we expect that these materials could be fabricated in the near future. Especially, for Sb and Bi, the puckered structures have been demonstrated by PVD growth down to single-layer thickness on suitable substrates (refs 26–32; W. Xu *et al.*, unpublished). Besides the topological properties, the presence of Dirac states is expected to endow these 2D materials with many intriguing properties for applications, such as the very high mobility, the half-quantised quantum Hall effect,[40] the universal optical absorption[41] and etc. Due to the highly anisotropic dispersions of these new Dirac points, the electronic transport properties such as the conductivities would show strong direction dependence. In addition, since there is no symmetry connection between the two types of Dirac points, when they are both present, it is possible to independently shift each type of points relative to the Fermi level, e.g., by strain engineering, leading to self-doping and even the interesting scenario with both electron-like and hole-like Dirac fermions in the same system. With the multiple Dirac points with different chiralities, it is possible to further control the carriers near different Dirac points for valleytronic applications.

In conclusion, based on first-principles calculations of 2D allotropes of group-Va elements with puckered lattice structure, we predict the coexistence of two different types of Dirac points: Dirac points on high-symmetry lines and at generic *k*-points. In particular, the 2D Dirac points at generic *k*-points are fully-unpinned, have highly anisotropic dispersions and are discovered here for the first time. Combined with low-energy effective modelling, we unveil the low-energy properties of these Dirac points. We show that their appearance is associated with the band-inversion process corresponding to a topological phase transition. The topology/symmetry protection of the Dirac nodes is analysed in detail. Interestingly, because of the unique lattice structure, there is a triple-protection of the nodes by three independent point group symmetries. This also implies versatile methods to control the locations, as well as the dispersions around the Dirac points. When SOC is strong, the Dirac nodes are gapped, and in the case of type-I points (such as for Bi) this drives the system into a QSH insulator phase. We further show that the topological phase transition and the novel unpinned Dirac points can be realized in strained phosphorene. Our work represents a significant conceptual advance in our fundamental understanding of 2D Dirac points. The result also suggests a new platform to explore novel types of 2D Dirac fermions both for their fascinating fundamental properties and for their promising electronic and valleytronic applications.

MATERIALS AND METHODS

First-principles calculations

Our first-principles calculations are based on the DFT implemented in the Vienna *ab initio* simulation package.[42] The projector augmented wave pseudopotential method is employed to model ionic potentials.[43] Kinetic energy cutoff is set to 400 eV and *k*-point sampling on the rectangular BZ is with a mesh size 20×20. The minimum vacuum layer thickness is >20 Å; which is large enough to avoid artificial interactions with system images. The structure optimisation process is performed including SOC with the local density approximation for the exchange-correlation energy[44] and with van der Waals corrections in the Grimme implementation.[45] The force convergence criteria is set to be 0.01 eV/Å. Hybrid functional (HSE06)[46] is used for the band structure calculations.

ACKNOWLEDGEMENTS

The authors thank D.L. Deng and Shengli Zhang for helpful discussions. This work was supported by NSFC (Grant No. 11374009, 61574123 and 21373184), the National Key Basic Research Program of China (2012CB825700), SUTD-SRG-EPD2013062, Singapore MOE Academic Research Fund Tier 1 (SUTD-T1-2015004), A*STAR SERC 122-PSF-0017 and AcRF R-144-000-310-112. H.L. acknowledges support by Singapore National Research Foundation under NRF Award No. NRF-NRFF2013-03. Y.L. acknowledges Special Program for Applied Research on Super Computation of the NSFC-Guangdong Joint Fund (the second phase). The authors gratefully acknowledge support from SR16000 supercomputing resources of the Center for Computational Materials Science, Tohoku University.

CONTRIBUTIONS

Y.L. and D.Z. performed the first-principles calculations. G.C. and S.G. helped with the data analysis and model fitting. H.L. and S.A.Y. did the analytical modelling and symmetry/topology analysis. W.C., Y.J., J.J. and X.-s.W. participated in the discussion and analysis. X.-s.W., Y.P.F., Y.K., Y.L., S.A.Y. and H.L. supervised the work. Y.L., S.G. and S.A.Y. prepared the manuscript. All authors reviewed the manuscript.

COMPETING INTERESTS

The authors declare no conflict of interest.

REFERENCES

1. Hasan, M. Z. & Kane, C. L. Colloquium: topological insulators. *Rev. Mod. Phys.* **82,** 3045 (2010).
2. Qi, X.-L. & Zhang, S.-C. Topological insulators and superconductors. *Rev. Mod. Phys.* **83,** 1057 (2011).
3. Wan, X., Turner, A. M., Vishwanath, A. & Savrasov, S. Y. Topological semimetal and fermi-arc surface states in the electronic structure of pyrochlore iridates. *Phys. Rev. B* **83,** 205101 (2011).
4. Young, S.-M. *et al.* Dirac semimetal in three dimensions. *Phys. Rev. Lett.* **108,** 140405 (2012).
5. Novoselov, K. S. *et al.* Electric field effect in atomically thin carbon films. *Science* **306,** 666–669 (2004).
6. Castro Neto, A. H., Guinea, F., Peres, N. M. R., Novoselov, K. S. & Geim, A. K. The electronic properties of graphene. *Rev. Mod. Phys.* **81,** 1095 (2009).
7. Cahangirov, S., Topsakal, M., Akturk, E., Sahin, H. & Ciraci, S. Two- and one-dimensional honeycomb structures of silicon and germanium. *Phys. Rev. Lett.* **102,** 236804 (2009).
8. Liu, C.-C., Feng, W. & Yao, Y. Quantum spin Hall effect in silicene and two-dimensional germanium. *Phys. Rev. Lett.* **107,** 076802 (2011).
9. Xu, Y. *et al.* Large-gap quantum spin Hall insulators in tin films. *Phys. Rev. Lett.* **111,** 136804 (2013).
10. See Wang J., Deng S., Liu Z. & Liu Z. The rare two-dimensional materials with Dirac cones. *Nat. Sci. Rev* 2015; **2:** 22.
11. Malko, D., Neiss, C., Vines, F. & Görling, A. Competition for graphene: graphynes with direction-dependent dirac cones. *Phys. Rev. Lett.* **108,** 086804 (2012).
12. Xu, L.-C. *et al.* Two dimensional Dirac carbon allotropes from graphene. *Nanoscale* **6,** 1113 (2014).
13. Zhou, X.-F. *et al.* Semimetallic two-dimensional boron allotrope with massless Dirac fermions. *Phys. Rev. Lett.* **112,** 085502 (2014).
14. Li, L. *et al.* Black phosphorus field-effect transistors. *Nat. Nanotechnol.* **9,** 372 (2014).
15. Xia, F., Wang, H. & Jia, Y. Rediscovering black phosphorus as an anisotropic layered material for optoelectronics and electronics. *Nat. Commun.* **5,** 4458 (2014).
16. Liu, H. *et al.* Phosphorene: An unexplored 2D semiconductor with a high hole mobility. *ACS Nano* **8,** 4033 (2014).
17. Qiao, J., Kong, X., Hu, Z.-X., Yang, F. & Ji, W. High-mobility transport anisotropy and linear dichroism in few-layer black phosphorus. *Nat. Commun.* **5,** 4475 (2014).
18. Rudenko, A. N. & Katsnelson, M. I. Quasiparticle band structure and tight-binding model for single-and bilayer black phosphorus. *Phys. Rev. B* **89,** 201408 (2014).
19. Liu, Q., Zhang, X., Abdalla, L. B., Fazzio, A. & Zunger, A. Switching a normal insulator into a topological insulator via electric field with application to phosphorene. *Nano Lett.* **15,** 1222 (2015).
20. Kim, J. *et al.* Observation of tunable band gap and anisotropic Dirac semimetal state in black phosphorus. *Science* **349,** 723 (2015).
21. Woomer, A. H. *et al.* Phosphorene: Synthesis, scale-up, and quantitative optical spectroscopy. *ACS Nano* **9,** 8869 (2015).
22. Kamal, C. & Ezawa, M. Arsenene: Two-dimensional buckled and puckered honeycomb arsenic systems. *Phys. Rev. B* **91,** 085423 (2015).

23. Zhu, Z. & Tomanek, D. Semiconducting layered blue phosphorus: a computational study. *Phys. Rev. Lett.* **112**, 176802 (2014).

24. Zhang, S., Yan, Z., Li, Y., Chen, Z. & Zeng, H. Atomically thin arsenene and antimonene: semimetal-semiconductor and indirect-direct band-gap transitions. *Angew. Chem. Int. Ed.* **54**, 3112 (2015).

25. Wang, G., Pandey, R. & Karna, S. P. Atomically thin group V elemental films: theoretical investigations of antimonene allotropes. *ACS Appl. Mater. Interfaces* **7**, 11490 (2015).

26. Wang, X.-S., Kushvaha, S. S., Yan, Z. & Xiao, W. Self-assembly of antimony nanowires on graphite. *Appl. Phys. Lett.* **88**, 233105 (2006).

27. Bianchi, M. *et al.* Surface states on a topologically nontrivial semimetal: the case of Sb (110). *Phys. Rev. B* **85**, 155431 (2012).

28. Strozecka, A. *et al.* Unconventional spin texture of a topologically nontrivial semimetal Sb (110). *N. J. Phys.* **14**, 103026 (2012).

29. Nagao, T. *et al.* Nanofilm allotrope and phase transformation of ultrathin Bi film on Si(111)-77. *Phys. Rev. Lett.* **93**, 105501 (2004).

30. Kowalczyk, P. J. *et al.* Electronic size effects in three-dimensional nanostructures. *Nano Lett.* **13**, 43 (2013).

31. Kokubo, I., Yoshiike, Y., Nakatsuji, K. & Hirayama, H. Ultrathin Bi (110) films on Si (111) $\sqrt{3} \times \sqrt{3}$-B substrates. *Phys. Rev. B* **91**, 075429 (2015).

32. Lu, Y. H. *et al.* Topological properties determined by atomic buckling in self-assembled ultrathin Bi(110). *Nano Lett.* **15**, 80 (2015).

33. Volovik, G. E.. *The Universe in a Helium Droplet* (Clarendon Press, 2003).

34. Schnyder, A. P., Ryu, S., Furusaki, A. & Ludwig, A. W. W. Classification of topological insulators and superconductors in three spatial dimensions. *Phys. Rev. B* **78**, 195125 (2008).

35. Yang, S. A., Pan, H. & Zhang, F. Dirac and Weyl superconductors in three dimensions. *Phys. Rev. Lett.* **113**, 046401 (2014).

36. Young, S. M. & Kane, C. L. Dirac semimetals in two dimensions. *Phys. Rev. Lett.* **115**, 126803 (2015).

37. Kane, C. L. & Mele, E. J. Quantum spin hall effect in graphene. *Phys. Rev. Lett.* **95**, 226801 (2005).

38. Fu, L. & Kane, C. L. Topological insulators with inversion symmetry. *Phys. Rev. B* **76**, 045302 (2007).

39. Wei, Q. & Peng, X. Superior mechanical flexibility of phosphorene and few-layer black phosphorus. *Appl. Phys. Lett.* **104**, 251915 (2014).

40. Zhang, Y., Tan, Y.-W., Stormer, H. L. & Kim, P. Experimental observation of the quantum Hall effect and Berry's phase in graphene. *Nature* **438**, 201 (2005).

41. Nair, R. R. *et al.* Fine structure constant defines visual transparency of graphene. *Science* **320**, 1308 (2008).

42. Kresse, G. & Furthmuller, J. Efficient iterative schemes for *Ab initio* total-energy calculations using a plane-wave basis set. *Phys. Rev. B* **54**, 11169 (1996).

43. Blochl, P. E. Projector augmented-wave method. *Phys. Rev. B* **50**, 17953 (1994).

44. Solovyev, I. V., Dederichs, P. H. & Anisimov, V. I. Corrected atomic limit in the local-density approximation and the electronic structure of d impurities in Rb. *Phys. Rev. B* **50**, 16861 (1994).

45. Grimme, S. Semiempirical GGA-type density functional constructed with a long-range dispersion correction. *J. Comput. Chem.* **27**, 1787 (2006).

46. Heyd, J., Scuseria, G. E. & Ernzerhof, M. Hybrid functionals based on a screened Coulomb potential. *J. Chem. Phys.* **118**, 8207 (2003).

14

A high-throughput technique for determining grain boundary character non-destructively in microstructures with through-thickness grains

Matteo Seita[1], Marco Volpi[2], Srikanth Patala[3], Ian McCue[4], Christopher A Schuh[1], Maria Vittoria Diamanti[2], Jonah Erlebacher[4] and Michael J Demkowicz[5]

Grain boundaries (GBs) govern many properties of polycrystalline materials. However, because of their structural variability, our knowledge of GB constitutive relations is still very limited. We present a novel method to characterise the complete crystallography of individual GBs non-destructively, with high-throughput, and using commercially available tools. This method combines electron diffraction, optical reflectance and numerical image analysis to determine all five crystallographic parameters of numerous GBs in samples with through-thickness grains. We demonstrate the technique by measuring the crystallographic character of about 1,000 individual GBs in aluminum in a single run. Our method enables cost- and time-effective assembly of crystallography–property databases for thousands of individual GBs. Such databases are essential for identifying GB constitutive relations and for predicting GB-related behaviours of polycrystalline solids.

INTRODUCTION

Polycrystalline materials are aggregates of differently oriented crystal grains joined along grain boundaries (GBs). Although they typically comprise a small fraction of the material's volume, GBs have tremendous impact on its properties, including strength and ductility,[1–3] thermal and electrical conductivity,[4,5] diffusion,[6,7] resistance to environment-assisted failure,[8,9] and radiation tolerance.[10,11] Although all grains may have identical crystal structure, GB structure depends at minimum on five crystallographic parameters:[12,13] three to describe the relative misorientation, **R**, of the adjoining grains and two for the GB normal vector, $\hat{n}$.

A growing body of evidence shows that many GB properties, ϕ, also depend on all five of these parameters.[14–21] Thus, we may write GB constitutive relations as $\phi = F(\mathbf{R}, \hat{n})$. Knowing these relations is key to predicting the behaviour of polycrystals and to designing materials with superior properties.[22] However, these relations are difficult to obtain experimentally because it is challenging to measure the property of interest, ϕ, and the full character—i.e., both **R** and $\hat{n}$—of numerous individual GBs. Measurement of GB properties, such as corrosion susceptibility, fracture strength or permeability, requires a physical test of the samples, which often irreversibly damages the GB or alters its character. Some techniques for determining GB character (GBC) involve the destruction of the sample, making it impossible to measure GB properties afterwards.

We have developed a new high-throughput method to measure GBC without altering the samples and thus allowing further testing to assess properties of individual GBs. Our method relies on the preparation of microstructures with through-thickness grains and integrates measurements of **R** through electron backscatter diffraction (EBSD)[23] with determinations of $\hat{n}$ using optical reflectance microscopy (ORM):[24] a hybrid technique we term EDOR (electron diffraction optical reflectance). We demonstrate and validate EDOR on GBs in polycrystalline aluminum (Al). EDOR may be employed to collect the large experimental data sets needed to establish GB constitutive relations, $\phi = F(\mathbf{R}, \hat{n})$.

Although the misorientation of GBs on free surfaces is easy to measure non-destructively using EBSD,[23] finding GB plane normal vectors is usually harder.[25] One approach to obtain them couples EBSD with focused ion-beam serial sectioning.[26] Although this is a versatile technique that may be employed on a wide range of microstructures and sample geometries, it requires the physical destruction of the GBs to measure their character, which prevents any further test aimed at measuring their physical properties. GBC may be found using transmission electron microscopy[20] or three-dimensional X-ray diffraction (3DXRD)[27–29] while leaving GBs available, in principle, for subsequent property measurements. However, transmission electron microscopy has limited throughput—it can only study a few GBs at a time—whereas 3DXRD is cost- and time-intensive—it requires advanced X-ray sources and significant computational effort to reconstruct GBC.[30] The advantage of EDOR over these techniques is its ability to assess GBC non-destructively, with high-throughput, and using commercially available equipment and software.

RESULTS

Figure 1 illustrates the elements of EDOR. The method is intended for samples with through-thickness grains. Many commonly

[1]Department of Materials Science and Engineering, Massachusetts Institute of Technology, Cambridge, MA, USA; [2]Department of Chemistry, Materials and Chemical Engineering 'G Natta', Politecnico di Milano, Milan, Italy; [3]Department of Materials Science and Engineering, North Carolina State University, Raleigh, NC, USA; [4]Department of Materials Science and Engineering, Johns Hopkins University, Baltimore, MD, USA and [5]Materials Science and Engineering, Texas A&M University, College Station, TX, USA.
Correspondence: M Seita (matteos@mit.edu)

A high-throughput technique for determining grain boundary character non-destructively in microstructures...

141

encountered types of samples have such microstructures, including films that have been epitaxially grown and then lifted off of their substrates[31] or polycrystals that have been mechanically thinned into foils with thickness smaller than the average grain size.[8] Not all such samples are representative of bulk microstructures. However, our work is not aimed at investigating bulk microstructure, but rather at relating the crystallographic character of numerous individual GBs to their physical properties. Samples with through-thickness grains provide rich data sets to study these relationships, provided that the GBs within them sample a broad range of GBCs. Here we illustrate EDOR using engineered Al foils with through-thickness grains created by repeated rolling and annealing (see the Methods section). The foils are ~600-μm thick and contain grains with average diameter of

~700 μm. EDOR consists of (1) the non-destructive characterisation of the GB plane normal vectors $\hat{n}$ using ORM, (2) assessment of the GB misorientation using EBSD and (3) integration of these two data sets into a list of complete GBCs for all GBs in the sample.

In the first step, the GB plane normal vector $\hat{n}$ in the laboratory frame is measured by analysing both sides of the sample through ORM—a process illustrated in Figure 2. To reveal the crystallographic facets of the constituent grains, both sides of the sample are mechanically polished and then chemically etched (see the Methods section). A sequence of micrographs is acquired from both sides of the engineered sample under variable incident light direction using a stereographic optical microscope, as shown in Figure 2a. The reflectance of a grain along a specific direction depends on the orientation of the crystallographic facets on its surface with respect to the incoming light.[24] Thus, different grains appear with different intensity through the optical microscope as a function of both their crystallographic orientation and the direction of the incident light (Figure 2a). Each optical micrograph is then digitally segmented using a MATLAB processing routine (described in detail in Supplementary Information) that finds the GB traces on the sample surfaces using a contrast-based edge detection algorithm,[32] as illustrated in Figure 2b. The algorithm identifies GB traces by detecting discontinuities in intensity that are larger than a pre-specified threshold. This image processing routine detects all GBs in the microstructure, where a GB is defined by a crystallographic misorientation of $\theta \geq 5°$, as in other techniques.[23]

Not all GBs appear in each optical micrograph. Acquiring multiple optical micrographs under different illumination conditions ensures that all GBs are detected in at least one of them. The digitalised micrographs are then summed into two cumulative 'TOP' and 'BOTTOM' micrographs that contain all detected GB traces from both sample sides, shown in Figure 2c. To remove noise arising from surface scratches or irregularities,

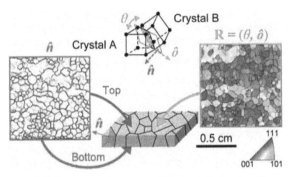

Figure 1. EDOR process to assess GB crystallography (top schematics) using through-thickness-grained microstructures. Left: optical reflectance micrographs taken from the top and bottom of the sample are used to evaluate the GB plane normal vector. Right: EBSD characterisation on the top side of the sample is used to measure GB misorientation.

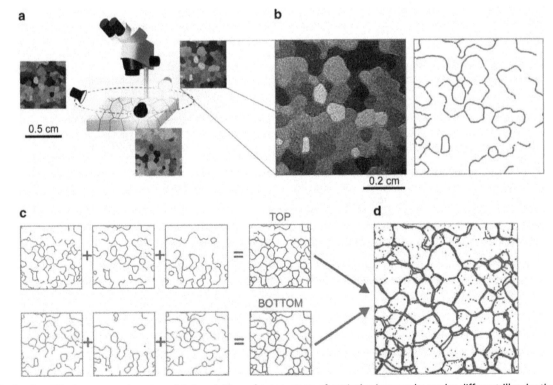

Figure 2. Evaluating GB plane normal vectors. (**a**) Acquisition of a sequence of optical micrographs under different illumination conditions. (**b**) Post-processing of each optical micrograph to detect GB traces using a contrast-based edge detection algorithm. (**c**) Reconstruction of the GB trace network from the top (blue) and bottom (red) of the sample by summing the sequence of processed optical micrographs. (**d**) Registration of the cumulative 'TOP' and 'BOTTOM' images to display the projected GB network (overlapping signal between TOP and BOTTOM displayed in purple). For interpretation of colours, the reader is referred to the web version of this article.

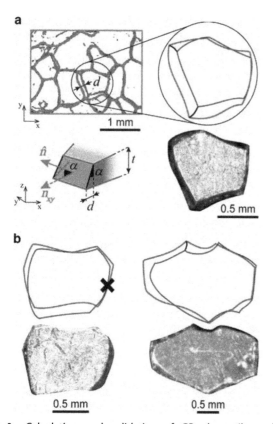

Figure 3. Calculation and validation of GB plane tilt angle a. (**a**) Evaluation of the distance d between top- and bottom-GB traces. The 3D schematic shows the orientation of the GB plane in the laboratory reference frame. (**b**) Comparison of grain shapes obtained through EDOR with direct images of the same grains after disaggregation of the Al microstructure via Ga permeation. The GB crossed in black is the only instance where the EDOR prediction does not match the actual plane tilt.

all features that have low cumulative signal are filtered out. This way, all GB traces from both top and bottom sides of the sample are found, with the exception of those that appear with low frequency in the cumulative images (Supplementary Information). TOP and BOTTOM are then registered in MATLAB using pre-fabricated fiducial markers: through-thickness notches engraved by electric discharge machining around the perimeter of the sample (Supplementary Information). Registration may be performed manually by selecting control points at the notch vertices or automatically using a built-in MATLAB function that uses an intensity-based registration algorithm. The result is a single micrograph, shown in Figure 2d, that contains GB traces from both top and bottom sides of the engineered sample.

To evaluate the normal vector $\hat{n}$, we follow the procedure illustrated in Figure 3a. The tilt angle a of the GB plane with respect to the top surface is computed from the sample thickness t at the GB location and the average distance d between the GB traces on the top and bottom sides:

$$a = \arctan\left(\frac{d}{t}\right). \tag{1}$$

t is estimated by interpolating the thickness values measured around the perimeter of the samples using a caliper, whereas d is computed automatically from the registered micrograph shown in Figure 2d (Supplementary Information). The GB plane normal vector $\hat{n}$ (in the lab frame) may then be written as:

$$\hat{n} = \left[\frac{n_x}{\sqrt{n_x^2+n_y^2}}\cos(a)\hat{x} \quad \frac{n_y}{\sqrt{n_x^2+n_y^2}}\cos(a)\hat{y} \quad |\sin(a)|\hat{z} \right] \tag{2}$$

n_x and n_y are measured from the linear fit to the top GB trace. We adopt the convention that the normal vector, $\hat{n}$, points from the bottom side to the top.

To test the accuracy of this calculation, we permeate our Al sample with liquid gallium (Ga; Supplementary Information). Permeation of Ga into Al is intergranular and leads to the disaggregation of the microstructure into constituent grains without altering the grain shapes.[33] We confirm the GB plane tilts calculated using EDOR by imaging individual grains using optical microscopy. Three illustrative grains are shown in Figure 3b. The measurement error, Δa, in the computed GB plane tilt may arise from neglecting possible out-of-plane curvature of the boundary plane, from uncertainties in the evaluation of the GB trace distance, Δd, and from local variations in sample thickness, Δt, that result from non-uniform grinding during sample preparation. We inspect selected grain facets by scanning electron microscopy and find that the out-of-plane curvature is negligible over the length of the GB (see Supplementary Figure S14 in the Supplementary Information). Following equation (1), we have computed an error of $\Delta a = \frac{\sqrt{t^2\Delta d^2 + d^2\Delta t^2}}{(d^2+t^2)} \approx 3°$ using conventional uncertainty propagation techniques (Supplementary Information).

In the second step of EDOR, we find GB misorientations. The top side of the sample is prepared for EBSD characterisation (see the Methods section). The EBSD data set is imported into MATLAB through MTEX,[34] a toolbox used to compute misorientation angles θ and axes $\hat{o}$ of each GB from the crystallographic orientation of the adjoining grains. In principle, EBSD or other electron-based microscopy techniques, such as orientation contrast imaging,[35] may also be conducted on both sides of the sample and then used to reconstruct the GB network—similar to how ORM operates in EDOR—or else performed on one side of the sample only and then combined with Monte Carlo simulations to infer the likeliest GB plane normal vector.[36] These approaches are advantageous in that they all use scanning electron microscopy and do not require registration of different data sets. However, they are mostly suitable for small-size samples and become impractical for large samples due to EBSD's and orientation contrast imaging's limited field of view, slow scanning speed and large image distortion.[37] For these reasons, EDOR relies on optical- rather than electron-based techniques to assess GB plane normal vectors.

Finally, in the third step of EDOR, the EBSD data (θ and $\hat{o}$ as well as the orientations of the two adjacent grains, O_A and O_B) are automatically assigned to the corresponding GB plane normal vector $\hat{n}$, such that for each GB both the misorientation and plane normal vector are specified. This procedure is detailed in the Supplementary Information. The grain orientations obtained with EBSD are used to rotate the GB plane normal vector $\hat{n}$ from the lab frame into the frames of the individual grains: $\hat{n}_A$ and $\hat{n}_B$:

$$\begin{aligned} \hat{n}_A &= O_A^{-1} \cdot \hat{n} \\ \hat{n}_B &= O_B^{-1} \cdot \hat{n} \end{aligned} \tag{3}$$

$\hat{n}_A$ and $\hat{n}_B$ are representations of the same GB plane, expressed with respect to reference frames fixed to the two adjoining crystal grains. In this paper, we use both $\hat{n}_A$ and $\hat{n}_B$ for representing each measured GB plane. We note that a unique representation of the GBC—considering both misorientation and boundary plane—can be found after considering all symmetry operations that stem from the underlying crystals.[38] The implications of using such a representation on the formulation of GB constitutive relations will be discussed in a future publication.

DISCUSSION

Using EDOR, we found the complete GBC of 976 GBs from three Al samples with cumulative area of ~ 1.4 cm². Figure 4 presents the disorientations—using the axis-angle parameterisation—(Figure 4a) and boundary plane orientations (Figure 4b) of these GBs,

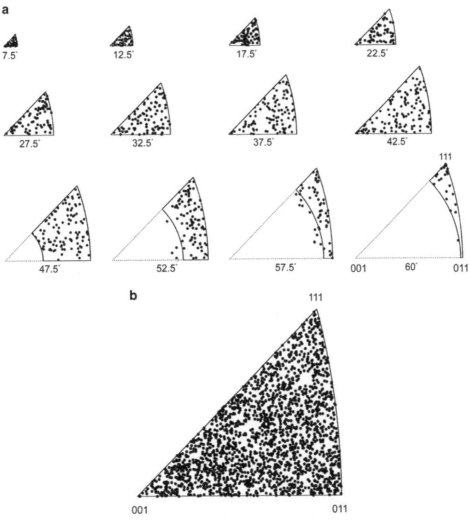

Figure 4. Visualisation of the GBC distribution as measured by EDOR in polycrystalline Al. (**a**) GB misorientation distribution. The plot shows progressive slices of constant θ through the asymmetric fundamental zone for misorientations in cubic crystals. Each section comprises GBs within the specified range of $\theta \pm 2.5°$ (indicated in degrees below each slice).[43] (**b**) Distribution of the GB plane normal vectors plotted in the standard stereographic triangle for cubic symmetry.

showing an approximately uniform and diverse sampling of GBC. The misorientation distribution function of one representative Al sample is also reported in the Supplementary Information. All three Al samples remained undamaged by these measurements and were therefore available for subsequent measurements of GB properties. These results demonstrate that EDOR is a reliable, high-throughput and non-destructive technique to assess the GBC of numerous individual GBs in polycrystalline materials. It therefore provides a critical capability for finding GB crystallography-property relationships, $\phi = F(\mathbf{R}, \hat{n})$. Owing to its hybrid nature—which combines ORM and EBSD—EDOR has higher versatility than other optical techniques and may therefore be applied to a wider range of materials.[39] As it relies on commercially available equipment and software, EDOR may be widely adopted by many groups to study GB crystallography–property relations. The MATLAB scripts for EDOR as well as the raw data presented in this paper are provided in the Supplementary Information. As a proof of concept, we employed EDOR on Al samples created by repeated rolling and annealing. However, EDOR may also be employed on other types of samples, with grain size resolvable by ORM and sample thickness smaller than the average grain size. Modified versions of EDOR may be employed to characterise nanostructured materials or thin films by means of transmission electron microscopy,[40,41] or as an *in situ* characterisation tool in

layer-by-layer additive manufacturing.[42] Further improvements to EDOR will aim at retrieving GB misorientations using ORM and at meshing the surface of curved GBs to extract variable GB plane normals.[27] These upgrades will significantly improve the efficiency of EDOR at sampling the GB crystallography space, by removing the need for EBSD—the most time-intensive step in EDOR—and maximising the number of detected GBs per sample.

MATERIALS AND METHODS

Al samples with through-thickness grains were produced from a 500-g high-purity Al (99.999%) master ingot. A ~20-mm-thick foil was cut from the master ingot and underwent a series of rolling and annealing cycles until the foil was ~600-μm thick. Each cycle consisted of a 50% thickness reduction, chemical etching in 1 mol/l NaOH solution at 80 °C for 15 min, and annealing at 620 °C for 20 min. The amount of thickness reduction was tailored to ensure complete recrystallisation upon annealing, while limiting texture formation and minimising work hardening. The chemical etching was used to remove the native oxide film that may form at each cycle and contaminate the rollers. The annealing temperature was chosen near the melting temperature of Al to maximise the kinetics of grain growth. Both sides of the Al samples with through-thickness grains were then mechanically polished using a sequence of SiC paper from 2,400 to 4,000 grit size. Subsequently, the samples were immersed in a bath of 1 N NaOH for 30 min in stirring condition to reveal the crystallographic facets of the constituent grains. After having characterised the microstructure by optical reflectance, one side of the sample (referred to as the top side) was further polished using a colloidal silica suspension (0.04 µm) to a mirror-like quality and then characterised by EBSD.

ACKNOWLEDGEMENTS

We acknowledge O.K. Johnson for providing help with the MATLAB code and for stimulating discussions. This work was supported by the US Department of Energy, Office of Basic Energy Sciences under award no DE-SC0008926. M.V. was supported by the MISTI Seed Fund 'Progetto Roberto Rocca.' I.M. and J.E. were supported by the National Science Foundation under grant DMR-1003901. Access to shared experimental facilities was provided by the MIT Center for Materials Science Engineering.

CONTRIBUTIONS

The project was planned and supervised by M.S., M.J.D., M.V.D. and C.A.S. EDOR was conceived by M.S. Data collection and analysis were performed by M.S. and M.V. The MATLAB code was written by M.S. and S.P. Samples were designed by J.E. and prepared by I.M. The manuscript was prepared by M.S., S.P. and M.J.D.

COMPETING INTERESTS

The authors declare no conflict of interest.

REFERENCES

1. Lu, K., Lu, L. & Suresh, S. Strengthening materials by engineering coherent internal boundaries at the nanoscale. *Science* **324**, 349–352 (2009).
2. Lu, L., Chen, X., Huang, X. & Lu, K. Revealing the maximum strength in nanotwinned copper. *Science* **323**, 607–610 (2009).
3. Wei, Y. J. *et al.* Evading the strength- ductility trade-off dilemma in steel through gradient hierarchical nanotwins. *Nat. Commun.* **5**, 3580 (2014).
4. Minnich, A. J., Dresselhaus, M. S., Ren, Z. F. & Chen, G. Bulk nanostructured thermoelectric materials: current research and future prospects. *Energy Environ. Sci.* **2**, 466–479 (2009).
5. Cook B. A., Kramer M. J., Wei X., Harringa J. L. & Levin E. M. Nature of the cubic to rhombohedral structural transformation in (AgSbTe2)(15)(GeTe)(85) thermoelectric material. *J. Appl. Phys.* **101**, 053715 (2007).
6. Suzuki, A. & Mishin, Y. Diffusion mechanisms in grain boundaries in conventional and nanostructured materials. *Minerals, Metals & Materials Soc.* 43–60 (2004).
7. Chen, Y. & Schuh, C. A. Diffusion on grain boundary networks: percolation theory and effective medium approximations. *Acta Mater.* **54**, 4709–4720 (2006).
8. Seita, M., Hanson, J. P., Gradecak, S. & Demkowicz, M. J. The dual role of coherent twin boundaries in hydrogen embrittlement. *Nat. Commun.* **6**, 6164 (2015).
9. Lehockey, E. M. *et al.* On improving the corrosion and growth resistance of positive Pb-acid battery grids by grain boundary engineering. *J. Power Sources* **78**, 79–83 (1999).
10. Shen T. D. *et al.* Enhanced radiation tolerance in nanocrystalline MgGa2O4. *Appl. Phys. Lett.* **90**, 263115 (2007).
11. Uberuaga, B. P., Vernon, L. J., Martinez, E. & Voter, A. F. The relationship between grain boundary structure, defect mobility, and grain boundary sink efficiency. *Sci. Rep.* **5**, 9095 (2015).
12. Sutton, A. P. & Balluffi, R. W. *Interfaces in crystalline materials.* (Oxford University Press, New York, NY, USA, 2007).
13. Rohrer, G. S. The distribution of grain boundary planes in polycrystals. *JOM* **59**, 38–42 (2007).
14. Kaur, I., Mishin, Y. & Gust, W. *Fundamentals of Grain and Interphase Boundary Diffusion.* (John Wiley & Sons, Chirenchester, UK, 1995).
15. Mishin, Y., Asta, M. & Li, J. Atomistic modeling of interfaces and their impact on microstructure and properties. *Acta Mater.* **58**, 1117–1151 (2010).
16. Rohrer, G. S. *et al.* Comparing calculated and measured grain boundary energies in nickel. *Acta Mater.* **58**, 5063–5069 (2010).
17. Olmsted, D. L., Holm, E. A. & Foiles, S. M. Survey of computed grain boundary properties in face-centered cubic metals-II: Grain boundary mobility. *Acta Mater.* **57**, 3704–3713 (2009).
18. Olmsted, D. L., Foiles, S. M. & Holm, E. A. Survey of computed grain boundary properties in face-centered cubic metals: I. Grain boundary energy. *Acta Mater.* **57**, 3694–3703 (2009).
19. Holm, E. A. & Foiles, S. M. How Grain Growth Stops: A Mechanism for Grain-Growth Stagnation in Pure Materials. *Science* **328**, 1138–1141 (2010).
20. Hugo, R. C. & Hoagland, R. G. The kinetics of gallium penetration into aluminum grain boundaries—In situ TEM observations and atomistic models. *Acta Mater.* **48**, 1949–1957 (2000).
21. Han, W. Z., Demkowicz, M. J., Fu, E. G., Wang, Y. Q. & Misra, A. Effect of grain boundary character on sink efficiency. *Acta Mater.* **60**, 6341–6351 (2012).
22. Vattre, A. J., Abdolrahim, N., Kolluri, K. & Demkowicz, M. J. Computational design of patterned interfaces using reduced order models. *Sci. Rep.* **4**, 6231 (2014).
23. Adams, B. L., Wright, S. I. & Kunze, K. Orientation imaging—the emergence of a new microscopy. *Metall. Trans. A* **24**, 819–831 (1993).
24. Weightman, P., Martin, D. S., Cole, R. J. & Farrell, T. Reflection anisotropy spectroscopy. *Rep. Prog. Phys.* **68**, 1251–1341 (2005).
25. Rohrer, G. S. The distribution of internal interfaces in polycrystals. *Z. Metallkd.* **95**, 197–214 (2004).
26. Khorashadizadeh, A. *et al.* Five-Parameter Grain Boundary Analysis by 3D EBSD of an Ultra Fine Grained CuZr Alloy Processed by Equal Channel Angular Pressing. *Adv. Eng. Mater.* **13**, 237–244 (2011).
27. Lieberman, E. J., Rollett, A. D., Lebensohn, R. A. & Kober, E. M. Calculation of grain boundary normals directly from 3D microstructure images. *Model. Simul. Mater. Sci.* **23**, 035005 (2015).
28. King, A., Johnson, G., Engelberg, D., Ludwig, W. & Marrow, J. Observations of intergranular stress corrosion cracking in a grain-mapped polycrystal. *Science* **321**, 382–385 (2008).
29. Poulsen, H. F. *et al.* Three-dimensional maps of grain boundaries and the stress state of individual grains in polycrystals and powders. *J. Appl. Crystallogr.* **34**, 751–756 (2001).
30. Suter R. M., Hennessy D., Xiao C. & Lienert U. Forward modeling method for microstructure reconstruction using x-ray diffraction microscopy: Single-crystal verification. *Rev. Sci. Instrum.* **77**, 123905 (2006).
31. Rupert, T. J., Gianola, D. S., Gan, Y. & Hemker, K. J. Experimental observations of stress-driven grain boundary migration. *Science* **326**, 1686–1690 (2009).
32. Canny, J. A computational approach to edge detection. *IEEE Trans. Pattern Anal. Mach. Intell.* **PAMI-8**, 679–698 (1986).
33. Joseph, B., Picat, M. & Barbier, F. Liquid metal embrittlement: a state-of-the-art appraisal. *Eur. Phys. J. Appl. Phys.* **5**, 19–31 (1999).
34. Bachmann, F., Hielscher, R. & Schaeben, H. Texture analysis with MTEX—free and open source software toolbox. *Solid State Phenom.* **160**, 63–68 (2010).
35. Prior, D. J., Trimby, P. W., Weber, U. D. & Dingley, D. J. Orientation contrast imaging of microstructures in rocks using forescatter detectors in the scanning electron microscope. *Mineral. Mag.* **60**, 859–869 (1996).
36. Sorensen, C., Basinger, J. A., Nowell, M. M. & Fullwood, D. T. Five-parameter grain boundary inclination recovery with EBSD and interaction volume models. *Metall. Mater. Trans. A* **45**, 4165–4172 (2014).
37. Pilchak, A. L., Shively, A. R., Shade, P. A., Tiley, J. S. & Ballard, D. L. Using cross-correlation for automated stitching of two-dimensional multi-tile electron backscatter diffraction data. *J. Microsc.* **248**, 172–186 (2012).
38. Patala, S. & Schuh, C. A. Symmetries in the representation of grain boundary-plane distributions. *Philos. Mag.* **93**, 524–573 (2013).
39. Heilbronner, R. P. & Pauli, C. Integrated spatial and orientation analysis of quartz c-axes by computer-aided microscopy. *J. Struct. Geol.* **15**, 369–382 (1993).
40. Ghamarian, I., Liu, Y., Samimi, P. & Collins, P. C. Development and application of a novel precession electron diffraction technique to quantify and map deformation structures in highly deformed materials-as applied to ultrafine-grained titanium. *Acta Mater.* **79**, 203–215 (2014).
41. Rauch, E. F. & Veron, M. Automated crystal orientation and phase mapping in TEM. *Mater. Charact.* **98**, 1–9 (2014).
42. Dinwiddie, R. B., Dehoff, R. R., Lloyd, P. D., Lowe, L. E. & Ulrich, J. B. Thermographic in-situ process monitoring of the electron beam melting technology used in additive manufacturing. *Proc. SPIE* **8705K**, 1–8 (2013).
43. Patala, S., Mason, J. K. & Schuh, C. A. Improved representations of misorientation information for grain boundary science and engineering. *Prog. Mater. Sci.* **57**, 1383–1425 (2012).

15

Predictive modelling of ferroelectric tunnel junctions

Julian P Velev[1,2], John D Burton[1,3], Mikhail Ye Zhuravlev[4,5] and Evgeny Y Tsymbal[1,3]

Ferroelectric tunnel junctions combine the phenomena of quantum-mechanical tunnelling and switchable spontaneous polarisation of a nanometre-thick ferroelectric film into novel device functionality. Switching the ferroelectric barrier polarisation direction produces a sizable change in resistance of the junction—a phenomenon known as the tunnelling electroresistance effect. From a fundamental perspective, ferroelectric tunnel junctions and their version with ferromagnetic electrodes, i.e., multiferroic tunnel junctions, are testbeds for studying the underlying mechanisms of tunnelling electroresistance as well as the interplay between electric and magnetic degrees of freedom and their effect on transport. From a practical perspective, ferroelectric tunnel junctions hold promise for disruptive device applications. In a very short time, they have traversed the path from basic model predictions to prototypes for novel non-volatile ferroelectric random access memories with non-destructive readout. This remarkable progress is to a large extent driven by a productive cycle of predictive modelling and innovative experimental effort. In this review article, we outline the development of the ferroelectric tunnel junction concept and the role of theoretical modelling in guiding experimental work. We discuss a wide range of physical phenomena that control the functional properties of ferroelectric tunnel junctions and summarise the state-of-the-art achievements in the field.

INTRODUCTION

Electron tunnelling refers to the ability of electrons to traverse potential barriers exceeding their energy.[1] This phenomenon is at the core of devices known as tunnel junctions, which consist of a nanometre-thick insulating layer separating two metallic electrodes. Significant interest in electron tunnelling has been triggered by the advent of spintronics—a branch of electronics using the electron spin in data storage and processing.[2] The magnetic tunnel junction (MTJ) is the staple of spintronics.[3] A MTJ exploits the switchable magnetisation of the two ferromagnetic (FM) metal electrodes. Changing their magnetic configuration from parallel to antiparallel, e.g., by an applied magnetic field, causes a large change in tunnelling resistance of the junction, an effect known as tunnelling magnetoresistance (TMR).[4]

In MTJs, the role of the barrier layer is passive: it separates the FM electrodes so that their magnetisations can be switched independently. Using the barrier layer as an active element of the device to control the charge and spin transport could be advantageous. This possibility is offered by complex oxide materials, which exhibit a wide range of properties, such as ferroelectricity, magnetoelectricity and metal–insulator transitions.[5,6] These properties are exploited in ferroelectric and multiferroic tunnel junctions (MFTJs).[7,8]

A ferroelectric tunnel junction (FTJ) consists of two metal electrodes separated by a nanometre-thick ferroelectric barrier layer, as illustrated in Figure 1a. The key feature of bulk ferroelectric materials is the spontaneous electric polarisation that can be switched between at least two stable orientations by applying an external electric field.[9] In most cases, the ferroelectric crystal structure represents a small distortion of a high-symmetry paraelectric structure, as in the case of $BaTiO_3$ (BTO), where the

ferroelectric state corresponds to displacement of the Ti atom from the centrosymmetric position (Figure 1b). The perovskite oxides of the BTO family have a simple crystal structure that makes them relatively easy to grow. Moreover, as by symmetry all ferroelectrics are also piezoelectric and pyroelectric, these materials find widespread use in technological applications such as actuators and transducers. There are several other families of more complex ferroelectric oxides, which, however, have been less studied.[9] Recently, organic ferroelectrics are also becoming mainstream.[10]

Although the FTJ concept was proposed long ago,[11] it was contingent on the possibility to grow nanometre-thick ferroelectric films. Until recently, it was commonly accepted that there is a macroscopic critical thickness for ferroelectricity and that the ferroelectric polarisation would be suppressed by the depolarising field in nanometre-sized films.[12,13] However, theoretical modelling[14] and experimental work[15-17] have demonstrated that when the ferroelectric is interfaced with a metal, the depolarising field is reduced due to screening of the polarisation charges and ferroelectricity can be maintained in nanometre-thick films. These developments paved the way for using ferroelectrics as barriers in tunnel junctions.

The signature property of the FTJ is that the reversal of the electric polarisation of the ferroelectric barrier in a FTJ produces a sizable change in resistance of the junction, as illustrated in Figures 1c and d. This phenomenon is known as the tunnelling electroresistance (TER) effect.[18-20] FTJs are interesting from the point of view of device applications. Contrary to the ferroelectric capacitors, where leakage currents are detrimental to the device performance, the conductance of a FTJ is the functional characteristic of the device. This property allows using FTJs in

[1]Department of Physics and Astronomy, University of Nebraska, Lincoln, NE, USA; [2]Department of Physics and Astronomy, University of Puerto Rico, San Juan, Puerto Rico, USA; [3]Nebraska Center for Materials and Nanoscience, University of Nebraska, Lincoln, NE, USA; [4]Kurnakov Institute for General and Inorganic Chemistry, Russian Academy of Sciences, Moscow, Russia and [5]Faculty of Liberal Arts and Sciences, St Petersburg State University, St Petersburg, Russia.
Correspondence: EY Tsymbal (tsymbal@unl.edu)

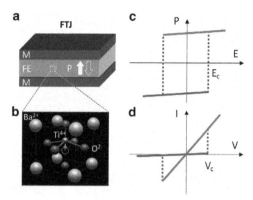

Figure 1. Schematic view of FTJ and the TER effect. (**a**) FTJ consisting of a nanometre-thick ferroelectric (FE) barrier between two metal (M) electrodes. (**b**) Crystal structure of the prototypical ferroelectric BaTiO₃ showing the two polarisation states. (**c**, **d**) TER effect showing the correlation between switching of the polarisation of the ferroelectric in applied electric field (**c**) with the resistance of the FTJ (**d**). The low (high)-resistance states are indicated by red (blue) lines.

non-volatile memory devices that are superior to the existing ferroelectric random access memories.[21] An extension of the functionality of FTJ is achieved by making the electrodes FM, which makes it into a MFTJ.[22] In a MFTJ, which can alternatively be thought of as a MTJ with a ferroelectric barrier, the TER and TMR effects coexist resulting in a four-state resistive device.[23,24] In the past decade, we have witnessed marked progress in the elaboration of the properties of FTJs and MFTJs.[25,26] This progress has been the result of a remarkable positive-feedback loop between theoretical modelling and experimental effort, during which theoretical predictions, based on both analytical and first-principles calculations, guided experimental effort, whereas experimental observations revealed novel phenomena stimulating theoretical work.

In this review, we outline this synergetic development with an emphasis on phenomena driven by the control of barrier polarisation in FTJs. In general, switching of the polarisation in the ferroelectric barrier affects the resistance of the FTJ in two principal ways: (i) modulation of the barrier height or (ii) modulation of the barrier width. All the mechanisms of TER that are discussed below fall in one of these two categories. The modulation of the barrier height typically results from asymmetry of the FTJ due to different electrode materials (section 'Effects of electrostatic screening'), different atomic structure of the two interfaces ('Effects of interface termination and interlayer') and/or finite bias ('Effects of finite bias'). In real materials, change in the barrier height is also associated with the complex band structure (CBS) modification ('Effects of the electronic structure'). The modulation of the barrier width can be realised through the Schottky barrier at the interface and barrier metallisation ('Effects of the Schottky barrier and barrier metallization') or phase transitions at the interface ('Effects of phase transitions at the interface'). We discuss all these mechanisms in some detail, providing links between theoretical modelling and experimental results where appropriate. In the section 'Multiferroic tunnel junctions', we briefly outline some results for FTJs with FM electrodes (i.e., MFTJs).

MECHANISMS OF TER AND THEIR EXPERIMENTAL MANIFESTATION

Effects of electrostatic screening

In thin ferroelectric films, the interface polarisation charges are not completely screened by the adjacent metal electrodes and therefore the depolarising electric field E in the ferroelectric is not

zero.[27] The electrostatic potential associated with this field depends on the direction of the ferroelectric polarisation. If a FTJ is made of metal electrodes with different screening lengths, the asymmetry in the electrostatic potential profile alters an effective barrier height when the ferroelectric polarisation is reversed. This leads to the TER effect.[7]

Potential profile. Quantitatively, the magnitude and the shape of the electrostatic potential profile across the junction are determined by the polarisation P, the background dielectric permittivity ε_{FE}[28] and the thickness d of the ferroelectric film, as well as the screening length λ of the electrodes. Within the Thomas–Fermi model, the ferroelectric polarisation charge at the interface is screened within the screening length $\lambda = \frac{1}{e}\sqrt{\varepsilon/\rho}$ (where ρ is the density of states at the Fermi energy and ε is the dielectric permittivity of the electrode), resulting in the exponential decay of the potential into the electrode.[18] The magnitude of the electrostatic potential at the left (L) and right (R) interfaces is given by[24]

$$\phi_{L,R} = \pm \frac{\gamma_{L,R}Pd}{d + \varepsilon_{FE}(\gamma_L + \gamma_R)}, \tag{1}$$

where $\varepsilon_{L,R}$ are the electrode dielectric permittivities, $\gamma_{L,R} = \lambda_{L,R}/\varepsilon_{L,R}$ is the normalised screening lengths and the positive (negative) sign corresponds to L (R) interfaces. For not too thick ferroelectric films, $d \sim \varepsilon_{FE}(\gamma_L + \gamma_R)$, as in the case of FTJs, the potential profile depends on characteristics of the whole tunnel junction.

Figures 2a,b shows a representative electrostatic potential energy ($e\phi$) profile, calculated for two opposite polarisation directions in a FTJ with two different metals electrodes, one being a 'good' metal and the other a 'bad' metal.[24] There is a notable difference in the magnitude of the electrostatic energy step at the interfaces, leading to a sign change in the average electrostatic potential across the barrier region when polarisation is reversed from right ($\rightarrow$) to left ($\leftarrow$). As a result, the average potential barrier height changes from $U_\rightarrow = U_0 + e(\phi_L + \phi_R)/2$ to $U_\leftarrow = U_0 - e(\phi_L + \phi_R)/2$. This is evident from Figures 2c,d showing the tunnelling potential energy profile for two opposite polarisation directions.

Conductance and TER. This difference in the effective barrier height leads to a change in the tunnelling conductance, G. The measure of this change is the TER ratio, which we define here as TER $= G_\leftarrow/G_\rightarrow$ (same as the ON/OFF resistance ratio). At the most basic level, the tunnelling current through an asymmetric barrier can be described within the Wentzel–Kramers–Brillouin approximation, in which the exact form of the potential barrier is approximated by its average value.[29,30] Within the Wentzel–Kramers–Brillouin approximation, TER can be written explicitly in terms of ferroelectric polarisation and screening lengths in the electrodes[31,32]

$$TER \approx \exp\left[\frac{\sqrt{2m}}{\hbar}\frac{\Delta U}{\sqrt{U_0}}\right] = \exp\left[\frac{e}{\hbar}\sqrt{\frac{2m}{U_0}}\frac{P(\gamma_L - \gamma_R)d^2}{d + \varepsilon_{FE}(\gamma_L + \gamma_R)}\right], \tag{2}$$

where m is the electron effective mass in the barrier and $\Delta U = U_\leftarrow - U_\rightarrow$ is the barrier height change upon polarisation reversal (derived assuming $\Delta U \ll U_0$). As seen from equation (2), TER is expected to depend stronger than exponentially on the barrier width d. For the parameters used in Figures 2a–d, we find $\Delta U \approx 0.4$ eV and a steep increase of TER as a function of d (Figure 2e), resulting in TER $\sim 10^3$ for a reasonable barrier width of ~ 2.5 nm and polarisation $P = 40$ μC/cm². Also from equation (2), TER depends exponentially on ferroelectric polarisation P, which dependence is also shown in Figure 2e. Therefore, the enhancement of polarisation magnitude and its stability is critical for observing a large TER in FTJs. The TER ratio increases with increasing 'asymmetry' of the electrodes as determined by γ_L/γ_R. In

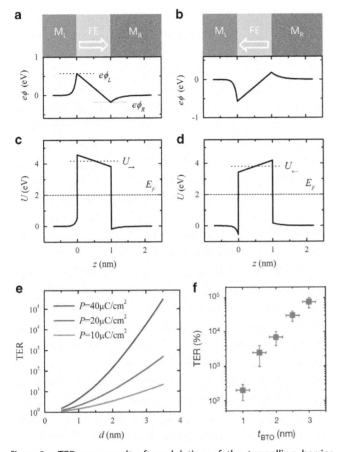

Figure 2. TER as a result of modulation of the tunnelling barrier height by ferroelectric polarisation switching. (**a, b**) Electrostatic potential energy profile across a FTJ for ferroelectric polarisation pointing right (**a**) and left (**b**) as shown on top panels by arrows. FTJ consists of two different metals (M_L and M_R, located at $z < 0$ and $z > d$, respectively) separated by a ferroelectric tunnelling barrier (FE), which is located at $0 < z < d$ (top panels). ϕ_L and ϕ_R denote potential steps at left and right interfaces, respectively. It is assumed that $\lambda_L = 0.07$ nm, $\varepsilon_L = \varepsilon_0$, $\lambda_R = 0.24$ nm, $\varepsilon_R = 10 \ \varepsilon_0$, $\varepsilon_{FE} = 50 \ \varepsilon_0$ and $P = 40 \ \mu C/cm^2$. (**c, d**) Potential energy profiles across the FTJ for ferroelectric polarisation pointing right (**c**) and left (**d**). $U_\rightarrow$ and $U_\leftarrow$ indicate average potential barrier heights for the two opposite polarisations. The dashed lines denote the Fermi energy E_F. (**e**) Calculated TER as a function of barrier thickness d for different polarisation P of the ferroelectric barrier. It is assumed that $d = 1$ nm, $E_F = 2.0$ eV and $U_0 = 2.0$ eV. (**f**) Measured TER of highly strained BaTiO$_3$ (BTO) films deposited on La$_{0.7}$Sr$_{0.3}$MnO$_3$ electrodes versus BTO film thickness (t_{BTO}) as obtained from conductive AFM measurements. After ref. 33 with permission.

the limit $\gamma_L/\gamma_R \gg 1$, for not too large thickness ($d \ll \varepsilon_{FE}\gamma_L$), the TER ratio approaches value of $\text{TER} \approx \exp\left[\left(e\sqrt{2mP}d^2\right)/\left(\varepsilon_{FE}\hbar\sqrt{U_0}\right)\right]$, which does not depend on electrode properties.

Experimentally, the TER effect was first observed with uncovered ferroelectric films, where local probe techniques were used to control the ferroelectric polarisation orientation and measure the tunnelling current through the film.[31,33–35] In particular, it was demonstrated that highly strained BTO films retain robust room-temperature ferroelectricity down to 1 nm.[33] A large resistance change was observed in these films correlated with ferroelectricity, as revealed from comparing images of ferroelectric domains and resistance maps of these domains. The resistance change between polarisation states was found to increase exponentially with BTO thickness, reaching 750 for 3-nm films at room temperature (Figure 2f). Fitting of the device I–V curves with the Brinkman model revealed a change of the average

barrier height of the order of tenths of eV associated with polarisation reversal,[31,34] which is consistent with the model predictions.

In addition, there have been several successful demonstrations of the TER effect using patterned ferroelectric films with top electrodes.[36,37] In particular, tunnel junctions based on 2-nm-thick BTO films grown on La$_{0.7}$Sr$_{0.3}$MnO$_3$ (LSMO) bottom electrodes with Au/Co top electrodes demonstrated robust ferroelectricity of BTO inside the junction.[37] The remanent resistance of the device was shown to change hysteretically with ON/OFF ratios of ~ 100 and coercive fields matching ferroelectric switching. Fitting of the I–V characteristics indicated that mechanisms based on the modulation of tunnel transmission through variation of the barrier height by ferroelectric polarisation are responsible for the TER effect.

Role of electrode work function. The model discussed above can be extended to include the possibility of different work functions of the two electrodes. In this case, different potential steps are expected at the two interfaces producing an electric field in the barrier even in the absence of ferroelectric polarisation. This field is screened by interface charges that are formed to balance the electrochemical potentials in the electrodes, as required by the short-circuit boundary condition, resulting in a change in the effective tunnelling barrier height from U_0 to

$$\tilde{U}_0 = U_0 + \frac{\Delta W}{2} \frac{\gamma_L - \gamma_R}{d + \varepsilon_{FE}(\gamma_L + \gamma_R)}, \tag{3}$$

where $\Delta W = W_L - W_R$ is the difference of the work functions of the left and right electrodes. It is seen that the effective barrier height can increase or decrease depending on sign of ΔW, which is expected to affect TER according to equation (2).

Experimentally, different potential steps at the two interfaces have been measured in Cr/BTO/Pt FTJs by reconstructing the electrostatic potential profile by hard X-ray photoemission spectroscopy.[38] The effect of electrode's work function on TER was explored for M/BiFeO$_3$/Ca$_{0.96}$Ce$_{0.04}$MnO$_3$ FTJs with different top metal electrodes (M = W, Co, Ni and Ir).[39] Fitting of the I–V characteristics indicated that the M/FE interfacial barrier height increases with the electrode work function. This led to an increased resistance in the OFF state and a larger ON/OFF ratio for top electrodes with a larger work function. The results, however, also indicated that the higher TER comes at the cost of deteriorated switching characteristics. A large difference in TER was also observed for M/BTO/LSMO (M = Au and Cu) FTJs.[40]

Effects of the Schottky barrier and barrier metallisation

Another approach to produce a sizable TER effect is to modulate the barrier width by switching ferroelectric polarisation in a FTJ. This may be achieved by a reversible depletion/accumulation of charge in an interfacial region of a semiconducting electrode (modulation of the Schottky barrier) and/or reversible metallisation of an interfacial region in the barrier itself.

Schottky barrier. In a FTJ with a semiconducting electrode, which is characterised by a small Fermi energy and large screening length, the barrier width can be controlled by ferroelectric polarisation through the Schottky barrier. For example, if one of the electrodes in a FTJ is an n-type semiconductor and ferroelectric polarisation is pointing away from the semiconductor, electron carriers are depleted in the semiconductor adding a Schottky barrier to the ferroelectric tunnel barrier (Figure 3b). On the other hand, when the polarisation is pointing towards the semiconductor, electron carriers are accumulated which only leads to a local increase of the chemical potential, but does not change the barrier width in the FTJ (Figure 3a).

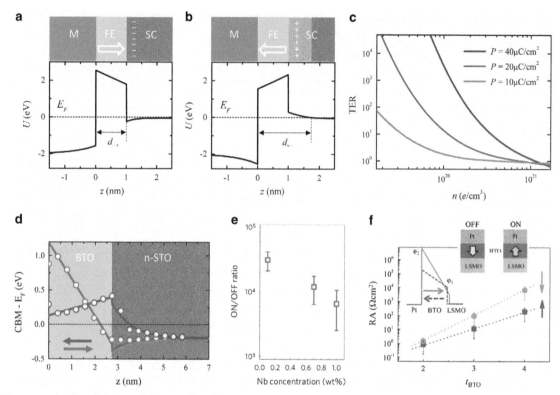

Figure 3. TER as a result of modulation of the tunnelling barrier width by ferroelectric polarisation switching. (**a, b**) Calculated potential energy profiles for ferroelectric polarisation pointing right (**a**) and left (**b**) as shown on top panels by arrows. FTJ consists of a metal electrode (M, located at $z < 0$) and a semiconducting electrode (SC, located at $z > d$), separated by a ferroelectric tunnelling barrier (FE), which is placed at $0 < z < d$ (top panels). The dashed lines denote the Fermi energy E_F. $d_\rightarrow$ and $d_\leftarrow$ indicate the average potential barrier width for the two opposite polarisations. It is assumed that $\lambda_M = 0.07$ nm, $\varepsilon_M = \varepsilon_0$, $d = 1$ nm, $n = 3.4 \times 10^{10}$ e/cm^3, $\varepsilon_{SC} = 10\,\varepsilon_0$, $\varepsilon_{FE} = 50\,\varepsilon_0$ and $P = 40$ µC/cm^2. (**c**) Calculated TER as a function of carrier concentration n in the semiconducting electrode for different polarisation P of the ferroelectric barrier. TER is defined as $G_\rightarrow/G_\leftarrow$. It is assumed that $d = 1$ nm, $E_F = 2$ eV and $U = 2$ eV. (**d**) Energy profile of the conduction band minimum (CMB) in SrRuO$_3$/BaTiO$_3$/n-SrTiO$_3$ tunnel junctions for two polarisation orientations (indicated by arrows) and doping concentration $n = 0.09$ e/f.u. calculated using a model (solid lines) and first-principles (dots) approach. After ref. 43. (**e**) ON/OFF ratio as a function of Nb doping concentration in Pt/BaTiO$_3$/Nb:SrTiO$_3$ FTJs. After ref. 44 with permission. (**f**) Average resistance of Pt/BaTiO$_3$/La$_{0.7}$ Sr$_{0.3}$MnO$_3$ (Pt/BTO/LSMO) tunnel junctions as a function of BTO barrier thickness. The top-left inset shows the barrier for ON and OFF states as deduced from the fits of the I–V curves. The top inset shows the relative orientation of the BTO polarisation (arrow) in the ON and OFF states. After ref. 46 with permission.

Model calculations of the conductance of a metal/ferroelectric/semiconductor tunnel junction show very large TER ratio due to modulation of the Schottky barrier. Figure 3c shows the calculated TER ratio as a function of electron concentration n in a semiconducting electrode. Here the electronic potential profile (Figures 3a,b) is obtained by solving self-consistently the Poisson equation for the junction, imposing the boundary conditions at interfaces and assuming that the local electron concentration is a linear function of the electrostatic potential.[41,42] As is seen from Figure 3c, the TER increases markedly when the carrier concentration is reduced. This reflects the increase of the Schottky barrier width for polarisation pointing into the semiconductor, as the result of the enhanced depletion region. The effect also increases with the polarisation due to a stronger electron depletion required to screen a larger polarisation charge. At a very high carrier concentration, $n \geqslant 10^{21}$ e/cm^3, a metallic regime is reached, which corresponds to the case where the change in the barrier height controls the TER ('Effects of electrostatic screening').

This result is corroborated by density functional theory (DFT) calculations for a SrRuO$_3$/BTO/n-SrTiO$_3$ FTJ, where n-doped SrTiO$_3$ (STO) serves as a semiconducting electrode.[43] When the polarisation is pointing away from n-STO, electron depletion and the associated band bending near the interface lead to an additional narrow barrier formed within the n-STO electrode (Figure 3d, blue curve and symbols). When the polarisation is pointing into n-STO,

however, the Schottky barrier is eliminated by electron accumulation (Figure 3d, red curve and symbols). Moreover, a conducting region within the BTO is formed near the interface, further reducing the tunnelling barrier width (the effect of barrier metallisation is discussed in the subsection 'Barrier metallization').

This physical mechanism for the enhanced TER explains the experimental data on Pt/BTO/Nb:STO FTJs, exhibiting resistance change as high as 10^4 at room temperature.[44] In these FTJs, Nb-doped STO single-crystal substrates with Nb varying from 0.1 to 1% were used as n-doped semiconducting electrodes. The large TER was interpreted by the depletion or accumulation of carriers in Nb:STO depending on polarisation orientation of the barrier. This mechanism was found to be consistent with the measured TER dependence on the dopant concentration: the lower Nb concentration led to the higher Schottky barrier and hence the larger TER (Figure 3e). Interesting results were obtained for Pt/La$_{1-x}$Sr$_x$MnO$_3$/BTO/Nb:STO FTJs, where Sr doping was used to tune the chemical potential of the manganite to optimise the TER effect, and surprisingly high TER ratios (~400%) were observed at room temperature for FTJs with BTO layer thickness down to 2 u.c.[45]

Barrier metallisation. In addition to the Schottky barrier in a semiconducting electrode, a metallic region within the ferroelectric barrier layer near interface may be formed, reducing the tunnelling barrier width in the ON state. The partial metallisation of the BTO barrier is seen in Figure 3d for polarisation pointing into the n-STO, where electrons are spilled from n-STO to BTO, due to their close electron affinities (3.9 and 4.0 eV, respectively). The effect of barrier metallisation was put forward to explain very large ON/OFF resistance ratios observed in Pt/BTO/LSMO[46] and Co/PbTiO$_3$ (PTO)/LSMO[47] tunnel junctions. In particular, ON/OFF ratios up to 300 were observed in Pt/BTO/LSMO tunnel junctions at room temperature.[46] On the basis of the measured capacitance change with reversal of ferroelectric polarisation of BTO, it was argued that there is an n-type semiconducting region at the BTO/LSMO interface. Upon biasing, the n-type region is driven to accumulation or depletion regimes with subsequent changes of the effective barrier width for tunnelling transport across the junction (Figure 3f). The measurement also indicated an exponential dependence of the resistance and TER on the BTO barrier width consistent with the expectation (Figure 3f). Similarly, it was found that the TER effect in a Co/PTO/LSMO tunnel junction is controlled by the change in the tunnelling barrier thickness upon reversal of polarisation of the ferroelectric PTO layer.[47] On the basis of first-principles calculations, it was argued that the ferroelectric layer exhibits a reversible metallisation at one of the interfaces.

Effects of the electronic structure

Simple, free-electron models have been essential for understanding the basic physics of FTJs and the TER effect. At the same time, they have limitations, both in the depth of the description as well as in the range of the physical mechanisms that are captured.

For example, the electrostatic profile in these models is determined by the screening length in the electrodes and the polarisation in the barrier, however, both of these are taken as parameters. Using first-principles DFT-based methods makes it possible to calculate self-consistently the distribution of the screening charge accounting for the density of states of the materials.[48] DFT calculations also give the electrostatic potential corresponding to the charge distribution, which goes beyond the Tomas–Fermi approximation used in the model calculations, and accounts for the realistic electronic structure of the materials, as well as for details of the atomic structure of the interfaces. The work functions of the electrodes are also parameters in the model; however, they are routinely calculated for different surface terminations using first-principles calculations, as the difference between the Fermi energy of the material and the vacuum energy.[49] The very existence of polarisation in ultrathin films cannot be predicted within simple models, but it is captured by first-principles calculations.[14] These calculations also capture changes in the polarisation profile in the barrier due to interface boundary conditions,[23] and take into account not only the electronic but also the ionic screening of the polarisation charge.[50,51] Furthermore, first-principles calculations capture details of the bonding and the atomic structure of the interface,[20] in particular the effect of different interface terminations and/or oxidation.[52,53] Finally, first-principles calculations account for the details of the electronic structure of the electrode and barrier materials, involving multiple bands whose orbital character and symmetry matching across the interface largely determine the transmission coefficient.

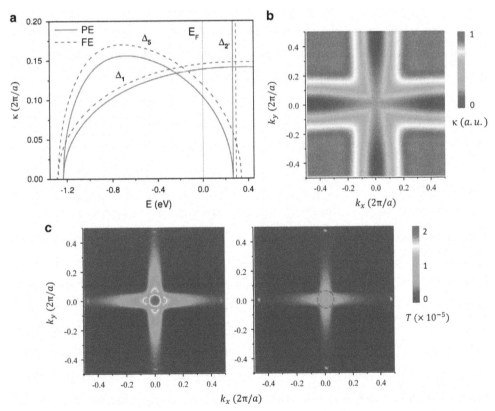

Figure 4. Complex band structure and conductance in the BaTiO$_3$-based FTJs. (**a**) Complex band structure of BaTiO$_3$ in paraelectric (PE) and ferroelectric (FE) states. The two lowest decay rates of the evanescent states of Δ_1 and Δ_5 symmetry are displayed. After ref. 20 with permission. (**b**) The lowest decay rate as a function of $\mathbf{k}_\| = (k_x, k_y)$ in the two-dimensional Brillouin zone calculated at the Fermi energy, which is indicated in **a** by a vertical line. After ref. 20 with permission. (**c**) Calculated transmission as a function of $\mathbf{k}_\|$ for a non-magnetic SrRuO$_3$/BaTiO$_3$/SrRuO$_3$ FTJ with asymmetric interfaces for polarisation pointing toward the RuO$_2$/BaO interface (left panel) and the TiO$_2$/BaO interface (right panel).

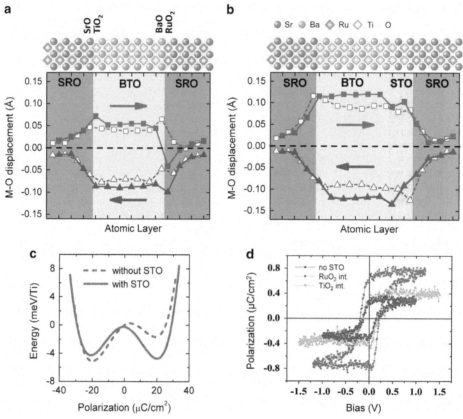

Figure 5. Atomic structure, polarisation stability and TER in $SrRuO_3/BaTiO_3/SrRuO_3$ FTJs. (**a**) Atomic structure of SRO/BTO(6 u.c.)/SRO FTJ. Displacements of the Ru and Ti cations are measured with respect to the O atoms in the same atomic plane. Red (blue) symbols correspond to polarisation of $BaTiO_3$ pointing to the right (left). Open symbols correspond to Ba–O and Sr–O displacements; closed symbols correspond to Ti–O_2 and Ru–O_2 displacements. After ref. 23. (**b**) Same as **a** for SRO/BTO(6 u.c.)/STO/SRO FTJ with a 2-u.c. STO layer at the interface. After ref. 73 with permission. (**c**) Total energy per Ti atom as a function of polarisation of the 5-u.c.-thick $BaTiO_3$ layer in SRO/BTO/SRO (dashed line) and SRO/BTO/STO/SRO (solid line) FTJs. After ref. 73. (**d**) Remanent P–V hysteresis loops in FTJs with no STO layer, STO inserted at the RuO_2/BaO interface and the TiO_2/BaO interfaces. After ref. 73 with permission.

An important effect of the electronic structure is that in complex insulators, such as STO and BTO, there is more than one evanescent state available for carriers in the bandgap.[20,54] These states have different decay rates and hence contribute differently to the tunnelling current. The CBS of the insulator determines the rate of decay of the carriers in the bandgap. Formally, CBS is an analytical continuation of the real band structure into the complex momentum space, which reveals both the propagating and evanescent states supported by the material.[55,56] The importance of the CBS and evanescent states was recognised as a powerful tool to analyse spin-transport properties of MTJs.[54,57,58]

In the case of ferroelectrics, the CBS has been calculated for BTO[20,23,52,59] and PTO.[60–63] The results show that there are multiple decay rates in the barrier for propagating states with different symmetry. The largest contribution to the transmission is expected from the bands that have lowest decay rates. For example, in the case of BTO, there are two evanescent states of Δ_1 and Δ_5 symmetry with the lowest decay rates in the bandgap (Figure 4a). For not too thick barriers, contributions to the tunnelling transmission are expected from both bands, provided that states compatible with these symmetries are available in the electrodes. It was also found that the bandgap and, respectively, the decay rates increase with the magnitude of the polarisation.[20,52] For example, comparing the decay rates for paraelectric and ferroelectric BTO (Figure 4a) shows that the structural change driven by spontaneous polarisation increases the bandgap and enhances the decay rates of both the Δ_1 and Δ_5 bands.

The CBS is useful to make qualitative conclusions about mechanisms of tunnelling conductance; however, as a bulk property of the barrier, it is only a part of the device description. Direct first-principles transport computations are required for a full quantitative analysis of the conductance and TER. These are performed using the Landauer formula,[64] which relates the conductance to the transmission probability between different propagating states in the electrodes

$$G = \frac{e^2}{h} \sum_{\sigma, \mathbf{k}_\|} T_\sigma(\mathbf{k}_\|). \tag{4}$$

Here $T_\sigma(\mathbf{k}_\|)$ is the transmission probability of the electron with spin σ at the Fermi energy and $\mathbf{k}_\| = (k_x, k_y)$ is the Bloch wave vector corresponding to the periodicity in the plane of the junction. The transmission probability is calculated either using the Green's function method[65] or the wave-function matching method[66,67] implemented in the context of DFT calculations. In both cases, the barrier with the interfaces is considered as a scattering region, connected to semi-infinite electrodes. The propagating states in one of the electrodes are scattered and transmitted into propagating states in the other electrode. Within the Green's function method, the transmission probability is expressed through the Green's function of the scattering region connected to the electrodes, G, as[68]

$$T_\sigma(\mathbf{k}_\|) = \mathrm{Tr}\left[\Gamma_L^\sigma G_{LR}^\sigma \Gamma_R^\sigma G_{RL}^{\sigma\dagger}\right] \tag{5}$$

where Γ_L (Γ_R) are the escape rates to the left (right) electrodes, which are calculated form the surface Green's function of the electrodes.[69] The trace is over all the orbital indices. Within wave-function matching method, the wave function in the electrodes is written as an incoming and reflected wave on the left and a transmitted wave to the right[67]

$$\Psi = \begin{cases} \psi_j + \sum_{i \in L} r_{ij}\psi_i, & z < 0 \\ \sum_{i \in R} t_{ij}\psi_i, & z > d \end{cases} \qquad (6)$$

where the scattering region is situated $0 < z < d$, and r_{ij} and t_{ij} are the reflection and transmission amplitudes between the Bloch states, respectively. By matching the electrode wave function with that of the scattering region at both interfaces ($z = 0$ and $z = d$), the transmission coefficients are obtained, and the transition probability is given as $T = \sum_{ij} |T_{ij}|^2$, where $T_{ij} = \sqrt{I_i / I_j} t_{ij}$ is the transmission amplitude normalised to unit flux.

An example of such full transport calculation is illustrated in Figure 4c, where the $k_\parallel$-resolved transmission in the two-dimensional Brillouin zone is plotted for a non-magnetic SrRuO$_3$/BTO/SrRuO$_3$ (SRO) FTJ for two polarisation directions. The transmission pattern seen for both polarisation states is determined by the CBS of the barrier, as follows from Figure 4b showing the $k_\parallel$-resolved lowest decay rate at the Fermi energy. It is evident from Figure 4c that, although features in the transmission distribution look similar for the two polarisation states, there is a notable difference in the overall contrast, indicating a sizable reduction in the total conductance and thus a large TER effect when polarisation is switched from pointing towards the RuO$_2$/BaO interface to the TiO$_2$/BaO interface (see section 'Interface termination' for further details). Full transport calculations have been carried out for FTJs featuring the perovskite ferroelectrics BTO[20,23,59] and PTO,[63] as well as organic ferroelectric barriers.[53,70]

Effects of interface termination and interlayer

Interface termination. The requirement of two different electrodes is not mandatory for observing TER in FTJs. Symmetry can be broken by different terminations at the two interfaces of the junction. For example, in FTJs with two SRO electrodes and BTO barrier, epitaxial unit cell by unit cell growth of the perovskite heterostructure preserves the AO–BO$_2$ (A = Sr, Ba; B = Ru, Ti) sequence of the atomic layers, resulting in SrO/TiO$_2$ termination at the left interface and BaO/RuO$_2$ at the right interface of the FTJ (Figure 5a, top). The asymmetric interfaces lead to a different polarisation profile when the ferroelectric is switched. As seen from Figure 5a, when polarisation is pointing to the left, polar displacements between metal (M) and oxygen (O) ions in BTO are larger than those when polarisation is pointing to right. This sizable change in the polarisation magnitude affects the decay rate of the states carrying tunnelling current (in this case, Δ_1 and Δ_5 symmetry states shown in Figure 4a), resulting in a large TER effect (Figure 4c).[23]

It is well known that the interface atomic relaxations impose boundary conditions on polarisation, affecting ferroelectric stability in thin films.[14,50] Asymmetric interfaces may produce a strong poling effect making one of the polarisation states preferential.[71] In the case of SRO/BTO/SRO FTJs, the built-in dipole at the BaO/RuO$_2$ interface (seen in Figure 5a from the Ru–O displacement at this interface) suppresses the polarisation when it is pointed towards this interface and for sufficiently thin BTO films leads to non-switchable polarisation.[23,72,73] The calculated total energy as a function of polarisation (Figure 5c, dashed line) displays an asymmetric double-well potential with one of the minima being very shallow due to the built-in interface dipole. Interestingly, inserting a thin layer of STO at the BaO/RuO$_2$-terminated interface counteracts this unfavourable interface dipole effect (Figure 5b). Both theory[72] and experiment[73] demonstrate that the associated change of the interface termination sequence to SrO/TiO$_2$ on both sides of the heterostructure leads to a restoration of bi-stability with a smaller critical thickness,

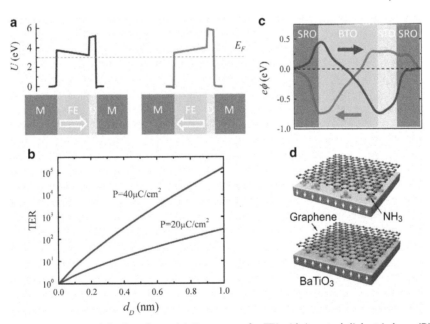

Figure 6. FTJs with a dielectric layer at one of the interfaces. (**a**) Geometry of a FTJ with inserted dielectric layer (D) and calculated potential energy profile for the two opposite polarisation orientations. The dielectric layer is described by the dielectric constant $\varepsilon_D = 10$, potential barrier height $U_D = 2.5$ eV and thickness $d_D = 0.5$ nm. Other parameters are as follows: $\lambda = 0.1$ nm, $d_{FE} = 2.5$ nm and $P = 20$ μC/cm^2. The Fermi energy (E_F) is indicated by the dashed line. (**b**) Calculated TER as a function of thickness of the dielectric interlayer d_D for two values of polarisation. Other parameters are the same as in **a**. (**c**) Electrostatic potential energy profile across SRO/BTO(6 u.c.)/STO(2 u.c.)/SRO FTJ calculated from first-principles for two polarisation orientations indicated by arrows. After ref. 76. The potential is averaged over the plane parallel to the interfaces. (**d**) Schematic of a FTJ with the top graphene electrode and the interfacial ammonia (NH$_3$) layer for ferroelectric polarisation of BaTiO$_3$ pointing up (bottom) and down (top). Courtesy of Alexey Lipatov.

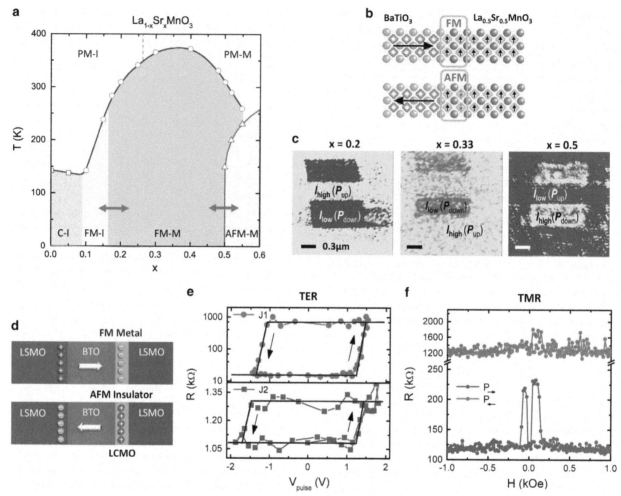

Figure 7. Polarisation controlled interfacial phase transitions. (**a**) Phase diagram showing the diversity of electronic and magnetic phases in bulk La$_{1-x}$Sr$_x$MnO$_3$. PM–I, PM–M, FM–I, FM–M, AFM–M and C–I denote paramagnetic insulator, paramagnetic metal, ferromagnetic insulator, ferromagnetic metal, antiferromagnetic metal and spin-canted insulator states, respectively. Red arrows show phase transitions that may be exploited in FTJs. After ref. 80. (**b**) Predicted FM–AFM phase transition at the interface between BTO and La$_{0.5}$Sr$_{0.5}$MnO$_3$ induced by ferroelectric polarisation reversal in BTO. After ref. 81 Horizontal arrows indicate the orientation of ferroelectric polarisation; vertical arrows show magnetic moments on Mn atoms. (**c**) Change in the tunnelling current across a PZT(10 nm)/La$_{1-x}$Sr$_x$MnO$_3$ (5 nm) bilayer at various phase boundaries of LSMO: $x = 0.2$ (left), $x = 0.33$ (middle) and $x = 0.5$ (right), measured by conductive AFM at room temperature. After ref. 87 with permission. (**d**) A schematic diagram of the LSMO/BTO/LCMO/LSMO FTJ with a thin layer (~1 nm) of La$_{0.5}$Sr$_{0.5}$MnO$_3$ (LCMO) inserted at the interface. Reversal of ferroelectric polarisation leads to hole charge depletion or accumulation in the LCMO layer resulting in its phase transition from FM metal to AFM insulator, respectively. After ref. 86 with permission. (**e**) Resistance (*R*) as a function of pulsed poling voltage (*V*$_{pulse}$) measured at 40 K for junction with (J1, TER ~ 100) and without (J2, TER ~ 1.3) LCMO interlayer. The arrows indicate the direction of pulse sequence. After ref. 86 with permission. (**f**) Resistance (*R*) as a function of magnetic field (*H*) for a junction with the LCMO interlayer in the high- (top) and low- (bottom) resistance states. After ref. 86 with permission.

along with enhancement of the barrier for polarisation reversal (Figure 5c, solid line). The effect is evident from Figure 5d showing the increase of the remanent polarisation of the FTJ when the BaO/RuO$_2$ interface is eliminated.[73]

Recently, it was demonstrated that the surface termination of the ferroelectric barrier layer in contact with a simple-metal electrode critically affects electroresistive properties.[74] Using Co/BTO/LSMO, FTJs with either TiO$_2$- or BaO-terminated BTO led to the opposite sign of TER. The TER ratio was found to be dependent on the fraction of the BaO termination at the interface, which was explained in terms of the termination-dependent depolarising field.[50]

Interface interlayer. At the same time, inserting a thin layer of dielectric at one of the interfaces in a FTJ provides asymmetry necessary for TER. Model calculations predict that TER is strongly enhanced due to the presence of the dielectric layer.[75] The effect

originates from the electrostatic potential in the dielectric being shifted up or down when the ferroelectric polarisation is switched. This changes substantially the tunnelling barrier height in the dielectric portion of the barrier (Figure 6a) resulting in a strong enhancement of TER. In the limit of perfect identical electrodes (λ_L, $\lambda_R \rightarrow 0$), we obtain within the Wentzel–Kramers–Brillouin approximation[31]

$$\text{TER} \approx \exp\left[\frac{e}{\hbar}\sqrt{\frac{2m}{U}}\frac{Pd_{FE}d_D(d_{FE}+d_D)}{(d_{FE}\varepsilon_D + d_D\varepsilon_{FE})}\right], \qquad (7)$$

where d_{FE} (d_D) and ε_{FE} (ε_D) are thickness and the electric permittivity of the ferroelectric (dielectric) layer, and U is the average height of the composite barrier when $P = 0$. It is seen from equation (7) that for small d_D TER increases exponentially, which is in agreement with the more rigorous calculation shown in Figure 6b.[75] These conclusions are consistent with the results of first-principles calculations showing that the inclusion of the STO interlayer at the BTO/SRO interface leads to the ferroelectric modulation of the electrostatic potential in the STO interlayer (Figure 6c).[76] This enhances the TER, whose magnitude is scaled with STO thickness.

Experimentally, the enhancement of TER is evident in FTJs with uncovered ferroelectric layers. Comparison of the TER results for FTJs with open and capped ferroelectric layers shows that the capped structures exhibit much more modest values of TER. This difference may be attributed to the presence of an additional dielectric layer between the local probe tip and the ferroelectric.[31,33,34] Surface chemistry also has an essential role in these and related films, and likely affects the observed switching phenomena.[77,78]

We also note that the presence of the dielectric layer at the interface strongly affects the ferroelectric polarisation stability and therefore special attention needs to be given to this issue. Clear evidence of the effect of the interfacial layer has been demonstrated recently for FTJs with graphene electrodes.[79] It was found that ferroelectric polarisation stability and resistive switching are strongly affected by a molecular layer at the graphene/BTO interface. For FTJs with an interfacial H_2O layer, only one polarisation state was stable and the TER effect was impossible to observe. In contrast, for FTJs with an interfacial NH_3 layer (Figure 6d), the polarisation was stable and switchable, and an enhanced TER effect reaching a magnitude of 6×10^3 was observed.

Effects of phase transitions at the interface

Using complex oxide materials, such as manganites $La_{1-x}A_xMnO_3$ (A = Ca, Sr or Ba), as electrodes or interlayers in FTJs brings into play new interesting phenomena involving the coupling between ferroelectricity of the barrier and magnetism of the manganite. Strongly correlated manganite materials display a rich phase diagram that consists of competing phases with different conducting properties and magnetic orderings (Figure 7a).[80] Normally, the control of different phases is achieved though chemical doping. Instead, a ferroelectric/manganite interface can be employed for electrostatic doping of the manganite with the advantage of being switchable by an applied electric field. The electrostatic doping occurs at such an interface due to polarisation charge on the ferroelectric side of the interface being screened by a build-up of opposite charge on the manganite side, effectively altering the hole concentration of the manganite near the interface. If the manganite has chemical composition x close to a phase boundary the reversal of polarisation can induce a magnetic (and/or metal–insulator) phase transition locally near the interface (red arrows in Figure 7a). For example, a FM to antiferromagnetic (AFM) phase transition was predicted theoretically for the $La_{0.5}Ba_{0.5}MnO_3$/BTO interface.[81] Because of a finite screening length in the manganite, the predicted magnetoelectric effect is confined within 2 u.c. near the interface (Figure 7b). Experimentally, the related effects have been observed in $La_{0.8}Sr_{0.2}MnO_3$/$PbZr_{0.2}Ti_{0.8}O_3$ (PZT) heterostructures.[82,83]

The electrically induced FM to AFM phase transition at the interface has important implications for FTJs. The FM metal phase in LSMO is half metallic with electronic density of states only in the majority spin channel.[84] Switching the magnetic order near the interface from FM to A-type AFM order blocks the majority spin conduction due to the spin valve effect. As the spin polarisation is nearly 100% this 'spin valve' exhibits a huge difference in conductance between parallel and antiparallel configurations, giving rise to the large TER effect. First-principles calculations for a junction consisting of BTO as a barrier with $La_{0.7}Sr_{0.3}MnO_3$ and $La_{0.6}Sr_{0.4}MnO_3$ as the left and right electrodes, predicts a factor of 18 change of conductance with polarisation reversal due to the interface magnetic phase transition.[85]

These findings were confirmed in experiments on LSMO/BTO/LCMO/LSMO FTJs with a thin layer of $La_{0.5}Ca_{0.5}MnO$ (LCMO) inserted at one of the interfaces.[86] The doping of the LCMO (x = 0.5) was intentionally chosen at the boundary between FM-metallic (x < 0.5) and AFM-insulating (x > 0.5) states. Reversal of ferroelectric polarisation was expected to produce hole charge depletion or accumulation in the LCMO layer resulting in a phase transition in the LCMO from FM metal to AFM insulator, respectively (Figure 7d). Indeed, it was found that ferroelectric polarisation switching caused the material to undergo the metal–insulator transition effectively changing the barrier width by converting part of the electrode into a barrier. This was accompanied by a giant TER ratio of about 100 measured at T = 40 K (Figure 7, top panel). For reference, LSMO/BTO/LSMO junctions without the LCMO spacer layer demonstrated a low TER ratio of just about 1.3 (Figure 7, bottom panel).

Interfacial phase transitions were also investigated experimentally using PZT/$La_{1-x}Sr_xMnO_3$ bilayers with a conductive-AFM tip as a top electrode.[87] A sign change of the TER was observed when the LSMO electrode composition was altered from x = 0.2 to 0.5 (Figure 7c). This result is consistent with the phase diagram of LSMO (Figure 7a), suggesting that at x = 0.2 hole accumulation (polarisation points away from LSMO) leads to the FM insulator to FM metal phase transition, whereas at x = 0.5 it leads to the FM metal to AFM insulator phase transition.

Effects of finite bias

The majority of theoretical effort has been directed towards elaborating the TER effect at zero bias, in which case structural asymmetry is necessary to break the degeneracy between the two polarisation directions. At the same time, applied bias is a simple and dynamically tunable way to create asymmetry in the junction.

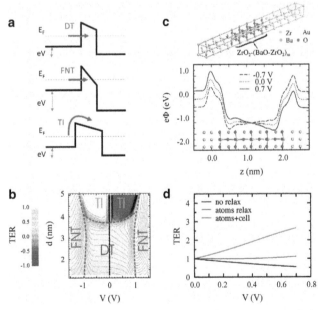

Figure 8. TER at finite bias. (**a**) Schematic representation of the different transport regimes at finite bias: direct tunnelling (DT), Fowler–Nordheim tunnelling (FNT) and thermionic injection (TI). (**b**) A contour plot of TER versus applied bias and barrier thickness for an FTJ with following parameters: $P = 3\ \mu C/cm^2$, $\varepsilon_{FE} = 60\ \varepsilon_0$, $\lambda_L = 0.8$ Å, $\lambda_R = 0.55$ Å, $\varepsilon_L = 2\varepsilon_0$, $\varepsilon_R = 8\varepsilon_0$ and $U_0 = 1.0$ eV. After ref. 88 with permission. (**c**) Schematic view of the symmetric Au/$BaZrO_3$/Au FTJ and the electrostatic potential profile in the unrelaxed junction for several voltages. After ref. 90 with permission. (**d**) ON/OFF ratio for the FTJ for the case of electronic screening, electronic and ionic screening, and screening plus piezoelectric response. After ref. 90 with permission.

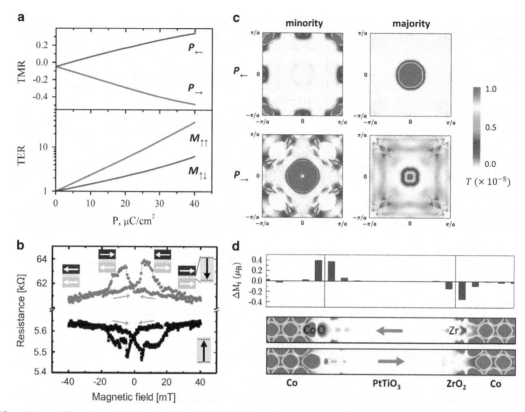

Figure 9. Multiferroic tunnel junctions. (**a**) Results of model calculations demonstrating the simultaneous presence of TER and TMR effects in a MFTJ: TMR for opposite ferroelectric polarisation orientations in the barrier (top panel), and TER for parallel and antiparallel magnetisation in the electrodes (bottom panel), as a function of ferroelectric polarisation P. After ref. 24 with permission. (**b**) Resistance as a function of magnetic field in as-grown Co/PZT(3.2 nm)/LSMO junction (black squares) and after polarisation switching (red circles) measured at $T = 50$ K. Magnetisation directions in each magnetic layer and polarisation states of the barrier are schematically shown. After ref. 101 with permission. (**c**) Spin- and $k_{\parallel}$-dependent transmission in Co/PTO–ZrO$_2$/Co tunnel junctions for the parallel configuration of magnetic Co electrodes and for the two opposite orientations of ferroelectric polarisation. After ref. 91 with permission. (**d**) Magnetoelectric coupling in a Co/PTO–ZrO$_2$/Co tunnel junction. Top: difference of the layer-resolved magnetic moments upon reversal of the ferroelectric polarisation in the barrier. Bottom: spin-resolved charge densities for the two ferroelectric polarisations (majority spin—red; minority spin—blue). After ref. 91 with permission.

Within simple models, TER was calculated at finite bias for several transport regimes: direct tunnelling, Fowler–Nordheim tunnelling and thermionic injection (Figure 8a).[88] It was found that TER is present in all three transport regimes with the Fowler–Nordheim active at high bias and thermionic injection at high barrier thickness (Figure 8b). Large TER is favoured in FTJs with thicker barriers in all three transport regimes, but at the cost of current density. The largest TER is predicted when polarisation switching changes the transport regime. Importantly, at finite bias, the TER was shown to be non-zero even for a symmetric junction.

This result is corroborated by calculations of TER based on a single-orbital tight-binding model, showing that finite bias produces TER in a symmetric junction.[89] The effect is due to the change of the effective screening lengths in the electrodes, as the result of the shift in the chemical potential by the applied bias. This conclusion is consistent with the first-principles calculations of TER in Au/BaZrO$_3$/Au FTJs (Figure 8c).[90] Under finite applied bias, the junction electrostatic potential, shown in Figure 8c, breaks the symmetry which makes the two polarisation states inequivalent. These results show that at finite bias, the TER is present even in symmetric junctions; however, the slope of the TER curve with bias, in addition to the electronic and ionic screening, depends on the piezoelectric response which appears to produce the largest contribution to TER (Figure 8d).

Multiferroic tunnel junctions

The functionality of FTJs can be extended by using FM electrodes. This kind of FTJ is known as a MFTJ, which can also be considered as a MTJ with a ferroelectric barrier. The key property of a MFTJ is the co-existence of the TMR and TER effects.[7,22] Similar to MTJs, the TMR effect arises from the band structure mismatch between the FM electrodes when the magnetisation orientation changes from parallel to antiparallel. The TER effect arises from the asymmetry of the junction breaking the degeneracy between the two polarisation directions. Equivalently these features of MFTJs can be stated as: (i) the presence of four resistance states determined by the mutual orientation of the magnetisation in the electrodes (parallel or antiparallel) and the direction of the polarisation in the barrier or (ii) the tunability of the TMR and TER effects by the complementary ferroic order, that is, there are two values of TMR dependent on ferroelectric polarisation orientation and two values of TER dependent on magnetisation orientation (Figure 9a).

A number of first-principles calculations have been performed for different MFTJs to elaborate these conclusions.[23,59,63,70,91] One of the important outcomes of these calculations is the demonstrated connection between the interfacial magnetoelectric effect and the ferroelectrically controlled TMR effect.[92] The reversible electric field induced by the ferroelectric polarisation is predicted to modify the magnetic structure of an adjacent ferromagnet.[93–96] On the other hand, the TMR effect is largely determined by the spin-polarised transmission across interfaces,[97] and thus by altering the interface spin structure the ferroelectric polarisation affects the spin-polarisation and TMR. For example, the calculated $k_{\parallel}$-resolved transmission for Co/PbTiO$_3$–ZrO$_2$/Co MFTJs[91] shows that the dominant spin contribution to the conductance is switched from majority to minority when the ferroelectric polarisation of PTO is reversed from pointing to the Co/PbO interface to the ZrO_2/Co interface (Figure 9c). This effect is explained by the spin density and magnetic moments at the interfaces dependent on ferroelectric polarisation orientation, which constitutes the interface magneto-electric effect (Figure 9d).[91]

Experimentally, there have been a number of demonstrations of the co-existence of the TMR and TER effects in MFTJs.[98–101] Sizable changes in the negative TMR were observed depending on the orientation of the ferroelectric polarisation in nanoscale Fe/BTO/LSMO and Co/BTO/LSMO MFTJs.[98,100] X-ray resonant magnetic scattering indicated that in these junctions there is an induced magnetic moment on the Ti atom that is coupled with the Fe electrode.[100] This result, as well as the measured negative spin polarisation of the Fe/BTO interface, is in agreement with first-principles calcuations.[93,95] A very large TMR was observed in a LSMO/BTO/LCMO/LSMO MFTJ (Figure 7f).[86] The value of TMR was found to be much larger for the junction in a low resistance state (100% at $T = 80$ K) than in a high-resistance state (hardly seen within the noise level). The significantly diminished TMR in the high-resistance state is explained by the LCMO phase transition to the AFM-insulating state.[86] Finally, it was shown that switching the ferroelectric polarisation in Co/PZT/LSMO tunnel junctions changes sign of the TMR effect from −3 to +4% (Figure 9b).[101] Although the TMR itself was not large, it indicates a possibility to control the sign of the spin polarisation by ferroelectric polarisation.[53] Overall, simultaneous TER and TMR effects and their tunability by the complementary ferroic order parameter in MFTJs are consistent with the original theoretical predictions.[23]

CONCLUSIONS AND OUTLOOK

Overall, the combined effort of theory and experiment has managed to predict, demonstrate and elucidate the fundamental properties of FTJs. The underling physical mechanisms behind the functionality of FTJs are now understood and the figures of merit that have been achieved are remarkable. In particular, the ON/OFF ratios can reach several orders of magnitude and significant scalability and reliability of FTJs have been demonstrated. Thus, FTJs possess a great deal of potential for non-volatile ferroelectric memory applications. In order for ferroelectric memory to become a reality, however, more work is needed in various areas, such as improving the stability of the ferroelectric polarisation in the barrier, controlling and understanding the role of defects such as oxygen vacancies in oxide materials, and understanding the finite bias behaviour of these devices. This is in addition to practical challenges, such as making the manufacturing process scalable and cost-effective for industrial applications and demonstrating the compatibility with existing electronic components.

In terms of future research, there are a number of avenues for development. First, current devices utilise oxide ferroelectrics, such as BTO, PTO and PZT. These materials often contain structural defects, such as oxygen vacancies. Understanding the role of these defects on ferroelectric switching as well as on electronic transport properties of FTJs is of critical importance. Second, the current theoretical methodology, in particular DFT-based calculations, describes very well the effect of the electronic screening of the polarisation and the interface bonding effect at zero bias. However, the finite bias regime and the effects appearing at finite bias are still largely unexplored. Thus, methodology improvements to target these regimes are necessary. Third, a potentially interesting research direction is the extension of the functionality of the devices using more complex materials with multiple and possibly coupled ferroic orders, such as single-phase magnetoelectric multiferroics. Finally, exploration of other ferroelectric materials, such as organic ferroelectrics that could preserve ferroelectricity at monolayer thickness, would be interesting.

ACKNOWLEDGEMENTS

This work was supported by the National Science Foundation (NSF) through Materials Research Science and Engineering Center (MRSEC, grant no. DMR-1420645) and the Semiconductor Research Corporation through Center for NanoFerroic Devices (CNFD). The work at UPR was supported by NSF (grants nos EPS-1010094 and DMR-1105474). We thank Verona Skomski for her help in preparation of the manuscript for submission.

COMPETING INTERESTS

The authors declare no conflict of interest.

REFERENCES

1. Frenkel, J. On the electrical resistance of contacts between solid conductors. *Phys. Rev.* **36**, 1604–1618 (1930).
2. Chappert, C., Fert, A. & Van Dau, F. N. The emergence of spin electronics in data storage. *Nat. Mater.* **6**, 813–823 (2007).
3. Jullière, M. Tunneling between ferromagnetic films. *Phys. Lett. A* **54**, 225–226 (1975).
4. Tsymbal, E. Y., Mryasov, O. N. & LeClair, P. R. Spin-dependent tunnelling in magnetic tunnel junctions. *J. Phys. Condens. Matter* **15**, R109–R142 (2003).
5. Zubko, P., Gariglio, S., Gabay, M., Ghosez, P. & Triscone, J.-M. Interface physics in complex oxide heterostructures. *Annu. Rev. Condens. Matter Phys.* **2**, 141–165 (2011).
6. Bibes, M., Villegas, J. E. & Barthélémy, A. Ultrathin oxide films and interfaces for electronics and spintronics. *Adv. Phys.* **60**, 5 (2011).
7. Tsymbal, E. Y. & Kohlstedt, H. Tunneling across a ferroelectric. *Science* **313**, 181–183 (2006).
8. Velev, J. P., Jaswal, S. S. & Tsymbal, E. Y. Multiferroic and magnetoelectric materials and interfaces. *Phil. Trans. R. Soc. A* **369**, 3069–3097 (2011).
9. Rabe, K. M., Dawber, M., Lichtensteiger, C., Ahn, C. H., Triscone, J.-M. in *Topics in Applied Physics* 105, 1–30 (eds Rabe K., Ahn C. H. & Triscone J.-M.) (Springer-Verlag, Berlin, Heidelberg, Germany, 2007).
10. Horiuchi, S. & Tokura, Y. Organic ferroelectrics. *Nat. Mater.* **7**, 357–366 (2008).
11. Esaki, L., Laibowitz, R. B. & Stiles, P. J. Polar switch. *IBM Tech. Discl. Bull.* **13**, 2161–2162 (1971).
12. Ahn, C. H., Rabe, K. M. & Triscone, J.-M. Ferroelectricity at the nanoscale: local polarization in oxide thin films and heterostructures. *Science* **303**, 488–491 (2004).
13. Dawber, M., Rabe, K. M. & Scott, J. F. Physics of thin-film ferroelectric oxides. *Rev. Mod. Phys.* **77**, 1083–1130 (2005).
14. Junquera, J. & Ghosez, P. Critical thickness for ferroelectricity in perovskite ultrathin films. *Nature* **422**, 506–509 (2003).
15. Fong, D. D. *et al.* Ferroelectricity in ultrathin perovskite films. *Science* **304**, 1650–1653 (2004).
16. Lichtensteiger, C., Triscone, J.-M., Junquera, J. & Ghosez, P. Ferroelectricity and tetragonality in ultrathin $PbTiO_3$ films. *Phys. Rev. Lett.* **94**, 047603 (2005).
17. Tenne, D. A. *et al.* Probing nanoscale ferroelectricity by ultraviolet raman spectroscopy. *Science* **313**, 1614 (2006).
18. Zhuravlev, M. Y., Sabirianov, R. F., Jaswal, S. S. & Tsymbal, E. Y. Giant electroresistance in ferroelectric tunnel junctions. *Phys. Rev. Lett.* **94**, 246802 (2005).
19. Kohlstedt, H., Pertsev, N. A., Rodríguez Contreras, J. & Waser, R. Theoretical current-voltage characteristics of ferroelectric tunnel junctions. *Phys. Rev. B* **72**, 125341 (2005).
20. Velev, J. P., Duan, C.-G., Belashchenko, K. D., Jaswal, S. S. & Tsymbal, E. Y. Effect of ferroelectricity on electron transport in $Pt/BaTiO_3/Pt$ tunnel junctions. *Phys. Rev. Lett.* **98**, 137201 (2007).
21. Tsymbal, E. Y. & Gruverman, A. Ferroelectric tunnel junctions: Beyond the barrier. *Nat. Mater.* **12**, 602–604 (2013).
22. Zhuravlev, M. Y., Jaswal, S. S., Tsymbal, E. Y. & Sabirianov, R. F. Ferroelectric switch for spin injection. *Appl. Phys. Lett.* **87**, 222114 (2005).
23. Velev, J. P. *et al.* Magnetic tunnel junctions with ferroelectric barriers: Prediction of four resistance states from first principles. *Nano Lett.* **9**, 427–432 (2009).
24. Zhuravlev, M. Y., Maekawa, S. & Tsymbal, E. Y. Effect of spin-dependent screening on tunneling electroresistance and tunneling magnetoresistance in multiferroic tunnel junctions. *Phys. Rev. B* **81**, 104419 (2010).
25. Tsymbal, E. Y., Gruverman, A., Garcia, V., Bibes, M. & Barthélémy, A. Ferroelectric and multiferroic tunnel junctions. *MRS Bull.* **37**, 138–143 (2012).
26. Garcia, V. & Bibes, M. Ferroelectric tunnel junctions for information storage and processing. *Nat. Commun.* **5**, 4289 (2014).
27. Mehta, R. R., Silverman, B. D. & Jacobs, J. T. Depolarization fields in thin ferroelectric films. *J. Appl. Phys.* **44**, 3379–3385 (1973).
28. Kim, D. J. *et al.* Polarization relaxation induced by a depolarization field in ultrathin ferroelectric $BaTiO_3$ capacitors. *Phys. Rev. Lett.* **95**, 237602 (2005).

29. Simmons, J. G. Generalized formula for the electric tunnel effect between similar electrodes separated by a thin insulating film. *J. Appl. Phys.* **34**, 1793–1803 (1963).

30. Brinkman, W. F., Dynes, R. C. & Rowell, J. M. Tunneling conductance of asymmetric barriers. *J. Appl. Phys.* **41**, 1915–1921 (1970).

31. Gruverman, A. *et al.* Tunneling electroresistance effect in ferroelectric tunnel junctions at nanoscale. *Nano Lett.* **9**, 3539–3543 (2009).

32. Sokolov, A., Bak, O., Lu, H., Tsymbal, E. Y. & Gruverman, A. Effect of epitaxial strain on tunneling electroresistance in ferroelectric tunnel junctions. *Nanotechnology* **25**, 305202 (2015).

33. Garcia, V. *et al.* Giant tunnel electroresistance for non-destructive readout of ferroelectric states. *Nature* **460**, 81–84 (2009).

34. Maksymovych, P., Jesse, S., Yu, P., Ramesh, R., Baddorf, A. P. & Kalinin, S. V. Polarization control of electron tunneling into ferroelectric surfaces. *Science* **324**, 1421–1425 (2009).

35. Li, Z. *et al.* An epitaxial ferroelectric tunnel junction on silicon. *Adv. Mater.* **26**, 7185–7189 (2014).

36. Pantel, D., Goetze, S., Hesse, D. & Alexe, M. Room-temperature ferroelectric resistive switching in ultrathin $Pb(Zr_{0.2}Ti_{0.8})O_3$ films. *ACS Nano* **5**, 6032–6038 (2011).

37. Chanthbouala, A. *et al.* Solid-state memories based on ferroelectric tunnel junctions. *Nat. Nanotechnol.* **7**, 101–104 (2012).

38. Zenkevich, A. *et al.* Electronic band alignment and electron transport in $Cr/BaTiO_3/Pt$ ferroelectric tunnel junctions. *Appl. Phys. Lett.* **102**, 062907 (2013).

39. Boyn, S. *et al.* Engineering ferroelectric tunnel junctions through potential. *APL Mater.* **3**, 061101 (2015).

40. Soni, R. *et al.* Giant electrode effect on tunneling electroresistance in ferroelectric tunnel junctions. *Nat. Commun.* **5**, 5414 (2014).

41. Liu, X., Wang, Y., Burton, J. D. & Tsymbal, E. Y. Polarization-controlled Ohmic to Schottky transition at a metal/ferroelectric interface. *Phys. Rev. B* **88**, 165139 (2013).

42. Liu, X., Burton, J. D., Zhuravlev, M. Y. & Tsymbal, E. Y. Electric control of spin injection into a ferroelectric semiconductor. *Phys. Rev. Lett.* **114**, 046601 (2015).

43. Liu, X., Burton, J. D. & Tsymbal, E. Y. Enhanced tunneling electroresistance in ferroelectric tunnel junctions due to the reversible metallization of the barrier. *Phys Rev. Lett.* **116**, 197602 (2016).

44. Wen, Z., Li, C., Wu, D., Li, A. & Ming, N. Ferroelectric-field-effect-enhanced electroresistance in metal/ferroelectric/semiconductor tunnel junction. *Nat. Mater.* **12**, 617–621 (2013).

45. Li, C. *et al.* Ultrathin $BaTiO_3$-based ferroelectric tunnel junctions through interface engineering. *Nano Lett.* **15**, 2568–2573 (2015).

46. Radaelli, G. *et al.* Large room-temperature electroresistance in dual-modulated ferroelectric tunnel barriers. *Adv. Mater.* **27**, 2602–2607 (2015).

47. Quindeau, A. *et al.* Origin of tunnel electroresistance effect in $PbTiO_3$-based multiferroic tunnel junctions. *Phys. Rev. B* **92**, 035130 (2015).

48. Martin, R. M. *Electronic Structure: Basic Theory and Practical Methods.* (Cambridge University Press, Cambridge, UK, 2008).

49. Singh-Miller, N. E. & Marzari, N. Surface energies, work functions, and surface relaxations of low-index metallic surfaces from first principles. *Phys. Rev. B* **80**, 235407 (2009).

50. Gerra, G., Tagantsev, A. K., Setter, N. & Parlinski, K. Ionic polarizability of conductive metal oxides and critical thickness for ferroelectricity in $BaTiO_3$. *Phys. Rev. Lett.* **96**, 107603 (2006).

51. Stengel, M., Vanderbilt, D. & Spaldin, N. A. Enhancement of ferroelectricity at metal-oxide interfaces. *Nat. Mater.* **8**, 392–397 (2009).

52. Shen, L. *et al.* Systematic study of ferroelectric, interfacial, oxidative, and doping effects on conductance of $Pt/BaTiO_3/Pt$ ferroelectric tunnel junctions. *Phys. Rev. B* **85**, 064105 (2012).

53. Velev, J. P., Lopez-Encarnacion, J. M., Burton, J. D. & Tsymbal, E. Y. Multiferroic tunnel junctions with poly(vinylidene fluoride). *Phys. Rev. B* **85**, 125103 (2012).

54. Velev, J. P. *et al.* Negative spin polarization and large tunneling magnetoresistance in epitaxial $Co/SrTiO_3/Co$ magnetic tunnel junctions. *Phys. Rev. Lett.* **95**, 216601 (2005).

55. Kohn, W. Analytic properties of Bloch waves and Wannier functions. *Phys. Rev* **115**, 809 (1959).

56. Heine, V. On the general theory of surface states and scattering of electrons in solids. *Proc. Phys. Soc.* **81**, 300 (1963).

57. Mavropoulos, P., Papanikolaou, N. & Dederichs, P. Complex band structure and tunneling through ferromagnet/insulator/ferromagnet junctions. *Phys. Rev. Lett.* **85**, 1088 (2000).

58. Butler, W. H., Zhang, X.-G., Schulthess, T. C. & MacLaren, J. M. Spin-dependent tunneling conductance of Fe/MgO/Fe sandwiches. *Phys. Rev. B* **63**, 054416 (2001).

59. Caffrey, N. M., Archer, T., Rungger, I. & Sanvito, S. Prediction of large bias-dependent magnetoresistance in all-oxide magnetic tunnel junctions with a ferroelectric barrier. *Phys. Rev. B* **83**, 125409 (2011).

60. Hinsche, N. F. *et al.* Strong influence of complex band structure on tunneling electroresistance: A combined model and ab initio study. *Phys. Rev. B* **82**, 214110 (2010).

61. Wortmann, D. & Blügel, S. Influence of the electronic structure on tunneling through ferroelectric insulators: Application to $BaTiO_3$ and $PbTiO_3$. *Phys. Rev. B* **83**, 155114 (2011).

62. Aguado-Puente, P. & Junquera, J. First-principles study of metal-induced gap states in metal/oxide interfaces and their relation with the complex band structure. *MRS Commun.* **3**, 191–197 (2013).

63. Dai, J.-Q., Zhang, H. & Song, Y.-M. Magnetoelectric coupling and spin-dependent tunneling in Fe/PbTiO₃/Fe multiferroic heterostructure with a Ni monolayer inserted at one interface. *J. Appl. Phys.* **118**, 054104 (2015).

64. Landauer, R. Electrical resistance of disordered one-dimensional lattices. *Philos. Mag.* **21**, 863–867 (1970).

65. Brandbyge, M., Mozos, J.-L., Ordejon, P., Taylor, J. & Stokbro, K. Density-functional method for nonequilibrium electron transport. *Phys. Rev. B* **65**, 165401 (2002).

66. Choi, H. J. & Ihm, J. Ab initio pseudopotential method for the calculation of conductance in quantum wires. *Phys. Rev. B* **59**, 2267–2275 (1999).

67. Smogunov, A., Dal Corso, A. & Tosatti, E. Ballistic conductance of magnetic Co and Ni nanowires with ultrasoft pseudopotentials. *Phys. Rev. B* **70**, 045417 (2004).

68. Caroli, C., Combescot, R., Nozieres, P. & Saint-James, D. Direct calculation of the tunneling current. *J. Phys. C* **4**, 916 (1971).

69. Velev, J. & Butler, W. On the equivalence of different techniques for evaluating the Green function for a semi-infinite system using a localized basis. *J. Phys. Cond. Mater.* **16**, R637 (2004).

70. López-Encarnación, J. M., Burton, J. D., Tsymbal, E. Y. & Velev, J. P. Organic multiferroic tunnel junctions with ferroelectric poly(vinylidene fluoride) barriers. *Nano Lett.* **11**, 599–603 (2011).

71. Gerra, G., Tagantsev, A. K. & Setter, N. Ferroelectricity in asymmetric metal-ferroelectric-metal heterostructures: a combined first-principles-phenomenological approach. *Phys. Rev. Lett.* **98**, 207601 (2007).

72. Liu, X., Wang, Y., Lukashev, P. V., Burton, J. D. & Tsymbal, E. Y. Interface dipole effect on thin film ferroelectric stability: First-principles and phenomenological modeling. *Phys. Rev. B* **85**, 125407 (2012).

73. Lu, H. *et al.* Enhancement of ferroelectric polarization stability by interface engineering. *Adv. Mater.* **24**, 1209–1216 (2012).

74. Yamada, H. *et al.* Strong surface-termination effect on electroresistance in ferroelectric tunnel junctions. *Adv. Funct. Mater.* **25**, 2708–2714 (2015).

75. Zhuravlev, M. Y., Wang, Y., Maekawa, S. & Tsymbal, E. Y. Tunneling electro-resistance in ferroelectric tunnel junctions with a composite barrier. *Appl. Phys. Lett.* **95**, 052902 (2009).

76. Caffrey, N. M., Archer, T., Rungger, I. & Sanvito, S. Coexistence of giant tunneling electroresistance and magnetoresistance in an all-oxide composite magnetic tunnel junction. *Phys. Rev. Lett.* **109**, 226803 (2012).

77. Chen, Y. & McIntyre, P. C. Effects of chemical stability of platinum/lead zirconate titanate and iridium oxide/lead zirconate titanate interfaces on ferroelectric thin film switching reliability. *Appl. Phys. Lett.* **91**, 232906 (2007).

78. Wang, R. V. *et al.* Reversible chemical switching of a ferroelectric film *Phys. Rev. Lett.* **102**, 047601 (2009).

79. Lu, H. *et al.* Ferroelectric tunnel junctions with graphene electrodes. *Nat. Commun.* **5**, 5518 (2014).

80. Dagotto, E., Hotta, T. & Moreo, A. Colossal magnetoresistant materials: the key role of phase separation. *Phys. Rep.* **344**, 1–153 (2001).

81. Burton, J. D. & Tsymbal, E. Y. Prediction of electrically induced magnetic reconstruction at the manganite/ferroelectric interface. *Phys. Rev. B* **80**, 174406 (2009).

82. Burton, J. D. & Tsymbal, E. Y. Giant tunneling electroresistance effect driven by an electrically controlled spin valve at a complex oxide interface. *Phys. Rev. Lett.* **106**, 157203 (2011).

83. Molegraaf, H. J. A. *et al.* Magnetoelectric effects in complex oxides with competing ground states. *Adv. Mater.* **21**, 3470–3474 (2009).

84. Park, J. H. *et al.* Direct evidence for a half-metallic ferromagnet. *Nature* **392**, 794–796 (1998).

85. Vaz, C. A. F. *et al.* Origin of the magnetoelectric coupling effect in $Pb(Zr_{0.2}Ti_{0.8})O_3/La_{0.8}Sr_{0.2}MnO_3$ multiferroic heterostructures. *Phys. Rev. Lett.* **104**, 127202 (2010).

86. Yin, Y. W. *et al.* Enhanced tunnelling electroresistance effect due to a ferroelectrically induced phase transition at a magnetic complex oxide interface. *Nat. Mater.* **12**, 397 (2013).

87. Jiang, L. *et al.* Tunneling electroresistance induced by interfacial phase transitions in ultrathin oxide heterostructures. *Nano Lett.* **13**, 5837–5843 (2013).

88. Pantel, D. & Alexe, M. Electroresistance effects in ferroelectric tunnel barriers. *Phys. Rev. B* **82**, 134105 (2010).

89. Useinov, A., Kalitsov, A., Velev, J. & Kioussis, N. Bias-dependence of the tunneling electroresistance and magnetoresistance in multiferroic tunnel junctions. *Appl. Phys. Lett.* **105**, 102403 (2014).

90. Bilc, D. I., Novaes, F. D., Íñiguez, J., Ordejón, P. & Ghosez, P. Electroresistance effect in ferroelectric tunnel junctions with symmetric electrodes. *ACS Nano* **6**, 1473–1478 (2012).

91. Borisov, V. S., Ostanin, S., Achilles, S., Henk, J. & Mertig, I. Spin-dependent transport in a multiferroic tunnel junction: Theory for Co/PbTiO$_3$/Co. *Phys. Rev. B* **92**, 075137 (2015).

92. Burton, J. D. & Tsymbal, E. Y. Magnetoelectric interfaces and spin transport. *Philos. Trans. A Math. Phys. Eng. Sci.* **370**, 4840–4855 (2012).

93. Duan, C. G., Jaswal, S. S. & Tsymbal, E. Y. Predicted magnetoelectric effect in Fe/BaTiO$_3$ multilayers: ferroelectric control of magnetism. *Phys. Rev. Lett.* **97**, 047201 (2006).

94. Yamauchi, K., Sanyal, B. & Picozzi, S. Interface effects at a half-metal/ferroelectric junction. *Appl. Phys. Lett.* **91**, 062506 (2007).

95. Fechner, M. *et al.* Magnetic phase transition in two-phase multiferroics predicted from first principles. *Phys. Rev. B* **78**, 212406 (2008).

96. Niranjan, M. K., Burton, J. D., Velev, J. P., Jaswal, S. S. & Tsymbal, E. Y. Magnetoelectric effect at the SrRuO$_3$/BaTiO$_3$ (001) interface: An *ab initio* study. *Appl. Phys. Lett.* **95**, 052501 (2009).

97. Belashchenko, K. D. *et al.* Effect of interface bonding on spin-dependent tunneling from the oxidized Co surface. *Phys. Rev. B* **69**, 174408 (2004).

98. Garcia, V. *et al.* Ferroelectric control of spin polarization. *Science* **327**, 1106–1110 (2010).

99. Hambe, M. *et al.* Crossing an interface: Ferroelectric control of tunnel currents in magnetic complex oxide heterostructures. *Adv. Funct. Mater.* **20**, 2436–2441 (2010).

100. Valencia, S. *et al.* Interface-induced room-temperature multiferroicity in BaTiO$_3$. *Nat. Mater.* **10**, 753 (2011).

101. Pantel, D., Goetze, S., Hesse, D. & Alexe, M. Reversible electrical switching of spin polarization in multiferroic tunnel junctions. *Nat. Mater.* **11**, 289–293 (2012).

16

Cybermaterials: materials by design and accelerated insertion of materials

Wei Xiong[1] and Gregory B Olson[1,2]

Cybermaterials innovation entails an integration of Materials by Design and accelerated insertion of materials (AIM), which transfers studio ideation into industrial manufacturing. By assembling a hierarchical architecture of integrated computational materials design (ICMD) based on materials genomic fundamental databases, the ICMD mechanistic design models accelerate innovation. We here review progress in the development of linkage models of the process–structure–property–performance paradigm, as well as related design accelerating tools. Extending the materials development capability based on phase-level structural control requires more fundamental investment at the level of the Materials Genome, with focus on improving applicable parametric design models and constructing high-quality databases. Future opportunities in materials genomic research serving both Materials by Design and AIM are addressed.

BACKGROUND: ICMD BLUEPRINT

The far-reaching multi-agency enterprise, Materials Genome Initiative (MGI),[1,2] highlights computational materials design techniques grounded in fundamental databases, which can support an ambition of decreasing the full development cycle of new materials from the present 10–20 years to $\leqslant 5$ years. As a subfield of the broader field of integrated computational materials engineering (ICME), which includes modelling of existing materials,[3,4] the MGI centres on design of new materials and their accelerated qualification through the inherent predictability of designed systems. Creation of the infrastructure of this technology has been a global activity as summarised in a recent series of viewpoint papers on materials genomics.[5–10] Particularly notable has been the design work of Bhadeshia[8] at the University of Cambridge and his former students including Harada[11,12] and Reed.[13] The highest achievements of full cycle compression have been demonstrated in US research, which will be the focus of this paper.

In the development of applied engineering materials design powered by fundamental thermodynamic and kinetic genomic databases, a hierarchical infrastructure called ICMD presents a proven scenario of materials genomic design for accelerated engineering innovation. As indicated in Figure 1, the backbone of the ICMD infrastructure is composed of Materials by Design and accelerated insertion of materials (AIM) techniques. The application of both techniques spans the entire course of materials innovation, which can be divided into three phases: concept implementation, materials and processing design, and material qualification. The quality of the materials innovation is determined by the mechanistic models applied in the ICMD framework, which follow the universal process–structure–property–performance paradigm in materials science.[14] In Materials by Design, both process–structure and structure–property models are evaluated and refined to maximise the model-predictability grounded in the Materials Genome, which is powered by fundamental research on

high-quality databases. Although the AIM technique also takes mechanistic models as its basis, it requires extra effort in uncertainty quantification, design sensitivity analysis, integrated component-level process simulation and probabilistic prediction of manufacturing variation.

A recent review of ICMD application highlights notable achievements made in ferrous materials.[15] This review provides further detail on the linkage models for Materials by Design, and methods of qualification adopted in AIM for a technology transfer from lab-scale materials innovation to industrial commercial practice. It is formulated as a snapshot of materials innovations on selected alloy designs for improving mechanical performance. Its foundation is sufficiently general that all levels of materials design should share the same principles, not only for metal and alloys, but also for oxides and polymers.[14,16,17]

Here we revisit some key process–structure and structure–property linkage models applied in Materials by Design followed by a technical review of the AIM method application in projects supported by DARPA (Defense Advanced Research Projects Agency) and National Aeronautics and Space Administration. Although the above linkage models are applied in AIM technology as well, related discussion on AIM will be more about its unique efforts on model integration, uncertainty quantification and design sensitivity control. Future research opportunities and perspectives are presented as concluding remarks in the last section.

MATERIALS BY DESIGN: DESIGN ENGINE

The paradigm of materials design guided by governing processing–structure–properties–performance is a universal principle for developing advanced materials.[14] Predictive-science-based computational modelling reinforced by experimental calibration allows us to perform multiscale modelling for various properties with temperature/time dependence. Therefore, the development

[1]Department of Materials Science and Engineering, Northwestern University, Evanston, IL, USA and [2]QuesTek Innovations LLC, Evanston, IL, USA.
Correspondence: W Xiong (wxiong@yahoo.com)

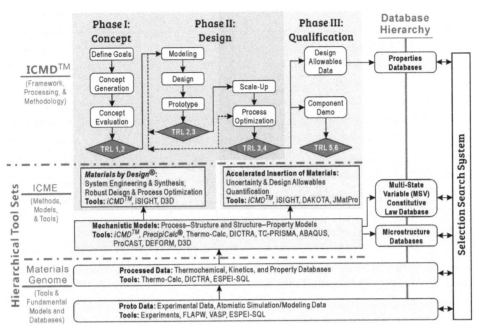

Figure 1. Overall hierarchical architecture of ICMD methods, tools and techniques are highlighted in green and databases are marked in yellow. Please note that iCMD is the toolkit used during application of ICME methods based on Materials Genome. Figure is reprinted with permission from Elsevier.

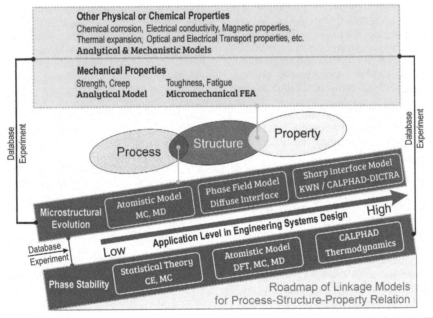

Figure 2. Materials design roadmap for linkage models of process–structure and structure–property relations. CE, cluster expansion; MC, Monte Carlo; MD, molecular dynamics.

of the process–structure and structure–property models will be beneficial to Materials by Design, and boost AIM applications.

As shown in Figure 2, process–structure models empower two levels of modelling: phase stability and microstructural evolution, which establish a fundamental basis for structure–property design models. With particular engineering design purpose, high-quality engineering models are integrated with science-based mechanistic models forming a hierarchical model system.

Process–structure relation and design models

In the course of systems design for advanced materials, a thorough understanding of thermodynamics and kinetics enables

processing design, which often starts with a parametric design for chemical composition. As shown in Figure 3, intrinsic design parameters for materials processing encompass chemical composition, temperature, pressure and external force fields, on which even a small adjustment may lead to significantly different evolution of microstructure driven by the thermodynamics and kinetics of the system. The CALPHAD (calculation of phase diagrams) method[18] and its correlated modelling techniques quantitatively express our mechanistic understanding of the process–structure linkage relations.

Thermodynamics is the core of the precipitation simulation. Considerable efforts have been made on integration of density functional theory (DFT)-based atomistic modelling, CALPHAD and

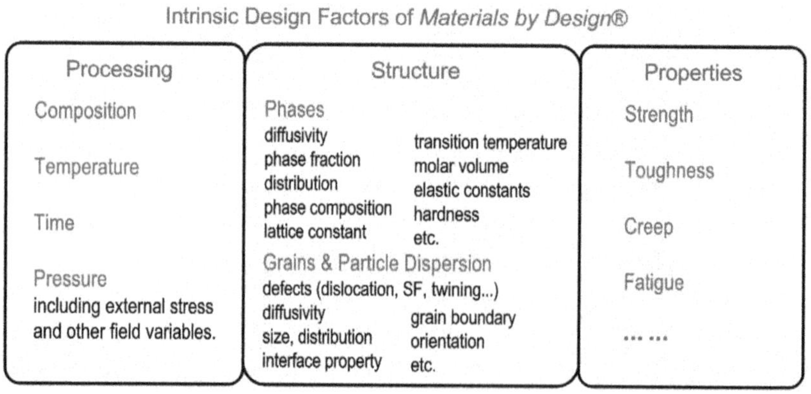

Intrinsic Design Factors of *Materials by Design*®

Figure 3. Intrinsic design factor of the process–structure–property-performance paradigm. The most common attributes of design factors for microstructure control are presented.

phase equilibrium measurements for assembling materials genomic databases. DFT calculations for zero-kelvin thermal–chemical properties, such as, enthalpy of formation, are often used as protodata to support CALPHAD thermodynamic modelling. This is now considered as a routine method to perform physically sound thermodynamic modelling, which can sometimes circumvent the restriction of calorimetry (e.g., radiative alloy systems),[19,20] and assist in determining reliable experimental data sets.[19,21–23] At present, there are many commercial software packages available based on the CALPHAD technique, e.g., Thermo-Calc, Pandat, FactSage, MTDATA.[24,25] However, not until the release of the DICTRA (diffusion controlled transformations) package by the Thermo-Calc software company in early 90s[26,27] for diffusion kinetic modelling, did the CALPHAD approach start to demonstrate its capability of handling sophisticated process–structure modelling. The significance of the CALPHAD method nowadays has exceeded its original definition as phase diagram computation. It is indisputable that the modern CALPHAD approach[28] serves as a fundamental technique for establishing the materials genomic databases, and widely recognised as a foundational tool for Materials by Design.

When related CALPHAD genomic databases are available, although thermodynamic and diffusion calculations can be readily performed, one should be aware of application limits and evaluate the quality of database predictions prior to a pragmatic design, or before adopting CALPHAD as an input for explicit microstructural simulation. One should also bear in mind that a fundamental CALPHAD database generally provides the Gibbs energy of bulk phases by neglecting the contribution from interfaces to the total energy of the system. However, when modelling precipitation strengthening, the Gibbs–Thompson effects, i.e., size-dependent capillary effects, need to be considered. When experimental data determined by atom probe tomography (APT) is available, a semi-log plot of composition versus inverse of particle size can often generate a good trajectory defining equilibrium tie-lines and initial critical nucleus size defining interfacial energy.[29,30] In addition, elastic coherency (misfit) energy, G_{el}, can be evaluated for a further energy penalty to the precipitation when particles are coherent. In the work by Olson et al.[31] the method of estimating coherent interfacial energy (G_{coh}) and elastic misfit energy was demonstrated using single sensor differential thermal analysis[32] combined with APT.

Another issue related to phase stability should be addressed when applying CALPHAD thermodynamics is evaluation of competitive growth behaviour of different phases. As CALPHAD thermodynamic database calculates phase equilibrium as a default option, some phases with sluggish formation should be carefully evaluated in a quantitative comparison with experimental data determined from materials processing, which are

generally nonequilibrium. For example, in the work performed by Wusatowska-Sarnek et al.[33] experimental phase fraction were initially compared with calculated results under full equilibrium. Good agreement was ultimately obtained when the calculation was performed using constrained equilibria, under which topologically close-packed phases are excluded because of their sluggish formation.

In view of the above, for the sake of parametric design, experimental validation is essential to assess the accuracy of a CALPHAD model-prediction. This can in turn contribute to database refinement in support of higher fidelity applications of Materials by Design and AIM methodologies. Similar to the case in thermodynamics, diffusional kinetic databases also require adjustments for the constrained case in a quantitative comparison with experimental results measured after specific processing. For developing dual microstructure processing turbine disk in a project supported by National Aeronautics and Space Administration Glenn Research Center,[31] a multicomponent CALPHAD diffusion database was constructed by National Institute of Standards and Technology, which demonstrates a good predictability after calibration on the atomic mobility of Nb, Al, Ti and Cr in fcc using multicomponent diffusion couples made between disc alloys and pure Ni. A need for calibration of CALPHAD diffusivity databases originates from the simplifications made by omitting the influence of fast diffusion path, e.g., dislocations, phase/grain boundaries. A diffusion scaling factor, D_{scale}, can be applied to the diffusivity matrix for calibration, particularly for a precipitation simulation.

A well-calibrated CALPHAD thermodynamic and kinetic database is a prerequisite of accurate parametric control of microstructural length and time scales. For example, effective precipitation strengthening relies on the size control of the particle precipitation, and thus requires predictive models for the kinetic process of particle size evolution. A good example is the extensive application of the coarsening model represented by the Lifshitz–Slyozov–Wagner (LSW) theory[34–38] for precipitation size evolution in a dilute binary system. To couple with the CALPHAD thermodynamic/kinetic databases for multicomponent and multiphase simulation, several models[39,40] were proposed based on extensions of the LSW theory. One of the simplified models was proposed by Lee et al.[39] based on the dilute solution case. A later work performed by Umantsev and Olson[40] removed the constraint of dilute solution thermodynamics, but neglected effects of capillarity on the precipitation composition. Further improvements have been made by Morral and Purdy[36,37] with a more general description of the coarsening in multicomponent alloys. Kuehmann and Voorhees[38] considered flux balance boundary conditions on the interface rather than local equilibrium, but off-diagonal terms in the diffusion matrix were omitted for simplification. For practical design purposes, it is found that the simplified dilute solution model proposed by Lee *et al.*[39] may be directly applied, if only an underdeveloped database is available. However, the Morral–Purdy model[36,37] overall provides coarsening rate with highest accuracy.

To maximise hardening effects by reducing initial strengthening particle size, a designed alloy usually preserves a high-solute supersaturation, i.e., high-driving force for nucleation of a second-phase precipitate particle, which can allow a bypass of the growth regime from nucleation directly to the coarsening regime. Therefore, the coarsening rate predicted by the aforementioned multicomponent coarsening models is often adopted as the principle design factor to control overall precipitation rate. Such a concept can be confirmed using a more rigorous approach based on the Kampmann and Wagner numerical (KWN) model[41] for concomitant nucleation, growth and coarsening of precipitates, extended to multicomponent alloy systems. A detailed discussion on the influence of supersaturation was made by Robson,[42] who analysed different scenarios with overlapping of nucleation, growth and coarsening occurs during precipitation. Another key factor to control particle size is interfacial energy, which also markedly influences kinetic behaviour of precipitation. As shown in Figure 4a,b, in the KWN modelling, both interfacial energy and supersaturation driving force can bring significant impact and cause the overlap of nucleation and coarsening regime. However, in the design process, supersaturation expressed as the thermodynamic driving force is the most accessible parameter using the CALPHAD genomic databases. Figure 4c is a kinetic map calculated by Robson[42] indicating the correlation between supersaturation and interface energy where nucleation and coarsening can be controlled by both parameters. Experiments are from studies carried out on Fe-Cu,[43] Cu-Co,[44] Cu-Ti[45,46] and Ni-Al[47] alloys.

Although a design model-chain from CALPHAD thermodynamics and kinetics to phase-level microstructural information is developed and can be directly used as a robust tool for Materials by Design, some constraints require further development of the sharp-interface model. For example, the current KWN model-based simulations assume a spherical geometry for the second-phase particle precipitation, which is an oversimplification for rod or plate shape precipitates, e.g., Q phase in the Al–Cu–Mg–Si-based alloy systems. A compromising method is to simulate precipitate kinetics by assuming spherical particles of equivalent particle volume. However, such a compromise could deviate significantly from experimental observation, and should be applied with caution for design purposes.

A less efficient but sometimes useful method which may circumvent the above difficulties by considering geometrical/crystal anisotropy is the so-called phase-field model (PFM) or diffuse interface model. For the sake of the process–structure modelling, the governing functionals of PFM are equivalent to the time-dependent Ginzburg–Landau equation[48] and Cahn–Hilliard equation.[49] Similar to the case of sharp-interface KWN model, both approaches take fundamental thermodynamic and kinetic parameters as the input, which can be read from the CALPHAD databases. In addition, PFM also experiences the difficulties of determining several key input parameters, such as, interfacial energy, gradient energy term and interfacial mobility, which are generally hard to measure, without even further considering their time and temperature dependence. It is noteworthy, even though PFM of anisotropic structure is feasible, the anisotropic input parameters, e.g., interfacial energy, often require a good quantitative estimation, which is still a daunting task for both atomistic modelling and experiments. Therefore, more often, these input parameters are still considered as the fitting parameter to perform the simulation to reach a good agreement with experimental microstructural evolution.

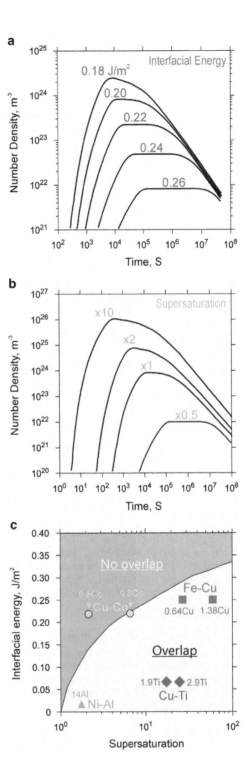

Figure 4. Predicted evolution of number density with time for a selected range of (**a**) interfacial energies and (**b**) supersaturation. (**c**) Kinetic map showing the overlap region of nucleation and coarsening influenced by supersaturation and interfacial energy. Simulations are all based on the work performed by Robson.[42] Alloy compositions are given in atomic percentage. Figure is reprinted with permission from Elsevier.

Figure 5 shows the relation among CALPHAD, experiments, sharp-interface (KWN) and diffuse interface models (PFM) for a typical process–structure simulation of solid-state precipitation phenomena. Although PFM can also be performed based on the CALPHAD method, nowadays it is still a significant challenge for effective coupling with CALPHAD database for multicomponent

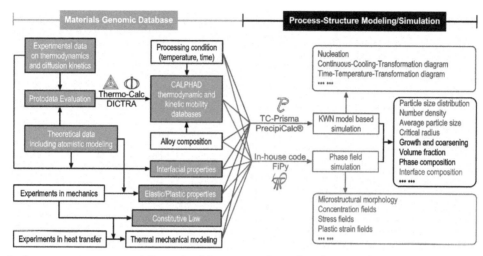

Figure 5. Flow chart of process–structure modelling of solid-state transformation for second-particle growth. Both KWN model-based simulation and phase-field simulation take materials genomic database for the model input. The unique outputs of these two different simulation techniques are highlighted in purple (KWN) and blue (phase-field).

and multiphase simulations[50,51] owing to the multiplicity of both bulk and interfacial properties. Another distinctive disadvantage of PFM is its limitation on simulating nucleation stage by generating nucleation sites using some ad hoc methods. Overall, in a practical simulation, comparing with KWN methods, PFM usually contains more adjustable parameters, which are not easy to determine, despite its capability of handling more complex phenomena. Nevertheless, PFM is often a useful science tool to enhance understanding of mechanism, it is generally less useful as an engineering tool than KWN-based methods that also provide better-defined statistical quantities.

A parametric design often relies on efficient computational design toolkits. Because the KWN model-based simulations can be implemented in an intuitive way by coupling with fundamental CALPHAD thermodynamic and kinetic parameters, they are widely adopted in the process–structure design. The PrecipCalc software developed by the QuesTek Innovations (Evanston, IL, USA) is a good example utilising the sharp-interface precipitation model for alloy design and process optimisation, which led the commercialising of several Ferrium high-performance alloys, e.g., S53, M54, C61 and C64.[52]

Structure–property relation and design models

As shown in Figure 3, strength, toughness, creep and fatigue are four elementary mechanical properties of importance in structural materials design. Although corrosion resistance is also important, it is commonly treated as a composition constraint while improving these four elementary properties. It is a formidable task to cover all of the structure–property linkage models in this review. For elucidating the application of structure–property models to linkage design, in this section, we selectively discuss some well-developed mechanical models, which are available for application, by citing some practical structural alloy designs. As indicated in Figure 2, whereas analytical models are available for improving strength and creep, toughness and fatigue are optimised by performing micromechanical finite element analysis (FEA).

Structure-strength. One of the most well-developed structure–property analytical models is probably the one for predicting mechanical strength based on alloy hardening mechanism. Table 1 shows a summary of elementary hardening mechanism of materials strength by Hornbogen.[53] Therefore, the microstructure from zero- to three-dimensional (3D) discontinuities are defined as the source of elementary hardening mechanism.

Table 1. Summary of elementary hardening mechanism of crystalline phases[53]

Symbol	Obstacles	Hardening mechanisms
σ_0	Lattice friction, order	Order hardening
$\Delta\sigma_S$	Solute atoms, vacancies	Solid solution hardening
$\Delta\sigma_d$	Dislocations	Work hardening
$\Delta\sigma_b$	Grain- or phase-boundaries	Grain boundary hardening
$\Delta\sigma_p$	Particles	Precipitation hardening
$\Delta\sigma_{ac}$	Crystal anisotropy	Texture hardening
$\Delta\sigma_{am}$	Microstructure anisotropy	Fibre hardening
$\Delta\sigma_m$	Crystal metastability	Pre-martensitic softening, transformation hardening

Excellent examples of applying such rules for materials strengthening can be found in the hardening enhancement design for high strength steels.[54,55] Typically, overall yield strength is described using the superposition law:

$$\sigma_Y = M\left[\tau_0 + \sum_i \tau_i\right] + \sigma_{HP} \qquad (1)$$

where M represents the Taylor factor, τ_0 is the Peierles–Nabarro stress contribution,[56–58] which often adopts the intrinsic value of the pure base alloy. τ_i takes into account contributions from work hardening τ_W, solid solution strengthening τ_{SS} and precipitation strengthening τ_P. The Hall–Petch effect σ_{HP} is the strengthening caused by grain refinement. Different contributions to the yield strength model is listed in Table 2. On the basis of the alloy composition and processing time, σ_P in Table 2 representing the dispersed particle strengthening contribution can be modelled based on the size of obstacle precipitates. As illustrated in Figure 6a, as a start, the small shearable particles follow the Friedel stress; with particle growth, the Orowan bypass stress becomes dominant, decreasing with the incremental particle size. However, in the case of ordered precipitate strengthening for superalloys, due to the antiphase boundary behaviour with paired dislocations, the peak hardening usually arrives earlier than the case with single-dislocation interaction with particle obstacles as shown in Figure 6a.[59–61] In this case, the peak-aged state corresponds to the intensity of pair dislocation interaction with obstacle particles.

A recent study performed by Wang *et al.*[54] demonstrated the application of the above superposition law on predicting the yield

Table 2. Contributions to yield strength by considering effects of solution strengthening, precipitation strengthening, work hardening, Hall–Petch effects based on intrinsic strength due to Peierles–Nabarro stress[54]

	Numerical expression	Constants and variables	
σ_0	$K_0 \cdot \left(\frac{2G}{1-\nu}\right) \cdot \exp\left(\frac{-4\pi w}{b}\right)$	G: shear modulus w: dislocation width K_0: Taylor coefficient	b: Burgers vector v: Poisson's ratio
σ_W	$K_W \cdot \frac{\sqrt{3}}{60} a_0 \left[3a_0\vartheta k_1 + (3a_0\vartheta k_1)^2 + \beta b \rho^{1/2} \right]$	ϑ: misorientation angles between two subgrains a_0: Taylor factor, ρ: dislocation density β: elastic modulus coefficient, k_1: subgrain interaction constant K_W: Taylor coefficient	
σ_{HP}	$K_{HP} \cdot d^{-1/2}$	d: grain size or martensite packet size. K_{HP}: Hall–Petch coefficient	
σ_{SS}	$K_{SS} \cdot \left(\sum_i k_{SS,i}^2 c_i \right)^{1/2}$	$k_{SS,i}$: strengthening coefficient c_i: mole fraction of element in matrix K_{SS}: model coefficient	
σ_P	$K_P \cdot \tau_P = K_P \cdot \min(\tau_{PF}, \tau_{PO})$	K_P: model coefficient τ_P: Precipitation stress by taking minimum value between Friedel τ_{PF} and Orowan stress τ_{PO}.	

strength of a secondary-hardening steel Blastalloy 160. A new transformation strengthening model accounting for Cu particle phase transformation was developed by associating with dislocation interaction with Cu of different crystal structure (bcc, 9R martensite and fcc). In addition, a modified predictive yield strength model was developed based on 3D APT of M_2C and bcc Cu-hardening precipitates. Especially, to directly link quantitative grain structure characterisation and yield strength, 3D precipitate size distribution and characterisation parameters (e.g., phase composition, phase fraction) measured by 3D APT were adopted in the strengthening model for predicting yield strength within 10% uncertainty. Such a method can directly bridge the aforementioned process–structure and structure–property models for improving yield strength as long as all input physical parameters can be well-determined with high accuracy. Therefore, a comprehensive scenario for modelling dispersed particle-hardening uses the process–structure model to obtain desired kinetic behaviour of the particle dispersion (see Figures 6c–e). The structure-strength model can determine the optimal particle radius to reach the maximum strengthening as shown in Figures 6a,b providing a clear microstructural objective for design.

Structure-toughness. Materials design is clearly a delicate compromise of a number of controlling parameters, which may be beneficial to one type of mechanical properties, but detrimental to the other. A classic example is the intrinsic inverse proportionality of toughness to materials strength. A specific example of designing Co-Ni ultra-high strength (UHS) steels is illustrated in Figures 7 and 8, which summarize a systems design effort on improving both strength and toughness synchronously. Refining grain size can be beneficial to both strength and toughness, but it is not always achievable. Therefore, understanding multiple ways to maximise both strength and toughness using structure-toughness models is crucial to a successful design.

Much research on toughness mechanism of high strength steels is devoted to avoidance of brittle fracture failure modes such as intergranular fracture. In a systems design for Co-Ni UHS steel, several methods were applied for improving both toughness and strength synergistically in the steel research group (SRG) consortium centred at Northwestern University.[15,62–68] The related design efforts for enhancing both strength and toughness are summarised in Figure 8, which focuses on (i) ductile–brittle transition temperature (DBTT),[69,70] (ii) enhancement of interphase chemical bonding to delay microvoid softening and (iii)

transformation toughening. Further, as hydrogen embrittlement can promote brittle intergranular fracture, grain boundary (GB) chemistry optimisation is also a primary requirement as indicated in Figures 7 and 8.

On the heuristic level, DBTT is often a primary design parameter to avoid brittle fracture at desired service temperature. An analytical 'master curve' model based on the shift of DBTT was constructed for designing the SRG UHS steels[69] by considering the influence of weight per cent of key components, hardness and grain size as primary model variables. Such a model directly correlating with alloy composition made it possible to create a criteria for optimising Ni and Co content in the Co-Ni UHS steel design by making trade-off among toughness, strength and Martensite start temperature.

In the course of the UHS SRG steel designs, the submicron microstructure design for improving ductile toughness and retain high strength is based on suppressing microvoid softening and introducing dispersed-austenite transformation toughening. It is well-established that microvoid nucleation during plastic deformation accelerates ductile fracture via a process of shear localisation, whereby sudden microvoid nucleation destabilises plastic flow by strain location into shear bands.[64,71] However, the quantitative role of fine ~0.1-μm scale secondary particle dispersion which nucleate the microvoids is less well-understood. In an optimised microstructure, these small particles correspond to the Zener-pinning particles[72–74] acting as the necessary grain refiners. Considerable efforts have been made by the SRG consortium developing several generations of multiscale ductile fracture simulators[75–79] to quantify the role of this level of microstructure. Extensive efforts were made on micromechanical FEA combined with the DFT calculation of interfacial strength,[75,76,80–83] which can evaluate ideal work of adhesion. For example, in the work by Hao et al.[75,76] a hierarchical constitutive model was derived with the plastic potential by considering DFT-calculated interfacial debonding and microvoid softening effects. Such an effort directly bridges atomistic modelling and continuum finite element simulation, which offers deeper insights into correlation between strength and fracture toughness. Furthermore, the interfacial adhesion energy can be enhanced by controlling dispersed phase compositions and geometric features in the design process using process–structure models.

On the basis of these simulations, a phenomenological model of toughness related to microvoid nucleation[84–86] was developed to

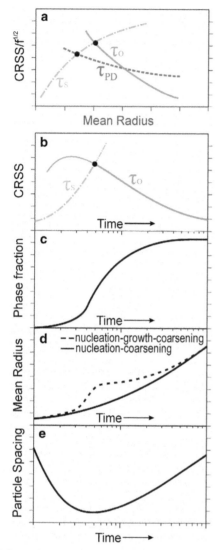

species along the GB and free surface (FS), $\Delta E_{GB} - \Delta E_{FS}$,[80,82,92] which indicates a species as a cohesion enhancer when this quantity is negative. Rice and Wang[92] directly correlated this quantity with measured embrittlement sensitivity quantities, relating intergranular DBTT and interfacial segregant atomic fraction. Accurate prediction of this quantity based on FLAPW were then performed by the SRG at the Northwestern University.[89,91] Combining initial FLAPW calculations and handbook quantities, such as elemental cohesive energies, a simplified model was developed to predict potencies represented in Figure 9.[89] To quantify the accuracy of this simplified model, several rigorous validation FLAPW calculations were performed (solid points in the inset of Figure 9), and confirmed that W and Re exhibit the highest cohesion enhancement. Eventually, a genomic surface thermodynamic database was constructed to enable design of the GB composition to entirely eliminate the intergranular mode of hydrogen embrittlement.[55,87,88]

In addition to these methods of increasing GB cohesion and interfacial bonding between matrix and particles, transformation plasticity was further introduced to improve toughness during plastic deformation in the mechanical Co-Ni UHS steels as shown in Figure 8. A key design parameter is the transition temperature called M_s^σ, below which stress-assisted martensitic transformation controls materials yielding. As quantified by the Olson–Cohen theory,[93,94] below M_s^σ, the transformation controlled yield stress rises due to the stability of the parent austenite increasing with temperature. Above M_s^σ, the slip-controlled yield stress decreases with temperature through thermal activation of dislocation motion. Controlling the stability of the austenite phase qualified by M_s^σ is the key to optimising transformation toughening. In the studies by Leal[95] and Stavehaug,[96] the M_s^σ temperature was further characterised by distinguishing between that defined by uniaxial tension $M_s^\sigma(ut)$ and by the crack tip stress state denoted as $M_s^\sigma(ct)$. As shown in Figure 10, the uniform ductility reaches a maximum between these two temperatures, and maximum fracture toughness occurs at $M_s^\sigma(ct)$. To quantitatively design transformation toughness in UHS steel, a model linked with thermodynamic database is proposed by Olson and his colleagues[67,97–101] in a series of works. The underlining approach is based on an austenite stability parameter (ASP),[97,98] which can be expressed as:

$$\Delta G_{ch}(x_i, T) + W_f(x_i, T) = -G_n - \Delta G_\sigma(\sigma, \Delta V/V) \qquad (2)$$

where the first term on the left side of equation is chemical contribution of the total energy, the second term is the interfacial friction energy, where both are functions of temperature T and composition x_i. The right side of the above equation is composed of defect potency, G_n, which can be approximated as a constant value for a certain material,[97,98] and mechanical driving force, ΔG_σ, which is a function of stress state and strength goal of the material. According to the Olson–Cohen model,[93] given the stress state of the material and desired M_s^σ temperature, it is possible to provide a critical value for the austenite stability parameter, and thus allow dispersed-austenite composition to be designed through the CALPHAD genomic databases.

Structure-creep/fatigue. Unlike strength and toughness, creep and fatigue are dynamically structure-sensitive properties, where predictive models are more complex. For creep modelling, many available phenomenological models require numerous empirical parameters, and thus generate a heavy load for experiments. However, there is a promising method for parametric design purposes when dynamic creep models are correlated to diffusion kinetic modelling. For example, in single-crystal Ni superalloys, under a high-temperature, low-stress condition, where dislocation climb at the matrix γ-precipitate γ' interface becomes the dominant rate-controlling step in the kinetic creep process, modelling vacancy flux can be directly employed for creep

Figure 6. Schematic representation of dispersed particle strengthening mechanism governed by process–structure–property-performance paradigm. (**a**) relation between normalised critical resolved shear stress (CRSS) using square root of the volume fraction for the second-phase strengthening particle $f^{1/2}$ and mean particle radius. Blue chain line denotes shear stress, red-dashed curve corresponding to the interaction between paired dislocation and obstacle, and the green curve is for Orowan looping effects. (**b–e**) show kinetic evolution of CRSS, particle phase fraction, mean radius and particle spacing.

aid parametric design of ductile toughness. According to the model developed by Wang and Olson,[84] the limit of ductile fracture toughness can be quantitatively related to the critical strain for shear localisation governing primary void coalescence.

As illustrated in Figure 8, another interface level design concept for avoiding environmental GB embrittlement is the GB cohesion enhancement, which is mainly applied to minimise hydrogen-induced intergranular embrittlement.[68,87,88] As a basis for establishing an understanding on the electronic level, the full-potential augmented planewave (FLAPW) method was applied in a series studies on UHS steel[68,87–91] based on a thermodynamic theory developed by Rice and Wang.[92] Their theory describes the mechanism of impurity-induced intergranular embrittlement via the competition between plastic crack blunting and brittle boundary separation. The derived 'embrittlement potency' is defined as the segregation energy difference between segregant

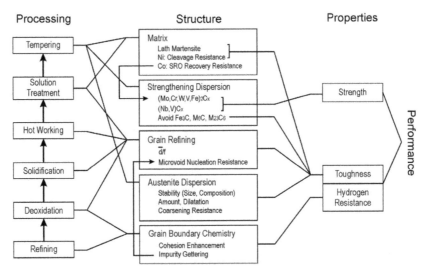

Figure 7. Systems design chart of the Co–Ni UHS steel.

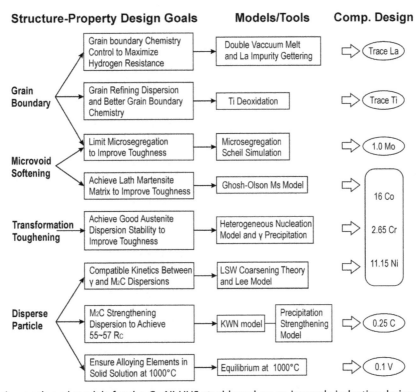

Figure 8. Summary of design goals and models for the Co–Ni UHS steel based on an inversely inductive design approach.

resistance design. Recent studies published by Dyson[102] and Zhu et al.[103] provide such promising models with vacancy diffusivity as a key design parameter, which can be calculated using the CALPHAD diffusion database for multicomponent alloy design. Further at very high temperatures, where rafting affects creep, structure–property models can also be related to the CALPHAD genomic database. In the viscoplastic regime described by the Socrate–Parks model,[104] the thermodynamic driving force for rafting is primariily controlled by the lattice mismatch of the matrix γ and the precipitate γ', which can be predicted through models[105] linking to the CALPHAD molar volume databases.[106,107]

With regard to structure–property models for fatigue, as indicated in Figure 2, besides conventional statistical analysis-based phenomenological models, more attention is now paid to micromechanical FEA as a foundation for a predictive probabilistic

approach. The state-of-the-art micromechanical FEA on fatigue is comprehensively reviewed by Pineau et al.[108] highlighting the seminal contributions of McDowell[108] and his former students. Fatigue is an intrinsically multiscale and multistage phenomenon, and thus highly sensitive to microstructure level design. Conventional fatigue modelling separates the total life into two stages: initiation life N_i and crack propagation life defined by number of cycles N_p. For low cycle fatigue, N_p dominates, and thus can be modelled by statistical methods via some parametric scaling law (e.g., the Coffin–Manson law[109]). In contrast, fatigue crack initiation dominates the total life of (ultra) high cycle fatigue. Therefore, it should be predictable by the micromechanical model with process-microstructure sensitivity. A recent thesis work by Moore[110] further developed the micromechanics-based method for multiscale fatigue nucleation prediction, which successfully

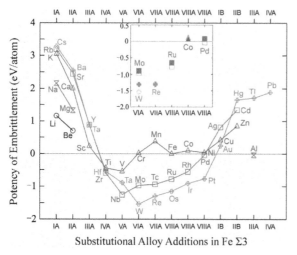

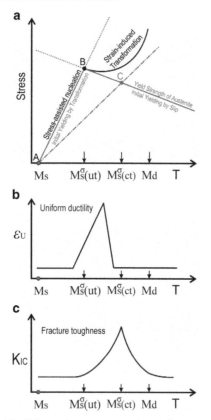

Figure 9. Embrittlement potency prediction made for substitutional elements in Fe grain boundaries.[89] Solid points in inset represent FLAPW calibration/validation results.[55]

Figure 10. (**a**) Schematic representation of interrelationships between stress-assisted and strain-induced martensitic transformation. (**b**) Uniform ductility ε_U as a function of temperature. (**c**) Fracture toughness K_{IC} as a function of temperature.

predicted the fatigue behaviour of the Ni–Ti shape memory alloy. The 3D image-based micromechanical FEA demonstrates that an alloy design with enhanced yield strength can significantly increase fatigue strength in addition to the effects of minimising inclusion size. It should be emphasised that such micromechanical FEA for ultra-high cycle fatigue simulation directly drove a shape memory biomedical alloy design[111,112] performed synchronously. It was determined that a 50% increase in Ni–Ti yield strength resulted in a 44% increase in ultra-high cycle fatigue limit at 10^9 cycles.

We note that, in a practical design, although a completely deductive process–structure–property model focusing on sequential cause and effects is difficult to obtain, an inversely inductive design approach directed by moving fundamental design parameters in the right direction is usually sufficient to implement a hierarchical design using specific means based on the available mechanistic models of Materials by Design.

AIM: ACCELERATING TECHNOLOGY TRANSITION

As indicated in the ICMD hierarchical architecture (see Figure 1), the AIM method completes the process of component-level materials development. Here, final design allowables bounded by the process–structure–property models are evaluated deliberately. To fully implement AIM, an integrated product team consisting of original equipment manufacturer, small company, university and government laboratory are often assembled determining design goals, sharing information and integrating available design models/tools within the framework now known as ICME.

Although the above described process–structure–property models in Materials by Design is utilised extensively in AIM as a basis, there are three additional unique thrusts in the AIM method, which includes (i) multidisciplinary engineering model integration, (ii) location-specific modelling throughout a component and (iii) uncertainty quantification and management.

Standard statistical methods are naturally vital for uncertainty quantification, which guides design tolerances. Such techniques were applied in the aforementioned design of the UHS steel, Ferrium S53 by the SRG consortium, run concurrently with the DARPA-AIM project. Design sensitivity analysis was performed using a combined method of Monte Carlo simulation and iCMD (an integrated computational materials design software package created by QuesTek Innovations LLC, Evanston, IL, USA), implemented through the commercial iSIGHT design optimization platform. For example, within 12 min on a Pentium IV 2.2 GHZ CPU (QuesTek Innovations LLC), a thousand parametric of structure–property iCMD calculations using the three-σ limits of the composition and process temperature tolerances generated the probability distribution of Charpy V-Notch toughness, hardness, Martensite start temperature and DBTT.

Under the DARPA-AIM 3-year project starting from 2000 using the example of IN100 superalloy turbine disks, the iSIGHT design integration system was adopted as the cross-disciplinary platform to link distributed software capabilities in Utah and Illinois as shown in Figure 11a. The turbine disk heat treatment process modelling performed using a FEA-based DEFORM Heat Treatment module (Scientific Forming Technologies Corporation, Columbus, OH, USA) determined the history of temperature profile at every location in a part, which were taken as the input of the precipitation simulator, PrecipiCalc (QuesTek Innovations LLC), for the process–structure modelling. In the final step, a yield strength structure–property model was applied to predict the strength at both room temperature and elevated temperatures by taking results of PrecipiCalc and a CALPHAD-prediction of antiphase boundary energies as the input. The efficiency and high fidelity inherited from the Materials by Design models for the AIM methodology can be reflected by the quantitative simulation of structure and yield strength for the disk centre location. By tuning parameters of interfacial property and diffusivity through model-calibration performed at the University of Connecticut, the dynamic behaviour of the strengthening particle can be quantitatively captured using PrecipiCalc simulation with a high accuracy.[113] Figures 11b,c shows the time evolution of mean sizes and fractions of three γ' populations under a three-step heat treatment cycle. To further perform a location-specific design, process–structure–property models were applied to an entire minidisk as shown in Figures 11d,e. The calculated spatial contours of secondary γ' size show high consistency with

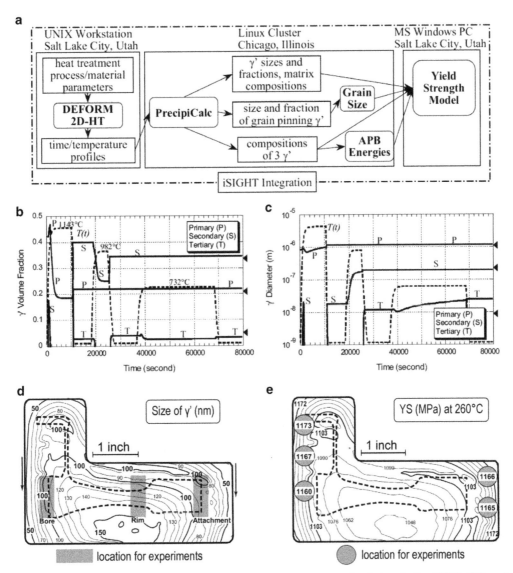

Figure 11. (a) Distribution iSIGHT integration for mechanistic process–structure–property simulations in the DARPA-AIM project, performed on different computer hardware and operating systems across network. (**b,c**): Kinetic simulation performed using PrecipiCalc for a centre location of an aero-turbine disk under the three-step heat treatment cycle. Time evolutions of mean sizes and fractions of three γ′ distributions are shown in solid curves overlapping with temperature profile as dashed curve. Experimental γ′ microstructural results are indicated by solid triangles on the right. (**d**) Calculated spatial contours of secondary γ′ size (nm) for a radially symmetric cross-section of the minidisk. The disk is immersed into quenching media from bottom to top denoted by side-arrows. (**e**) Calculated spatial contour of yield strength (MPa) at 260 °C for minidisk. Figure is reproduced based on the work performed by Jou *et al.*[113]

experimental observation on bore, rim and attachment parts indicated with shaded rectangles in Figure 11d. Using the structure-strength model, the tensile yield strength at five different locations was predicted within quantified uncertainty ranges by comparing with experimental results on specific locations (see circles highlighted in Figure 11d.

As a follow-on to the DARPA-AIM programme, a National Aeronautics and Space Administration-funded project at QuesTek Innovations in 2005–2008 further extended the above AIM techniques for a hierarchical calibration/validation uncertainty management approach for all available process–structure models[31] for a Dual Microstructure Heat Treatment (DMHT) process design indicated in Figure 12. As different location-specific properties require different microstructures, the design sensitivity of the process–structure model is crucial to meet property requirements. Table 3 summarises the sequential characterisation protocol performed for standard uncertainty

quantification of individual inputs in the process–structure models. Subsequently, a hierarchical uncertainty management strategy was adopted by balancing model sensitivity against database intrinsic accuracy. Rigid shifts to second-phase free-energy functions according to Gibbs–Thompson effects were applied for a precise description of the secondary strengthening particles. Meanwhile, DICTRA atomic mobility databases were recalibrated by diffusion couple experiments, which achieved an improved accuracy of diffusivity prediction. Regarding the high sensitivity of interfacial energy in precipitation simulation, PrecipiCalc simulations were carried out by adjusting interfacial energy to match the critical nucleation undercooling accurately determined by the single sensor differential thermal analysis technique, which measures a precise thermal history of a rapidly-quenched 3-mm diameter pins.[31,32] Afterwards, the PrecipiCalc simulated microstructural attributes were further compared with the APT characterisation to evaluate uncertainty limits.

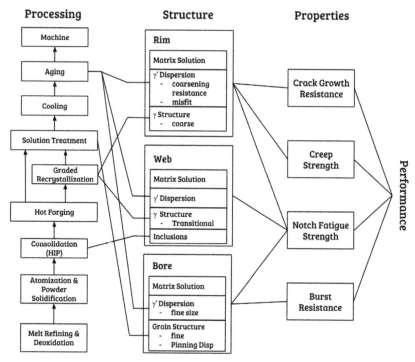

Figure 12. Systems design chart of Ni turbine disk alloy using the AIM method. Hierarchical microstructure are designed for specific locations in the turbine disk.

Table 3. Summary of database, model, calibration and validation protocol for precipitation modelling adopted in the AIM method

Database	Quantity	Model	Model parameters	Calibration and validation
Thermodynamics	Bulk energy	CALPHAD	Free energy	XRD/EDS/TEM/APT
	Interfacial energy	Gibbs–Thompson	G_{el}, σ_{coh}	APT/SSDTA
Molar volume	Lattice mismatch	CALPHAD	G_{el}, R	XRD/TEM/DFT
Kinetics	Atomic mobility	CALPHAD	Mobility	Diffusion couple
	D_{scale}		D_{scale}	EPMA or WDS

Abbreviations: APT, atom probe tomography; DFT, density functional theory; EDS, energy dispersive spectroscopy; G_{el}, estimated elastic coherency energy; R, coherency transition size from coherent to incoherent state; SSDTA: single sensor differential thermal analysis;[32] TEM, transmission electron microscopy; XRD, X-ray diffraction.

It should be noted that, having demonstrated the importance of the AIM technique, a suite of advanced 3D tomographic multiscale characterisation tools were further developed under the support of ONR and DARPA. Thus, the AIM technique is now supported by both novel experiments and 3D process–structure–property models reaching a new level of fidelity, which is greatly beneficial to the enterprise of materials cycle compression.

SUMMARY: CYBERMATERIALS INNOVATION BASED ON MATERIALS GENOME

The continued development of the ICMD techniques, Materials by Design and AIM, is already accelerating materials innovation, thus leveraging the current design techniques in materials science and mechanical engineering, which foster synergistic interactions among disciplines.

Materials by Design and AIM techniques serve as the two major pillars in the ICMD hierarchical architecture. As reviewed in the foregoing sections, sustained efforts have made grand achievements in transferring science-based ideas to industrial production. However, a sustainable development strategy is still needed to fulfil the potential of the ICMD blueprint in Figure 1 for cybermaterials innovations. It is clear that Materials Genome database as the cornerstone has a decisive role.

First, the systems design models of process–structure–property–performance paradigm can all benefit from further improvement. For instance, the preceding discussion on Materials by Design has highlighted needs of model development for non-spherical particles in the framework of the KWN model-based simulation. In view of the high sensitivity of nucleation to interfacial energy, more rigorous atomistic modelling of interfacial thermodynamics would enhance predictive capability.

Second, because process–structure linkage design relies heavily on the quality of the fundamental materials genomic databases assembled by computational thermodynamics and kinetics, more dedicated research on constructing high-quality genomic database should be motivated and promoted. Importantly, these research activities also include experiments providing protodata. The so-called high-throughput experiment represented by diffusion multiples[114] is a good example to make such contributions. Meanwhile, more efforts should be made on bridging the current DFT-based atomistic models and CALPHAD method by extending their temperature scales.[115] A recent work on the Fe–Cr system[22] demonstrated that the CALPHAD models extended to 0 K will garner valuable insights into anomalous phenomenon, i.e., negative enthalpy of mixing on Fe-rich side, found in enthalpy of formation generated by DFT calculations.[116]

Expending the impact of the structure–property model development, genomic databases of mechanistic constitutive laws for mechanical FEA should be initiated. Such efforts need intimate collaborations between experimentalists and modelling experts in both materials science and mechanical engineering. As mentioned in the above discussion, the state-of-the-art materials design usually considers experimental calibration as an important part in design iterations. Therefore, technical improvement of such experiments for fundamental model-calibration purposes should be a primary effort.

Materials genomic database construction requires corresponding efforts on standardisation and database management as indispensable support. Taking DBTT as an example, although DBTT can be evaluated using Charpy impact energy tests, precise modelling of DBTT is currently data intensive.[117,118] If a standard of efficient DBTT measurement is available, limitations affecting the accuracy of DBTT will be significantly diminished, and thus improve the DBTT model predictability. Further, we note that related topics are often much broader than the research field of materials science, driving the call for collaborative efforts by various scientific and engineering communities.

Last but not least, perhaps the greatest promise of materials development cycle compression is the new opportunity of concurrent design of materials and devices.[119] Notable achievements have already been demonstrated in the field of consumer electronics which would never have been possible under trial-and-error empirical materials development.

Overall, it is evident that the next generation of cybermaterials innovation applicable across all materials classes will be grounded in ICMD hierarchical architecture powered by an expanded system of materials genomic databases.

ACKNOWLEDGEMENTS
This work was performed within the CHiMaD Center for Hierarchical Materials Design at the Northwestern Univeristy, funded by the National Institute of Standards and Technology, USA. ICMD and Materials by Design are trademarks of QuesTek Innovations. We are grateful to Professor John Ågren (KTH, Sweden), Dr Qing Chen (Thermo-Calc Software AB, Sweden), Dr Ikumu Watanabe (NIMS, Japan), and Mr Abhinav Saboo (Northwestern University) for stimulating discussions.

COMPETING INTERESTS
The authors declare no conflict of interest.

REFERENCES
1. Apelian, D. et al. Accelerating Techology Transition: Bridge the Valley of Death for Materials and Processes in Defense Systems. National Research Council of The National Academies, The National Academes Press, (2004). Available online at: http://www.nap.edu/catalog.php?record_id = 11108.

2. Holdren, J. P. Materials Genome Initiative for Global Competitiveness. (Executive office of the president national science and technology council, 2011).

3. Pollock, T. M. et al. Integrated Computational Materials Engineering: A Transformational Discipline for Improved Competitiveness and National Security. (National Research Council of the National Academies, The National Academies Press, 2008) Available online at: http://www.nap.edu/catalog/12199.html.

4. Horstemeyer, M. F.. An Introduction to Integrated Computational Materials Engineering (ICME). Integr Comput Mater Eng Met. (John Wiley & Sons, Inc., 2012).

5. Olson, G. B. Preface to the viewpoint set on: the Materials Genome. Scr. Mater. 70, 1–2 (2014).

6. Kaufman, L. & Ågren, J. CALPHAD, first and second generation—birth of the materials genome. Scr. Mater. 70, 3–6 (2014).

7. Campbell, C. E., Kattner, U. R. & Liu, Z. K. File and data repositories for Next Generation CALPHAD. Scr. Mater. 70, 7–11 (2014).

8. Bhadeshia, H. K. D. H. Computational design of advanced steels. Scr. Mater. 70, 12–17 (2014).

9. Christensen, S. & D'Oyen, R. Computational formulation of a new composite matrix. Scr. Mater. 70, 18–24 (2014).

10. Olson, G. B. & Kuehmann, C. J. Materials genomics: from CALPHAD to flight. Scr. Mater. 70, 25–30 (2014).

11. Yuan, Y. et al. A new method to strengthen turbine disc superalloys at service temperatures. Scr. Mater. 66, 884–889 (2012).

12. Yuan, Y. et al. Deformation mechanisms in a new disc superalloy at low and intermediate temperatures. Scr. Mater. 67, 137–140 (2012).

13. Reed, R. C., Tao, T. & Warnken, N. Alloys-By-Design: Application to nickel-based single crystal superalloys. Acta. Mater. 57, 5898–5913 (2009).

14. Olson, G. B. Computational design of hierarchically structured materials. Science 277, 1237–1242 (1997).

15. Olson, G. B. Genomic materials design: the ferrous frontier. Acta. Mater. 61, 771–781 (2013).

16. Martinetti, L. et al. A critical gel fluid with high extensibility: the rheology of chewing gum. J. Rheol. 58, 821–838 (2014).

17. D'Oyen, R. M.. Systems Design of Case Hardened Polymers for Gears, Doctoral thesis Northwestern University, (1997).

18. Saunders, N. & Miodownik, A. P. CALPHAD (Calculation of Phase Diagrams): A Comprehensive Guide Pergamon Mater Ser. Pergamon, (1998). Available online at: http://dx.doi.org/10.1016/S1470-1804(98)80019-9.

19. Xiong, W., Xie, W., Shen, C. & Morgan, D. Thermodynamic modelling of the U-Zr system—a revisit. J. Nucl. Mater. 443, 331–341 (2013).

20. Xiong, W., Xie, W. & Morgan, D. Thermodynamic evaluation of the Np-Zr system using CALPHAD and ab initio methods. J. Nucl. Mater. 452, 569–577 (2014).

21. Xiong, W., Du, Y., Lu, X., Schuster, J. C. & Chen, H. Reassessment of the Ce-Ni binary system supported by key experiments and ab initio calculations. Intermetallics 15, 1401–1408 (2007).

22. Xiong, W. et al. An improved thermodynamic modelling of the Fe-Cr system down to zero kelvin coupled with key experiments. CALPHAD 35, 355–366 (2011).

23. Xiong, W. et al. Construction of the Al-Ni-Si phase diagram over the whole composition and temperature ranges: thermodynamic modelling supported by key experiments and first-principles calculations. Int. J. Mater. Res. 99, 598–612 (2008).

24. Hallstedt, B. & Liu, Z.-K. Software for thermodynamic and kinetic calculation and modelling. CALPHAD 33, 265 (2009).

25. Lukas, H., Fries, S. G. & Sundman, B. Computational Thermodynamics: The Calphad Method (Cambridge University Press, 2007).

26. Andersson, J.-O., Höglund, L., Jönsson, B., Ågren, J. in Proc Metall Soc Can Inst Min Metall (ed. Purdy G.) 153–163 (Pergamon, 1990). Available online at: http://dx.doi.org/10.1016/B978-0-08-040412-7.50023-2.

27. Borgenstam, A., Engstrom, A., Hoglund, L. & Agren, J. DICTRA, a tool for simulation of diffusional transformations in alloys. J. Phase Equilibria. 21, 269–280 (2000).

28. Xiong, W. CALPHAD-based integrated computational materials engineering research for materials genomic design. JOM 67, 1864–1865 (2015).

29. Bender, M. D. & Olson, G. B. Designing a precipitation-strengthened, superelastic, TiNi-based alloy for endovascular stentsin Int. Conf. Martensitic Transform 159–166 (John Wiley & Sons, Inc., 2010).

30. Bender, M. D. & Olson, G. B. Computational thermodynamics-based design of nanodispersion-strengthened shape memory alloys. in SMST-2007—Proc. Int. Conf. Shape Mem. Superelastic Technol. 115–122 (2008).

31. Olson, G. B. et al. Precipitation model validation in 3rd generation aeroturbine disc alloysin 11th Int. Symp. Superalloys 923–932 (Minerals, Metals and Materials Society, 2008).

32. Alexandrov, B. T. & Lippold, J. C. Single sensor differential thermal analysis of phase transformations and structural changes during welding and postweld heat treatment. Weld World 51, 48–59 (2007).

33. Wusatowska-Sarnek, A. M., Ghosh, G., Olson, G. B., Blackburn, M. J. & Aindow, M. Characterization of the microstructure and phase equilibria calculations for the powder metallurgy superalloy IN100. J. Mater. Res. 18, 2653–2663 (2003).

34. Lifshitz, I. M. & Slyozov, V. V. The kinetics of precipitation from supersaturated solid solutions. J. Phys. Chem. Solids 19, 35–50 (1961).

35. Wagner, C. Theorie der der Alterung von Niederschlägen durch Umlösen (Ostwald-Reifung). Zeitschrift für Elektrochemie, Berichte der Bunsengesellschaft für Phys Chemie 65, 581–591 (1961).

36. Morral, J. E. & Purdy, G. R. Particle coarsening in binary and multicomponent alloys. Scr. Metall. Mater. 30, 905–908 (1994).

37. Morral, J. E. & Purdy, G. R. Thermodynamics of particle coarsening. J. Alloys Compd. 220, 132–135 (1995).

38. Kuehmann, C. J. & Voorhees, P. W. Ostwald ripening in ternary alloys. Metall. Mater. Trans. A 27, 937–943 (1996).

39. Lee, H. M., Allen, S., Grujicic, M. Stability and coarsening resistance of M2C carbides in secondary hardening steelsin Innov Ultrahigh-Strength Steel Technol Sagamore Army Mater. Res. Conf. (eds Olson G. B., Azrin M. & Wright E. S.) 127–146 (U.S. Government Printing Office, 1990).

40. Umantsev, A. & Olson, G. B. Ostwald ripening in multicomponent alloys. Scr. Metall. Mater. 29, 1135–1140 (1993).

41. Wagner, R., Kampmann, R., Voorhees, P. W. in *Phase Transform Mater* (ed. Kostorz G.) 309–407 (Wiley-VCH Verlag GmbH and Co. KGaA, 2005).

42. Robson, J. D. Modelling the overlap of nucleation, growth and coarsening during precipitation. *Acta. Mater.* **52**, 4669–4676 (2004).

43. Kampmann, R., Wagner, R. in *At Transp Defects Met by Neutron Scatt SE—12* (eds Janot C., Petry W., Richter D. & Springer T.) 10, 73–77 (Springer, 1986).

44. Aaronson, H. I. & LeGoues, F. K. Assessment of studies on homogeneous diffusional nucleation kinetics in binary metallic alloys. *Metall. Trans. A, Phys. Metall. Mater. Sci.* **23A**, 1915–1945 (1992).

45. Alvensleben, L. V., Wagner, R.FIM-atom probe studies of early stage decomposition in Cu-Ti alloysin *Proc 2nd Acta-Scripta Metall. Conf.* (eds Haasen P., Gerold V., Wagner R. & Ashby M. F.) 143–148 (Pergamon, 1984) Available online at: http://dx.doi.org/10.1016/B978-0-08-031651-2.50026-4.

46. Eckerlebe, H., Kampmann, R., Wagner, R. in *At Transp Defects Met by Neutron Scatt SE—11* (eds Janot C., Petry W., Richter D. & Springer T.) 10, 66–72 (Springer, 1986).

47. Wendt, H. & Haasen, P. Nucleation and growth of γ'-Precipitates in Ni-14 at.% Al. *Acta. Metall.* **31**, 1649–1659 (1983).

48. Allen, S. M. & Cahn, J. W. A microscopic theory for antiphase boundary motion and its application to antiphase domain coarsening. *Acta. Metall.* **27**, 1085–1095 (1979).

49. Cahn, J. W. & Hilliard, J. E. Free energy of a nonuniform system. I. interfacial free energy. *J. Chem. Phys.* **28**, 258–267 (1958).

50. Kitashima, T. Coupling of the phase-field and CALPHAD methods for predicting multicomponent, solid-state phase transformations. *Philos. Mag.* **88**, 1615–1637 (2008).

51. Larsson, H. & Höglund, L. A scheme for more efficient usage of CALPHAD data in simulations. *CALPHAD* **50**, 1–5 (2015).

52. Sebastian, J. & Olson, G. B. Examples of QuesTek Innovations' Application ICME to Materials Design, Development, and Rapid Qualificationin *55th AIAA/ASMe/ASCE/AHS/SC Struct Struct Dyn Mater Conf* 1–7 (American Institute of Aeronautics and Astronautics, 2014).

53. Hornbogen, E. Design alloys for >3 GPa steelsin *Innov Ultrahigh-Strength Steel Technol Sagamore Army Mater Res Conf* (eds Olson G. B., Azrin M. & Wright E. S.) 113–126 (U.S. Government Printing Office, 1990).

54. Wang, J. S., Mulholland, M. D., Olson, G. B. & Seidman, D. N. Prediction of the yield strength of a secondary-hardening steel. *Acta. Mater.* **61**, 4939–4952 (2013).

55. Kantner, C. D. *Designing Strength, Toughness, and Hydrogen Resistance*, Doctoral Thesis (Northwestern University, 2002).

56. Nabarro, F. R. N. Fifty-year study of the Peierls-Nabarro stress. *Mater Sci Eng A* **234–236**, 67–76 (1997).

57. Nabarro, F. R. N. Dislocations in a simple cubic lattice. *Proc. Phys. Soc.* **59**, 256 (1947).

58. Wang, J. N. A new modification of the formulation of Peierls stress. *Acta. Mater.* **44**, 1541–1546 (1996).

59. Nembach, E. *Particle Strengthening of Metals and Alloys.* (John Wiley, 1997).

60. Nembach, E., Schänzer, S., Schröer, W. & Trinckauf, K. Hardening of nickel-base superalloys by high volume fractions of γ'-precipitates. *Acta. Metall.* **36**, 1471–1479 (1988).

61. Huether, W. & Reppich, B. Interaction of dislocations with coherent, stress-free, ordered particles. *Zeitschrift fuer Met. Res. Adv. Tech.* **69**, 628–634 (1978).

62. Tiemens, B., Sachdev, A. & Olson, G. B. Cu-precipitation strengthening in ultrahigh-strength carburizing steels. *Metall. Mater. Trans. A* **43**, 3615–3625 (2012).

63. Tiemens, B. L., Sachdev, A. K., Mishra, R. K. & Olson, G. B. Three-dimensional (3-D) atom probe tomography of a Cu-precipitation- strengthened, ultrahigh-strength carburized steel. *Metall. Mater. Trans. A Phys. Metall. Mater. Sci.* **43**, 3626–3635 (2012).

64. Cowie, J. G., Azrin, M. & Olson, G. B. Microvoid formation during shear deformation of ultrahigh strength steels. *Metall. Trans. A, Phys. Metall. Mater. Sci.* **20A**, 143–153 (1989).

65. Lippard, H. E. *et al.* Microsegregation behavior during solidification and homogenization of AerMet100 steel. *Metall. Mater. Trans. B Process Metall. Mater. Process. Sci.* **29**, 205–210 (1998).

66. Olson, G. B. Transformation plasticity and toughening. *J. Phys. IV JP* **6**, C1-407–C1-418 (1996).

67. Haidemenopoulos, G. N., Grujicic, M., Olson, G. B. & Cohen, M. Thermodynamics-based alloy design criteria for austenite stabilization and transformation toughening in the Fe-Ni-Co system. *J. Alloys Compd.* **220**, 142–147 (1995).

68. Wu, R., Freeman, A. J. & Olson, G. B. First principles determination of the effects of phosphorus and boron on iron grain boundary cohesion. *Science* **265**, 376–380 (1994).

69. Olson, G. B. *Corrosion Resistant Steels for Structural Applications in Aircraft* (Strategic Environmental Research and Development Program (SERDP), Department of Defense, 2005). Available online at: https://www.serdp-estcp.org/content/download/6302/84489/file/PP-1224-FR-01.pdf.

70. Moskovic, R. & Flewitt, P. E. J. An overview of the principles of modeling charpy impact energy data using statistical analyses. *Metall. Mater. Trans. A* **28**, 2609–2623 (1997).

71. Cowie, J. G. & Tuler, F. R. The influence of second-phase dispersions on shear instability and fracture toughness of ultrahigh strength AISI 4340 steel. *Mater. Sci. Eng. A* **141**, 23–37 (1991).

72. Hillert, M. On the estimation of the Zener drag on grain boundaries. *Scr. Metall.* **18**, 1431–1432 (1984).

73. Hillert, M. Inhibition of grain growth by second-phase particles. *Acta. Metall.* **36**, 3177–3181 (1988).

74. Wörner, C. H., Cabo, A. & Hillert, M. On the limit for particle attachment in Zener drag. *Scr. Metall.* **20**, 829–831 (1986).

75. Hao, S., Moran, B., Kam Liu, W. & Olson, G. B. A hierarchical multi-physics model for design of high toughness steels. *J. Comput. Mater. Des.* **10**, 99–142 (2003).

76. Hao, S., Liu, W. K., Moran, B., Vernerey, F. & Olson, G. B. Multi-scale constitutive model and computational framework for the design of ultra-high strength, high toughness steels. *Comput. Methods Appl. Mech. Eng.* **193**, 1865–1908 (2004).

77. Vernerey, F. *et al.* The 3-D computational modeling of shear-dominated ductile failure in steel. *JOM* **58**, 45–51 (2006).

78. Vernerey, F. J., Liu, W. K., Moran, B. & Olson, G. B. A micromorphic model for the multiple scale failure of heterogeneous materials. *J. Mech. Phys. Solids* **56**, 1320–1347 (2008).

79. Tian, R. *et al.* A multiresolution continuum simulation of the ductile fracture process. *J. Mech. Phys. Solids* **58**, 1681–1700 (2010).

80. Zhang, S., Kontsevoi, O. Y., Freeman, A. J. & Olson, G. B. First principles investigation of zinc-induced embrittlement in an aluminum grain boundary. *Acta. Mater.* **59**, 6155–6167 (2011).

81. Medvedeva, N. I., Gornostyrev, Y. N., Kontsevoi, O. Y. & Freeman, A. J. Ab-initio study of interfacial strength and misfit dislocations in eutectic composites: NiAl/Mo. *Acta. Mater.* **52**, 675–682 (2004).

82. Zhang, S., Kontsevoi, O. Y., Freeman, A. J. & Olson, G. B. Cohesion enhancing effect of magnesium in aluminum grain boundary: a first-principles determination. *Appl. Phys. Lett.* **100**, 231904 (2012).

83. Lee, J. H., Shishidou, T., Zhao, Y. J., Freeman, A. J. & Olson, G. B. Strong interface adhesion in Fe/TiC. *Philos. Mag.* **85**, 3683–3697 (2005).

84. Wang, J.-S. & Olson, G. B. Prediction of ductile fracture toughness. in *12th Int. Conf. Fract. 2009, ICF-12* **1**, 145–154 (2009).

85. Pardoen, T. & Hutchinson, J. W. An extended model for void growth and coalescence. *J. Mech. Phys. Solids* **48**, 2467–2512 (2000).

86. Pardoen, T. & Hutchinson, J. W. Micromechanics-based model for trends in toughness of ductile metals. *Acta. Mater.* **51**, 133–148 (2003).

87. Geng, W.-T., Freeman, A. J., Olson, G. B., Tateyama, Y. & Ohno, T. Hydrogen-promoted grain boundary embrittlement and vacancy activity in metals: Insights from ab initio total energy calculatons. *Mater. Trans.* **46**, 756–760 (2005).

88. Zhong, L., Wu, R., Freeman, A. J. & Olson, G. B. Charge transfer mechanism of hydrogen-induced intergranular embrittlement of iron. *Phys. Rev. B* **62**, 13938–13941 (2000).

89. Geng, W. T., Freeman, A. J. & Olson, G. B. Influence of alloying additions on grain boundary cohesion of transition metals: first-principles determination and its phenomenological extension. *Phys. Rev. B* **63**, 165415 (2001).

90. Wu, R., Freeman, A. J. & Olson, G. B. On the electronic basis of the phosphorus intergranular embrittlement of iron. *J. Mater. Res.* **7**, 2403–2411 (1992).

91. Wu, R., Freeman, A. J. & Olson, G. B. Effects of carbon on Fe-grain-boundary cohesion: first-principles determination. *Phys. Rev. B* **53**, 7504–7509 (1996).

92. Rice, J. R. & Wang, J.-S. Embrittlement of interfaces by solute segregation. *Mater. Sci. Eng. A* **107**, 23–40 (1989).

93. Olson, G. B., Cohen, M. Martensitic transformations as a deformation processin *Mechanical properties and phase transformations in engineering materials-Earl R Parker Symposium on Structure Property Relationships* (eds Antolovich S. D., Ritchie R. O. & Gerberich W. W.) 367–390 (The Metallurgical Society, Inc., 1986).

94. Olson, G. B., Feinberg, Z. D. in *Phase Transform Steels Fundam Diffus Control Transform Vol 2* (eds Arnold S. & Pereloma E.) 59–82 (Woodhead Publishing Ltd., 2012).

95. Leal, R. H. *Transformation Toughening of Metastable Austenitic Steels*, Doctoral Thesis (Massachusetts Institute of Technology, 1984).

96. Stavehaug, F. *Transformation Toughening of gamma-strengthened Metastable Austenitic Steels*, Doctoral Thesis Massachusetts Institute of Technology, (1990).

97. Ghosh, G. & Olson, G. B. Kinetics of f.c.c. → b.c.c. heterogeneous martensitic nucleation—I. The critical driving force for athermal nucleation. *Acta. Metall. Mater.* **42**, 3361–3370 (1994).

98. Ghosh, G. & Olson, G. B. Kinetics of F.c.c. → b.c.c. heterogeneous martensitic nucleation—II. Thermal activation. *Acta. Metall. Mater.* **42**, 3371–3379 (1994).

99. Olson, G. B. & Cohen, M. A general mechanism of martensitic nucleation: Part I. General concepts and the FCC → HCP transformation. *Metall. Trans. A* **7**, 1897–1904 (1976).

100. Olson, G. B. & Cohen, M. A general mechanism of martensitic nucleation: Part II. FCC → BCC and other martensitic transformations. *Metall. Trans. A* **7**, 1905–1914 (1976).

101. Olson, G. B., Cohen, M. in *Dislocations Solids* (ed. Nabarro F. R. N.) (Elsevier Science B.V., 1986).

102. Dyson, B. F. Microstructure based creep constitutive model for precipitation strengthened alloys: theory and application. *Mater. Sci. Technol.* **25**, 213–220 (2009).

103. Zhu, Z., Basoalto, H., Warnken, N. & Reed, R. C. A model for the creep deformation behaviour of nickel-based single crystal superalloys. *Acta. Mater.* **60**, 4888–4900 (2012).

104. Socrate, S. & Parks, D. M. Numerical determination of the elastic driving force for directional coarsening in Ni-superalloys. *Acta. Metall. Mater.* **41**, 2185–2209 (1993).

105. Jung, J., Ghosh, G. & Olson, G. B. A comparative study of precipitation behavior of Heusler phase (Ni2TiAl) from B2-TiNi in Ni-Ti-Al and Ni-Ti-Al-X (X = Hf, Pd, Pt, Zr) alloys. *Acta. Mater.* **51**, 6341–6357 (2003).

106. Lu, X.-G., Selleby, M. & Sundman, B. Theoretical modeling of molar volume and thermal expansion. *Acta. Mater.* **53**, 2259–2272 (2005).

107. Lu, X.-G., Selleby, M. & Sundman, B. Assessments of molar volume and thermal expansion for selected bcc, fcc and hcp metallic elements. *CALPHAD* **29**, 68–89 (2005).

108. Pineau, A., McDowell, D. L., Busso, E. P. & Antolovich, S. D. Failure of metals II: Fatigue. *Acta. Mater.* http://dx.doi.org/10.1016/j.actamat.2015.05.050 (in press).

109. Sornette, D., Magnin, T. & Brechet, Y. The physical origin of the Coffin-Manson law in low-cycle fatigue. *Europhys. Lett.* **20**, 433 (1992).

110. Moore, J. A. *A Micromechanics-Based Method for Multiscale Fatigue Prediction* (Doctoral Thesis Massachusetts Institute of Technology, 2015).

111. Frankel, D. J. *Design of Fatigue Resistant Heusler-strengthened PdTi-based Shape Memory Alloys for Biomedical Applications* (Doctoral Thesis, Massachusetts Institute of Technology, 2015).

112. Frankel, D. & Olson, G. Design of Heusler precipitation strengthened NiTi- and PdTi-Base SMAs for cyclic performance. *Shape Mem. Superelasticity* **1**, 162–179 (2015).

113. Jou, H.-J., Voorhees, P. W. & Olson, G. B. Computer simulations for the prediction of microstructure/property variation in aeroturbine disks. in *Superalloys* (eds Green, K. A. *et al.*) 877–886 (2004).

114. Zhao, J.-C., Jackson, M. R., Peluso, L. A. & Brewer, L. N. A diffusion multiple approach for the accelerated design of structural materials. *MRS Bull.* **27**, 324–329 (2002).

115. Palumbo, M. *et al.* Thermodynamic modelling of crystalline unary phases. *Phys. Status Solidi.* **251**, 14–32 (2014).

116. Xiong, W., Selleby, M., Chen, Q., Odqvist, J. & Du, Y. Phase equilibria and thermodynamic properties in the Fe-Cr System. *Crit. Rev. Solid State Mater. Sci.* **35**, 125–152 (2010).

117. Squires, D. R. & Wilson, E. A. Effect of cobalt on impact toughness of steels. *Mater. Sci. Technol.* **10**, 52–55 (1994).

118. Shibayama, T., Yamagata, I., Kayano, H. & Namba, C. Effect of small additional elements on DBTT of V-4Cr-4Ti irradiated at low temperatures. *J. Nucl. Mater.* **258–263**, 1361–1368 (1998).

119. McDowell, D. L. & Olson, G. B. Concurrent design of hierarchical materials and structures. *Sci. Model. Simul. SMNS* **15**, 207–240 (2008).

17

Recent developments in computational modelling of nucleation in phase transformations

Lei Zhang[1], Weiqing Ren[2,3], Amit Samanta[4] and Qiang Du[5]

Nucleation is one of the most common physical phenomena in physical, chemical, biological and materials sciences. Owing to the complex multiscale nature of various nucleation events and the difficulties in their direct experimental observation, development of effective computational methods and modeling approaches has become very important and is bringing new light to the study of this challenging subject. Our discussions in this manuscript provide a sampler of some newly developed numerical algorithms that are widely applicable to many nucleation and phase transformation problems. We first describe some recent progress on the design of efficient numerical methods for computing saddle points and minimum energy paths, and then illustrate their applications to the study of nucleation events associated with several different physical systems.

INTRODUCTION

The recent call of Materials Genome Initiative (MGI) exemplifies the use of computational modelling in new materials design.[1] One of the most effective ingredients to design materials with certain desired properties is through the control of their phase transformations and microstructure evolution. These processes often start with the nucleation of nanoscale new-phase particles, followed by growth and particle impingement or coarsening.

Generically, nucleation of a new phase requires overcoming a minimum thermodynamic barrier, which leads to a saddle point configuration along the minimum energy path on the energy landscape. Being a rare event, nucleation is difficult to observe directly in physical experiments as a critical nucleus typically appears transiently at very fast time scales. Therefore, there have been many theoretical studies of the nucleation event. Early classical nucleation theories mainly study phase changes in fluids, e.g., nucleation of a liquid droplet from a vapour phase. The thermodynamic properties of a nucleus are assumed to be the same as in the corresponding bulk. Consequently, the size of a critical nucleus is determined as a result of bulk free-energy reduction and interfacial energy increase, $r^* = -2\gamma/\Delta G_\nu$ where γ is the interfacial energy per unit area between a nucleus and the parent matrix and ΔG_ν is the free-energy-driving force per unit volume. The nucleation rate then depends on the height of the critical energy barrier ΔE^*, i.e., $I = I_0 \exp(-\Delta E^*/k_B T)$ with the pre-exponential factor I_0 calculated from fundamental statistical approaches, k_B and T being the Boltzmann's constant and the absolute temperature, respectively. However, nucleation process in general phase transformation problems is much more challenging to characterise due to the possible complex geometry and structure of critical nuclei. Therefore, much computational effort has been called for to study the nucleation events in various applications.[2–9] Being a saddle point of the free energy, the critical nucleus satisfies the Euler–Lagrange equation of the energy.

However, solving the Euler–Lagrange equation directly or classical optimisation methods are inefficient for this problem due to the unstable nature of the saddle point. Our goal here is to provide a review of some relatively new progress in this direction to the computational materials science community.

This review is by no means a comprehensive treatment of nucleation modelling and transition state search. We refer to[10–12] for other excellent reviews on the subject. In preparing for this article, over a hundred papers on the topic were researched, about half of them are included as references. The main objective of this paper is to provide a sampler of some relatively new progress on the development of numerical algorithms that are applicable to general nucleation and phase transformation problems. Our discussions are focused on a few approaches developed in the past decade, and our reviews of the literature are narrowed to those highly relevant works. This serves as our particular search criterion.

We organise the review by describing some developments of algorithmic works first, followed by illustrations of their applications to nucleation events in various material systems. On the algorithmic side, we first introduce some recent developments of methods for finding saddle points and minimum energy paths. One of the popular approaches is the class of surface walking methods. These methods locate the saddle point starting from an initial state. Here we mainly consider methods based on the dimer method,[13,14] the gentlest ascent method such as the recent works on the gentlest ascent dynamics[15–18] and the shrinking dimer dynamics.[19] Another approach is categorised as path-finding methods that are to compute the minimum energy path (MEP). Some of the representative methods include the nudged elastic band (NEB) method[20,21] and the string method[22–27] with the latter being a focus of our discussion for this type of approaches. For complex and rough energy landscapes, we review some methods that can be used to compute either the transition tubes[28–30] or the mean free-energy path in the space of collective variables.[31,32]

[1]Beijing International Center for Mathematical Research, Center for Quantitative Biology, Peking University, Beijing, China; [2]Department of Mathematics, National University of Singapore, Singapore; [3]Institute of High Performance Computing, Agency for Science, Technology and Research, Singapore; [4]Physics Division, Lawrence Livermore National Laboratory, Livermore, CA, USA and [5]Department of Applied Physics and Applied Mathematics, Columbia University, New York, NY, USA.
Correspondence: Q Du (qd2125@columbia.edu)

To illustrate how the numerical methods can be utilised in the nucleation studies, we present their applications to the nucleation processes in three different systems. In the first application, we consider solid-state phase transformation, and present predicted morphologies of critical nuclei with long-range elastic interactions.[8,33,34] In the second example, we consider the search of transition pathways in micromagnetics, and show the application of the string method in the study of thermally activated switching and the energy landscape of submicron-sized magnetic elements.[3] The last example is on the solid melting problem, as an illustration of nucleation in solid-fluid phase transition, the multiple barrier-crossing events within the solid basin are discussed, which reveal the importance of nonlocal behaviour.[5]

COMPUTING SADDLE POINTS AND TRANSITION PATHS

The classical transition state theory gives a sufficiently accurate description of the transition process for systems with smooth energy landscapes.[35,36] For such systems, the transition state is a saddle point with the lowest energy that connects two neighbouring local minima. Here we focus on the numerical algorithms for computing saddle points and MEPs in systems with smooth potential energy. Generally speaking, there are two distinct classes of numerical methods: (1) surface walking methods for finding saddle points starting from a single state; and (2) path-finding methods for computing MEPs, which involve two end states. In this subsection, we illustrate the ideas behind these approaches via some recently developed methods.

Finding saddle points/surface walking methods

Several surface walking methods have been developed to locate saddle points. An important character of such methods is to perform a systematic search for a saddle point starting from a given initial state, without knowing the final states. Here we focus on a special class of surface walking methods, the so-called minimum mode following methods, where only the lowest eigenvalue and the corresponding eigenvector of the Hessian are calculated and subsequently used together with the energy gradient (often referred as the force) to locate the saddle point. The representative methods include the dimer method,[13,14] the gentlest ascent method/dynamics,[15-17] and the shrinking dimer dynamics.[19,37] Besides the minimum mode following methods, there are some other surface walking methods as well, such as the eigenvector-following method,[38] the minimax method,[39,40] the activation-relaxation technique,[41-43] the step and slide method,[44] to name a few.

The gentlest ascent method/dynamics. The gentlest ascent method was first developed by Crippen and Scheraga[15] as a numerical algorithm to search the transition path from a local minimum to a neighbouring minimum via an intervening saddle point on an energy surface. Later, E and Zhou[16] reformulated the gentlest ascent method as a dynamical system and proposed the gentlest ascent dynamics (GAD).

Consider a system with N degrees of freedom (DOF) contained in a vector $x \in \mathbb{R}^N$, the GAD refers to the following dynamic system

$$\begin{cases} \dot{x} = -\nabla V(x) + 2\frac{(\nabla V, v)}{(v,v)}v, \\ \dot{v} = -\nabla^2 V(x)v + \frac{(v, \nabla^2 Vv)}{(v,v)}v, \end{cases} \quad (1)$$

where ∇V is the gradient of the potential energy V, $\nabla^2 V$ is the Hessian of V, and $(\cdot, \cdot)$ denotes the standard inner product. The second equation in Equation (1) determines the orientation vector v to be the eigenvector that corresponds to the smallest eigenvalue of $\nabla^2 V$, which is used in the first equation in Equation

(1) as an ascent direction to find the saddle point. The stable fixed points of this dynamical system were proved to be the index-1 saddle points in ref. 16. The GAD has also be applied to non-gradient systems.

The dimer method. A dimer system consists of two nearby points x_1 and x_2 separated by a small distance, that is, with a small dimer length $l = \|x_1 - x_2\|$. The dimer orientation is given by a unit vector v so that $x_1 - x_2 = lv$. The (rotating) center of the dimer is usually taken as the midpoint of the dimer, i.e., $x_c = \frac{x_1}{2} + \frac{x_2}{2}$. The dimer method developed by Henkelman and Jonsson[13] uses only first-order derivatives of the energy to calculate the forces F_1 and F_2 on the two end points of the dimer.

The dimer method proceeds by alternately performing the rotation and translation steps. The rotation step is to find the lowest eigenmode at the center of the dimer. This is done by a single rotation towards the configuration which minimises the dimer energy with its center fixed. In practice, a conjugate gradient algorithm is used to choose the plane of rotation, whereas the minimisation of the force on the dimer within the plane is carried out by using the modified Newton method. In the translation step, the potential force is first modified so that its component along the dimer is reversed. Then the dimer is translated using the modified force with either the steepest descent algorithm or the conjugate gradient method.

Much effort has been made to improve the dimer method. In ref. 14, it was argued that the method is more effective when the rotation step is proceeded until convergence, compared with using a single rotation after every translation step. More recently, in the work of Gould, Ortner and Packwood,[45] a preconditioner technique is used in the dimer iteration and a line search technique is applied for finding the step size to achieve better efficiency and convergence. A recent attempt was made in ref. 46 to unify various techniques for accelerating the dimer methods though most of the approaches discussed there do not share superlinear convergence property. In contrast, an interesting development was given by Kastner and Sherwood in ref. 47, which used the Limited-memory Broyden–Fletcher–Goldfarb–Shanno (L-BFGS) algorithm for the dimer translation to greatly improve the convergence of the dimer method. Moreover, the numbers of gradient calculations per dimer iteration are reduced through an extrapolation of the gradients during repeated dimer rotations.

The shrinking dimer dynamics. The shrinking dimer dynamics (SDD) was proposed by Zhang and Du in ref. 19 to find the index-1 saddle points based on the original dimer method.[13] It follows a dynamic system as

$$\begin{cases} \mu_1 \dot{x}_a = (I - 2vv^T)((1-a)F_1 + aF_2), \\ \mu_2 \dot{v} = (I - vv^T)\frac{(F_1 - F_2)}{l}, \\ \mu_3 \dot{l} = -\nabla E_{dimer}(l), \end{cases} \quad (2)$$

where μ_1, μ_2, μ_3 are nonnegative relaxation constants, and a is a constant parameter between 0 and 1, which determines the rotating center of the dimer.

The first two equations of SDD in Equation (2) represent the translation step and the rotation step, respectively, which are essentially same with the classical dimer method. The operator $(I - 2vv^T)$ is the Householder mirror reflection that reverses the component of the force along v. The operator $(I - vv^T)$ is a projection that makes v of unit length. Instead of using a fixed small dimer length in the classical dimer method, the third equation of SDD in Equation (2) follows a gradient flow of the energy function $E_{dimer}(l)$, which is generally taken as a monotonically increasing function in l such that it allows the shrinking of the dimer length over time by forcing it to approach zero at the steady state.

The dynamic system (Equation (2)) is similar to Equation (1), but avoids the calculation of the second-order derivatives by requiring only the evaluation of the natural forces. Rigorous analysis for both the continuous dynamic system and its time discretisation were carried out in ref. 19 to demonstrate the linear stability and convergence, in particular, the importance of shrinking the dimer length for the guaranteed convergence.

In terms of numerical schemes, the SDD employs either explicit or modified Euler method to obtain the linear convergence in ref. 19. In ref. 37, a constrained SDD has been proposed as an variant of SDD to handle the saddle point search on a constrained manifold. The use of preconditioner has also been alluded to in refs 37,48 but without implementation. To accelerate the convergence and further improve the efficient of the dimer method, the optimisation-based shrinking dimer (OSD) method was recently proposed by Zhang et al. in ref. 49 as the generalised formulation of SDD. The OSD method translates the rotation and translation steps of the dimer in Equation (2) to the corresponding optimisation problems such that the efficient optimisation methods can be naturally employed to substantially speed up the computation of saddle points. In ref. 49, the Barzilai–Borwein gradient method was used as an implementation of OSD and showed a superlinear convergence.

Finding minimum energy path

In the case when the object of interest is the most probable transition path between metastable states of the smooth potential energy, it is known that for overdamped Langevin dynamics the most probable path for the transition is the MEP, which is the path in the configuration space such that the potential force is parallel to the tangents along the path, i.e.,

$$(\nabla V)^\perp(\varphi) := \nabla V - (\nabla V, \hat{\tau})\hat{\tau} = 0, \tag{3}$$

where $\hat{\tau}$ is the unit tangent vector to the curve. Several numerical methods have been developed for finding MEPs. Below we review two popular ones, the (zero temperature) string method[24,25,50] and the nudged elastic band method,[20,21,51] for computing the MEPs. Once the MEP is found, the transition states can be identified from the maxima of the energy along the MEP.

The (zero-temperature) string method. The string method was first proposed by E et al. in ref. 24, and it proceeds by evolving a string in the configuration space by using the steepest descent dynamics. Let $\varphi(a,t)$ denote the string at the time t with parameterisation $a \in [0,1]$, then the string evolves according to

$$\dot{\varphi} = -\nabla V(\varphi) + \lambda\hat{\tau}, \quad 0 < a < 1, \tag{4}$$

where λ is the Lagrange multiplier to impose the equal arc-length constraint, and $\tau = \varphi'/|\varphi'|$ is the unit tangent vector to the string. Here we use $\dot{\varphi}$ and φ' to denote the temporal and spatial derivatives, respectively. The above evolution equation is supplemented with the boundary conditions:

$$\varphi(0,t) = x_a, \quad \varphi(1,t) = x_b, \tag{5}$$

where x_a and x_b are the two minima of the potential energy $V(x)$.

In the numerical implementation, the discretised string is composed of a number of images $\{\varphi_i(t), i = 0, 1, \cdots, N\}$, where $\varphi_i(t) = \varphi(i/N, t)$. Equation (4) is solved using a time-splitting scheme, and the string method iterates the following two steps:

String evolution updates the images on the string over some time interval Δt according to the potential force:

$$\dot{\varphi}_i = -\nabla V(\varphi_i), \quad i = 1, 2, \cdots, N-1. \tag{6}$$

This equation can be integrated by any ODE solver, e.g., the Euler method, or Runge–Kutta methods.

String reparametrisation is applied to redistribute the images along the string using linear or cubic spline interpolation according to equal arc-length parameterisation.

The string method is a simple but an effective technique for finding MEPs. It only requires an ODE solver and an interpolation scheme, thus it is easy to implement and can be readily incorporated into any existing code as long as the force evaluation is provided.

In the case that the exact locations of the minima x_a and/or x_b are unknown beforehand, the two end points can be computed on-the-fly by following the relaxation dynamics as the string evolves towards the MEP, i.e.,

$$\dot{\varphi}_0 = -\nabla V(\varphi_0), \quad \dot{\varphi}_N = -\nabla V(\varphi_N). \tag{7}$$

At the steady state, the end points converge to the minima x_a and x_b, as long as they initially lie in the basins of attraction of these minima, respectively.

In ref. 26, the dynamics of the final point φ_N is modified so that it converges to a saddle point. This can be used for saddle point search. Specifically, the dynamics of φ_N follows the modified potential force

$$\dot{\varphi}_N = -\nabla V(\varphi_N) + 2(\nabla V(\varphi_N), \hat{\tau}_N)\hat{\tau}_N, \tag{8}$$

where $\hat{\tau}_N$ is the unit tangent vector to the string at φ_N. In Equation (8), the component of the potential force in the direction along the string is reversed. This makes φ_N climb uphill towards a saddle point.

In practice, an initial string is constructed in the basin of the initial state, e.g., the linear interpolation of the initial state and its small perturbation. Then following the dynamics in Equations (4) and (8), the final state converges to a saddle point, and the string converges to the MEP connecting the initial state and the saddle point.

The string method can also be generalised to compute the mean free-energy path (MFEP) in collective variable spaces. This will be discussed later in this review.

The nudged elastic band method. The NEB method is another approach to compute MEPs. It connects two minima x_a and x_b by a chain of states, then evolves this chain by using a combination of the potential force and a spring force:

$$\dot{x}_i = -(\nabla V(x_i))^\perp + (F_i^s \cdot \hat{\tau}_i)\hat{\tau}_i, \quad i = 1, \cdots, N-1, \tag{9}$$

where $-(\nabla V)^\perp = -\nabla V + (\nabla V, \hat{\tau})\hat{\tau}$ denotes the projection of the potential force in the hyperplane perpendicular to the chain, $F_i^s = k(x_{i+1} - 2x_i + x_{i-1})/\Delta a$ is the spring force with $\Delta a = 1/N$ and $k > 0$ being a parameter, and $\hat{\tau}_i$ denotes the unit tangent vector along the chain at x_i.

The NEB is an improvement upon the elastic band method.[52-54] The elastic band method, which evolves the chain using the total spring forces, fails to converge to the MEP as the spring force tends to make the chain straight, which leads to corner-cutting. The NEB overcomes this difficulty by using only the normal component of the potential force and the tangential component of the spring force.

Compared with the string method, which evolves a curve with intrinsic arc-length parameterisation, the NEB uses a spring force to connect the states along the chain. This requires the prescription of an additional parameter k. The value of k controls the strength of the spring force: if it is too large, then the dynamics in Equation (9) becomes stiff, which may limit the time step; on the other hand, a small k leads to weak spring force then the states may not be evenly distributed along the chain.

COMPLEX ENERGY LANDSCAPES

The concept of transition states and MEPs become inappropriate when the energy landscape is rough with many saddle points, and most of these saddle points have potential-energy barriers that are less then or comparable to the strength of the noise, thus do not act as barriers for the transition. In this case, one approach is to compute the so-called transition tubes. The transition tube carries most of the current of the transition in the configuration space. The other way is to consider the free-energy surface (FES) instead of potential energy surface. One first selects a set of collective variables, then computes the transition state or the MFEP in the space of these collective variables. The FES is typically much more smooth than the potential energy surface. Below we review these two approaches.

Computing transition tubes

The transition tube is defined by considering the committor function and the current of reaction trajectories. The committor function specifies, at each point of the configuration space, the probability that the reaction or transition will succeed if the system is initiated at that point. The isosurfaces of the committor function, called the iso-committor surfaces, foliate the space between the metastable states under consideration. Assume the reaction current is localised in the configuration space, then it intersects with the isocommittor surfaces in one or a few isolated regions. The collection of these regions form one or a few tubes, which are called transition tubes (Figure 1).[28–30]

Under the assumptions that the transition tube is thin compared with the local radius of the curvature of the centerline and the isocommittor surfaces can be approximated by hyperplanes within the transition tube, it was shown that the centerline of the transition tube, which is defined as

$$\varphi(a) = \langle x \rangle_{P_a} \equiv \frac{\int_{P_a} x e^{-\beta V} d\sigma(x)}{\int_{P_a} e^{-\beta V} d\sigma(x)}, \tag{10}$$

is normal to the isocommittor surfaces, i.e.,

$$\hat{n}(a) \| \varphi'(a), \tag{11}$$

where the centerline φ is parameterised by a (e.g., the arc length), $\varphi'(a)$ is the tangent vector to the centerline, and $\hat{n}$ is the normal vector of the hyperplanes.

The finite-temperature string method is an iterative procedure to solve Equations (10) and (11).[30,55,56] Let φ^n denote the centerline of the tube at the nth step, the new configuration of the centerline is computed following two steps:

Expectation step: sample on the hyperplanes perpendicular to the current configuration of the centerline and compute the center of mass $\langle x \rangle_{P_a}$ on each hyperplane.

Relaxation step: evolve the centerline to the new configuration according to

$$\varphi^{n+1} = \varphi^n + \Delta t \left(\langle x \rangle_{P_a} - \varphi^n \right). \tag{12}$$

In practice, each image is evolved according to Equation (12), followed by a reparameterisation step using interpolation as in the zero-temperature string method. The centerline of mass $\langle x \rangle_{P_a}$ can be estimated using constrained simulations in the hyperplanes[30,55] or Voronoi cells.[56] At convergence, the Voronoi cells also have the remarkable property that they form a centroidal Voronoi tessellation or CVT—a concept introduced in ref. 57.

Exploring the free-energy surface

We consider the case when the transition of interest can be described by the collective variables $\xi(x) = (\xi_1(x), \cdots, \xi_d(x))$. These collective variables correspond to the slow manifold along which the transition occurs, therefore, the dimension of the collective variable space is usually much smaller than the total DOF in the system. The free energy associated with $\xi(x)$ is given by

$$F(z) = -k_B T \ln \left(\frac{1}{Z} \int_{R^N} e^{-\beta V(x)} \prod_{i=1}^{d} \delta(\xi_i(x) - z_i) dx \right), \tag{13}$$

where $z = (z_1, \cdots, z_d)$ are the coordinates in the collective variable space, $\beta = 1/k_B T$, and Z is the partition function. On the FES, the metastable and transition states are given by the local minima of $F(z)$ and the saddle points between them, respectively. The path of maximum likelihood for the transition follows the MFEP,[31] which satisfies

$$(M(\varphi) \nabla F(\varphi))^{\perp} = 0, \tag{14}$$

where $(\cdots)^{\perp}$ denotes the projection on the hyperplane perpendicular to the path φ, and M is an $d \times d$ tensor whose entries are given by

$$M_{ij}(z) = \frac{1}{Z} e^{\beta F(z)} \int_{R^N} \sum_{k=1}^{N} \frac{\partial \xi_i(x)}{\partial x_k} \frac{\partial \xi_j(x)}{\partial x_k} e^{-\beta V(x)} \tag{15}$$

$$\times \delta(z_1 - \xi_1(x)) \cdots \delta(z_d - \xi_d(x)) dx. \tag{16}$$

If one is interested in the transition between two metastable states z_a and z_b, then Equation (14) is supplemented with the boundary conditions such that the path φ connects the two minima.

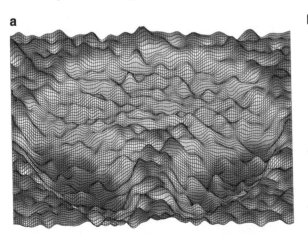

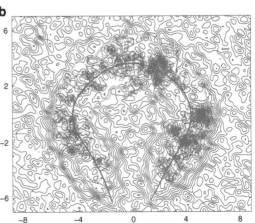

Figure 1. Example of the rough energy landscape (**a**) and the transition tube (**b**) identified by the string method. Figure modified from ref. 55. Permission has been granted by the publisher to use the original figures.

The numerical techniques in 'Finding saddle points/surface walking methods' and 'Finding minimum energy path' can be readily used to compute the saddle points and the MFEP on the FES. For example, the MFEP can be computed using the zero-temperature string method by evolving a string $\varphi(a)$ according to

$$\gamma\dot{\varphi}(a,t) = -M(\varphi(a,t))\nabla F(\varphi(a,t)) + \lambda(a,t)\hat{\tau}, \tag{17}$$

where the boundary conditions $\varphi(0,t) = z_a$ and $\varphi(1,t) = z_b$. At the steady state, the string converges to the MFEP with the prescribed parameterisation.

In the above algorithm or the methods for computing the saddle points, one needs to compute the mean force $-\nabla F$. When the dimension of the free-energy space is low, e.g., $d=2$ or 3, one can first construct the free-energy function, for example, using a variational reconstruction scheme as in the single sweep method.[5,58] The method follows two steps: (a) first a sufficiently long temperature-accelerated molecular dynamics (TAMD)/adiabatic free-energy dynamics (AFED) trajectory is obtained and a set of centers in the space of the collective variables and; (b) the free energy is expressed as a superposition of Gaussian radial basis functions (RBFs) placed at these centers. The optimal values of the coefficients of the RBFs are obtained by optimising the cost function, which is the standard mean-squared error in the gradient of the free energy.

The reconstruction of the FES is only possible when the the number of collective variables is small. When the number is large, the mean force can be computed on the fly without the requirement of a globally explicit form of the free-energy function a priori. This can be done by averaging over long trajectories with the collective variables fixed at their current locations.[31] Alternatively, one may couple the microscopic dynamics with the evolution of the string simultaneously.[32] Specifically, each image along the string is coupled with two independent microscopic systems, denoted by $x^{(1)}$ and $x^{(2)}$, respectively, where the first system is for the computation of M and the second system is for the computation of the mean force $-\nabla F$. These systems evolve by molecular dynamics or the Langevin dynamics at the temperature T in the extended potential

$$U_\kappa(x;\varphi(a,t)) = V(x) + \frac{1}{2}\kappa|\xi(x)-\varphi(a,t)|^2. \tag{18}$$

The second term in the extended potential is to constrain the system at the current location of the string, so that dynamics of the microscopic systems are slaved to the evolution of the string. The tensor M and the mean force $-\nabla F$ are computed from the instantaneous configurations of $x^{(1)}$ and $x^{(2)}$:

$$M_{ij} \approx \sum_{k=1}^{N} \frac{\partial\xi_i}{\partial x_k}(x^{(1)})\frac{\partial\xi_j}{\partial x_k}(x^{(1)}), \tag{19}$$

$$\nabla F \approx -\kappa(\xi(x^{(2)})-\varphi(a,t)). \tag{20}$$

To ensure the microscopic systems to have enough time to equilibrate before the string move significantly, the dynamics of the string in Equation (17) is slowed down by taking $\gamma 1$ in Equation (17).[32]

APPLICATIONS IN MATERIALS SCIENCE

Nucleation in solid-state phase transformations
Following the seminal work by Cahn and Hilliard,[59] the phase field (diffuse interface) model has been successfully employed to investigate nucleation and microstructure evolution in phase transformations.[60] In the diffuse interface description, a set of conserved field variables $c_1,c_2,...,c_m$ and non-conserved field

variables $\eta_1,\eta_2,...\eta_n$ are often used to describe the compositional/structural domains and the interfaces, and the total free energy of an inhomogeneous microstructure system is formulated as

$$E_{total} = \int [\sum_{i=1}^{m} a_i(\nabla c_i)^2 + \sum_{i=1}^{3}\sum_{j=1}^{3}\sum_{k=1}^{n}\beta_{ij}\nabla_i\eta_k\nabla_j\eta_k \tag{21}$$

$$+ f(c_1,c_2,...,c_m,\eta_1,\eta_2,...\eta_n)]dx$$
$$+ \int\int G(x-x',\vec{c},\vec{\eta})dxdx', \tag{22}$$

where the gradient coefficient a_i and β_{ij} can be used to reflect the interfacial energy anisotropy and the function f corresponds to the local free-energy density. The last integral in the above equation represents a nonlocal term that includes a general long-range interaction such as elastic interactions in solids.

In solid-state phase transformations, the local free-energy function f is often described by a polynomial of order parameters with a conventional Landau-type of expansion, for instance, a simple double-well potential,

$$f(\eta) = \frac{\eta^4}{4} - \frac{\eta^2}{2} + \lambda\frac{\eta^3-3\eta}{4}, \tag{23}$$

with two energy wells at $\eta = \pm 1$ and λ determines the well depth difference.

Furthermore, the lattice mismatch between solid phases and domains are accommodated by elastic displacements, so the computation of the long-range elastic energy is needed. For the case that the elastic modulus is anisotropic but homogeneous, the microscopic elasticity theory of Khachaturyan[61] is often used in phase field simulation and the total elastic energy of a microstructure can be given by

$$E_{elastic} = \frac{1}{2}\int_\Omega C_{ijkl}\,\epsilon_{ij}^{el}\,\epsilon_{kl}^{el}dx, \tag{24}$$

where the elastic strain ϵ^{el} is the difference between the total strain and stress-free strain as stress-free strain does not contribute to the total elastic energy.

Examples of predicted critical profiles in the presence of long-range elastic energy for a cubic crystal are shown in Figure 2a. By combining the diffuse-interface approach with the minimax technique,[39] it demonstrated that the elastic interactions can markedly change the critical nucleus morphology, thus revealing the fascinating possibility of nuclei with non-convex shapes, together with the phenomenon of shape-bifurcation and the formation of critical nuclei whose symmetry is lower than both the new phase and the original parent matrix.[8,33,62]

For the conserved solid field with profile $c = c(x)$, a combination of diffuse-interface description and a constrained string method[23] is able to predict both the critical nucleus and equilibrium precipitate morphologies simultaneously (Figure 2b).[34] Using the cubic to cubic or cubic to tetragonal transformation as examples, simulations showed that the morphology of a critical nucleus can be markedly different from the equilibrium one due to the elastic energy contributions.[34,63]

The general framework of diffuse interface model has been greatly extended to investigate complex nucleation phenomena, such as heterogenous nucleation,[48,64,65] homogeneous/heterogenous crystal nucleation[66–68] and nucleation dynamics.[69–72] In particular, to investigate the nucleation and growth kinetics in real alloys, the thermodynamic properties of critical nuclei with the chemical free energy and interfacial energy can be assessed from thermodynamic calculation and atomistic simulation, and then efficient numerical method of computing saddle points can be applied to quantitatively predict the critical nucleus, nucleation energy barriers and growth kinetics.[73]

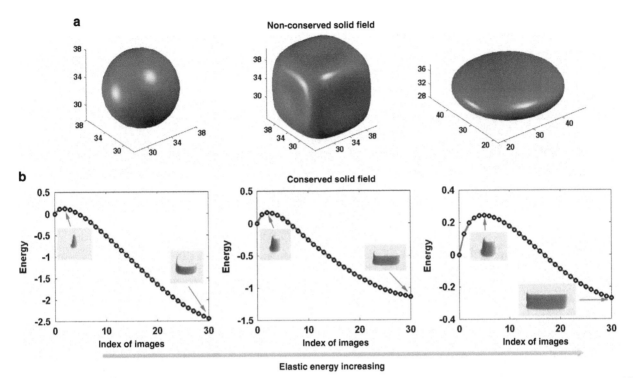

Figure 2. Critical nuclei with increasing of elastic energy in non-conserved solid field (**a**) and conserved solid field (**b**). Figure reproduced from refs 33,34. Permission has been granted by the publisher to use the original figures.

Transition pathways in micromagnetics

Submicron-sized magnetic elements have found a wide range of applications in science and technology, particularly as storage devices. As the elements get smaller, the effect of thermal noise and the issue of data retention time become a concern. In ref. 3, the string method was applied to study thermally activated switching and the energy landscape of submicron-sized magnetic elements. The free energy of the system is modelled by the Landau–Lifshitz functional:

$$F[\mathbf{M}] = \frac{1}{2} \int_\Omega \left\{ \frac{C_{ex}}{M_s^2} |\nabla \mathbf{M}|^2 + \phi \left(\frac{\mathbf{M}}{M_s} \right) \right\} dx \qquad (25)$$

$$- \frac{1}{2} \int_\Omega 2\mu_0 \mathbf{H}_e \cdot \mathbf{M} \, dx + \frac{\mu_0}{2} \int_{R^3} |\nabla U|^2 dx, \qquad (26)$$

where $\mathbf{M}$ is the magnetisation distribution normalised so that $|\mathbf{M}| = 1$, is the domain occupied by the element, $C_{ex} |\nabla \mathbf{M}|^2 / M_s^2$ is the exchange interaction energy the the spins, $\phi(\mathbf{M}/M_s)$ is the energy due to material anisotropy, $-2\mu_0 \mathbf{H}_e \cdot \mathbf{M}$ is the energy due to the external applied field. The last term is the energy due to the field induced by the magnetisation distribution inside the material. This induced field $-\nabla U$ can be computed by solving

$$\Delta U = \begin{cases} \nabla \cdot \mathbf{M}, & \text{in ,} \\ 0, & \text{outside ,} \end{cases} \qquad (27)$$

with the jump conditions

$$[U] = 0, \quad \left[\frac{\partial U}{\partial \nu} \right] = -\mathbf{M} \cdot \nu, \qquad (28)$$

at the boundary of the element. In the above conditions, $[\cdot]$ denotes the jump across the boundary, and the vector ν is the outward unit normal on the boundary of .

The string method was used to compute the MEPs and analyse the energy landscape of the Landau–Lifshitz free energy. It is found that the switching proceeds by two generic scenarios: Domain-wall propagation and reconnection followed by edge domain switching, or vortex nucleation at the boundary followed by vortex propagation through the element. These two pathways are shown in Figure 3. The second pathway is preferred for thicker films, whereas the first is preferred for thin films.

Nucleation in solid–liquid phase transition

Atomic processes associated with complex systems often exhibit a variety of time scales. The origins of this disparity in time scales can often be traced to the spatial scale associated with different process. Localised processes involve small subsets of the total degrees of freedom (DOF) of the system while collective processes involve large number of DOF; the latter typically occur at a slower rate. However, when collective processes are initiated through a series of one or more localised events, then the processes are inherently multiscale in nature. An important example of type of physical process is the melting of a solid.[5,74–78]

As a model system, copper is used to study the melting of a solid. The interatomic interactions were modelled using the embedded atom method potential for Cu developed by Mishin et al.[79] To efficiently explore the relevant parts of the configuration space and study the microscopic mechanisms of melting, the AFED,[80–82] a recently developed exploration technique, was used. Techniques like the AFED and the TAMD[82,83] exploit the adiabatic time scale separation between the evolution of the atomic DOF and the collective variables (CVs) by assigning fictitious masses to the CVs that are much larger than the atomic masses, and at the same time, maintaining the CVs at a temperature much higher than the physical temperature. We have used the volume (V) of the system and the Steinhardt orientation order parameters Q_4, Q_6 as the collective variables. The volume of the system captures the changes in the density, whereas the orientation order parameters capture the changes in the symmetry of the atomic structure.

The dynamic sampling of the FES in AFED/TAMD schemes is obtained by using the equations of motions obtained from the potential-energy surface $U_\kappa(x, z)$ that spans over an extended space of atomic and coarse-grained DOF. The TAMD scheme

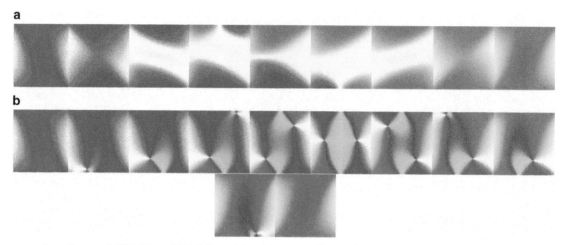

Figure 3. Two generic pathways (MEPs) (**a**) and (**b**) followed by the magnetisation during the switching of submicron-sized ferromagnetic elements. The figures show the successive minima and saddle points. The colour code represents the in-plane components of the magnetisation: blue = right, red = left, yellow = up, green = down. Figure modified from ref. 24. Permission has been granted by the publisher to use the original figures.

within the isobaric–isothermal conditions was used to sample the FES of Cu at different physical temperatures.[82] From these sampling trajectories, a set of CVs $\{\mathbf{z}_k\}_{k=0}^{m}$, was selected to reconstruct the Gibbs FES. The mean forces $\{\mathbf{f}_k\}_{k=0}^{m}$ on these CVs were obtained from an average of the instantaneous forces on these CVs along the sampling trajectories.

The reconstruction of the FES, using the knowledge of the mean forces at the set of CVs selected above, is essentially an inverse problem. The free-energy was expressed as a superposition of Gaussian RBFs placed at the chosen centers:[58]

$$\mathcal{F}(\mathbf{z}) = \sum_{k=1}^{m} a_k \phi_\sigma\big(|\mathbf{z} - \mathbf{z}_k|^2\big), \tag{29}$$

where $|\cdot|$ is the L_2 norm, $\phi_\sigma(r) = e^{-r/2\sigma^2}$ and σ is the width of the Gaussian RBFs. The optimal values of the coefficients a_k were obtained by optimising the following cost function

$$\mathcal{E}_r = \frac{1}{m} \sum_{k=1}^{m} |\nabla \mathcal{F}(\mathbf{z}_k) + \mathbf{f}_k|^2 + \lambda |\mathbf{a}|^2. \tag{30}$$

Here, λ is a regularisation parameter and $\lambda|\mathbf{a}|^2$ is a smoothness constraint introduced to stabilise the solution. From a statistical point of view, a proper handling of noisy data entails proper training as well as testing of the fitting model. To this end, multifold cross validation is used for model validation.[85]

The FES of Copper calculated close to the melting temperature, shown in Figure 4, illustrates the multitude of barriers that a system has to overcome to travel from the solid super-basin to the liquid basin. Some of the barriers inside the solid basin are associated with the formation of isolated point defects and are on the order of 1–5 eV. In contrast, the barriers to cross from these isolated point defect states to metastable states corresponding to extended defects like defect cluster, are on the order of 20 eV. The barrier to melting, on the other hand, is on the order of 120 eV. This disparity in the Gibbs free-energy barriers results in time scale disparities between the processes. In addition, the presence of a large number of metastable states inside the solid basin suggests that contrary to the tenets of the classical nucleation theory melting is not a simple, single thermally activated barrier crossing event but involves multiple thermally activated transition events.[5]

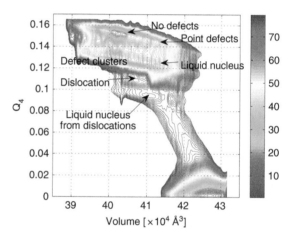

Figure 4. The free-energy surface of copper close to the melting temperature as a function of the collective variables—volume and Steinhardt orientation order parameter. The presence of multiple locally stable states illustrates that melting is a complex thermally activated transition event. Figure reproduced from ref. 5. Permission has been granted by the publisher to use the original figures.

SUMMARY AND OUTLOOK

Nucleation is a complex multiscale problem. Recent development of new numerical algorithms and modelling approaches on MEP calculation and saddle point search as well as transition path theory have brought new light to this challenging subject. The computational methods reviewed in 'Computing saddle points and transition paths' and 'Complex energy landscapes' are generic methods. We classify them according to the different purposes they serve, such as saddle point search, computing MEPs/MFEPs or transition tubes. Subject to the properties and structures of the problems, one is free to use different methods to study the same nucleation event. The examples in 'Applications in materials science' provided some illustrative applications of these new approaches.

There are naturally many other relevant and important issues that are not addressed here. For example, it is an interesting subject to see how global optimisation techniques may be adopted to efficiently identify saddle points with a low and/or minimum energy barrier. For complex models in high dimensions, it is worthwhile to mention the importance of data analysis techniques to our understanding of such systems. As it is difficult to *a priori* predict the relevant order parameters, one may combine FES exploration techniques in conjunction with dimension reduction techniques like diffusion maps[84] to obtain quantitative information about the relevant low-dimensional manifolds.

Furthermore, metastable states, saddle points, obtained from GAD or SDD or dimer like methods, in the space of collective variables, can be used to construct a weighted graph that is representative of the transition events taking place in a system. Such a network can have its own trapping regions, critical saddles and so on. One may also explore how to utilise the information encoded in such a network to conduct multiscale dynamic simulation of microstructure evolution that involves fast nucleation and slow coarsening processes. Preliminary studies along this direction can be found in ref. 71 These are only a few examples of the wide spectrum of multiscale analysis that is possible from the huge trove of that is obtained from saddle-point/MEP search and FES sampling schemes presented here.

ACKNOWLEDGEMENTS

We thank Professors Long-Qing Chen, Weinan E and Eric Vanden-Eijnden for fruitful discussions and collaborations on the subject. The work by L.Z. was supported by China NSFC No. 11421110001 and 91430217. The work of W.R. was partially supported by AcRF Tier-1 grant R-146-000-216-112. The work by A.S. was performed under the auspices of the U.S. Department of Energy by Lawrence Livermore National Laboratory under Contract DE-AC52-07NA27344. The work of Q.D. was supported in part by NSF-DMS1318586.

COMPETING INTERESTS

The authors declare no conflict of interest.

REFERENCES

1. Materials Genome Initiative for Global Competitiveness. *National Science and Technology Council, Office of Science and Technology Policy*. (Washington DC, 2011).
2. Cheng, X., Lin, L., E, W. Zhang, P. & Shi, A.-C. Nucleation of ordered phases in block copolymers. *Phys. Rev. Lett.* **104**, 148301 (2010).
3. E, W., Ren, W. & Vanden-Eijnden, E. Energy landscape and thermally activated switching of submicron-sized ferromagnetic elements. *J. Appl. Phys.* **93**, 2275–2282 (2003).
4. Li, T., Zhang, P. & Zhang, W. Nucleation rate calculations for the phase transition of diblock copolymers under stochastic Cahn-Hilliard dynamics. *SIAM Multi. Model. Simul.* **11**, 385–409 (2013).
5. Samanta, A., Tuckerman, M. E., Yu, T.-Q. & E, W. Microscopic mechanisms of equilibrium melting of a solid. *Science* **346**, 729–732 (2014).
6. Schlegel, H. Exploring potential energy surfaces for chemical reactions: an overview of some practical methods. *J. Comput. Chem.* **24**, 1514–1527 (2003).
7. Wales, D. Energy landscapes: calculating pathways and rates. *Int. Rev. Phys. Chem.* **25**, 237–282 (2006).
8. Zhang, L., Chen, L.-Q. & Du, Q. Morphology of critical nuclei in solid state phase transformations. *Phys. Rev. Lett.* **98**, 265703 (2007).
9. Zhang, W., Li, T. & Zhang, P. Numerical study for the nucleation of one-dimensional stochastic Cahn-Hilliard dynamics. *Commun. Math. Sci.* **10**, 1105–1132 (2012).
10. W. E. & Vanden-Eijnden, E. Transition-path theory and path-finding algorithms for the study of rare events. *Annu. Rev. Phys. Chem.* **61**, 391–420 (2010).
11. Laaksonen, A., Talanquer, V. & Oxtoby, D. W. Nucleation: measurements, theory, and atmospheric applications. *Annu. Rev. Phys. Chem.* **46**, 489–524 (1995).
12. Xu, X., Ting, C. L., Kusaka, I. & Wang, Z. G. Nucleation in polymers and soft matter. *Annu. Rev. Phys. Chem.* **65**, 449–475 (2014).
13. Henkelman, G. & Jónsson, H. A dimer method for finding saddle points on high dimensional potential surfaces using only first derivatives. *J. Chem. Phys.* **111**, 7010 (1999).
14. Olsen, R., Kroes, G., Henkelman, G., Arnaldsson, A. & Jonsson, H. Comparison of methods for finding saddle points without knowledge of the final states. *J. Chem. Phys.* **121**, 9776–9792 (2004).
15. Crippen, G. & Scheraga, H. Minimization of polypeptide energy XI. The method of gentlest ascent. *Arch. Biochem. Biophys.* **144**, 462–466 (1971).
16. E, W. & Zhou, X. The gentlest ascent dynamics. *Nonlinearity* **24**, 18311842 (2011).
17. Gao, W., Leng, J. & Zhou, X. An iterative minimization formulation for saddle-point search. *SIAM J. Numer. Anal.* **53**, 1786–1805 (2015).
18. Samanta, A., Chen, M., Yu, T. Q., Tuckerman, M. & E, W. Sampling saddle points on a free energy surface. *J. Chem. Phys.* **140**, 164109 (2014).
19. Zhang, J. Y. & Du, Q. Shrinking dimer dynamics and its applications to saddle point search. *SIAM J. Numer. Anal.* **50**, 1899–1921 (2012).
20. Henkelman, G. & Jónsson, H. Improved tangent estimate in the nudged elastic band method for finding minimum energy paths and saddle points. *J. Chem. Phys.* **113**, 9978–9985 (2000).
21. Henkelman, G., Uberuaga, B. P. & Jónsson, H. A climbing image nudged elastic band method for finding saddle points and minimum energy paths. *J. Chem. Phys.* **113**, 9901–9904 (2000).
22. Carilli, M. F., Delaney, K. T. & Fredrickson, G. H. Truncation-based energy weighting string method for efficiently resolving small energy barriers. *J. Chem. Phys.* **143**, 054105 (2015).
23. Du, Q. & Zhang, L. A constrained string method and its numerical analysis. *Commun. Math. Sci.* **7**, 1039–1051 (2009).
24. E, W., Ren, W. & Vanden-Eijnden, E. String method for the study of rare events. *Phys. Rev. B.* **66**, 052301 (2002).
25. E, W., Ren, W. & Vanden-Eijnden, E. Simplified and improved string method for computing the minimum energy paths in barrier-crossing events. *J. Chem. Phys.* **126**, 164103 (2007).
26. Ren, W. & Vanden-Eijnden, E. A climbing string method for saddle point search. *J. Chem. Phys.* **138**, 134105 (2013).
27. Samanta, A. & E, W. Optimization-based string method for finding minimum energy path. *Commun. Comput. Phys.* **14**, 265–275 (2013).
28. E, W., Ren, W. & Vanden-Eijnden, E. Transition pathways in complex systems: reaction coordinates, isocommittor surface, and transition tubes. *Chem. Phys. Lett.* **413**, 242–247 (2005).
29. E, W. & Vanden-Eijnden, E. Towards a theory of transition paths. *J. Stat. Phys.* **123**, 503–523 (2006).
30. Ren, W., Vanden-Eijnden, E., Maragakis, P. & E, W. Transition pathways in complex systems: application of the finite-temperature string method to the alanine dipeptide. *J. Chem. Phys.* **123**, 134109 (2005).
31. Maragliano, L., Fischer, A., Vanden-Eijnden, E. & Ciccotti, G. String method in collective variables: minimum free energy paths and isocommittor surfaces. *J. Chem. Phys.* **125**, 024106 (2006).
32. Maragliano, L. & Vanden-Eijnden, E. On-the-fly string method for minimum free energy paths calculation. *Chem. Phys. Lett.* **446**, 182–190 (2007).
33. Zhang, L., Chen, L.-Q. & Du, Q. Diffuse-interface description of strain-dominated morphology of critical nuclei in phase transformations. *Acta Mater.* **56**, 3568–3576 (2008).
34. Zhang, L., Chen, L.-Q. & Du, Q. Simultaneous prediction of morphologies of a critical nucleus and an equilibrium precipitate in solids. *Commun. Comput. Phys.* **7**, 674–682 (2010).
35. Eyring, H. The activated complex and the absolute rate of chemical reactions. *Chem. Rev.* **17**, 6577 (1935).
36. Wigner, E. The transition state method. *Trans. Farad. Soc.* **34**, 29–41 (1938).
37. Zhang, J. Y. & Du, Q. Constrained shrinking dimer dynamics for saddle point search with constraints. *J. Comput. Phys.* **231**, 4745–4758 (2012).
38. Cerjan, C. J. & Miller, W. H. On finding transition states. *J. Chem. Phys.* **75**, 2800–2806 (1981).
39. Li, Y. & Zhou, J. A minimax method for finding multiple critical points and its applications to semilinear PDEs. *SIAM J. Sci. Comput.* **23**, 840–865 (2001).
40. Rabinowitz, P. *Minimax Methods in Critical Point Theory with Applications to Differential Equations*. (American Mathematical Society, 1986).
41. Cances, E., Legoll, F., Marinica, M.-C., Minoukadeh, K. & Willaime, F. Some improvements of the activation-relaxation technique method for finding transition pathways on potential energy surfaces. *J. Chem. Phys.* **130**, 114711 (2009).
42. Machado-Charry, E. *et al.* Optimized energy landscape exploration using the ab initio based activation-relaxation technique. *J. Chem. Phys.* **135**, 034102 (2011).
43. Mousseau, N. & Barkema, G. T. Traveling through potential energy landscapes of disordered materials: The activation-relaxation technique. *Phys. Rev. E.* **57**, 2419–2424 (1998).
44. Miron, R. & Fichthorn, K. The step and slide method for finding saddle points on multi-dimensional potential surfaces. *J. Chem. Phys.* **115**, 8742–8750 (2001).
45. Gould, N., Ortner, C. & Packwood, D. An Efficient Dimer Method With Pre-conditioning And Linesearch. Preprint at http://arxiv.org/abs/1407.2817 (2014).
46. Zeng, Y., Xiao, P. & Henkelman, G. Unification of algorithms for minimum mode optimization. *J. Chem. Phys.* **140**, 044115 (2014).
47. Kastner, J. & Sherwood, P. Superlinearly converging dimer method for transition state search. *J. Chem. Phys.* **128**, 014106 (2008).

48. Zhang, L., Zhang, J. Y. & Du, Q. Finding critical nuclei in phase transformations by shrinking dimer dynamics and its variants. *Commun. Comput. Phys.* **16**, 781–798 (2014).

49. Zhang, L., Du, Q. & Zheng, Z. Optimization-based shrinking dimer method for finding transition states. *SIAM J. Sci. Comput.* (in the press).

50. Ren, W. Higher order string method for finding minimum energy paths. *Comm. Math. Sci* **1**, 377–384 (2003).

51. Jónsson, H., Mills, G. & Jacobsen, K. W. in *Classical and Quantum Dynamics in Condensed Phase Simulations*, (eds Berne B. J., Ciccoti G. & Coker D. F.) (World Scientific, 1998).

52. Elber, R. & Karplus, M. A method for determining reaction paths in large molecules: application to myoglobin. *Chem. Phys. Lett.* **139**, 375 (1987).

53. Gillilan, R. E. & Lilien, R. H. Optimization and dynamics of protein-protein complexes using b-splines. *J. Comput. Chem.* **25**, 1630 (2004).

54. Ulitsky, A. & Elber, R. A new technique to calculate steepest descent paths in flexible polyatomic systems. *J. Chem. Phys.* **92**, 1510 (1990).

55. W. E., Ren, W. & Vanden-Eijnden, E. Finite temperature string method for the study of rare events. *J. Phys. Chem. B.* **109**, 6688–6693 (2005).

56. Vanden-Eijnden, E. & Venturoli, M. Revisiting the finite-temperature string method for the calculation of reaction tubes and free energies. *J. Chem. Phys.* **130**, 194103 (2009).

57. Du, Q., Faber, V. & Gunzburger, M. Centroidal Voronoi tessellations: applications and algorithms. *SIAM Rev.* **41**, 637–676 (1999).

58. Maragliano, L. & Vanden-Eijnden, E. Single-sweep methods for free energy calculations. *J. Chem. Phys.* **128**, 184110 (2008).

59. Cahn, J. & Hilliard, J. Free energy of a nonuniform system. III. Nucleation in a two-component incompressible fluid. *J. Chem. Phys.* **31**, 688–699 (1959).

60. Chen, L.-Q. Phase-field models for microstructure evolution. *Annu. Rev. Mater. Res.* **32**, 113–140 (2002).

61. Khachaturyan, A. G. *Theory of Structural Transformations in Solids.* (Wiley, 1983).

62. Zhang, L., Chen, L. Q. & Du, Q. Mathematical and numerical aspects of phase-field approach to critical morphology in solids. *J. Sci. Comput.* **37**, 89–102 (2008).

63. Zhang, L., Chen, L.-Q. & Du, Q. Diffuse-interface approach to predicting morphologies of critical nucleus and equilibrium structure for cubic to tetragonal transformations. *J. Comput. Phys.* **229**, 6574–6584 (2010).

64. Gránásy, L., Pusztai, T., Saylor, D. & Warren, J. A. Phase field theory of heterogeneous crystal nucleation. *Phys. Rev. Lett.* **98**, 035703 (2007).

65. Laurila, T., Carlson, A., Do-Quang, M., Ala-Nissila, T. & Amberg, G. Thermo-hydrodynamics of boiling in a van der Waals fluid. *Phys. Rev. E* **85**, 026320 (2012).

66. Backofen, R. & Voigt, A. A phase-field-crystal approach to critical nuclei. *J. Phys. Condens. Matter* **22**, 364104 (2010).

67. Backofen, R. & Voigt, A. A phase field crystal study of heterogeneous nucleation—application of the string method. *Eur. Phys. J. Special Topics* **223**, 497–509 (2014).

68. Gránásy, L., Podmaniczky, F., Tóth, G. I., Tegze, G. & Pusztai, T. Heterogeneous nucleation of/on nanoparticles: a density functional study using the phase-field crystal model. *Chem. Soc. Rev.* **43**, 2159–2173 (2014).

69. Elder, K. R., Drolet, F., Kosterlitz, J. M. & Grant, M. Stochastic eutectic growth. *Phys. Rev. Lett.* **72**, 677 (1994).

70. Gránásy, L. *et al.* Phase-field modeling of polycrystalline solidification: from needle crystals to spherulites: a review. *Metall. Mater. Trans. A* **45**, 1694–1719 (2014).

71. Heo, T., Zhang, L., Du, Q. & Chen, L.-Q. Incorporating diffuse-interface nuclei in phase-field simulations. *Scripta Mater.* **63**, 8–11 (2010).

72. Roy, A., Rickman, J. M., Gunton, J. D. & Elder, K. R. Simulation study of nucleation in a phase-field model with nonlocal interactions. *Phys. Rev. E.* **57**, 2610–2617 (1998).

73. Li, Y., Hu, S., Zhang, L. & Sun, X. Non-classical nuclei and growth kinetics of Cr precipitates in FeCr alloys during aging. *Model. Simul. Mater. Sci. Eng.* **22**, 025002 (2014).

74. Brillouin, L. On thermal dependence of elasticity in solids. *Physical Review* **54**, 916–917 (1938).

75. Cahn, R. W. Crystal defects and melting. *Nature* **273**, 491–492 (1978).

76. Gorecki, T. Vacancies and changes of physical properties of metals at the melting point. *Z. Metallk.* **65**, 426–431 (1974).

77. Lindemann, F. A. The calculation of molecular natural frequencies. *Phys. Z.* **11**, 609–612 (1910).

78. Mott, N. F. Theories of the liquid state. *Nature* **145**, 801–802 (1940).

79. Mishin, Y., Mehl, M. J., Papaconstantopoulos, D. A., Voter, A. F. & Kress, J. D. Structural stability and lattice defects in copper: Ab initio, tight-binding, and embedded-atom calculations. *Phys. Rev. B* **63**, 224106 (2001).

80. Abrams, J. B. & Tuckerman, M. E. Efficient and direct generation of multidimensional free energy surfaces via adiabatic dynamics without coordinate transformations. *J. Phys. Chem. B* **112**, 15742 (2008).

81. Rosso, L., Mináry, P., Zhu, Z. & Tuckerman, M. On the use of the adiabatic molecular dynamics technique in the calculation of free energy profiles. *J. Chem. Phys.* **116**, 4389–4402 (2002).

82. Yu, T. Q. *et al.* Order-parameter-aided temperature-accelerated sampling for the exploration of crystal polymorphism and solid-liquid phase transitions. *J. Chem. Phys.* **140**, 214109 (2014).

83. Maragliano, L. & Vanden-Eijnden, E. A temperature accelerated method for sampling free energy and determining reaction pathways in rare events simulations. *Chem. Phys. Lett.* **426**, 168–175 (2006).

84. Coifman, R. R. *et al.* Geometric diffusion as a tool for harmonic analysis and structure definition of data, part I: diffusion maps. *Proc. Natl Acad. Sci. USA* **102**, 7426–7431 (2005).

85. Stone, M. Cross-validatory choice and assessment of statistical predictions. *J. R. Stat. Soc. B* **36**, 111–147 (1974).

18

Quasiparticle approach to diffusional atomic scale self-assembly of complex structures: from disorder to complex crystals and double-helix polymers

Mykola Lavrskyi[1], Helena Zapolsky[1] and Armen G Khachaturyan[2,3]

A self-organisation is an universal phenomenon in nature and, in particular, is highly important in materials systems. Our goal was to develop a new theory that provides a computationally effective approach to this problem. In this paper a quasiparticle theory of a diffusional self-organisation of atoms in continuum space during the diffusional time scale has been introduced. This became possible due to two novelties, a concept of quasiparticles, fratons, used for a description of dynamic degrees of freedom and model Hamiltonian taking into account a directionality, length and strength of interatomic bonds. To illustrate a predictive power and achievable level of complexity of self-assembled structures, the challenging cases of self-assembling of the diamond, zinc-blende, helix and double-helix structures, from a random atomic distribution, have been successfully modelled. This approach opens a way to model a self-assembling of complex atomic and molecular structures in the atomic scale during diffusional time.

INTRODUCTION

In the last 25 years, significant progress has been made by using molecular dynamic (MD) and Monte Carlo (MC) modelling for study of evolution of multi-atomic systems.[1–4] These methods supplement each other. The MD, is a straightforward approach of to a numerical solution of equations of motions for 6N dynamic variables where N is the number of atoms. However, a sheer number of variable limits the size of studied systems to $N \sim 10^5$ and time of its evolution to $t \sim 10^{-8}$ s. Recent modifications of MD by coarse graining improve the addressable time but at expense of the lost spatial resolution.[5,6] However, the MD is still not well-suited to study slow evolving systems with the typical diffusion time scale.

The MC alternative complements the MD since it can be applicable to the diffusional time scale. The MC approximates the mechanics of atomic motion by a stochastic dynamics of the Markov chain evolution.[3,4,7] A stochastic sampling in the MC dynamics requires a generation of a Markov chain that takes a significant time because it requires a search and update of databases, time scale separation and one process-at-a time execution.

Therefore, there are still significant difficulties of atomic scale prototyping of a slow diffusional self-organisation of atoms in complex structures. This is especially the case if the evolution is in the continuum space, the system consists of comparatively large number of atoms, and evolution time is long, ranging from a fraction of seconds to years. In this paper, we propose such an approach that may be supplemental to MD and MC that addresses their aforementioned limitations of MD and MC in computationally very effective way.

This development turned out to be possible because we (i) introduced a characterisation of a multi-atomic system in terms of quasiparticles named fratons, (ii) proposed a new simple form of phenomenological model potentials describing a directionality, length and strength of atom–atom bonding and (iii) used the kinetic equations of the atomic density field (ADF) theory describing the atomic scale diffusion. The latter was obtained by extending the ADF theory that was first developed for the Ising lattice gas model[8,9] and for the Ising lattice sites diffusion.[10] Recently, the extension of the ADF theory to the relevant case of atomic diffusion in the continuum space was also done.[11]

To check how this theory works, we chose examples of diffusional self-assembling of several atomic systems. Criteria for these choices are the following:

(i) To estimate a predictive power of the theory, we should choose the initial atomic configuration that has no any resemblance to the expected self-assembled atomic structure. This condition was satisfied by a choice of an initial state described by a completely random distribution of fratons in continuum space. The 'condensation' of fratons into the desirable final atomic structure should be provided solely by the input parameters of the model potentials, which characterise the directional atomic bonding.

(ii) The model potentials should have sufficient flexibility to be fitted not only to provide a desirable crystallography of a system, but also to approximate its thermodynamic and mechanic properties.

(iii) To make an illustration of potentiality of the approach sufficiently convincing, we have to demonstrate that it allows one to successfully simulate a self-assembly of high-complexity structures.

[1]GPM, UMR 6634, University of Rouen, Saint-Etienne du Rouvray, France; [2]Department of Materials Science & Engineering, Rutgers University, Piscataway, NJ, USA and [3]Department of Materials Science & Engineering, University of California, Berkeley, Berkeley, CA, USA.
Correspondence: H Zapolsky (helena.zapolsky@univ-rouen.fr)

We did successfully model a self-assembly of a randomly distributed atoms into complex atomic configurations of increasing complexity. The structures are ranged from single-component and multi-component crystals to single- and double-helix polymers. As far as we know, a self-assembling of the most complex of them that would started from the completely random state, has been never modelled before.

Model

A central part of the proposed theory is an introduction of non-traditional dynamic variables. Unlike the conventional approach describing the configuration of a classic multi-atomic system by coordinates of atomic centres, the proposed theory describes the atomic configurations by occupation numbers, $c(r)$, of quasiparticles that we call fratons where $c(r)$ is a stochastic function describing two possible events: $c(r)$ is equal to 1 if the point, r, reside inside any atomic sphere and is equal to zero otherwise. In other words, these two events indicate two possible states of each point of the continuous space, r. In this description, the dynamics of the system is described by a creation or annihilation of a fraton at each point of the continuous system: a creation of a fraton at a point, r, indicates that atomic movements resulted in a situation wherein the point r, which was previously outside of any atomic spheres, turned out inside of one of them; annihilation of a fraton describes an opposite process wherein the point, r, which initially is within of an atomic sphere, becomes outside of it.

We also introduced an analogue to the Pauli exclusion principle assuming that two fratons cannot occupy the same point. This exclusion automatically forbids interpenetration of the atoms and thus provides a dynamic 'exchange' repulsion preventing the atomic overlap. The introduction of fratons, in many respect, is conceptually similar to a transition to the secondary quantisation for multi-particle Fermi systems.[12]

A m-component system characterised in terms of fratons is described by m stochastic numbers $c_a(r)$ at each site r where $a = 1,2,..m$ labels the fratons related to the corresponding atom of the kind a. The averaging over the time-dependent Gibbs ensemble gives the occupation probability, $\rho_a(r, t) = <c_a(r, t)> T \leq 1$, where the symbol $< ... >$ implies averaging over the ensemble at temperature, T, and time, t. In this definition, the function $\rho_a(r, t)$ is an occupation probability, that point, r, is at any point inside of atomic sphere of any atom of the kind a at the time t.

The temporal evolution of the density function of fratons of the multi-component system is described by the atomic scale kinetic equation of the ADF theory[10] extended to the continuum space:

$$\frac{d\rho_a(\mathbf{r}, t)}{dt} = \sum_{\mathbf{r}'} \sum_{\beta=1}^{\beta=m} \frac{L_{a\beta}(\mathbf{r} - \mathbf{r}')}{k_B T} \frac{\delta G}{\delta \rho_\beta(\mathbf{r}', t)} \qquad (1)$$

where indices, a and β, label fratons describing different kinds of atoms ($a = 1, 2, ..., m$), k_B is the Boltzmann constant, G is the non-equilibrium Gibbs free energy functional, $L_{a\beta}(\mathbf{r})$ is a kinetic coefficients matrix. Summation is carried out over all points, $\mathbf{r}'$ of the computational grid approximating the continuum space. The kinetic parameters employed in our microscopic model are related to the phenomenological diffusion coefficients in the continuum model. It was shown previously[9] that in the continuous model (where $k \rightarrow 0$) the kinetic coefficients can be expressed as: $L(\mathbf{k}) = -M_{ij}k_ik_j$ where M is a diffusional mobility. The condition $\sum_{\mathbf{r}} L_{a\beta}(\mathbf{r}) = 0$ guarantees the conservation of the total number of fratons of each kind during evolution (and thus the conservation of the total volume of the corresponding atoms).

The simplest Gibbs free energy functional entering equation (1) is:

$$G = \frac{1}{2} \sum_{\mathbf{r},\mathbf{r}'} \sum_{a=1}^{a=m} \sum_{\beta=1}^{\beta=m} w_{a\beta}(\mathbf{r} - \mathbf{r}') \rho_a(\mathbf{r}) \rho_\beta(\mathbf{r}')$$

$$+ k_B T \sum_{\mathbf{r}} \left[\sum_{a=1}^{a=m} \rho_a(\mathbf{r}) \ln \rho_a(\mathbf{r}) + \left(1 - \sum_{a=1}^{a=m} \rho_a(\mathbf{r}) \right) \right.$$

$$\left. \ln \left(1 - \sum_{a=1}^{a=m} \rho_a(\mathbf{r}) \right) \right] - \sum_{\mathbf{r}} \sum_{a=1}^{a=m} \mu_a \rho_a(\mathbf{r}) \qquad (2)$$

where $w_{a\beta}(\mathbf{r} - \mathbf{r}')$ is the model potential of interaction of a pair of fratons of the components a and β, respectively, separated by a distance, $\mathbf{r} - \mathbf{r}'$, μ_a is the chemical potential of fratons of the kind a. Summation over $\mathbf{r}$ and $\mathbf{r}'$ in equation (2) is carried out over all N_0 sites of the computational grid lattice introduced to discretise the continuous space. The free energy (see equation (2)) corresponds to the mean field approximation[9]. It is asymptotically accurate at low and high temperatures, and its accuracy asymptotically increases if the interaction radius is much greater than the distance between interacting particles[13]. The latter condition is automatically satisfied in our case because interacting particles are fratons, and the computational grid increment, which is the minimum permitted distance between fratons, is much smaller than the atomic radius. The latter is also a requirement of accuracy of a description of a continuous atomic movement. The computational grid can be also interpreted as an Ising lattice and a spacing of the grid as a crystal lattice parameter of this 'Ising lattice'.

Equation (2) uses the Connolly–Williams approximation[14] that maps the fraton–fraton interaction into the chosen model potential, $w_{a\beta}(\mathbf{r} - \mathbf{r}')$. The Fourier transform (FT) of such a potential is:

$$\tilde{w}_{a\beta}(\mathbf{k}) = \frac{1}{N_0} \sum_{\mathbf{r}} w_{a\beta}(\mathbf{r}) e^{-i\mathbf{k}\mathbf{r}} \qquad (3)$$

where the summation is carried out over all sites of the computational grid, and the wave vector, $\mathbf{k}$, is defined at all quasi-continuum points, $\mathbf{k}$, of the first Brillouin zone of the computational grid, that is, at all N_0 the points in the k-space permitted by the periodical boundary conditions.

Structures that are really complex are usually formed in systems with directional covalent bonds between atoms. Therefore a theory whose goal is to study such systems by using the phenomenological model Hamiltonian should formulate the atom–atom interaction potentials, $\mathbf{k}$, as directional functions of atomic separation distance, $\mathbf{r} - \mathbf{r}'$. This function should describe the directionality of these bonds, their length and strength. The parameters of the potentials can be fitted to those calculated in quantum chemistry, Density Function formalism and/or by fitting to the observed crystallographic, mechanic and thermodynamic properties of the system. To make modelling sufficiently efficient, the approximation of the potentials should be as simple as possible without sacrificing a predictive power of the model.

In this paper, such model potentials are proposed. We tested their validity for the most challenging cases of self-assembling, some of which are not being modelled before. A central idea of the formulation of these potentials is based on the explicit use of what we call a bonding star, a group of vectors with the common beginning that describes directions and lengths of interatomic bonds.

A structural cluster designated as $a\beta$, is determined by a star of vectors numbered by the index $j_{a\beta}$. The vectors, $j_{a\beta}$, have the same origin, parallel to the corresponding bonds, $j_{a\beta}$, between the atoms of the kinds a and β, and have the lengths of these bonds. Therefore, the cluster, $a\beta$, is characterised by a set of geometrical

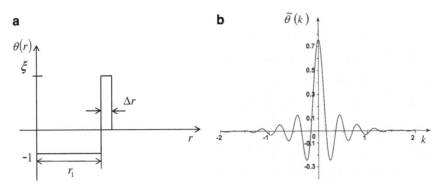

Figure 1. (**a**) Schematic representation of short-range potential $\theta(r)$. (**b**) Example of FT of $\theta(\mathbf{r})$ with the following input parameters: $\xi = 4$, $\hat{\Delta r} = 0.25$.

points determined by the ends of the vectors of the star and its centre. The geometry of the stars, $\alpha\beta$, and the strength of the bonds characterising the star describe interaction in the system and should be the only factor determining the structure of a self-assembled multi-atomic system.

These definitions allow us to introduce a cluster function, $\Psi_{\alpha\beta}^{\text{clstr}}(\mathbf{r})$ whose FT is:

$$\tilde{\Psi}_{\alpha\beta}^{\text{clstr}}(\mathbf{k}) = \sum_{j_{\alpha\beta}} \omega(j_{\alpha\beta}, \mathbf{k}) e^{-i\mathbf{k}\mathbf{r}_{j_{\alpha\beta}}} \qquad (4)$$

where summation is carried out over all vectors, $\mathbf{r}_{j_{\alpha\beta}}$, of the $\alpha\beta$ star of the cluster, $\omega(j_{\alpha\beta}, \mathbf{k})$ is a function characterising a strength of the bond $j_{\alpha\beta}$. The coefficients, $\omega(j_{\alpha\beta}, \mathbf{k})$ determine the thermodynamic and mechanical properties of the simulated system and can be used as fitting parameters to reproduce them.

Using these definitions, we present the FT of the model potential as sum of what we call the short-range and long-range interactions:

$$\tilde{w}_{\alpha\beta}(\mathbf{k}) = \lambda_1 \tilde{\theta}_a(\mathbf{k})\delta_{\alpha\beta} + \lambda_2(\mathbf{k})\Psi_{\alpha\beta}^{\text{cltr}}(\mathbf{k})\ \Psi_{\alpha\beta}^{\text{cltr}}(\mathbf{k})^* \qquad (5)$$

The first term in equation (5) describes the spherically symmetrical short-range fraton–fraton pair interaction. The function $\theta_a(\mathbf{r})$ is schematically presented in Figure 1a, where r_1 is a length parameter determining atomic radius, $\Delta\mathbf{r}$ is the width of repulsion part, $\xi = \frac{|\theta_{\max}(\mathbf{r})|}{|\theta_{\min}(\mathbf{r})|}$ is the ratio between the modules of minimum and maximum values of the shape function and λ_1 is a constant determining the strength of the short-range atomic repulsion. In particular, the parameter λ_1 characterises the rigidity of the atomic spheres and its value can be determined by a fitting of the elastic properties of a given system. The FT of the function, $\tilde{\theta}_a(\mathbf{k})$, schematically shown in Figure 1b is:

$$\tilde{\theta}_a(\mathbf{k}) = \frac{4\pi}{k^3}(-\sin(\mathbf{k}r_1) - \mathbf{k}r_1\cos(\mathbf{k}r_1)) + \xi(\sin(\mathbf{k}(r_1 + \Delta\mathbf{r}))$$
$$-\mathbf{k}(r_1 + \Delta\mathbf{r})\cos(\mathbf{k}(r_1 + \Delta\mathbf{r}) - \sin(\mathbf{k}r_1) + \mathbf{k}r_1\cos(\mathbf{k}r_1))) \qquad (6)$$

The second term of equation (5) is the long-range part of the fraton–fraton interaction describing a directional bonding of atoms of the kind α and β.

In fact, the long-range interaction in equation (5) is presented as a bilinear expansion in cluster functions, $\Psi_{\alpha\beta}^{\text{clstr}}(\mathbf{k})$, $\lambda_2(\mathbf{k})$ is a fitting parameter determining a strength of the long-range interaction.

The indexes, α and β can be dropped for a single-component system. Then equation (5) is simplified to:

$$\tilde{w}(\mathbf{k}) = \lambda_1 \tilde{\theta}(\mathbf{k}) + \lambda_2(\mathbf{k})\left|\Psi^{\text{clstr}}(\mathbf{k})\right|^2 \qquad (7)$$

In this paper we consider a particular case of application of the fratonic theory allowing to obtain a single crystalline state. This is a case of specifically oriented clusters. The constraint lifts the angular isotropy of the system and thus allows us to prevent the formation of a 'polycrystalline state' that is an atomic aggregate of grains with the same atomic structure but different orientation.

A self-assembling producing such a 'polycrystal' would make it difficult to identify the equilibrium atomic structure. However, in the general case, in which the angular isotropy is not lifted, the potential described by equation (7) describes a growth of polycrystal (see Supplementary Figure 1S in Supplementary Information). In this case the interaction energy of a pair of fratons is independent from its orientation. It can be achieved using a rotational averaging of a cluster function $\Psi_{\alpha\beta}^{\text{clstr}}(\mathbf{k})$.

RESULTS

To illustrate the versatility and effectiveness of the fraton theory, we tested its application to the modelling of the self-assembly of three groups of three-dimensional structures of increasing complexity. They are single-component crystals, two-component crystals and a polymer with a double-helix structure mimicking biological macromolecules. The modelling was carried out by numerical solution of the FT representation of the kinetic equation (1).

In our simulations, we used the reduced parameters, and, in particular, average density, defined as $\hat{\bar{\rho}}_a = \rho_a^{\text{at}}\frac{4\pi R_a^3}{3}$ where $\rho_a^{\text{at}} = \frac{N_a}{V}$ is the density of a atoms in the ground state, N_a is number of the atoms of sort a, V is the total volume of the system and R_a is the atomic radius of this atom. According to this definition, the reduced density, $\hat{\bar{\rho}}_a$, is also a fraction of all computational grid sites occupied by fratons of the kind a. The input parameter ξ of the energy $\theta_a(\mathbf{r})$ is measured in units of $k_B T_o$, where T_o is the solidification temperature. The lengths are measured in units of r_1, which is very close to the atomic radius; the grid lattice increment, $\hat{l}$ (the spacing of the underlying Ising lattice), is defined as a fraction of the atomic radius. The temperature $\hat{T}$ is also measured in units of T_o. The reduced time, $\hat{t}$, is measured in units of typical atomic migration time, τ_o. The reduced kinetic coefficients, $\hat{L}(\mathbf{r})$, are measured in units of τ_0^{-1} and $\tilde{L}(\mathbf{k}) = Dk^2$, where D is a constant. The numerical solution of the reduced form of kinetic equation (1) was obtained by using the semi-implicit Fourier-spectral method[15]. The number of computational grid sites in the simulation box was chosen in compromise between a necessity to reasonably reproduce the atomic structure and to optimise the simulation time. A time step of $\Delta t = 0.01$ was used. The periodic boundary condition was imposed for all simulations. The first example that we will consider in this article is the self-assembly of a single-component crystal with several atoms in a Bravais lattice unit cell. As an example, we tested whether the model Hamiltonian describing tetrahedral orientation of the interatomic bonds results in a self-assembling producing the diamond structure. An image of the structural cluster in this case is convenient to choose as a star of eight bonding vectors whose ends are located at the sites of a cubic unit cell of the diamond lattice: (000), $\left(0\frac{1}{2}\frac{1}{2}\right)$, $\left(\frac{1}{2}0\frac{1}{2}\right)$, $\left(0\frac{1}{2}\frac{1}{2}\right)$, $\left(\frac{1}{4}\frac{1}{4}\frac{1}{4}\right)$, $\left(\frac{3}{4}\frac{1}{4}\frac{1}{4}\right)$, $\left(\frac{1}{4}\frac{3}{4}\frac{3}{4}\right)$ and $\left(\frac{3}{4}\frac{3}{4}\frac{1}{4}\right)$.

The function $\Psi^{\text{cltr}}(\mathbf{k})$ was constructed by using its definition (equation (4)) and the coordinates of the chosen structural cluster points:

$$\Psi^{\text{cltr}}(\mathbf{k}) = \left(1 + e^{-i\frac{a}{4}(k_x + k_y + k_z)}\right)\left(1 + e^{-i\frac{a}{2}(k_x + k_y)} + e^{-i\frac{a}{2}(k_y + k_z)} + e^{-i\frac{a}{2}(k_x + k_z)}\right) \tag{8}$$

where a is a lattice constant of diamond structure, $k_i = \frac{2\pi m_i}{aN}$ (where $i = x, y$ or z, $m_i = 1 \ldots N$, N is the number of simulation grid in a given direction). With this definition, the FT of the long-range interaction, $\tilde{w}_{\text{LR}}(\mathbf{k})$, can be written as:

$$\tilde{w}_{\text{LR}}(\mathbf{k}) = \lambda_2(\mathbf{k})\tilde{\Omega}_D(\mathbf{k}) \tag{9}$$

where

$$\tilde{\Omega}_D(\mathbf{k}) = \Psi^{\text{cltr}}(\mathbf{k})\left(\Psi^{\text{cltr}}(\mathbf{k})\right)^*$$
$$= \left(2 + 2\cos\left(\frac{a}{4}(k_x + k_y + k_z)\right)\right) \times \left(4 + 4\left(\cos\left(\frac{k_x a}{2}\right)\right.\right.$$
$$\cos\left(\frac{k_y a}{2}\right) + \cos\left(\frac{k_y a}{2}\right)\cos\left(\frac{k_z a}{2}\right) + \cos\left(\frac{k_x a}{2}\right)$$
$$\left.\left.\cos\left(\frac{k_z a}{2}\right)\right)\right) \tag{10}$$

To choose for the model potential the form of $\lambda(\mathbf{k})$, we used the following consideration. The function $|\Psi(\mathbf{k})|^2$ by definition (see equation (10)) is periodical in k-space. Its longest period is along the [111] direction and equal to $k_1 = \frac{8\pi}{a}$. We assume that the function, $\lambda_2(\mathbf{k})$, is a step function. It is defined as:

$$\lambda_2(\mathbf{k}) = \begin{cases} \lambda_2 & \text{if} \quad 0 \le (k_x + k_y + k_z) \le \frac{8\pi}{a} + \delta \\ 0 & \text{otherwise} \end{cases} \tag{11}$$

where δ is a positive constant.

Then the FT of the model potential can be written as:

$$\tilde{w}(\mathbf{k}) = \lambda_1 \tilde{\theta}(\mathbf{k}) + \lambda_2(\mathbf{k})\tilde{\Omega}_D(\mathbf{k}) \tag{12}$$

where the first term describes the short-range interaction. The energy parameters λ_1 and λ_2 were normalised by the absolute value of a difference between the maximum and minima values of the function $\tilde{\Omega}_D(\mathbf{k})$ and $\tilde{\theta}(\mathbf{k})$, respectively.

In this simulation, the initial configuration was an embryo consisting of the small variation of the fratons density at the sites of the structural cluster of diamond structure embedded in the gas of disordered fratons (Figure 2a). However, the initial state can be also chosen as local random 'infinitesimals' statics fluctuations with respect to homogeneous state. The spontaneous self-organisation of fratons into the diamond structure is shown in Figure 2. The intermediate step in the pattern formation dynamics at the reduced time $t = 60,000$ is shown in Figure 2b. A very interesting aspect of this growth is the development of the transient bcc structure (Figure 2c) in the early stages of growth. Its lattice parameter approximately is half that of the diamond structure.

A probability of occupation of each sites of the obtained transient bcc lattice is less than unity. The latter indicates that the bcc lattice sites, in fact, are randomly occupied by atoms and their vacancies and form the disordered distribution in the bcc lattice. This result is fully consistent with the thermodynamics dictating a mandatory presence of a disordered distribution of vacancies in the bcc (or any other) lattice at finite temperature, and thus dictating the greater number of lattice sites than the number of occupying them atoms. It should be also noted that the total number of atoms in our simulation is automatically conserved because the kinetic equation (or more specifically, a choice of matrix of kinetic coefficients) guarantees the atomic conservation.

The formation of the diamond lattice, in fact, is a fratonic theory description of the bcc → B32 (NaTl-type) ordering of atoms and their vacancies over the preferential sites of the previously 'condensed' bcc host lattice. A result of such an ordering is a doubling of the crystal lattice parameters of the underlying bcc lattice. It should be also mentioned that a possibility of the crystallisation of the diamond structure through the transient bcc structure is a result that could hardly be predicted in advance for a system with tetrahedral atomic bonding.

To better visualise the final structure, presented in Figure 2e, the links between the first neighbours are shown. The growth dynamic of the diamond structure is shown in Supplementary Movie 1 (see Supplementary Information). To better characterise the different steps of the growth dynamic, the diffraction patterns at different time steps have been calculated. The diffraction pattern, which is a distribution of intensity of scattered radiation in the three-dimensional reciprocal space of the wave vectors, $\mathbf{k} = (k_x, k_y, k_z)$, was determined as the squared modulus of the FT of the density function. The strongest diffraction peaks that characterise the diamond structure are: (220), (111), (311) and (400). The peaks {200}, which are forbidden for the diamond lattice, are present in the diffraction pattern of the bcc structure. Then, following these peaks it is possible to distinguish the different stage of the growth dynamic. The evolution of the diffraction pattern during the growth dynamic of the diamond structure is shown in Supplementary Movie 2 (see Supplementary Information).

Continuing a gradual increase in the complexity of the modelled structure, we also considered the formation of a two-component crystalline phase in a system with tetrahedral direction of bonds expecting the formation of the zinc-blende structure. A two-component systems atomic arrangement is formed by a 'condensation' of two kinds of fratons, belonging to type A and B. This condensation should produce atoms A and B, respectively. The spontaneous arrangement caused by an equilibration of a disordered distribution of one sort of fratons is described by the kinetic equation (1). For a two-component system, equation (1) is reduced to two equations for two fraton densities $\rho_A(\mathbf{r})$ and $\rho_B(\mathbf{r})$. Thus in the reciprocal space these equations can be written as:

$$\frac{\partial \tilde{\rho}_A(\mathbf{k}, t)}{\partial t} = L_{AA}(\mathbf{k})\left(\tilde{w}_{AA}(\mathbf{k})\tilde{\rho}_A(\mathbf{k}, t) + \tilde{w}_{AB}(\mathbf{k})\tilde{\rho}_B(\mathbf{k}, t)\right.$$
$$+ \left(\ln\frac{\rho_A(\mathbf{r}', t)}{1 - \rho_A(\mathbf{r}', t) - \rho_B(\mathbf{r}', t)}\right)_{\mathbf{k}}\right)$$
$$+ L_{AB}(\mathbf{k})(\tilde{w}_{BB}(\mathbf{k})\tilde{\rho}_B(\mathbf{k}, t) + \tilde{w}_{AB}(\mathbf{k})\tilde{\rho}_A(\mathbf{k}, t)$$
$$+ \left(\ln\frac{\rho_B(\mathbf{r}', t)}{1 - \rho_A(\mathbf{r}', t) - \rho_B(\mathbf{r}', t)}\right)_{\mathbf{k}}\right) \tag{13a}$$

$$\frac{\partial \tilde{\rho}_B(\mathbf{k}, t)}{\partial t} = L_{BB}(\mathbf{k})\left(\tilde{w}_{BB}(\mathbf{k})\tilde{\rho}_B(\mathbf{k}, t) + \tilde{w}_{AB}(\mathbf{k})\tilde{\rho}_A(\mathbf{k}, t)\right.$$
$$+ \left(\ln\frac{\rho_B(\mathbf{r}', t)}{1 - \rho_A(\mathbf{r}', t) - \rho_B(\mathbf{r}', t)}\right)_{\mathbf{k}}\right)$$
$$+ L_{AB}(\mathbf{k})(\tilde{w}_{AA}(\mathbf{k})\tilde{\rho}_A(\mathbf{k}, t) + \tilde{w}_{AB}(\mathbf{k})\tilde{\rho}_B(\mathbf{k}, t)$$
$$+ \left(\ln\frac{\rho_A(\mathbf{r}', t)}{1 - \rho_A(\mathbf{r}', t) - \rho_B(\mathbf{r}', t)}\right)_{\mathbf{k}}\right) \tag{13b}$$

where A and B designate two sorts of fratons.

The FTs of the interaction energies, $\tilde{w}_{\alpha\beta}(\mathbf{k})$, determined by equation (5), are:

$$\tilde{w}_{AA}(\mathbf{k}) = \lambda_{1A}\tilde{\theta}_A(\mathbf{k}) + \lambda_{2A}(\mathbf{k})\tilde{\Omega}_{ZB}^{AA}(\mathbf{k}) \tag{14a}$$

$$\tilde{w}_{BB}(\mathbf{k}) = \lambda_{1B}\tilde{\theta}_B(\mathbf{k}) + \lambda_{2B}(\mathbf{k})\tilde{\Omega}_{ZB}^{BB}(\mathbf{k}) \tag{14b}$$

$$\tilde{w}_{AB}(\mathbf{k}) = \lambda_{2AB}(\mathbf{k})\tilde{\Omega}_{ZB}^{AB}(\mathbf{k}) \tag{14c}$$

where $\tilde{\Omega}_{ZB}^{\alpha\beta}(\mathbf{k}) = \Psi_\alpha(\mathbf{k})\Psi^*_\beta(\mathbf{k})$

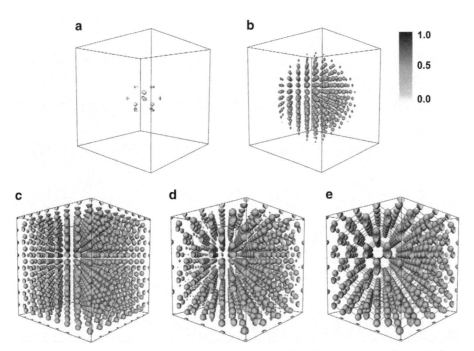

Figure 2. Example of a self-assembly of fratons into diamond structure at reduced times $\hat{t}$ of (**a**) $\hat{t} = 0$, (**b**) $\hat{t} = 60{,}000$, (**c**) $\hat{t} = 100{,}000$, (**d**) $\hat{t} = 280{,}000$ and (**e**) $\hat{t} = 300{,}000$. The parameters in these simulations are $\hat{\lambda}_1 = 14.085$, $\hat{\lambda}_2 = -7.042$, $\hat{a} = 4.57$, $\xi = 2$, $\hat{D} = 1$, $\hat{\bar{\rho}} = 0.07$, $\hat{l} = 0.286$, $\Delta\hat{r} = 0.17$ and $\hat{T} = 0.732$. The initial configuration is the atomic cluster of the diamond structure placed in the centre of the simulation box. The size of the simulation box is $64 \times 64 \times 64$.

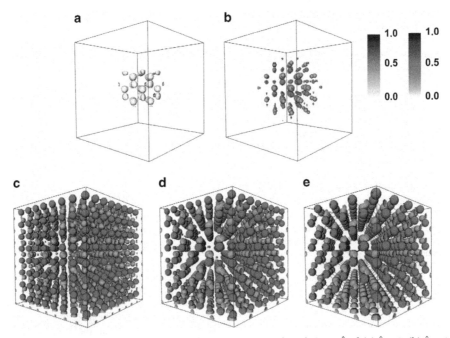

Figure 3. Example of a self-assembly of fratons into zinc-blende structure at reduced times $\hat{t}$ of (**a**) $\hat{t} = 0$, (**b**) $\hat{t} = 160{,}000$, (**c**) $\hat{t} = 190{,}000$, (**d**) $\hat{t} = 280{,}000$ and (**e**) $\hat{t} = 3{,}000{,}000$. The parameters in this simulation are $\xi = 2$, $\hat{D}_{AA} = \hat{D}_{BB} = 1$, $\hat{D}_{AB} = -0.5$, $\hat{\lambda}_{1A} = 3.77$, $\hat{\lambda}_{2A} = -1.88$, $\hat{\lambda}_{1B} = 5.84$, $\hat{\lambda}_{2B} = -2.92$, $\hat{\lambda}_{2AB} = -2.26$, $\hat{l} = 0.25$, $r_{1A} = 1.143\, r_{1B}$, $\Delta\hat{r} = 0.17$, $\hat{a} = 4.0$, $\hat{\bar{\rho}}_A = 0.07$, $\hat{\bar{\rho}}_B = 0.045$ and $\hat{T} = 0.235$. The initial configuration is the atomic cluster of a diamond structure placed in the centre of the simulation box. The size of the simulation box is $64 \times 64 \times 64$. Two sorts of atoms with different atomic sizes are indicated in red and green.

The zinc-blende structure has two atoms, A and B in a primitive unit cell of the fcc Bravais lattice with positions (000) and $\left(\frac{1}{4}\frac{1}{4}\frac{1}{4}\right)$, correspondingly. To describe the model Hamiltonian providing evolution to this structure, we needed two structural clusters, which is, the clusters of type A and B. The cluster A consists of four points: the point (000) and its nearest neighbours in the fcc lattice,

$a\left(\frac{1}{2}\frac{1}{2}0\right)$, $a\left(0\frac{1}{2}\frac{1}{2}\right)$, $a\left(\frac{1}{2}0\frac{1}{2}\right)$. The cluster B also consists of four point. They are obtained from the four points of the cluster A by the shift, $a\left[\frac{1}{4}\frac{1}{4}\frac{1}{4}\right]$. With this definition, the Ψ-functions for the two structural clusters are:

$$\Psi_A^{\text{cltr}}(\mathbf{k}) = \left(1 + e^{-i\frac{a}{2}\left(k_x + k_y\right)} + e^{-i\frac{a}{2}\left(k_y + k_z\right)} + e^{-i\frac{a}{2}\left(k_x + k_z\right)}\right) \tag{15a}$$

$$\Psi_B^{\text{cltr}}(\mathbf{k}) = e^{-\frac{ia}{4}(k_x+k_y+k_z)}\left(1 + e^{-\frac{ia}{2}(k_x+k_y)} + e^{-\frac{ia}{2}(k_y+k_z)} + e^{-\frac{ia}{2}(k_x+k_z)}\right)$$

(15b)

Using this definition in equation (7) and assuming that the functions $\lambda_{2A}(\mathbf{k}) = \lambda_{2B}(\mathbf{k}) = \lambda_{2AB}(\mathbf{k}) = \lambda_2(\mathbf{k})$, where $\lambda_2(\mathbf{k})$ is defined by equation (11), we have:

$$\tilde{\Omega}_{ZB}^{AA}(\mathbf{k}) = \tilde{\Omega}_{ZB}^{BB}(\mathbf{k}) = 4 + 4\left(\cos\left(\frac{k_x a}{2}\right)\cos\left(\frac{k_y a}{2}\right)\right.$$
$$\left. + \cos\left(\frac{k_y a}{2}\right)\cos\left(\frac{k_z a}{2}\right) + \cos\left(\frac{k_x a}{2}\right)\cos\left(\frac{k_z a}{2}\right)\right)$$

(16a)

$$\tilde{\Omega}_{ZB}^{AA}(\mathbf{k}) = 2\cos\left(\frac{a}{4}(k_x+k_y+k_z)\right) \times \left(4 + 4\left(\cos\left(\frac{k_x a}{2}\right)\cos\left(\frac{k_y a}{2}\right)\right.\right.$$
$$\left.\left. + \cos\left(\frac{k_y a}{2}\right)\cos\left(\frac{k_z a}{2}\right) + \cos\left(\frac{k_x a}{2}\right)\cos\left(\frac{k_z a}{2}\right)\right)\right)$$

(16b)

The difference in size of different species of atoms has been taken into account in the short-range potential. In these simulations, the ratio of two atomic radii was chosen to be 0.875. The size of the simulation grid, $\hat{l} = 0.25$, was measured in the unities of r_{1A}. Therefore, the value of r_{1B} was chosen equal to $3.5\hat{l}$. The growth of the zinc-blende structure is shown in Figure 3 and in Supplementary Movie 3 (see Supplementary Information).

Molecules with a helix architecture are observed in organic materials[16], helix-shaped graphite nanotubes,[17,18] liquid crystal,[19,20] proteins, and, of course, DNA and RNA polymeric molecules.[21,22] However, the most challenging test of the potency of the fraton theory would be its ability to describe a spontaneous self-assembly for the most interesting case relevant to biology, that is, the self-assembly of a double-helix polymer from a 'soup' of randomly distributed monomers. We chosen this system because, as far as we know, a self-assembly of randomly distributed monomers into double-thread helix polymers was a too complex phenomenon to prototype by the existing methods.

To model a spontaneous self-assembling of a helix polymer consisting of two complimentary threads, we considered two types of mutually complementary fratons whose 'condensation' should produce two kinds of mutually complementary monomers.

The model fraton–fraton potential producing a helix structure should be directional and have a built-in chirality. This is achieved by a choice of a bonding star of the cluster that has the chirality corresponding to the desired helix geometry. The geometrical parameters of the configuration of the structural cluster are the pitch length, P, the number of coils per pitch, $n_0 = 6$, the distance between coils in z-direction, h, and the radius of the coil, u (see Figure 4). Then the coordinates of points of the helix occupied by molecules are:

$$\mathbf{r}_s = \left(u\cos\left(\frac{2\pi}{n_0}s\right), u\sin\left(\frac{2\pi}{n_0}s\right), \frac{h}{n_0}s\right)$$

(17)

where s runs from 0 to $n_0 - 1$, n_0 is the number of coils in the pitch.

We again start from a construction of the model Hamiltonian that should lead to a condensation of a randomly distributed fratons into the helix structure. To do this, the structural cluster that consists of two pitches has been chosen. We had to choose a two-period cluster because fratons in this model have no orientational degrees of freedom and thus have no built-in chirality. The second-pitch segment of the cluster is needed to introduce a chirality into the long-range part of the model potential. The size of the structural cluster would be drastically reduced if the chirality were built-in in the short-range part of the fraction–fraton interaction. This can be done by a straightforward modification of the short-range interaction.

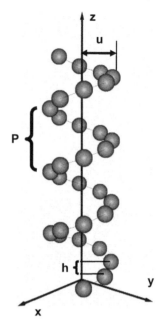

Figure 4. Illustration of the geometrical parameters of the helical structure: P is a pitch, h is a distance between the nearest coils along the z-axis and helix radius u.

Using this definition of cluster for the formulation of the function $\Psi(\mathbf{k})$ for the helical structure, presented in Figure 4 gives:

$$\Psi^{\text{cltr}}(\mathbf{k}) = \left(1 + e^{-n_0 h k_z}\right)\left(\sum_{n=0}^{n_0-1} e^{-i(x_n k_x + y_n k_y + z_n k_z)}\right)$$

(18)

Then function $\tilde{\Omega}_H(\mathbf{k})$ is:

$$\tilde{\Omega}_H(\mathbf{k}) = (2 + 2\cos(hn_0 k_z))\left(6 + 2\sum_{n>m=0}^{n_0-1}\cos(\phi(n,m))\right)$$

(19)

where

$$\phi(n,m) = k_x u\left(\cos\left(\frac{2\pi n}{n_0}\right) - \cos\left(\frac{2\pi m}{n_0}\right)\right)$$
$$+ k_y u\left(\sin\left(\frac{2\pi n}{n_0}\right) - \sin\left(\frac{2\pi m}{n_0}\right)\right)$$
$$+ k_z h(n-m)$$

(20)

We also assumed that the desired helix has a single monomer in each coil and approximate these monomers by a spherical shape. The latter is not a critical assumption for the theory. It is just a simplification that reduces the computational time. The short-range interaction was described by equation (6). The parameters of the $\tilde{\theta}(\mathbf{k})$ function were chosen the same as for the diamond structure. The size of simulation box was $210 \times 32 \times 32$. The initial embryo lifting the spatial and rotational energy degeneration was a one pitch inhomogeneity introduced in the centre of the simulation box. As was discussed previously, to construct a model potential we should choose the function $\lambda_2(\mathbf{k})$. For the same reasons as before, we assumed that the function, $\lambda_2(\mathbf{k})$, is a step function defined as:

$$\lambda_2(\mathbf{k}) = \begin{cases} \lambda_2 & \text{if } -\delta \leq \phi(n,m) \leq 2\pi + \delta \quad \text{for } n-m=1, \\ 0 & \text{otherwise} \end{cases}$$

(21)

The condition $n - m = 1$ defines the longest period of the function $\tilde{w}_{\text{LR}}(\mathbf{k})$ for helix.

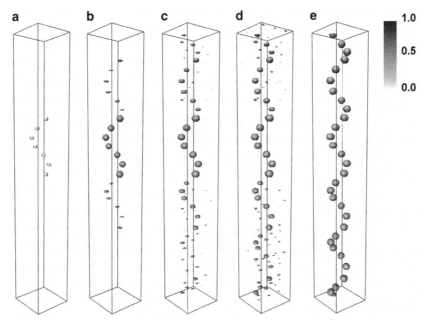

Figure 5. Example of self-assembly of the fratons into the helix structure at reduced time of (**a**) $\hat{t} = 0$, (**b**) $\hat{t} = 200,000$, (**c**) $\hat{t} = 250,000$, (**d**) $\hat{t} = 300,000$ and (**e**) $\hat{t} = 700,000$. The input parameters in this simulation are: $\hat{\lambda}_1 = 61.14$, $\hat{\lambda}_2 = -69.87$, $\hat{h} = \hat{u} = 1.56$, $n_0 = 6$, $\xi = 2$, $\hat{D} = 1$, $\hat{\bar{\rho}} = 0.0096$, $\hat{l} = 0.22$, $\Delta\hat{r} = 0.17$ and $\hat{T} = 0.568$. The size of the simulation box is $32 \times 32 \times 210$. The initial configuration shown in **a** is n_0+1 coils in the helix structure.

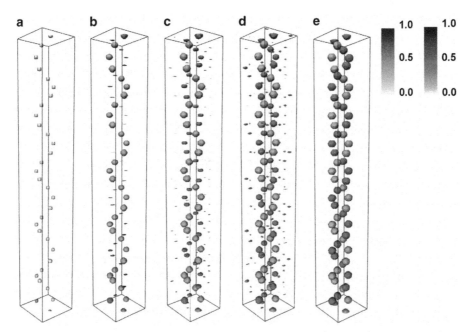

Figure 6. Self-assembly of the fratons into a double-helix structure at reduced time of (**a**) $\hat{t} = 0$, (**b**) $\hat{t} = 150,000$, (**c**) $\hat{t} = 200,000$, (**d**) $\hat{t} = 300,000$ and (**e**) $\hat{t} = 1,500,000$. The input parameters in this simulation are $\hat{\lambda}_{1A} = \hat{\lambda}_{1B} = 4.07$, $\hat{\lambda}_{2A} = \hat{\lambda}_{2B} = -4.07$, $\hat{\lambda}_{2AB} = -1.78$, $r_{1A} = r_{1B}$, $\hat{h} = \hat{u} = 1.56$, $n_0 = 6$, $\xi = 2$, $\hat{D}_{AA} = \hat{D}_{BB} = 1$, $\hat{D}_{AB} = -0.5$, $\hat{\bar{\rho}}_A = \hat{\bar{\rho}}_B = 0.0096$, $\hat{l} = 0.22$, $\Delta\hat{r} = 0.17$, and $\hat{T} = 0.033$. The size of the simulation box is $32 \times 32 \times 210$. The initial configuration shown in **a** is one helix and one coil of the second helix.

In the first step, we considered a random distribution of fratons of one kind. The solution of equation (1) with the model potential given by equations (19) and (20) in this case describes an evolution that eventually produces the single-stranded helix shown in Figure 5.

The introduction of the second kind of fratons, which is complementary to the first kind, results in their self-assembly of a complementary strand and the formation of a double-helix molecule. To model double-helix growth, we have to introduce (as for the zinc-blend structure) complimentary fratons of two types, A and B that form each helix thread. We used the same type of the structural clusters for the fratons of the kind A and B. Each of them consists of two pitches of helix structure. However, the clusters are rotated with respect to each other about z-axis by

$\varphi = \pi$. Then, using the definition of the functions $\Psi_a(\mathbf{k})$ by equation (4) for $a = A, B$, gives:

$$\Psi_A^{cltr}(\mathbf{k}) = \left(1 + e^{-i\frac{n_0 h}{2}k_z}\right)\left(\sum_{n=0}^{n_0-1} e^{-i\left(x_n k_x + y_n k_y + z_n k_z\right)}\right) \qquad (22a)$$

$$\Psi_B^{cltr}(\mathbf{k}) = e^{-i\frac{n_0 h}{2}k_z}\left(1 + e^{-i\frac{n_0 h}{2}k_z}\right)\left(\sum_{n=0}^{n_0-1} e^{-i\left(x_n k_x + y_n k_y + z_n k_z\right)}\right) \qquad (22b)$$

We assume that, where $\tilde{\Omega}_{DH}^{AA}(\mathbf{k}) = \tilde{\Omega}_{DH}^{BB}(\mathbf{k}) = \tilde{\Omega}_H(\mathbf{k})$, $\tilde{\Omega}_H(\mathbf{k})$ is defined by equation (19). Then the function $\tilde{\Omega}_{DH}^{AB}(\mathbf{k})$ is:

$$\tilde{\Omega}_{DH}^{AB}(\mathbf{k}) = 2\cos\left(h\frac{n_0}{2}k_z\right)(2 + 2\cos(hn_0 k_z))$$
$$\left(6 + 2\sum_{n>m=0}^{n_0-1}\cos\left(\phi(n,m)\right)\right) \qquad (23)$$

The simulation box size and input parameters for two kinds of complimentary fratons were chosen the same as for a single-thread helix and the interaction between helix is defined by $\hat{\lambda}_{2AB} = -1.78$, $\hat{T} = 0.033$.

In spite of all these oversimplifications, this model describes some generic features relevant to the spontaneous formation of single-stranded polymeric molecule and the growth of the complementary strand of the monomers eventually producing a double-stranded helix configuration (see Figure 6). In this case, the first single-stranded helix is a template for the aggregation on it of complementary monomers to form a double-stranded helix. For clarity, we show the clusters of fratons (monomers) of the second strand in red. The spontaneous growth of the helix and double-helix structure is also shown in Supplementary Movie 4 and 5 (see Supplementary Information).

DISCUSSION

In this paper, we selected the most difficult cases wherein the initial system is atomically disordered so that its configuration 'knows' nothing about the final atomic pattern that should be spontaneously self-assembled. This self-assembling is driven only by the chosen model Hamiltonian, and, specifically, by mutual orientation, length and strength of interatomic bonds. On a top of that, we considered situations wherein the self-assembling is diffusional and usually takes a long time, which may range from a fraction of a second to years. A typical time of this evolution is dictated by the typical time of evolution of time-dependent ensemble rather than typical times of atomic dynamics like time of atomic vibrations. Difficulty in addressing such slow evolving systems probably was a reason why a spontaneous formation of some of them (crystals and polymers) from a liquid solution of atoms or monomers has not been modelled yet in the diffusional time scale.

The developed approach opens a way to answer numerous outstanding questions concerning the atomistic mechanisms of the formation of defects (dislocations, grain boundaries, etc.), nucleation in solid–solid transformations, the formation of polymers due to aggregation of monomers in their solution, folding and crystallisation of polymers, and their responses to external stimuli. This list can be significantly extended.

Especially interesting are the modelling results describing the spontaneous self-assembly of monomers into a single-stranded polymeric helix and the formation of a double-helix structure obtained by aggregation of complementary monomers on the single-stranded helix playing the role of a template. This result may be also considered as an attempt to formulate and execute the simplest prototyping of the spontaneous formation of homopolymeric DNA from a liquid solution of monomers playing the role of nucleotides.

Finally, the use of the new model Hamiltonian formulated in terms of the structural clusters and proposed fraton model provide already a ready tool to address a general problem of spontaneous pattern formation by self-assembling of any randomly distributed building elements in the time scale ranging from sub-seconds to years. This approach can be also straightforwardly extended for the prototyping of self-assembly of elementary building block monomers with more complex molecular structures. In the latter case, we have to generalise the concept of fratons of atoms by introducing fraton of molecules and modify accordingly the model short-range part of the model fraton–fraton Hamiltonian. Then this modification should provide a 'condensation' of the molecular fratons into molecules and subsequent self-assembly of these molecules. In principle, this approach can be even used for the description of three-dimensional pattern formation by any macroscopic objects and optimisation of their properties. The 'fratons' in this case being fragments of these objects are also macroscopic.

MATERIALS AND METHODS

The proposed fraton theory rests on two novel conceptual premises: (a) the introduction of interacting pseudoparticles that we call fratons that described two configurational states of each point of continuum space. One is an event in which the point is inside the atomic sphere of any atom and the other is an event in which the point is outside of atomic sphere; the fratons are considered as a non-ideal gas whose 'condensation' describes a diffusional self-assembling of atomic system, and (b) a concept of a structural cluster function describing the directions, length and strength of interatomic bonds. The latter allows us to formulate a new and simple model Hamiltonian that is proportional to a bilinear expansion in these cluster functions. This model Hamiltonian provides the formation of a predetermined atomic structure and has a sufficient flexibility to describe the desired mechanic and thermodynamic properties of this structure.

The numerical solution of kinetic equation (1) for the density function of atomic fractons was carried out by using the semi-implicit Fourier-spectral method in which the time variable is discretised using semi-implicit schemes and space variables are discretised using the Fourier-spectral method[15].

ACKNOWLEDGEMENTS

We thank R Patte for helpful advices and discussions during code optimisation. This work was supported in part by the grant from the French National Agency for the Research (ANR) project 'Spiderman'. The simulations have been performed at the Centre de Ressources Informatiques de Haute-Normandie (CRIHAN) and at the IDRIS of CNRS.

CONTRIBUTIONS

All authors contributed extensively to the work presented in this paper.

COMPETING INTERESTS

The authors declare no conflict of interest.

REFERENCES

1. Klein, M. L. & Shinoda, W. Large-scale molecular dynamics simulations of self-assembling systems. *Science* **321**, 798–800 (2008).
2. Rapaport, D. C. *The Art of Molecular Dynamics Simulation*. Cambridge Univ. Press, (2004).
3. Landau, D. P. & Binder, K. *A guide to Monte Carlo simulations in statistical physics* (Cambridge Univ. Press, 2015).
4. Chatterjee, A. & Vlachos, D. G. An overview of spatial microscopic and accelerated kinetic Monte Carlo methods. *J. Comput. Aided Mater. Des.* **14**, 253–308 (2007).
5. Rudd, R. E. & Broughton, J. Q. Coarse-grained molecular dynamics and the atomic limit of finite elements. *Phys. Rev. B* **58**, R5893–R5897 (1998).
6. Tozzini, V. Coarse-grained models for proteins. *Curr. Opin. Struct. Biol.* **15**, 144–150 (2005).

7. Metropolis, N., Rosenbluth, A. W., Rosenbluth, M. N., Teller, A. H. & Teller, E. Equation of state calculations by fast computing machines. *J. Chem. Phys.* **21**, 1087–1092 (1953).

8. Khachaturyan, A. G. Ordering in substitutional and interstitial solid solutions. *Prog. Mater. Sci.* **22**, 1–150 (1978).

9. Khachaturyan, A. G. *Theory of Structural Transformations In Solids* (Wiley, 1983).

10. Khachaturyan, A. G. Microscopic theory of diffusion in crystalline solutions and the time evolution of X-ray and thermal neutron diffuse scattering. *Fiz. Tverd. Tela* **9**, 2595–2601 (1967).

11. Lavrskyi, M., Zapolsky, H. & Khachaturyan., A. G. Fraton Theory and Modelling of Self-Assembling of Complex Structures. Preprint at < http://http://arxiv.org/abs/1411.5587 > (2014).

12. Mahan, G. D. *Many Particle Physics* (Springer, 1981).

13. Suris, R. A. The application of functional methods to the theory of solid solutions. *Fiz. Tverd. Tela* **4**, 1154–1161 (1962).

14. Connolly, J. W. D. & Williams, A. R. Density-functional theory applied to phase transformations in transition-metal alloys. *Phys. Rev. B* **27**, 5169–5173 (1983).

15. Chen, L. Q. & Shen, J. Applications of semi-implicit Fourier-spectral method to phase field equations. *Comp. Phys. Commun.* **108**, 147–158 (1998).

16. Meyers, M. A., Chen, P.-Y., Lin, A. Y. & Seki, Y. Biological materials: structure and mechanical properties. *Prog. Mater. Sci.* **53**, 1–206 (2008).

17. Iijima, S. Single-shell carbon nanotubes of 1-nm diameter. *Nature* **354**, 56–58 (1991).

18. Ajayan, P. M. Nanotubes from carbon. *Chem. Rev.* **99**, 1787–1800 (1999).

19. Dierking, I. *Texture of Liquid Crystals* (Wiley-VCH Verlag, 2003).

20. Singh, S. Phase transitions in liquid crystals. *Phys. Rep.* **324**, 107–269 (2000).

21. Watson, J. D. & Crick, F. H. C. *Molecular structure of nucleic acidsNature* **171**, 737–738 (1953).

22. Seeman, N. C. DNA in a material world. *Nature* **421**, 427–431 (2003).

Permissions

The contributors of this book come from diverse backgrounds, making this book a truly international effort. This book will bring forth new frontiers with its revolutionizing research information and detailed analysis of the nascent developments around the world.

We would like to thank all the contributing authors for lending their expertise to make the book truly unique. They have played a crucial role in the development of this book. Without their invaluable contributions this book wouldn't have been possible. They have made vital efforts to compile up to date information on the varied aspects of this subject to make this book a valuable addition to the collection of many professionals and students.

This book was conceptualized with the vision of imparting up-to-date information and advanced data in this field. To ensure the same, a matchless editorial board was set up. Every individual on the board went through rigorous rounds of assessment to prove their worth. After which they invested a large part of their time researching and compiling the most relevant data for our readers.

The editorial board has been involved in producing this book since its inception. They have spent rigorous hours researching and exploring the diverse topics which have resulted in the successful publishing of this book. They have passed on their knowledge of decades through this book. To expedite this challenging task, the publisher supported the team at every step. A small team of assistant editors was also appointed to further simplify the editing procedure and attain best results for the readers.

Apart from the editorial board, the designing team has also invested a significant amount of their time in understanding the subject and creating the most relevant covers. They scrutinized every image to scout for the most suitable representation of the subject and create an appropriate cover for the book.

The publishing team has been an ardent support to the editorial, designing and production team. Their endless efforts to recruit the best for this project, has resulted in the accomplishment of this book. They are a veteran in the field of academics and their pool of knowledge is as vast as their experience in printing. Their expertise and guidance has proved useful at every step. Their uncompromising quality standards have made this book an exceptional effort. Their encouragement from time to time has been an inspiration for everyone.

The publisher and the editorial board hope that this book will prove to be a valuable piece of knowledge for researchers, students, practitioners and scholars across the globe.

List of Contributors

Zhen-Yu Ye
Department of Physics, State Key Laboratory for Silicon Materials, Zhejiang University, Hangzhou, China State Key Laboratory of Superlattices and Microstructures, Institute of Semiconductors, Chinese Academy of Sciences, Beijing, China

Hui-Xiong Deng, Shu-Shen Li and Jun-Wei Luo
State Key Laboratory of Superlattices and Microstructures, Institute of Semiconductors, Chinese Academy of Sciences, Beijing, China

Hui-Zhen Wu
Department of Physics, State Key Laboratory for Silicon Materials, Zhejiang University, Hangzhou, China

Su-Huai Wei
National Renewable Energy Laboratory, Golden, CO, USA

Fahhad H Alharbi, Nouar Tabet and Sabre Kais
College of Science and Engineering, Hamad Bin Khalifa University, Doha, Qatar
Qatar Environment and Energy research Institute (QEERI), Hamad Bin Khalifa University, Doha, Qatar

Sergey N Rashkeev and Fedwa El-Mellouhi
Qatar Environment and Energy research Institute (QEERI), Hamad Bin Khalifa University, Doha, Qatar

Hans P Lüthi
Department of Chemistry and Applied Bioscience, ETH Zurich, Zurich, Switzerland

Alexander Urban
Department of Materials Science and Engineering, University of California, Berkeley, CA, USA

Dong-Hwa Seo
Department of Materials Science and Engineering, Massachusetts Institute of Technology, Cambridge, MA, USA

Gerbrand Ceder
Department of Materials Science and Engineering, University of California, Berkeley, CA, USA
Materials Science Division, Lawrence Berkeley National Laboratory, Berkeley, CA, USA

Yongmei M Jin and Yu U Wang
Department of Materials Science and Engineering, Michigan Technological University, Houghton, MI, USA

Yang Ren
X-Ray Science Division, Advanced Photon Source, Argonne National Laboratory, Argonne, IL, USA

Scott Kirklin, James E Saal, Bryce Meredig, Alex Thompson, Jeff W Doak, Muratahan Aykol and Chris Wolverton
Department of Materials Science and Engineering, Northwestern University, Evanston, IL, USA

Stephan Rühl
FIZ Karlsruhe—Leibniz Institute for Information Infrastructure, Eggenstein-Leopoldshafen, Germany

Thomas P Senftle and Michael J Janik
Department of Chemical Engineering, Pennsylvania State University, University Park, PA, USA

Sungwook Hong, Md Mahbubul Islam, Yun Kyung Shin, Chad Junkermeier and Adri CT van Duin
Department of Mechanical and Nuclear Engineering, Pennsylvania State University, University Park, PA, USA

Sudhir B Kylasa and Ananth Grama
Department of Computer Science and Engineering, Purdue, West Lafayette, IN, USA

Yuanxia Zheng and Roman Engel-Herbert
Department of Materials Science and Engineering, Pennsylvania State University, University Park, PA, USA

Hasan Metin Aktulga
Department of Computer Science and Engineering, Michigan State University, East Lansing, MI, USA

Toon Verstraelen
Center for Molecular Modeling (CMM), Ghent University, Zwijnaarde, Belgium

Yi Wang, Shun-Li Shang, Huazhi Fang, Zi-Kui Liu and Long-Qing Chen
Department of Materials Science and Engineering, The Pennsylvania State University, University Park, Pennsylvania, PA, USA

Ligang Sun and Xiaoqiao He
Department of Architecture and Civil Engineering, City University of Hong Kong, Kowloon, Hong Kong, China

Jian Lu
Department of Mechanical and Biomedical Engineering, City University of Hong Kong, Kowloon, Hong Kong, China
Centre for Advanced Structural Materials, City University of Hong Kong, Shenzhen Research Institute, Shenzhen Hi-Tech Industrial Park, Shenzhen, China

Shuozhi Xu
Woodruff School of Mechanical Engineering, Georgia Institute of Technology, Atlanta, GA, USA

Liming Xiong
Department of Aerospace Engineering, Iowa State University, Ames, IA, USA

Youping Chen
Department of Mechanical and Aerospace Engineering, University of Florida, Gainesville, FL, USA

David L McDowell
Woodruff School of Mechanical Engineering, Georgia Institute of Technology, Atlanta, GA, USA
School of Materials Science and Engineering, Georgia Institute of Technology, Atlanta, GA, USA

Shiva Rudraraju
Department of Mechanical Engineering, University of Michigan, Ann Arbor, MI, USA

Anton Van der Ven
Materials Department, University of California, Santa Barbara, CA, USA

Krishna Garikipati
Department of Mechanical Engineering, University of Michigan, Ann Arbor, MI, USA
Department of Mathematics, University of Michigan, Ann Arbor, MI, USA

Narumasa Miyazaki and Masato Wakeda
Department of Mechanical Science and Bioengineering, Graduate School of Engineering Science, Osaka University, Osaka, Japan

Yun-Jiang Wang
State Key Laboratory of Nonlinear Mechanics, Institute of Mechanics, Chinese Academy of Sciences, Beijing, China

Shigenobu Ogata
Department of Mechanical Science and Bioengineering, Graduate School of Engineering Science, Osaka University, Osaka, Japan
Center for Elements Strategy Initiative for Structural Materials (ESISM), Kyoto University, Kyoto, Japan

Jiong Yang and Lihua Wu
Materials Genome Institute, Shanghai University, Shanghai, China

Lili Xi, Xun Shi and Lidong Chen
State Key Laboratory of High Performance Ceramics and Superfine Microstructure, Shanghai Institute of Ceramics, Chinese Academy of Sciences, Shanghai, China

Wujie Qiu
State Key Laboratory of High Performance Ceramics and Superfine Microstructure, Shanghai Institute of Ceramics, Chinese Academy of Sciences, Shanghai, China
Department of Physics, East China Normal University, Shanghai, China

Jihui Yang
Material Science and Engineering Department, University of Washington, Seattle, WA, USA

Wenqing Zhang
Materials Genome Institute, Shanghai University, Shanghai, China
State Key Laboratory of High Performance Ceramics and Superfine Microstructure, Shanghai Institute of Ceramics, Chinese Academy of Sciences, Shanghai, China

Ctirad Uher
Department of Physics, University of Michigan, Ann Arbor, MI, USA

David J Singh
Department of Physics and Astronomy, University of Missouri, Columbia, MO, USA

Yunhao Lu, Yinzhu Jiang and Jianzhong Jiang
School of Materials Science and Engineering, Zhejiang University, Hangzhou, China
State Key Laboratory of Silicon Materials, Zhejiang University, Hangzhou, China

Di Zhou
School of Materials Science and Engineering, Zhejiang University, Hangzhou, China

Guoqing Chang and Hsin Lin
Centre for Advanced 2D Materials and Graphene Research Centre, National University of Singapore, Singapore, Singapore
Department of Physics, National University of Singapore, Singapore, Singapore

Shan Guan and Shengyuan A Yang
Research Laboratory for Quantum Materials, Singapore University of Technology and Design, Singapore, Singapore

Weiguang Chen
College of Physics and Electronic Engineering, Zhengzhou Normal University, Zhengzhou, China

Xue-sen Wang and Yuan Ping Feng
Department of Physics, National University of Singapore, Singapore, Singapore

Yoshiyuki Kawazoe
New Industry Creation Hatchery Center, Tohuku University, Sendai, Japan
Institute of Thermophysics, Siberian Branch of Russian Academy of Sciences, Novosibirsk, Russia

Matteo Seita and Christopher A Schuh
Department of Materials Science and Engineering, Massachusetts Institute of Technology, Cambridge, MA, USA

Marco Volpi and Maria Vittoria Diamanti
Department of Chemistry, Materials and Chemical Engineering 'G Natta', Politecnico di Milano, Milan, Italy

Srikanth Patala
Department of Materials Science and Engineering, North Carolina State University, Raleigh, NC, USA

Ian McCue and Jonah Erlebacher
Department of Materials Science and Engineering, Johns Hopkins University, Baltimore, MD, USA

Michael J Demkowicz
Materials Science and Engineering, Texas A&M University, College Station, TX, USA

Julian P Velev
Department of Physics and Astronomy, University of Nebraska, Lincoln, NE, USA
Department of Physics and Astronomy, University of Puerto Rico, San Juan, Puerto Rico, USA

John D Burton and Evgeny Y Tsymbal
Department of Physics and Astronomy, University of Nebraska, Lincoln, NE, USA

Nebraska Center for Materials and Nanoscience, University of Nebraska, Lincoln, NE, USA

Mikhail Ye Zhuravlev
Kurnakov Institute for General and Inorganic Chemistry, Russian Academy of Sciences, Moscow, Russia Faculty of Liberal Arts and Sciences, St Petersburg State University, St Petersburg, Russia

Wei Xiong
Department of Materials Science and Engineering, Northwestern University, Evanston, IL, USA

Gregory B Olson
Department of Materials Science and Engineering, Northwestern University, Evanston, IL, USA
QuesTek Innovations LLC, Evanston, IL, USA

Lei Zhang
Beijing International Center for Mathematical Research, Center for Quantitative Biology, Peking University, Beijing, China

Weiqing Ren
Department of Mathematics, National University of Singapore, Singapore
Institute of High Performance Computing, Agency for Science, Technology and Research, Singapore;

Amit Samanta
Physics Division, Lawrence Livermore National Laboratory, Livermore, CA, USA

Qiang Du
Department of Applied Physics and Applied Mathematics, Columbia University, New York, NY, USA

Mykola Lavrskyi and Helena Zapolsky
GPM, UMR 6634, University of Rouen, Saint-Etienne du Rouvray, France

Armen G Khachaturyan
Department of Materials Science & Engineering, Rutgers University, Piscataway, NJ, USA
Department of Materials Science & Engineering, University of California, Berkeley, Berkeley, CA, USA

Index